Hanford

A conversation about nuclear waste and cleanup

Roy E. Gephart

The front cover was designed by Chris Walling (copyright 2002).

The back cover photo was taken by Alan Gephart and manipulated using graphics design software.

Regarding the photo of Love Canal, featured in Chapter 8, every effort has been made to contact copyright holders, and any omission is regretted.

Library of Congress Cataloging-in-Publication Data

Gephart, R.E.
Hanford: a conversation about nuclear waste and cleanup/by Roy Gephart.
p. cm.
Includes bibliographical references and index.
ISBN 1-57477-134-5 (pbk. : alk. paper)
1. Hanford Site (Wash.) 2. Nuclear weapons plants—Waste disposal—Environmental aspects—Washington (State)—Hanford Site. 3. Hazardous waste site remediation—Washington (State)—Hanford Site. 4. Radioactive waste sites—Cleanup—Washington (State)—Hanford Site. I. Title

TD898.12.W2 G46697 2003
363.72'89'0979751—dc21

2002038578

Printed in the United States of America

Battelle Press
505 King Avenue
Columbus, Ohio 43201-2693, USA
614-424-6393 or 1-800-451-3543
Fax: 614-424-3819
E-mail: press@battelle.org
Website: www.battelle.org/bookstore

Acknowledgments

The idea for this book took root in 1998 when I gave a tour of Hanford for Catholic bishops and parishioners from across the Pacific Northwest. They wrote a pastoral letter about caring for the Columbia River as part of Creation, for the common good of all generations. That visit energized my thinking about nuclear waste issues. Over the next 4 years, ideas matured as I wrote several drafts, culminating in the book you are now holding.

Along the way, I consulted with dozens of people inside and outside of the Hanford community. While we did not always share the same opinions, we did hold a common goal: protecting people and the environment. Their critical comments on drafts and candid discussions were essential in breathing life into this publication. Any errors are mine and do not reflect the expert knowledge and guidance of others.

I'm indebted to the professional editorial and graphic staff of the Pacific Northwest National Laboratory, especially for the keen eyes, skills, and patience of my lead editor Kristin L. Manke. The book was designed by Rose M. Watt and illustrated by Michael C. Perkins and Amy E. Madden. They helped craft images from words and my paper scribbles. Julie M. Gephart expertly completed the final editing.

I also wish to thank the staff at the U.S. Department of Energy's Public Reading Room, located at Washington State University Tri-Cities. They searched historical references and checked citations. Thanks, too, to Media Services of Lockheed Martin Information Technology, Richland, who preserve invaluable historical photographs of Hanford and have the knack for finding just the right ones.

The Fundamental Science and Environmental Technology Directorates of the Pacific Northwest National Laboratory provided funding for editorial, graphic, and photographic services.

Thanks also to Joseph E. Sheldrick of Battelle Press, Columbus, Ohio, for his encouragement and publication skills and to the U.S. Department of Energy for use of their many fine photographs.

The book's cover design and original color painting is the creative work of artist Chris Walling of Richland, Washington.

The views offered do not represent the opinions of any single individual, agency, or institution. Nonetheless, all writing reflects, in part, how the author views the world.

Amy Madden

To my sons Alan and Ian
who, along with other children,
will inherit the choices
our generation makes.

Foreword

"To understand today you have to search yesterday."
—Pearl S. Buck, writer

Thirty years have passed since I first walked these jagged rocks atop Gable Mountain. This ridge of pancake-layered basalt, lifted above the surrounding desert, lies in the middle of the Hanford Site in Washington State. Geologists call this ridge the Umtanum-Gable Mountain anticline. Indigenous people named it Nooksiah, for the mountain resembles an otter gliding across water. Here visitors gaze across 600 square miles of Hanford landscape. Except for a few blocky buildings, standing like silent relics from an ancient culture, the land appears untouched.

The Hanford Site created plutonium for nuclear weapons. It was born in a world war and grew during the darkened era of superpower distrust. Hanford was a benchmark of global anxiety where strength was measured in megatons. We stand in the shadow of that history, inheriting its contaminants. Our wisdom in caring for this land and our commitment to environmental reconciliation will be judged for generations.

Our wisdom in caring for this land will be judged for generations.

Hanford played a critical role in securing the military strength of the United States. However, today it is judged by a new social consciousness, one critical of how waste was handled after the first unambiguous concerns were voiced. Yes, Hanford had its failings. But therein lies a lesson: when powerful interests are involved, no generation is immune to pretense.

Decades of restricted access have preserved much of the area's natural beauty. Hanford provides sanctuary for a rich variety of plants and animals, including many that are threatened or endangered. The last free-flowing stretch of the magnificent Columbia River rests within its boundaries. A portion of this land is now designated the Hanford Reach National Monument, a treasure to be wisely used and passed to future generations. Without knowing, one would never suspect that pockets of radioactive and chemical contaminants exist onsite in underground tanks, landfills, soil, and groundwater. Some contaminants were also released into the air and Columbia River, exposing those who lived downwind and downstream.

When powerful interests are involved, no generation is immune to pretense.

Hanford is a site of contrast. It contains a portion of the nation's most dangerous radioactive waste while preserving some of the most unique desert ecology within the Pacific Northwest. It houses the world's oldest full-scale nuclear reactors and reprocessing plants while supporting the newest laboratories for studying environmental science. Hanford contrasts yesterday's certainty for achieving plutonium production goals with today's uncertainty about how to reach cleanup objectives. Plutonium production was done in secrecy; cleanup is an open book being written through an evolving social consensus and understanding of the problems faced.

Nuclear issues are socially amplified and stigmatized.

The problem of radioactive waste polarizes discussions like few other topics. It cuts deep across our emotional, cultural, and theological fabric. Nuclear issues are socially amplified and stigmatized. Discourse easily slips into skirmishes over language—word games ranging from the exactness of science understood by some to sensationalized images capturing the attention of many. Beliefs seem hardwired. Simply put, people do not like radioactive waste. It guarantees conversation gridlock. Many feel the "all clear" signal is never sounded; the books are never closed on nuclear issues. In such an environment, one begins to understand why the public is conservative and prescriptive in their approach to cleanup.

Compared to our daily lives, the complexity of dealing with radioactive and chemical waste is beyond comprehension. Potential long-term dangers are real yet commonly abstract, shrouded in probabilities and assumptions. It's unfamiliar terrain lying beyond our life spans where facts, mixed with beliefs, drive issues whose outcomes are hidden from view. This challenge is not being taken lightly. Yet our desire for quick-fix solutions may be mismatched to this multi-generation task.

Cleanup of nuclear waste is the largest environmental restoration project ever undertaken by the United States, perhaps the world. Yet, outside of local communities, does anyone care? It's increasingly difficult to hold the attention of policy makers on problems located in isolated areas of the country when environmental benefits and investment outcomes are sometimes ill-defined and unrecognized for years.

I step along the steep southern edge of Gable Mountain, stop, and look. This afternoon's sun accents the land's weathered contours. It reflects off the waters of a distant pond. Now, only shadows slip across the mountain. The sky is a continuum of blue. Clear and clean. Though alone, I feel part of a larger whole.

These rocks allow us to interrogate the past and anticipate the future. Like a multi-colored fabric, streams, lakes, and wind laid down blankets of interwoven sediments atop most of the surrounding basalt. Traces of ancestral rivers lie visible in abandoned channels braided across the land. Slowly, invisible forces folded the underlying basalt into long, linear ridges trending eastward from the Cascade Mountains. Several ridges contour the horizon. Beyond them, the earth curves from view.

The Hanford landscape evolved over eons. Each event in its geologic history, from the deep scouring of swift currents to the settling of clay in quiet backwaters, sculptured the land's character. I see an ancient, earth-born language that could have spoken many futures but chose this one.

There is no substitute for visiting Hanford, talking with people who work here, listening to those exposed to past contaminant releases, and understanding cleanup from a range of voices. Hanford does not live between the hard edges of viewgraphs, three-ring binders, and notebooks. It rests inside buildings, underground, and in the lives of people.

Today, I view Hanford not through the lens of numbers but as a father of two children. In many ways, environmental restoration is for the young, for those not yet born. Each generation is obligated to act beyond itself, to shape the opportunities others will face. How we deal with today's problems will gauge the humanness of our generation.

For years, Hanford officials sought to understand the potential hazards from the release of contaminants. Hanford health science was world class. However, it was first carried out when the environment was used as a sink for industrial contaminants, when the public trusted what it was told. Independent oversight, unshackled by preserving plutonium production goals, came years later.

I view Hanford not through the lens of numbers but as a father.

Nearby, a sign cautions visitors to remain on the dirt path. These are sacred rocks cradling ancient memories. Countless generations walked these outcrops. Their paths have vanished. Hanford will shadow this land for times studied only by geologists and astronomers.

Can Hanford be cleaned? Yes—how much depends upon what cleanup means and what cost, risk, and schedule are acceptable. In many ways, the word "cleanup" is a misnomer. It implies what can't be delivered. Perhaps "stabilize" or "contain" are more accurate. Existing knowledge and capability were adequate to begin cleanup. However, they're inadequate for completing it and understanding the legacy left to others.

To the extent cleanup efforts are ethically wise, scientifically sound, and socially acceptable, we are building an enduring foundation. When will cleanup end? When social expectations, physical reality, and technical capabilities coincide. Afterwards, land and contaminant stewardship remains. How much will this cost? No one knows.

The consequences of what we do will far outlive the very institutions now planning and executing cleanup work. Waste treatment, containment, and monitoring are multi-century commitments. It's a challenge built on an economy where value is measured in more than dollars, where policy must balance liabilities and benefits across generations.

I view Hanford cleanup as an experiment bounded by uncertainty, where outcomes remain shrouded in issues unknown, by expectations yet expressed. It's an experiment where researchers strive to understand contaminant behavior and nature's response to it, where some predictions tease at the far edge of science. Cleanup is an experiment where regulators strive to enforce well-considered policies that protect this and future generations. It's an experiment requiring risk taking and contingencies to handle the inevitable surprises, for we are rarely wise enough to set a course of action that doesn't change, an agreement that isn't revised, or a finding that doesn't contain unknowns.

Many positive steps have moved Hanford into a new future. Most work accomplished has focused on ensuring safer, more effective waste management. Progress was achieved by workers guided by a far-reaching cleanup agreement and public involvement. Yet nearly all waste remains where it has been for decades, and most actions undertaken are temporary steps rather than permanent solutions. The first years of cleanup were a honeymoon—money was abundant, promises were easy, and obvious problems were addressed. Now comes the long haul, the tough year-by-year grueling work.

Progress was achieved by workers guided by a far-reaching cleanup agreement.

We are now in the second decade of cleanup. Some people are satisfied with the progress made; others are not. Studies point to administrative, budgetary, and business constraints creating powerful obstacles to progress. People ask why cleanup is so difficult, why problems are identified but change not embraced, whether we are reducing health risks. The cold truth is that onsite radioactivity is naturally decaying away quicker than the treatment benefits gained from all cleanup actions.

Concerns are expressed about debates bogged down in management detail rather than elevated to frame public policy, political ideologies masquerading as objective principles, intimidation versus persuasion, and discourse tailored to achieve private advantage. It's a world where knowledge persuades less than news-catching high-road rhetoric, where public confidence is shaken with each problem uncovered and milestone missed. Acrimonious arguments are the mainstay of many debates.

Still, the public desires candor and openness without minimizing or exaggerating hazards. They want solutions that are safe, fair, and suitable to the task.

This crisis is deeply rooted in distrust. Workers are dismayed that it takes so much effort to do anything and yet so little to stop it. Projects appear to inch forward while completion slips into the future. Meaningful actions require the intervention of too many competitive interests.

Have we reached the point of re-examining the fundamentals driving cleanup?

Have necessary safeguards and unnecessary administrative burdens become inseparably entwined? Have we reached the point for re-examining the fundamentals driving cleanup? The whole process screams for simplification. There is a growing belief that Hanford needs a firmer path, a course biased towards action, streamlined work practices, joint accountability, flexibility, and sustained commitment. People want drive not drift, problem solving not repackaged agendas. It's time to demonstrate substantial progress.

Like the history that shaped this landscape, many courses of action are possible. Reasonable people can and do disagree. Perhaps our greatest social challenge is achieving consensus where trust is still sought. Perhaps our greatest policy challenge is understanding how to apply high-quality knowledge to decision-making. Perhaps our greatest political challenge is forging solutions without proliferating second-guessing.

Information and involvement are central to exercising responsibility in a democratic society. Shared power fosters cooperation. Conviction preserves commitment. The American people, especially those in the communities surrounding and downstream of Hanford, are the customers for the cleanup accomplished and the impression others have about our environment.

We are in this together.

A promise was made to reclaim this land. Each of us has a commitment in that outcome. And regardless of our beliefs and where we live, we are in this together.

This afternoon, it's quiet atop Gable Mountain. Occasionally, visitors come to glance across the landscape and to share their stories. We stand at the doorstep of cleanup and will pass along some heritage. What tomorrow's visitors are told depends on us.

Roy Gephart

Introduction

"It means nothing to be open to a proposition we don't understand."

—Carl Sagan,
The Demon-Haunted World: Science as a Candle in the Dark (1996, p. 306)

This book is about the largest and most complex environmental cleanup project in the United States—the U.S. Department of Energy's Hanford Site.[1] It's a macrocosm of restoration problems facing the federal government at dozens of places once used for uranium mining, nuclear material production, research, and weapons testing. The challenge of cleaning up industrial contaminants pales in comparison.

As the title implies, this book is a conversation—a conversation by the author as well as the many people and organizations quoted. It's an information resource, written for the general reader as well as the technically trained person wanting an overview of Hanford's contaminant history and the multi-generation cleanup challenges faced. The book is also an idea guide that encourages readers to be informed consumers of Hanford news, and to recognize that critical thinking and personal values are at the heart of understanding Hanford.

This book is an information resource and idea guide that encourages readers to be informed consumers of Hanford news.

Remaining informed is important whether you are a taxpayer (paying for cleanup), politician (allocating funds), scientist or engineer (creating knowledge and capabilities), regulator (enforcing environmental laws), contractor (performing cleanup), or a member of the regional community (living with consequences).

No single publication can do justice to all opinions or critical events in Hanford's history. And no publication stands alone, especially one covering such a controversial topic. This is why readers are encouraged to seek information from other sources and to maintain a healthy degree of skepticism about what they read and hear—regardless of the source. Whether acknowledged or not, all publications contain points of view.

Sometimes, point-counterpoint perspectives are posed to encourage readers to examine issues from different angles. The author is not shy about asking questions or questioning answers. He provides a broad context while avoiding some traditionally strident beliefs. However, what one person thinks is a middle-of-the-road opinion may be considered an extreme viewpoint by someone else. Welcome to nuclear issues; welcome to Hanford.

(1) Over the years, several terms were used to identify what is now officially known as the Hanford Site. Past examples include Site W, Hanford Engineering Works, and the Hanford Reservation. For the purpose of this book, the "Hanford Site" is shortened to the more commonly used term "Hanford."

Words are hooks for hanging judgments. What is written echoes with both actual and inferred meaning. This book is no exception. Nonetheless, it's useful to recall Hanford's history, reframe problems, and examine the potential merits of past choices and future decisions. After all, most cleanup lies years in the future. We are only the first generation to face it.

Most cleanup lies years in the future.

Useful information is gained by reading the book in whole or in parts. Hundreds of references are cited. Sidebars and quotes are liberally used, providing mini-stories and insight through the experiences of others. Sometimes popular words, rather than technically perfect phrases, are used to communicate. Also for easier understanding by most readers, English units are used. Sections are numbered for quicker cross-reference.

The world is not perfectly known. This admission should keep us humble when trying to predict the future. For example, there are many sources of information about the inventory, distribution, and behavior of contaminants at Hanford. Some are measured values; most are estimates. None are satisfactory to answer all questions. A few values survive like urban legends. The numbers shared are best estimates built from references, conversations, and experience. Depending upon context, nearly every number should be prefaced with "approximately," "about," or "on the order of."

Contents

Chapter 1

Hanford—In the Beginning

"There was a period…when the problem of radioactive waste was considered to be nonexistent."

—Dr. Abel Wolman, congressional testimony
(U.S. Congress 1959, p. 12)

During the Christmas season of 1938, 9 months before the beginning of World War II, an Austrian physicist, Lise Meitner, and two German chemists, Otto Hahn and Fritz Strassmann, discovered nuclear fission—the splitting of an atom's nucleus into pieces with the simultaneous release of energy—lots of energy. This discovery culminated 40 years of research into radioactivity.

"By order of the Secretary of War, and effective August 16, 1942, a new engineer district, without territorial limits, to be known as the Manhattan District, is established" (Bessell 1942).

In 1896, Henri Becquerel discovered the effect of radiation by exposing photographic plates to invisible particles emitted from uranium crystals. Becquerel called the emissions "uranic rays." A few years later, Marie Curie recognized this effect as being nuclear rather than chemical in nature and named the effect "radioactivity." She and her husband Pierre studied the mysterious glow given off from the uranium-rich mineral called "pitchblende." They were the first to chemically separate radioactive elements, such as radium from uranium, and to study their properties.

After the Englishman James Chadwick discovered neutrons in 1932, physicists intentionally bombarded uranium with this new particle. Neutrons were very effective at causing nuclear changes within other atoms. They thought uranium was capturing a neutron and transforming it into a heavier element, elements beyond uranium in the periodic table (also known as transuranic elements). Hahn, Strassmann, and Meitner discovered that when the chemical compound uranium nitrate was hit with neutrons, uranium atoms were actually split into pieces and released energy plus other neutrons. Fission was discovered.

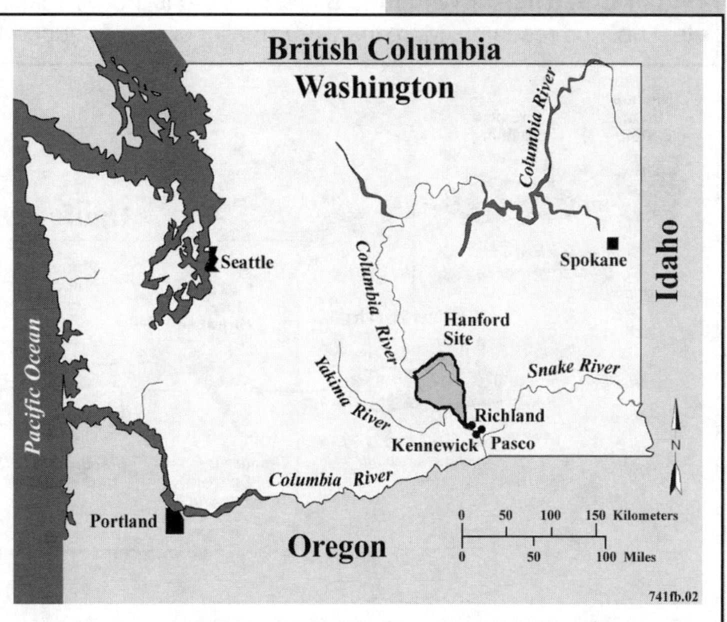

Hanford's Location. Hanford covers 586 square miles along the banks of the Columbia River in the desert of southeastern Washington State. The site is half the size of the state of Rhode Island.

The Danish physicist Niels Bohr brought word of this discovery to the United States. Upon hearing this news, physicist J. Robert Oppenheimer dismissed fission as impossible. But he soon changed his mind when briefed by fellow researchers. Bohr and John Wheeler of the United States developed the first detailed model of the fission process (Bodansky 1996, p. 59). Reports in the British publication *Nature* confirmed the discovery and spurred global interest in nuclear research by industrial nations (Rhodes 1995, p. 27).

Soon, other researchers verified that extra neutrons were ejected during nuclear fission, making it possible to have a self-sustaining chain reaction.

Chemical reactions involve the interplay of electrons surrounding the core—nucleus—of atoms. Nuclear reactions involve changes to the core.

This discovery was stunning, for the energy released was millions of times greater than the energy set free during chemical reactions, such as found in common explosives. Here was a new source of energy. Two uses sprang to mind: producing power or building weapons of mass destruction. The latter did not go unnoticed in a world entering the second war of the century.

In the early months of World War II, the basic design of atoms was well enough understood for countries to contemplate harnessing this new energy. Germany began work on the military use of fission in 1939 (Serber 1992). Japan also began to explore

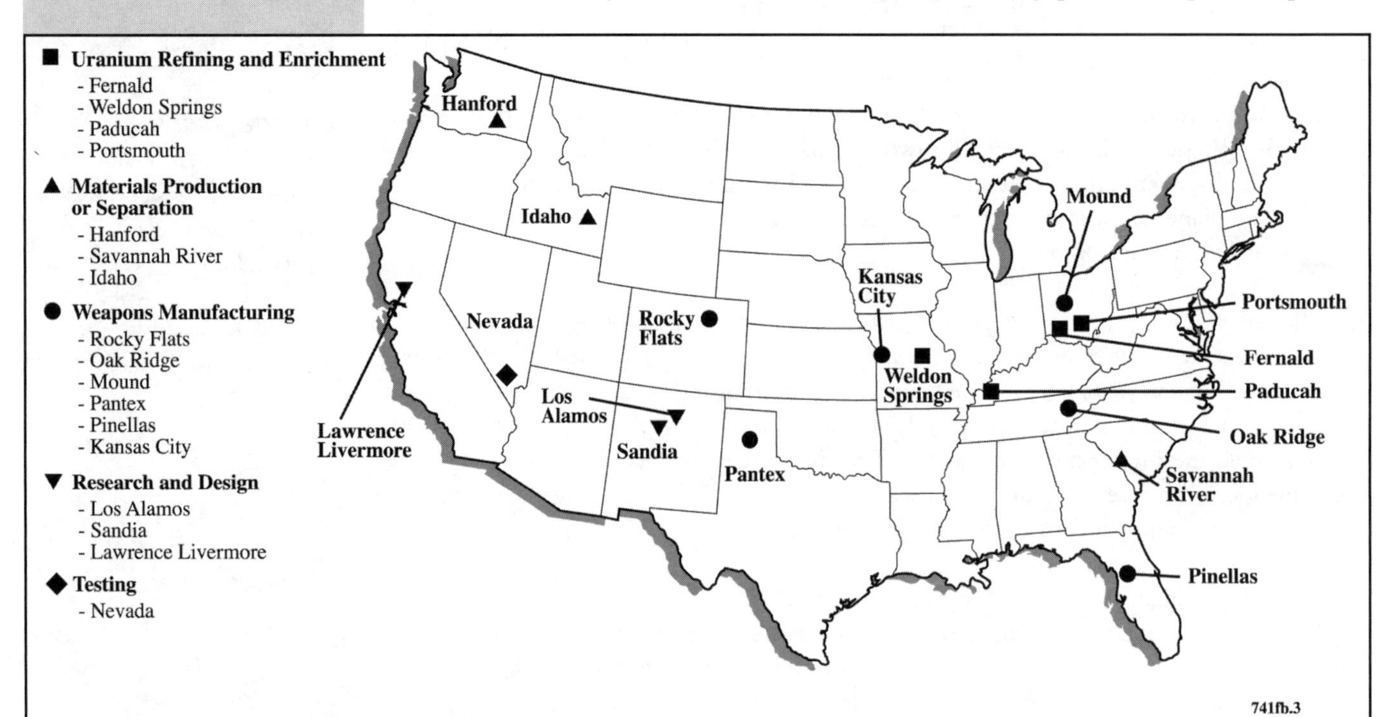

Major U.S. Nuclear Weapons Sites. Hanford is 1 of 17 major sites once used for nuclear material production, testing, and research. More than 100 smaller sites also supported nuclear material and weapon development.

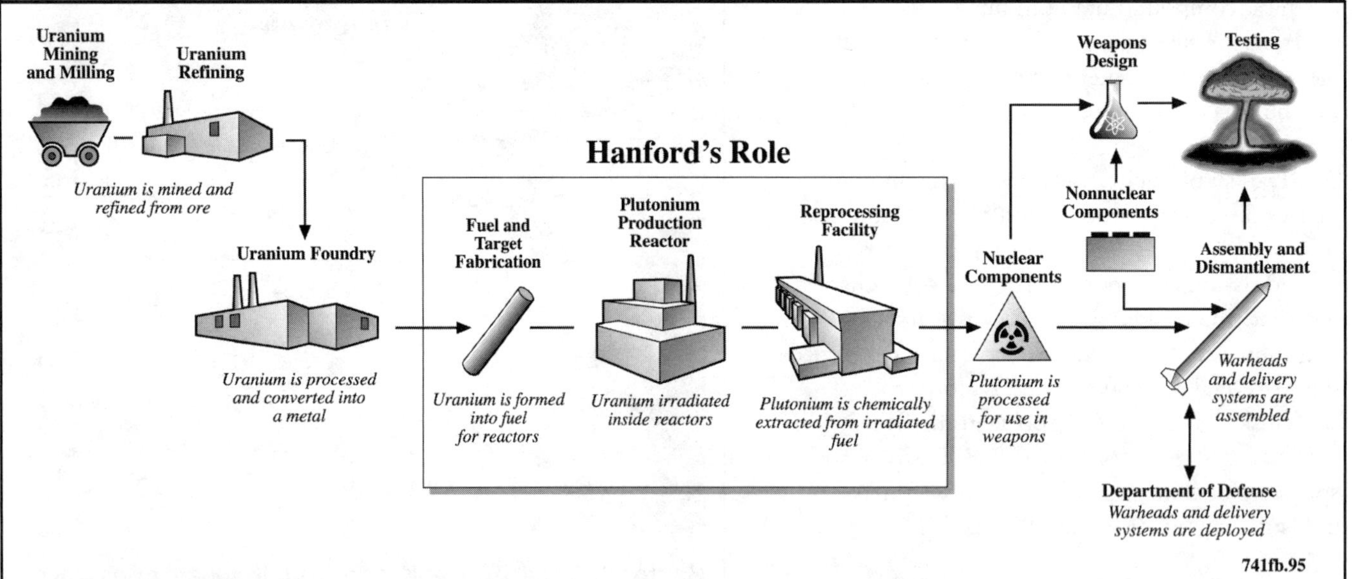

Hanford Role in U.S. Weapons Complex. The role of Hanford in U.S. nuclear weapons production was to create and process plutonium (DOE 1995a).

military applications. In October 1939, President Franklin Roosevelt was informed about the German effort. The Advisory Committee on Uranium was formed, and $6000 was allocated to study the fission process (Murray 1961). In April 1940, some scientists proposed voluntary censorship on publications about fission. These articles soon vanished from publicly available literature. In 1942, the United States began the Manhattan Project, under the jurisdiction of the Army Corps of Engineers, and started building nuclear research laboratories and factories. It quickly grew into the largest construction project in human history, driven by the rapid incorporation of new scientific discoveries into factory designs and operations. The British and Americans began cooperating in the Manhattan Project because "The U.S. had to develop an atomic bomb before the Germans did. Such weapons in the hands of Hitler would be the ultimate catastrophe for the world" (Makhijani 1995).

Though most industrialized nations held enough information to initiate some sort of atomic bomb project, only the United States had the industrial strength and human resources to undertake this mammoth effort (Fermi and Samra 1995). It was also the one industrial country whose home front industries were not under attack by invading forces. As it turned out, Japanese work never proceeded past the theoretical stage. Germany's progress stumbled as a result of Allied bombing, sabotage, lack of resources, and possibly a flawed approach to reactor design (Rhodes 1986). Russia undertook a modest nuclear research program coupled with aggressive spying to learn all they could from the United States and Britain.

If the Germans were able to produce the first nuclear weapons, it might "be necessary for us to stand the first punishing blows [of German atomic bombs] before we...[were] in the position to destroy the enemy" (Military Policy Committee 1943).

For the first time in human history, large quantities of naturally occurring radioactive elements were concentrated, and new elements created. The alchemist's dream of altering the fundamental makeup of nature had come true. Elements were not immutable. Common metals could be turned into valuable, rare materials, having unique properties. At Hanford, one of the first two nuclear material production sites of the Manhattan Project, uranium was transformed into plutonium—plus a host of unwanted radioactive elements. Plutonium was at the heart of a new approach to warfare—atomic bombs.

The world's first industrial-scale production of plutonium and release of radionuclides into the environment took place in the desert of Washington State, along the banks of the Columbia River. The time was 1944, and the place was Hanford.

The Smyth Report and the Soviet Union

The U.S. government published *A General Account of the Development of Methods of Using Atomic Energy for Military Purposes Under the Auspices of the United States Government, 1940-1945* (Smyth 1945) a few days after the second atomic bomb (Fat Man) was dropped. It detailed much of the basic science behind the Manhattan Project. Copies immediately went to Soviet intelligence. Its information was nearly equal to all the secret atomic documents gained in the previous 4 years by Soviet spies (Rhodes 1995, p. 182). The Soviet Union's first reactor was a carbon copy of Hanford's B Reactor, and their first atomic bomb, tested in Kazakhstan on August 29, 1949, was a duplicate of Fat Man (Rhodes 1995, p. 268).

There was a long-standing Soviet espionage rule that agents should only steal "conservative, tested technology" (Rhodes 1995, p. 332). It worked.

1.1 Hanford Site Selection

"I thought the Hanford site was perfect the first time I saw it ... we found the only place in the country that could match the requirements for a desirable site."

—Colonel Franklin Matthias,
Working on the Bomb:
An Oral History of WWII Hanford (Sanger 1995, p. 19)

The Manhattan Project was organized under the Manhattan Engineer District of the Corps of Engineers. It was named after one of the Corps' New York offices to avoid arousing suspicion and any association with technology.

The goal of the Manhattan Project was to build the first atomic bombs in the shortest possible time. Could it be done in just a few years? Some advisors to General Leslie Groves, Army commander of the Manhattan Project, thought "the entire project seemed beyond human capability" (Groves 1962, p. 47).

Project security and the potentially hazardous nature of nuclear operations were concerns from the earliest days. Scientists and government officials including General Groves and the staff of E.I. du Pont de Nemours and Company[(1)] recognized the potential danger from radioactive byproducts. High levels of exposure were an immediate safety problem. Radionuclide releases were a potential long-term health and environmental threat. Yet, our understanding of the potential biological danger from these byproducts was in its infancy.

U.S. Department of Energy D3065 741fb.84

General Leslie Groves. General Leslie Groves, head of the Manhattan Project, addresses Hanford workers in 1944.

The possible hazards from reprocessing of plutonium were thought too great to warrant building a semi-works[(2)] outside Chicago. The first alternative site was in eastern Tennessee at the Clinton Engineer Works (now known as the Oak Ridge Reservation) where a rare type of uranium was being separated for the first atomic bomb. However, if a nuclear accident occurred near Knoxville, the "loss of life and the damage to health of the area might be catastrophic" (Groves 1962, p. 69). Besides, any serious accident near Chicago or Knoxville would eliminate project secrecy, contaminate large numbers of people, and likely halt work on plutonium—resulting in a "Congressional investigation to end all Congressional investigations" (Groves 1962, p. 70). Therefore, only sites in isolated areas of the western United States were considered.

Though Groves believed the possibility of serious problems was small, no one was certain what would happen, if anything, when a chain reaction was first attempted in a fully powered nuclear reactor or when plutonium was reprocessed inside a huge chemical plant. Safety protocols were nonexistent. "Nonetheless, there were many decisions that had to be made when the unknown factors far

(1) E.I. du Pont de Nemours and Company, known today as DuPont, was responsible for the design, construction, and operation of the Manhattan Project facilities.

(2) A semi-works is a manufacturing plant built on a small scale to provide final tests for all processes to be used in a full-scale plant.

outweighed the known" (Groves 1962, p. 81). This was frontier science where some personnel operated unofficially "on the theory that a scientist may do anything he desires as long as he kills no one but himself" (Williams 1948, p. 5).

Every month, the frontier of nuclear physics was advanced with new discoveries. However, radiation health sciences lagged behind.

Colonel Franklin Matthias, an associate of General Groves, headed site selection. Matthias later became the officer-in-charge of Hanford construction. Before reconnaissance visits began in late December 1942, several locations were discussed including western Montana, central Oregon (near Bend), southern California, and eastern Washington (Matthias 1942). Montana was quickly eliminated for the lack of adequate electrical power. On December 20, Colonel Matthias and his team visited the area around Mansfield, Washington, located 100 miles west of Spokane. Two days later, Matthias flew south along the Deschutes River to Madras, Oregon, and then northward to Hanford. The Oregon site would require pumping Columbia River water 2000 feet high over hilly terrain, had limited space, and "would cut into productive wheat land" (Matthias 1942). On the other hand, Matthias wrote that the "Hanford site [was] far more promising." The team traveled to California and visited two sites east of Los Angeles along the California-Nevada border. A third location, northeast of Redding, was considered but not visited.

Upon his return to Washington D.C., Matthias recommended Hanford to General Groves. Hanford offered several advantages over other sites: electricity from the newly constructed Grand Coulee Dam, high-quality river water for cooling the reactors, a railroad line, vast sand deposits, a mild and dry climate, and a small population. In addition, the Columbia River furnished a "convenient dispersal mechanism" (disposal site) for reactor cooling waters containing contaminants (Honstead 1954). This "wasteland with sagebrush vegetation" (Grady 1943), as it was described in one report, was perfect.

A directive was issued under the Second War Powers Act for the federal government to acquire 670 square miles of land in southeastern Washington State (Manhattan Project 1977).[3] In March 1943, the families of about 1300 people received letters telling them to pack up and move. They were given 2 weeks to 3 months (Parker 1986, p. 375). The government paid what it considered a fair-market value for the land—cents to a few tens of dollars for each acre (Heriford in Sanger 1995). The cost for all land acquired was $5.1 million (Manhattan Project 1977). Groves thought this price was "exorbitant" (Groves 1962, p. 77).[4]

While some viewed the area as a "wasteland with sagebrush and vegetation" (Grady 1943), others called it home. Today, a number of towns and cities are within 50 miles of the Hanford Site.

Soon, concept turned into reality. Construction began on the secret Hanford location identified only as Site W. The enriched uranium[5] production facility at the Clinton Engineer Works, Tennessee, was known as Site X, and the research and weapon fabrication facilities at Los Alamos, New Mexico, were designated Site Y.

(3) In 2002, Hanford covers 586 square miles—nearly half the size of Rhode Island.
(4) M.B. Parker (1986, p. 376) wrote, "Records show that people who fought the government about the value of their lands and took their cases to jury trial received as much or more than twice the amount that the government had set as fair market value. This court action took as long as five years."
(5) Enriched uranium contains more of the uranium-235 isotope than occurs in natural uranium.

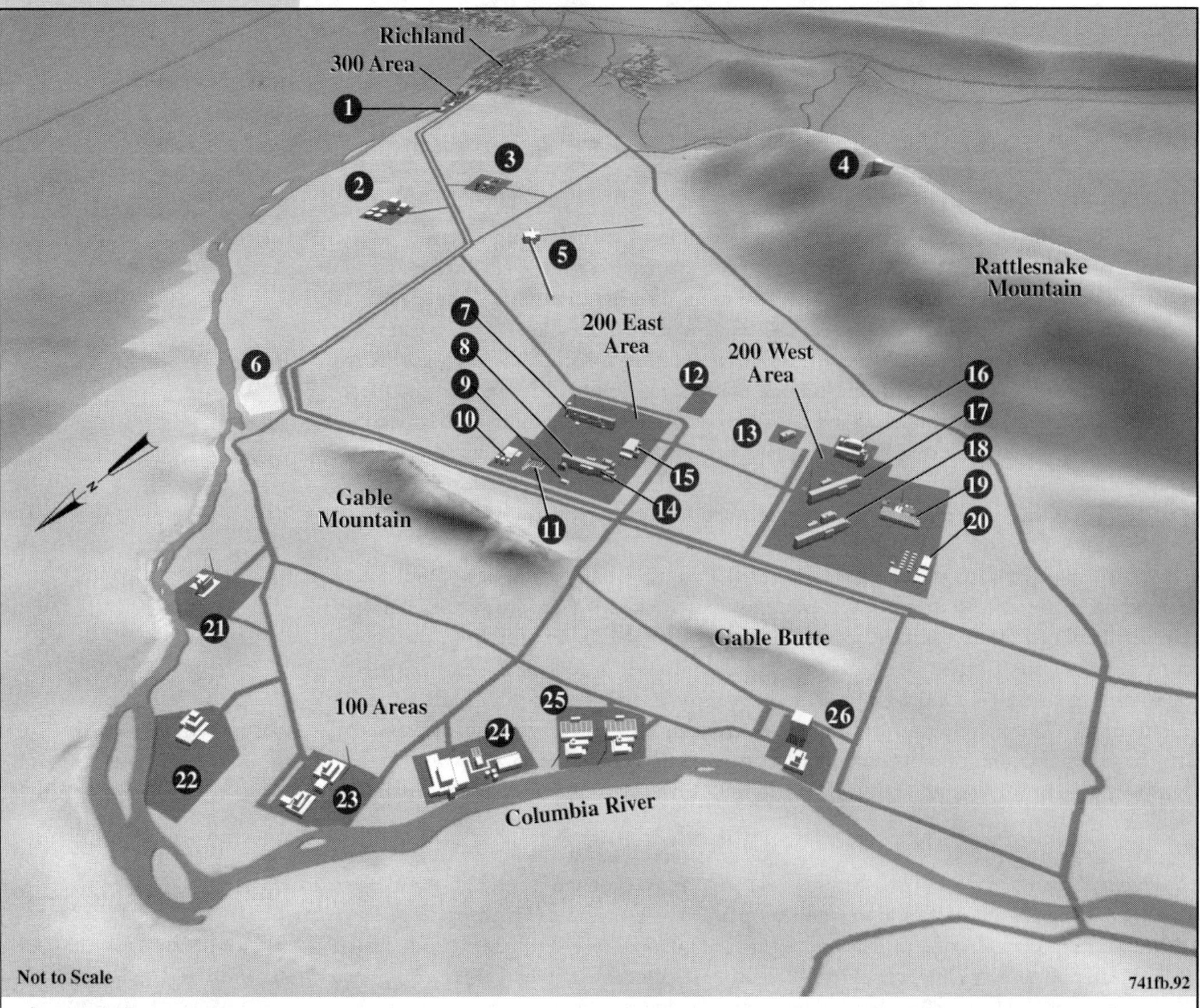

1. 300 Area Liquid Effluent Treatment Facility
2. Commercial Operating Nuclear Power Plant
3. Fast Flux Test Facility
4. Observatory
5. Laser Interferometer Gravitational Wave Observatory (LIGO)
6. Old Hanford Townsite
7. Plutonium-Uranium Extraction (PUREX) Plant
8. B Plant
9. Prototype Surface Engineered Barrier
10. 200 Area Liquid Effluent Treatment Facility
11. Submarine Burial
12. U.S. Ecology Commercial Solid Waste Site
13. Environmental Restoration Disposal Facility (ERDF)
14. Waste Encapsulation and Storage Facility (WESF)
15. Canister Storage Facility
16. Reduction-Oxidation (REDOX) Plant
17. U Plant
18. T Plant
19. Plutonium Finishing Plant
20. Waste Receiving and Processing (WRAP) Facility
21. F Reactor
22. H Reactor
23. D and DR Reactors
24. N Reactor
25. KE and KW Reactors; Cold Vacuum Drying Facility
26. B and C Reactors

Major Hanford Facilities and Geographic Features. Many of the Hanford facilities and onsite geographic features mentioned in this book are identified. This south-looking view is given to better display the layout of facilities in the northern half of Hanford.

1.2 Construction and Plutonium Production

"The first major construction was Camp Hanford, a sprawling temporary city for construction workers...The camp was a full-service town with banks, hospital, eight mess halls, taverns, movies, an auditorium for sports and dances, fire and police departments, baseball fields and a swimming pool."

—Steven Sanger, *Working on the Bomb: An Oral History of WWII Hanford* (1995, p. 68)

Hanford's prime contractors employed 31,000 workers in 1943 and 22,000 workers in 1944 (DOE 1997c). The height of Hanford employment came in mid-1944 when nearly 45,000 personnel worked onsite (Marceau et al. 2002). All aspects of Hanford's construction and purpose were secret. This bred rumors ranging from the plausible (manufacturing high-powered aircraft fuel) to the intentionally bizarre (summer home for President Roosevelt and his wife).

General Groves told his onsite manager, Colonel Matthias, "if the reactor blows up, jump in the middle of it, and save yourself a lot of trouble" (Sanger 1995, p. 77).

The first reactor (known as B Reactor) and reprocessing plant (known as T Plant) were built and first operated in late 1944[(6)]—less than 2 years after Enrico Fermi and his team at the University of Chicago proved that a controlled nuclear chain reaction was possible. DuPont officials believed that the Hanford reactor had a 60% chance of working.[(7)] Nonetheless, the pressure was intense to succeed with these first-of-a-kind facilities. General Groves told his onsite manager, Colonel Matthias, "If the reactor blows up, jump in the middle of it, and save yourself a lot of trouble" (Sanger 1995, p. 77).

Most incremental scale-up tests were bypassed to build these first facilities. Time-consuming engineering studies and pilot plants were sidestepped. Groves (1962, p. 72) wrote, "nothing like this had ever been attempted before...the great risks involved in designing, constructing, and operating plants such as these without extensive laboratory research and semi-works experience simply had to be accepted."

This caused problems. For instance, shortly after B Reactor went critical in September 1944, it began to lose power and shut down. The reactor was restarted over the next 3 days, but each time the power level dropped again. Something was absorbing the neutrons emitted from the uranium-235 in the nuclear fuel.[(8)] Physicists Enrico Fermi, John Wheeler, and Leona Marshall calculated that short-lived[(9)] xenon-135 was capturing the neutrons and not letting them participate in sustaining reactor criticality—that is, a condition in which a repeating succession of nuclear reactions (a chain reaction) could support itself.

(6) The first attempt to power up B Reactor was on September 26, 1944. Full power was not achieved until a few months later.

(7) Conversation with retired B Reactor operator William McCue, November 11, 1998.

(8) The number "235" following the word uranium is the element's atomic weight; that is, the total number of neutrons and protons inside the uranium nucleus.

(9) Half-lives are the average time for 50% of the atoms in a radioactive sample to decay away. By the time a radionuclide passes through 10 half-lives, it has essentially disappeared. The half-life of xenon-135 is 9.1 hours. See Appendix A for the half-lives of other radionuclides mentioned in this book.

Building the Hanford Site was the largest construction in human history. Why? Because each of the three key steps in creating plutonium call for constructing buildings with specialized missions. First, mined uranium was refined and fabricated into metallic uranium fuel rods that met precise engineering standards. At Hanford, machine shops and laboratories were built for fuel fabrication. Next, the fuel was shipped to a nuclear reactor, where it was exposed to neutrons in the reactor's core. Nuclear reactors and support facilities were constructed. Last, the plutonium contained inside the fuel rods was chemically removed in a separations or reprocessing plant. Mammoth separations plants were constructed. These steps were carried out in the 300, 100, and 200 Areas of Hanford, respectively.

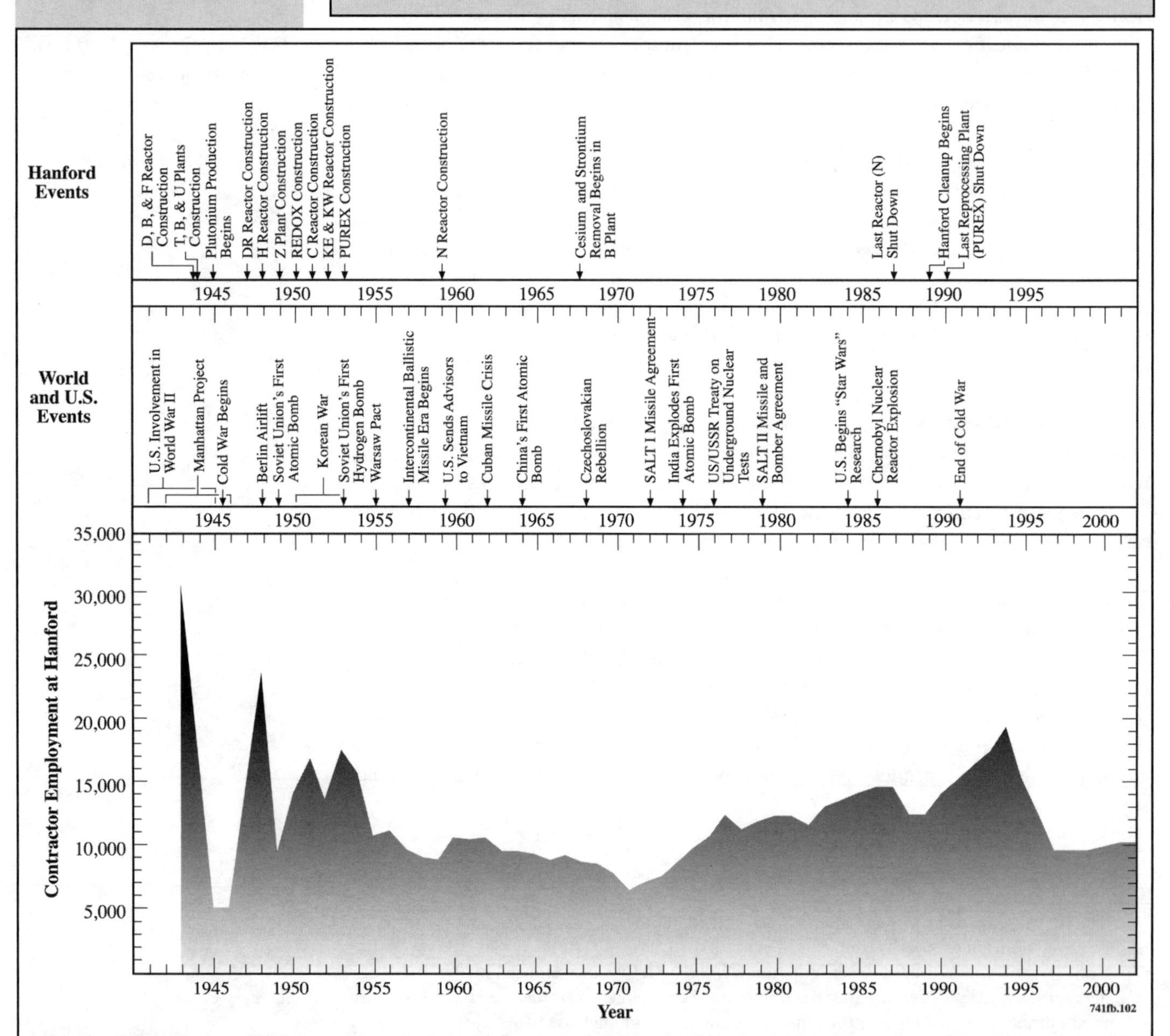

Hanford Employment. The number of prime contractor employees at Hanford varied over the years. At times, large numbers of subcontractor and military personnel also worked on the site. Plutonium production reflected national security and world events (Michael J. Scott, June 20, 2002, personal communication; Stapp 2002b).

Eventually, the problem was overcome by inserting extra uranium fuel into the reactor. This was possible because DuPont had over-designed the reactor by adding 500 extra process tubes that could hold more fuel rods. (The original reactor design contained 1500 tubes layout in a circular configuration.) During the design of B Reactor, physicist John Wheeler had insisted that the reactor's front be squared off and these extra 500 tubes inserted because of a nagging feeling he had about the possibility for some new fission product "poisoning" the nuclear reaction. His insistence cost millions of dollars and more months building the reactor. However, this extra uranium, and the neutrons it emitted, was enough to overcome the xenon poisoning. If B Reactor had been constructed to minimal specifications, it would have failed. Rebuilding would have been necessary, delaying plutonium production.

During the design of B Reactor, physicist John Wheeler had the nagging feeling that some new fission product would "poison" the nuclear reaction. He was right.

Engineering work was pragmatic and focused on results. Plutonium production goals were clear; jobs were well defined. A strong, integrated scientific underpinning for technology application and plutonium production was established at other Manhattan sites—Los Alamos (New Mexico), Berkeley (California), University of Chicago, and Clinton Engineer Works (Tennessee)—and then brought to Hanford. Originally, there was a "no research" attitude onsite. Science and technology advancements were imported. After the war, onsite research could be conducted "appropriate to the solution for local problems" (Parker 1952).

Originally, there was a "no research" attitude at Hanford. Science and technology were imported.

Key facilities were designed to handle specific jobs and built to run to failure. For the most part, major technology advances were carried out in the next-generation plant.

The reprocessing of the first batch of spent fuel[10] began on December 26, 1944, in T Plant located in the 200 West Area. In early February 1945, two military officials placed an inconspicuous wooden box containing a flask of syrupy plutonium solution into a car and drove to Portland, Oregon (Sanger 1995). From there, they traveled by train to Los Alamos and delivered 4 ounces of Hanford's first product. By May, a small armed convoy began carrying larger shipments.

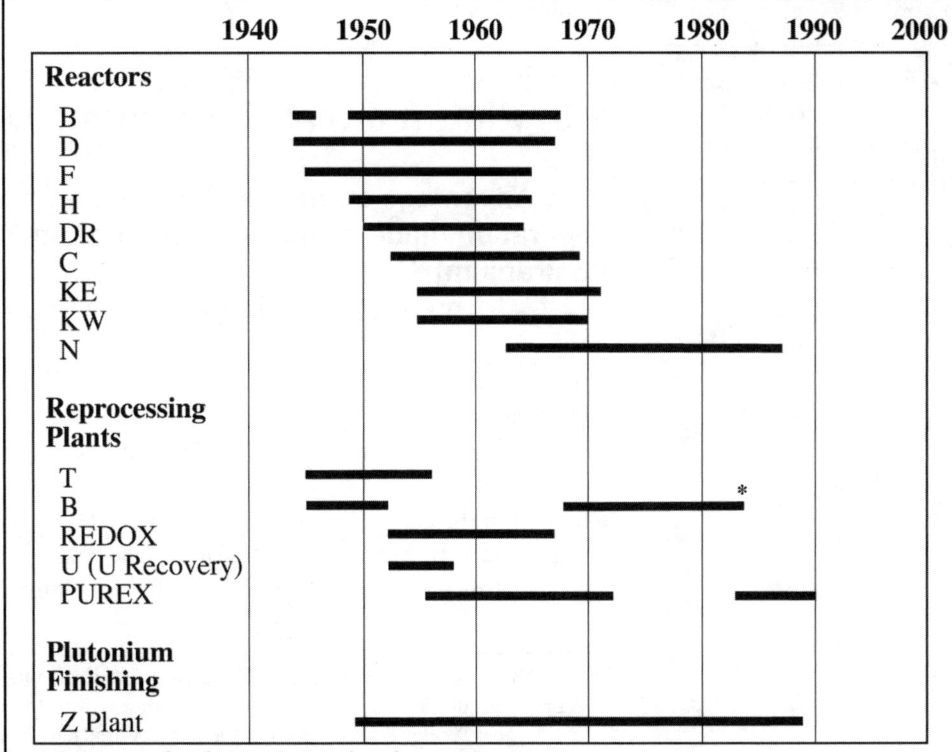

Operating History of Hanford Facilities. Nine nuclear reactors, five reprocessing plants, and other support facilities once operated at Hanford. Most were shut down by the early 1970s.

(10) Spent fuel is uranium fuel rods that have been removed from a reactor after exposure to neutrons.

The United States spent $2 billion (1940s dollars) on the Manhattan Project (Goldberg in Hevly and Findlay 1998). Accounting for inflation, this equals $20 billion dollars in 2002 value.

"To nearly everyone the news of what Richland was helping to make came as a complete surprise. Even those who may have been in the know would not admit it. The old habit of secrecy was strong upon them" (*The Villager* 1945, p. 1).

Hanford plutonium was used in the world's first nuclear test explosion, code named Trinity, at Alamogordo, New Mexico, in July 16, 1945, and in the atomic bomb dropped on Nagasaki, Japan, the following month.(11) The bomb released the explosive power carried by about 2000 B-29 bombers loaded with conventional weapons (Rhodes 1986, p. 737). The use of two atomic bombs led to the surrender of Japan and the end of World War II.

The United States spent $2 billion (1940s dollars) on the Manhattan Project (Goldberg in Hevly and Findlay 1998). About 20% ($350 million to $400 million) went to build Hanford and produce the first pounds of plutonium. Hanford was completed a year ahead of schedule and 11% over budget (Thayer 1996). In 1941 when the decision was made to begin a nationwide complex to build an atomic bomb, the initial (and likely informal) cost estimate was $133 million.

In the end, Groves guided the Manhattan Project in the direction of developing a vast infrastructure to create atomic bombs for use in World War II as well as national defense afterwards. This industrial complex was intended to turn out large numbers of bombs to provide the United States a "first strike" capability for the new Cold War (Rhodes 1995, p. 224). After the war and passage of the Atomic Energy Act of 1946, control of Hanford operations passed from the military to the civilian-run Atomic Energy Commission,(12) predecessor of today's U.S. Department of Energy (known by its initialism DOE).

1.3 Plutonium and Uranium: Metals of Choice

"The advantage of chemical extraction rather than isotope separation made element 94 [plutonium-239] a worthy competitor [to uranium]."

—Richard Rhodes, *The Making of the Atomic Bomb* (1986, p. 389)

Plutonium and uranium are the metals of choice for nuclear weapons. Both elements contain atoms that under the right conditions can fission and release large amounts of energy. Of the two, plutonium is the most fissionable.

Research in the early 1940s aimed at (1) separating a rare kind of uranium (uranium-235) from its more abundant counterpart (uranium-238) and (2) creating the new element plutonium-239. The Advisory Committee on Uranium focused

---

(11) The first atomic bomb, dropped August 6 on Hiroshima, used a special kind (isotope) of uranium (uranium-235) separated from other types of uranium in facilities at the Clinton Engineer Works. It was called "Little Boy." Three days later, the "Fat Man" atomic bomb was dropped on Nagasaki. It contained Hanford plutonium. This second bomb was physically larger because its plutonium core was surrounded by chemical explosives to implode upon the plutonium, thus creating a nuclear explosion.

(12) In 1974, Congress determined that the public interest was best served by separating the nuclear power licensing and related responsibilities of the Atomic Energy Commission from its nuclear material production and research activities. Under provisions of the Energy Reorganization Act, the Atomic Energy Commission was split into two agencies. Regulation of the commercial nuclear power industry went to the newly formed Nuclear Regulatory Commission. Material production stayed with the Energy Research and Development Administration (ERDA). Two years later, ERDA became the U.S. Department of Energy.

initially on the feasibility of using uranium-235 for nuclear weapons. However, after Glenn Seaborg and his co-workers created and synthesized microscopic amounts of plutonium at the University of California at Berkeley in 1941, the committee also sponsored research on its use.

In the early months of 1941, Glenn Seaborg, Emilio Segré, and fellow researchers produced milligram quantities of plutonium-239. The first pure plutonium metal was synthesized in late 1943. Thus, in less than 4 years, the nuclear physics, metallurgy, chemistry, full-scale industrial production, weapon design, and testing of plutonium-239 was accomplished.

Following a meeting at the University of Chicago with some of the world's top physicists, General Groves felt that "the plutonium process seemed to offer us the greatest chance for success in producing bomb material. Every other process then under consideration depended upon the physical separation of materials having almost infinitesimal differences in their physical properties" (Groves 1962, p. 41). Because of the cost and difficulty of separating uranium-235 from uranium-238, creating plutonium was considered a shortcut to building the first atomic bombs. Besides, for a given amount, plutonium-239 was more explosive.

Manhattan Project personnel believed that no more than about one bomb's worth of uranium-235 could be produced in 1 to 2 years (Serber 1992). Few people thought a single nuclear weapon would make a decisive difference in World War II. After all, the Allies were fighting on multiple fronts. This drove interest in creating a plutonium weapon.

Production of both materials was pursued. Uranium-235 was separated at the Clinton Engineer Works, and plutonium-239 was created at Hanford. At the same time, scientists at the Los Alamos National Laboratory explored the theory of nuclear fission to establish the knowledge for developing workable weapons of mass destruction.

Creating plutonium was considered a shortcut to building the first atomic bombs.

Natural uranium is made of 99.3% (by mass) uranium-238 and 0.7% uranium-235. These are two of the nearly two dozen kinds of uranium that can exist (Lide 1995). Uranium-235 is very fissile. It can support a chain reaction, caused when it captures a neutron, briefly transforms into uranium-236, splits into fragments, and releases more neutrons that cause other uranium-235 atoms to capture a neutron and split.

On the other hand, uranium-238 is not fissile but is fertile. That means it absorbs neutrons and transforms itself into other isotopes.[13] Plutonium-239 is formed when uranium-238 captures a neutron released from the fission of uranium-235. The

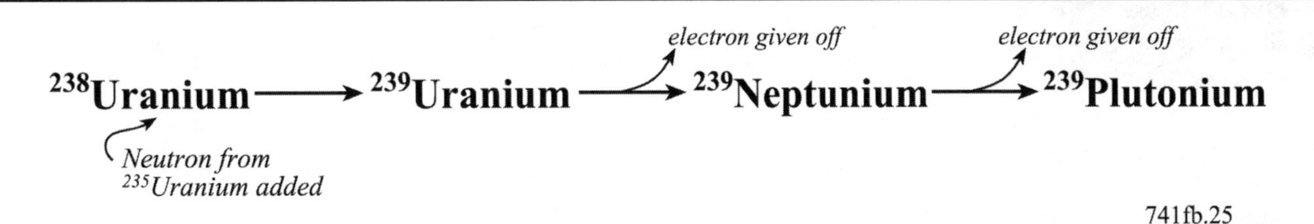

Creating Plutonium from Uranium. This nuclear reaction creates plutonium inside a reactor. When uranium-235 fragments (fissions), it releases neutrons. A neutron is captured by an atom of uranium-238 and converts it to uranium-239. The uranium-239 changes to neptunium-239 and then to plutonium-239. Two electrons (beta particles) are given off during this process.

(13) Isotopes are different types of the same element distinguished by having different numbers of neutrons in their nucleus. A single element may have many isotopes. Some isotopes are radioactive; others are not.

Each time a nucleus splits, it releases some of the energy that binds it together.

uranium-238 is briefly changed into uranium-239. It then decays by nuclear reactions to neptunium-239 and then to plutonium-239. This process takes 2.4 days and occurs inside a nuclear reactor. Longer exposure allows more opportunity for an atom of plutonium-239 to capture another neutron and change into plutonium-240, which is deleterious to the efficient detonation of a nuclear weapon. Therefore, to create a purer kind of plutonium, irradiation times were short (few weeks near the center of a reactor core), and uranium throughput was high.

Each time a nucleus splits, it releases some of the energy that binds it together. If enough energy is released suddenly in a highly pressurized and confined condition, an explosion occurs. This is what takes place inside a nuclear weapon.

Only a small baseball-sized, hollow sphere of plutonium-239 metal, weighing about 13 pounds, was needed for the first plutonium bomb (Rhodes 1995, p. 174). This sphere was imploded using 5000 pounds of chemical explosives. About 130 pounds of uranium-235, kept in two separate, sub-critical grapefruit-sized masses, were required in the first uranium bomb (Schwartz 1998, p. 548). The two masses were rapidly slammed together in a gun-type device inside the weapon.

The United States produced 114 tons of plutonium in government reactors at Hanford and the Savannah River Site in South Carolina (Usdin 1996).

Separating uranium-235 from the more abundant uranium-238 is difficult. Several processes were used in Tennessee. These relied on minute differences in how different uranium atoms diffused through porous membranes or behaved inside a strong electromagnetic field. Uranium-235 is less massive than uranium-238 because it has fewer neutrons in its nucleus. Therefore, uranium-235 diffused more rapidly and was more easily deflected.

Plutonium is classified into different grades depending on the amount of plutonium-239 present. Weapons-grade plutonium has 94% or greater (by mass) of plutonium-239 (DOE 1994a). The remaining 6% is plutonium-240 plus minor amounts of other isotopes such as plutonium-241 and -242. Power reactor fuel-grade plutonium contains 82% to 94% plutonium-239.

Between 1944 and the early 1990s, the United States produced 114 tons of plutonium in government reactors at Hanford and the Savannah River Site in South Carolina (Usdin 1996). This was in response to such world events as the first detonation of nuclear weapons by Russia (1949), the Korean War (1950-1953), Sputnik (1957), introduction of Russian intercontinental ballistic missiles (1957), the Cuban missile crisis (1962), the detonation of nuclear weapons by China (1964), and the Cold War (1945-1991).

Having a Russian satellite passing overhead every 90 minutes struck fear in the hearts of many Americans. Oceans no longer ensured safety from immediate attack. The story of Sputnik and the U.S. response is found in *Sputnik: The Shock of the Century* by Paul Dickson (2001).

Hanford generated 65% (74 tons) of the nation's plutonium tonnage (DOE 1996d; Schwartz 1998).[14] This consisted of 60 tons of weapons-grade plutonium and 14 tons of fuel-grade plutonium reprocessed from 106,600 tons of uranium fuel.[15] (For comparison, 40 tons of weapons-grade plutonium were produced at the Savannah River Site.) Based on this information, one can estimate that on an average, for each ton of fuel reprocessed, 1.4 pounds of plutonium were recovered at Hanford.

(14) If all 74 tons of plutonium were combined, they would form a dense metal cube measuring 5 feet on a side. In reality, such a combination could not take place without causing a runaway chain reaction and nuclear explosion.

(15) If all 106,600 tons of uranium fuel were combined, they would form a cube 70 feet on a side.

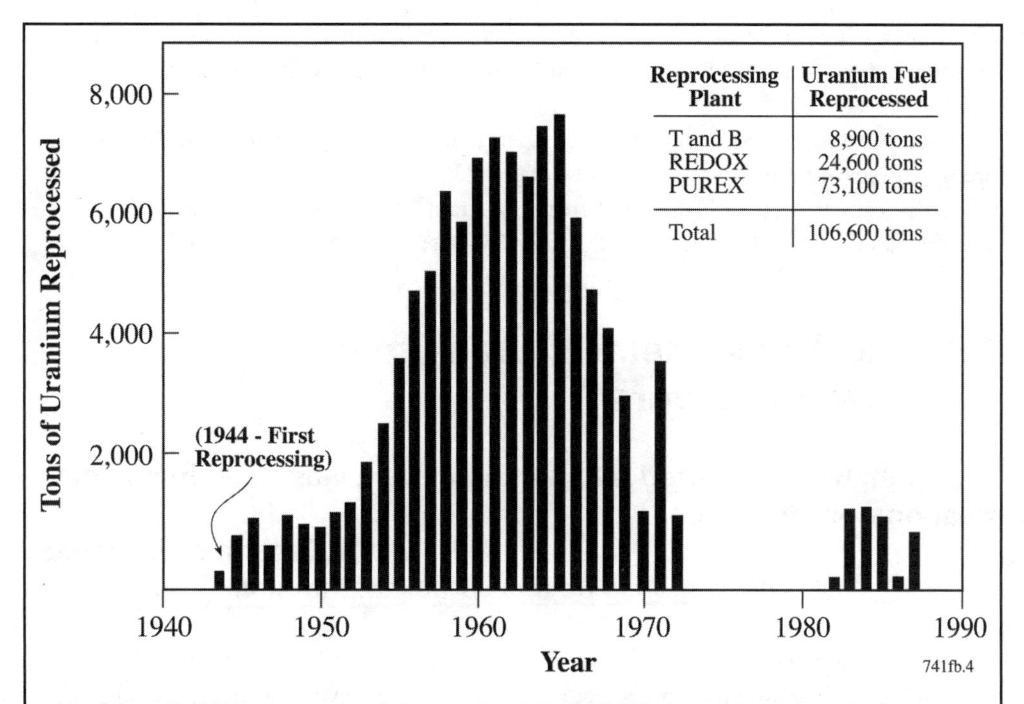

Uranium Fuel Reprocessed. Approximately 106,600 tons of uranium fuel were reprocessed at Hanford. Most of the reprocessing took place in the 1950s and 1960s. If all this uranium were combined, it would form a cube measuring 70 feet on each side. Most uranium was reprocessed in the PUREX Plant.

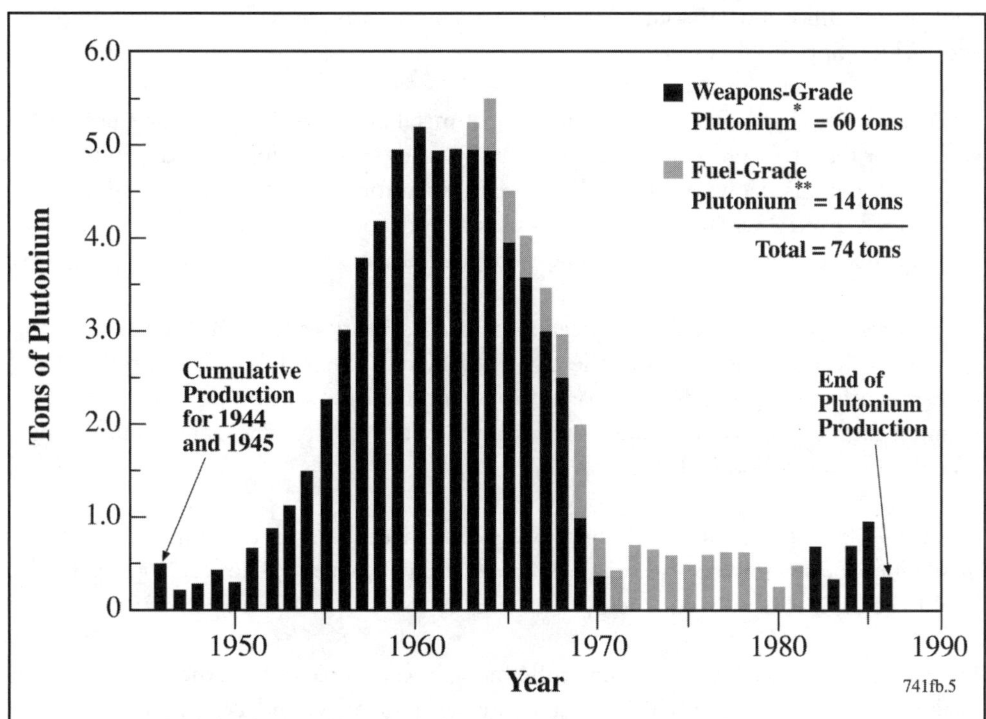

* Contained greater than 94% plutonium-239

** Contained 82% to 94% plutonium-239

Plutonium Produced. Seventy-four tons of plutonium were created at Hanford. If this plutonium could be combined, it would form a dense metal cube measuring 5 feet on each side. Most plutonium was high-grade weapon material (DOE 1996d).

For each ton of uranium fuel reprocessed, 1.4 pounds of plutonium were recovered.

Ninety percent of Hanford's plutonium was created before 1971. Plutonium production formally ended in 1990 when the last reprocessing plant was shut down.

Three steps were carried out to produce plutonium. First, metallic uranium fuel was prepared. Next, the fuel was exposed to neutrons inside a reactor. Last, the plutonium was chemically treated and separated from the uranium fuel in a reprocessing plant. These steps took place in the 300, 100, and 200 Areas of Hanford, respectively.

1.4 Fuel Rods: Metallic Uranium to Power Reactors

"Uranium was converted into metal before it was used in nuclear weapons production."

—U.S. Department of Energy,
Closing the Circle on the Splitting of the Atom (1995a, p. 15)

The first step in creating plutonium is mining uranium ore and then refining it into uranium metal. The Manhattan Project acquired ore from several sources, including the Belgian Congo, Canada (Northwest Territories and Ontario), Colorado, and Utah (DOE 1997a). Post-war sources included Australia and South Africa plus several western states such as Oregon, Idaho, and Washington. Much of the refined uranium metal used at Hanford during the Manhattan Project was shipped from facilities in New Jersey, Ohio, and Missouri. After the war, key sources shifted to Fernald, Ohio, and Weldon Spring, Missouri.

About 250 tons of uranium ore were needed to create 1 pound of plutonium.

Uranium fuel, called slugs, was made into short metal cylinders. Most of the slugs created for the first eight reactors were about 8 inches long, 1.4 inches wide, and weighed 8 pounds. However, those later used in Hanford's dual-purpose N Reactor were larger, up to 26 inches long and 2.4 inches in diameter and were of a double tube (tube-in-tube) design (Miller 1976). These weighed about 55 pounds each.

Slugs were prepared in the fuel fabrication part of Hanford (300 Area) built next to the Columbia River and just north of the city of Richland. The 300 Area was located closer to residential areas than any other part of the plutonium production process because uranium fuel contains only natural amounts of radioactivity and the fuel manufacturing process was not considered dangerous.[16]

Hanford is home to nine reactors and five reprocessing plants.

About 80% of the uranium used onsite was unenriched; that is, it contained the same mass ratio of uranium-238 and uranium-235 as found in nature. The rest, such as used in N Reactor, was slightly enriched (containing up to 1.2% uranium-235 by mass).

Initially, uranium metal was received at the fuel fabrication facilities in the form of short, thick metallic blocks called billets about 4.5 inches in diameter and 12 to 20 inches in length. Each weighed 250 to 325 pounds (Sanger 1995, p. 174). These were heated and

(16) The 300 Area did contain a small water-cooled test reactor for studying experimental plutonium fuels and recycling. Its official title was the 309 Plutonium Recycle Test Reactor. The reactor operated from 1960 to 1968 at a power level of 70 megawatts. The reactor's history is summarized at http://www.hanford.gov/history/300area/300-8th.htm#300-8-1.

U.S. Department of Energy

741fb.52
2052

300 Area. Uranium fuel preparation and research took place in the 300 Area, located along the Columbia River in the south end of Hanford. This 1953 photograph looks northeast and shows two large liquid seepage ponds along the riverbank used for temporary wastewater storage.

extruded into long uranium rods and then cut into shorter pieces. A water and oil mixture was used during uranium lathing and cutting to keep any fine particles from sparking or igniting.

Fuel manufacturing advanced over the years to lower costs and improve performance. In general, slugs used in the first reactors were dipped into various molten metals with a final dip in an aluminum-silicon alloy blend. They were then inserted into thin aluminum tubes, which were no thicker than a few sheets of paper, held inside steel jackets. The entire process was called canning. The canned uranium improved heat conduction, protected the uranium from corrosion,[17] and contained radionuclides created when the uranium was later exposed to neutrons. The canned uranium was machined to length, the ends were welded shut, and the final slugs were cleaned with chemical solutions. The slugs, now called uranium elements, were ready for railcar shipment to a reactor.

U.S. Department of Energy

1727-1
741fb.54

Uranium Fuel Elements. Uranium fuel irradiated inside reactors was machined into short cylinders and coated with metal. This 1958 photograph shows examples of uranium slugs used during the early years.

(17) Uranium chemically reacts (oxidizes) if in direct contact with hot water as found inside Hanford's water-cooled reactors.

Initially, the slugs were solid. By the mid-1950s, a variety of hollow slugs were made to permit water flow both around and through them. This improved their thermal behavior when heated by fission.

In N Reactor, zirconium-coated fuel was used. This fuel was made from a co-extrusion process wherein a zirconium-covered uranium billet, drilled through the center, was heated and extruded into a long rod. The metal was cut to length, and caps were welded to the ends.

Between 1955 and 1964, as many as eight reactors were operating at once.

Between 1955 and 1964 when as many as eight reactors were operating at once, about 30,000 slugs were canned each week. Nearly 70,000 slugs were loaded inside the first reactor—B Reactor—at any one time.

Before irradiation, the slugs were slightly radioactive, relatively safe to handle, and easily transported. Following irradiation, their radioactivity increased several million times (Parker 1959b). Now, they were very dangerous and were handled with great care and behind protective shielding.

1.5 Reactors: Where Plutonium Was Born

"All [reactor] design was governed by three rules: 1, safety first against both known and unknown hazards; 2, certainty of operation—every possible chance of failure was guarded against; and 3, the utmost saving of time in achieving full production."

—General Leslie Groves,
Now It Can Be Told:
The Story of the Manhattan Project (1962, p. 83)

Following irradiation, the radioactivity of the uranium fuel slugs increased several million times.

The transformation of uranium into plutonium took place inside reactors. Their designs included (Stapp 2002a):

Fuel: a few hundred tons of uranium fuel were machined into short tubes, coated with a layer of metal, and placed inside process tubes running through the core of a reactor. Different kinds of uranium provided the neutrons to sustain a chain reaction and to make plutonium.

Moderator: graphite blocks slowed down neutrons so that they could interact with atoms of uranium.

Control System: neutron-absorbing control and safety rods were removed or inserted through the graphite blocks to initiate or slow down the chain reaction.

Cooling Water (Coolant): river water was filtered and chemically treated before it was pumped through the process tubes to cool the reactors.

A cubical stack of 75,000 graphite blocks formed the 36 feet wide, 36 feet tall, and 28 feet deep interior of B Reactor.

Shielding: thick walls made of concrete, iron, and wood surrounded the reactor's core to protect workers from high levels of radiation.

Nine reactors were built between 1943 and 1963. Each was located along the shore of the Columbia River from 5 to nearly 15 miles north of the chemical reprocessing plants. This separated the reactors and reprocessing plants in case of an accident. In addition, each reactor, or group of two reactors, was separated 2 to 3 miles from the next closest reactor site.

741fb.57

B Reactor. B Reactor was the world's first industrial-scale reactor. This reactor created the plutonium used in the first atomic bomb test in July 1945 and the second bomb detonated over Japan.

B, D, and F Reactors were constructed during the Manhattan Project. B Reactor was the first completed. Construction began in August 1943. It began partial operation 13 months later in September 1944.

H, DR, and C Reactors came online during the years of the first Russian atomic bomb test (1949) and the Korean War (1950-1953). Hanford's largest reactors, KE and KW, began operating in 1955 while N Reactor started in 1963. These were the

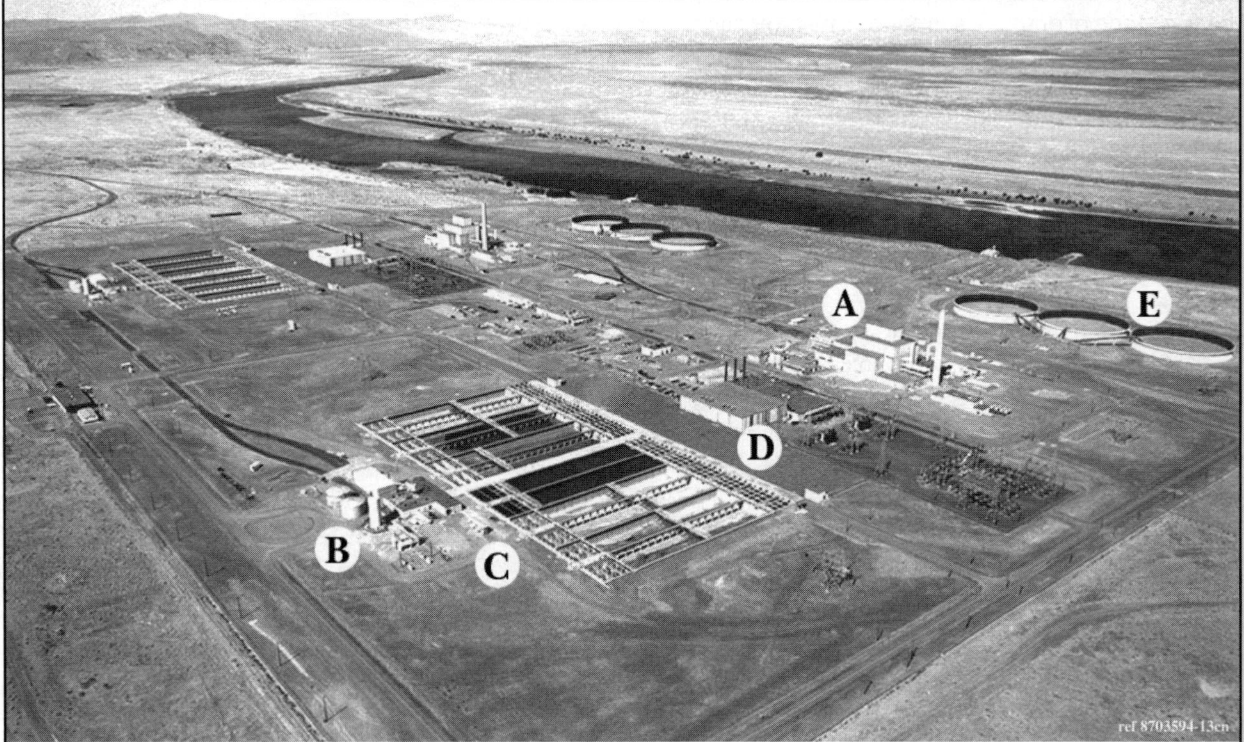

U.S. Department of Energy

741fb.13

A. Reactor Building
B. Water Treatment Plant
C. Treated Water Storage Basins
D. Pump House Supplying Cooling Water to Reactor
E. Water Retention Tanks for Radiation and Heat Decay

KW and KE Reactors. Uranium fuel was exposed to neutrons inside reactors built along the banks of the Columbia River. The river supplied cooling water. This 1987 photograph, looking upstream on the Columbia River, shows the two reactors (large block-shaped buildings near chimneys) alongside their support facilities.

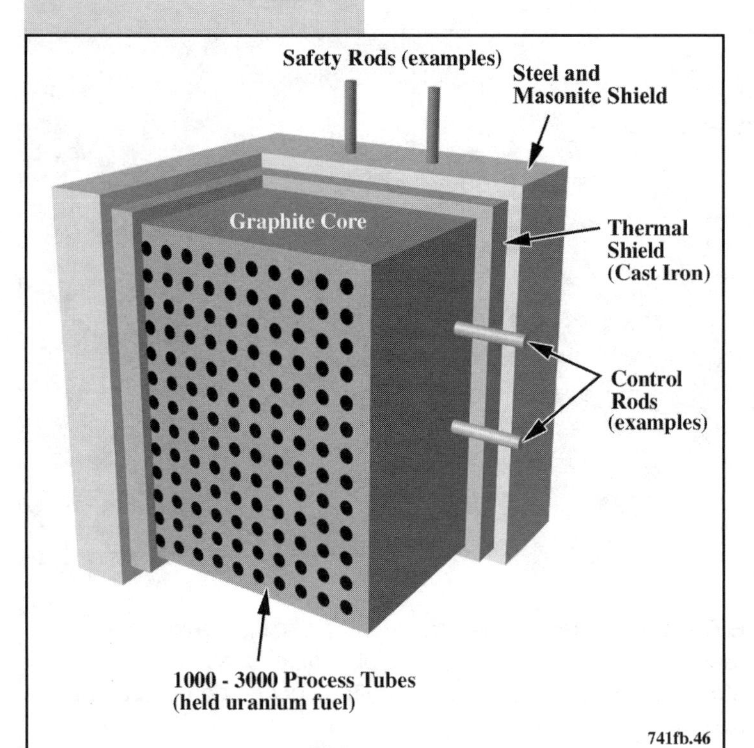

Nuclear Reactor Core. Each reactor contained a cube-shaped center made of graphite surrounded by protective walls of metal and Masonite®. Cores were the size of small block-shaped houses. Graphite slowed down (moderated) fast-moving neutrons to sustain the chain reaction and increase the chance for an atom of uranium-238 to capture a neutron and transform it into an atom of plutonium-239 (DOE 1997a).

years of Intercontinental Ballistic Missile deployment, fear of a bomber gap between the Americans and the Soviets, signing of the Warsaw Pact,(18) the beginning of U.S. involvement in Vietnam, and the Cuban missile crisis.

N Reactor served a dual purpose of generating electricity and creating plutonium. All of Hanford's fuel-grade plutonium was produced in N Reactor because the uranium slugs remained inside the reactor for a longer time than needed to create just weapons-grade plutonium-239. This caused additional neutrons to be captured by uranium-238, transforming more of it into plutonium-240.

Hanford reactors were graphite moderated because their interior cores or "piles" were built from thousands of finely machined and tightly stacked graphite blocks. Graphite is a kind of carbon. The graphite used for blocks was hardened.(19)

Inside a reactor, the graphite slowed down (moderated) neutrons given off by the fission of uranium-235 by bouncing neutrons off the carbon atoms. A cubical stack of 75,000 graphite blocks formed the 36 feet wide, 36 feet tall, and 28 feet deep interior of B Reactor. The block weighed 2200 tons. About 2000 small aluminum process tubes were inserted through holes in the pile to hold the fuel slugs. This block was enclosed within a 10-inch-thick cast-iron thermal shield and a 4-foot-thick biological shield made of sandwiched layers of steel and a pressed wood fiber (Masonite®). The thermal shield absorbed most of the reactor's heat not removed by water. Materials in the biological shield also absorbed neutrons and gamma rays to reduce radiation levels outside of the block. Water was pumped from the Columbia River, treated in purification plants, and passed through the reactor. Chemicals were added to adjust pH (lime), prohibit algal growth (chlorine), remove dissolved solids (ferric or aluminum sulfate and activated silica), and lessen metal corrosion (sodium dichromate) (Foster et al. 1954). Dissolved and suspended minerals were reduced by filters. Helium gas was circulated through the sealed pile to remove moisture-containing air and reactor-generated gases that would absorb neutrons. Later, carbon dioxide was added to the helium to reduce radiation-induced swelling of the graphite. Swelling caused jamming of the slugs and control rods plus it created unwanted gaps in the shielding walls through which radiation could egress (Parker 1948b).

The fissioning of uranium-235 atoms and the resulting radiation releases heated the reactor's cooling water to near boiling temperatures.

(18) A treaty of friendship, cooperation, and mutual assistance signed in 1955 between Albania, Bulgaria, Hungary, East Germany, Poland, Romania, the Union of Soviet Socialist Republics, and the Czechoslovak Republic.

(19) A soft, dark, and much less pure kind of carbon is used for making pencil "lead."

In the first eight reactors built, water was pumped through the reactor, heated, discharged to aboveground tanks, and released to the Columbia River. These were called single-pass reactors, because water made one pass through the core. Some water, contaminated with high levels of chemicals and radionuclides, was discharged to the nearby soil and the river. The possibility of avoiding contamination of the river was remote unless "drastic revisions" were made to the cooling of these reactors—such as re-circulating the water and cooling it by heat exchangers (Gamertsfelder 1947). Water contamination was a natural consequence of using single-pass reactors because as dissolved chemicals passed through the reactor, they absorbed neutrons and became radioactive. Examples included arsenic-76, manganese-56, sodium-24, and zinc-65. In the first days and weeks following release, these newly formed radionuclides raised the amount of radiation received by people and other organisms living downstream (see Chapter 3). After a couple of months, few of these radionuclides remain because of their short half-lives.

Water contamination was a natural consequence of using the single-pass reactors because chemicals absorbed neutrons and became radioactive.

A host of new radionuclides, called fission products, also formed in the fuel rods after uranium-235 captured a neutron and its nucleus fragmented. Sometimes, the metal coating bonded to the uranium slug would fail, releasing these fission products into the cooling water and into the river.

Two fission products, strontium-90 and cesium-137, created most of the radioactivity now found at Hanford.

In the early 1960s, water re-circulation was built into N Reactor. Its water was recycled within a closed-loop system and not discharged to the river. When a fuel slug ruptured, contaminated cooling water was diverted from the reactor and released underground to the soil.

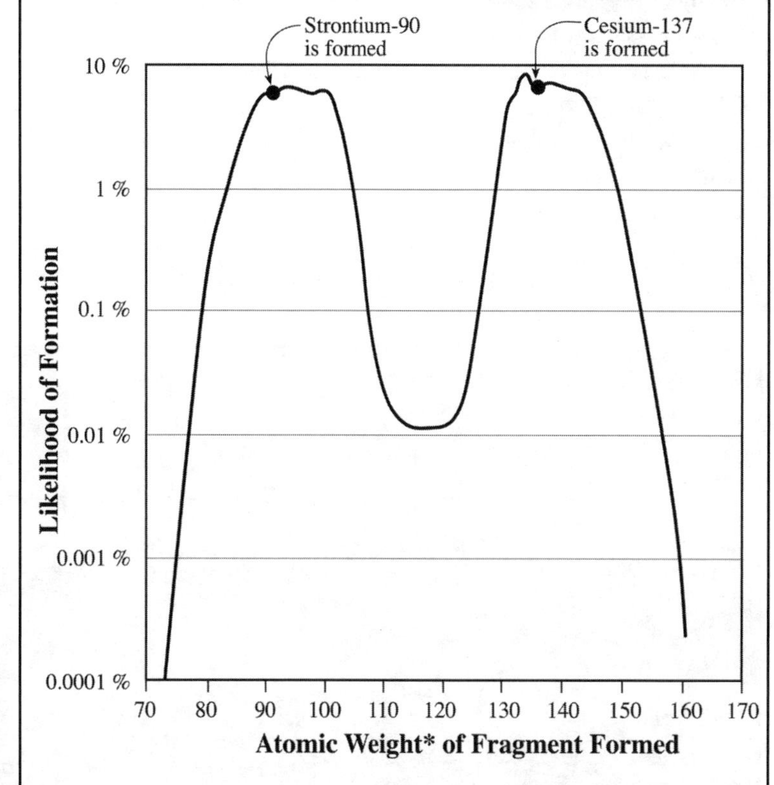

Creating Unwanted Radionuclides. When atoms of uranium-235 absorb neutrons, they split into fragments. This illustration shows the likelihood of different radioisotopes, having different atomic weights, being created during that fission process. The two peaks correspond to atomic weights of about 90 and 137—the atomic weights of strontium-90 and cesium-137. This high likelihood of formation, plus their nearly 30-year half-life, is why most radiation at Hanford is now dominated by cesium and strontium.

"Present plans call for holding the cooling water effluent from the pile [reactor] in a retention basin for nearly 3-1/2 hours, which is appreciably in excess of the time computed to be necessary for the normally expected noxious properties to die away. Nonetheless, because of uncertainty concerning the tolerable dosage for fish…it has been deemed advisable to consider means for ensuring rapid and effective mixing of the effluent with the river."

—T.B. Drew, *Admixture of Pile Effluent With River* (1943)

Hanford's reactors had an average operating life of 18 years.

The first reactor, B Reactor, had an initial power level of 250 megawatts, significantly larger than the initial level of the first small-scale test reactor at the University of Chicago with a power level of just one-half watt (Rhodes 1986, p. 440).

Hanford's reactors had an average operating life of 18 years. DR Reactor operated the shortest time (14 years), while N Reactor operated the longest (23 years).

The reactors were of different sizes and power levels. Between 220 and 430 tons of uranium fuel were inserted into 1003 to 3220 aluminum tubes running through the reactor's core (Miller 1976; Miller and Steffes 1987). (The tonnage and the number of process tubes varied between generations of reactors.) Reactor piles measured 23 feet to 36 feet on each side. The first reactor built had an initial power level of 250 megawatts[20] (DuPont 1946). Power levels increased to 4400 megawatts when Hanford's largest reactors (KE and KW), built in the mid-1950s, were upgraded in the 1960s.

The critical mass for Hanford's first reactor was reached when 65 tons of uranium fuel were loaded (Miller 1976). Critical mass is the amount of uranium needed to sustain a low-powered chain reaction. Full power was reached when 220 tons of uranium were loaded. For the larger KE and KW Reactors, the critical mass increased to about 195 tons of uranium. Full power was reached when 430 tons of uranium were inserted.

Once irradiated, the fuel slugs were dangerously radioactive. Workers used high-pressure water to push slugs through the process tubes and into a water-filled storage basin located on the backside of the reactor's core. This pool of water cooled the fuel and protected workers by absorbing most of the radiation emitted.

Using long tongs, the irradiated slugs were sorted, loaded inside large metal mesh-buckets, and kept underwater until it was time for reprocessing.[21] One-half to one ton of slugs were held in each bucket. The buckets were placed inside larger lead-lined, water-filled casks. Heavy lids were placed atop each cask. The casks were lifted out of the water and loaded in water-filled tanks (called cask cars) resting atop flatbed railroad cars. The casks were then moved to a reprocessing plant unless temporarily stored in the lag storage buildings halfway between the reactors and reprocessing plants. Three of these storage sites were used from 1945 to the mid-1950s (Freer and Conway 2002).

Normally, transportation took place at night when fewer workers were onsite. Nearby roads were barricaded to lessen the chance of people receiving unwanted radiation exposure called "shine." Once inside a plant, an overhead crane removed a cask's lid, lifted the bucket, and dumped the fuel into the

741fb.55

Railroad Entry into PUREX. Locked fences remind visitors that Hanford no longer produces plutonium. This is the gate where irradiated uranium fuel once entered the PUREX Plant in shielded railroad cars. The tracks are curved and concrete walls were built near the entrance to block direct radiation "shine" to workers who were outside the entrance when the fuel was remotely unloaded inside the building. Increased radiation exposure was possible because the entry doors were thinner than the building's walls.

(20) 250 megawatts is 250 million watts.
(21) Conversation with Steve Buckingham, retired Hanford reprocessing operator, November 10, 2000.

dissolver chute. This began the first step to chemically dissolve and separate plutonium from uranium and other radioactive byproducts.

Ruptured slugs, broken by tumbling into the reactor's back basin, were handled in much the same manner described above except "they were manually placed in cans with a screw-type cap prior to loading into the storage buckets" (Irish 1959). Because fine particles of bare irradiated uranium can burn, workers kept ruptured slugs away from the air. Nonetheless, when ruptured slugs were ejected from a reactor, sparks were sometimes observed during their free fall into the water basin.

In early 1964, President Johnson decided to begin closing down the older single-pass reactors (ERDA 1975). By the end of 1971, only N Reactor remained operating.

Few features were built into the first eight reactors to reduce the amount of radioactivity and chemicals released into the Columbia River. On the other hand, during the first 12 years of Hanford operations, the fuel reprocessing plants underwent major redesigns to improve operation efficiency, increase safety, and reduce the amount of high-level radioactive waste generated.

Normally, the transportation of irradiated fuel took place at night when fewer workers were onsite. Nearby roads were barricaded to lessen the chance of people receiving unwanted radiation exposure.

1.6 Reprocessing Plants: Where Plutonium Was Recovered

"The scale of the ultramicrochemistry work at [University of] Chicago was increased by a factor of a million at Hanford. Two huge chemical separation plants [T and B Plants] were built and nicknamed 'Queen Marys' after the famous ocean liner."

—Rachel Fermi and Esther Samra, *Picturing the Bomb* (1995, p. 74)

Large, windowless, concrete-gray buildings called reprocessing plants used chemical techniques to separate uranium and plutonium from unwanted byproducts contained in irradiated uranium fuel. These facilities accounted for more than 85% of the radioactivity stored or released during the production of material for nuclear weapons (DOE 1997a).

Five reprocessing plants were built in the 1940s and 1950s: T Plant, B Plant, U Plant, REDOX (for REDuction-OXidation) Plant, and PUREX (for Plutonium-URanium EXtraction) Plant.[22] Multiple plants were constructed to meet plutonium production goals, to implement process efficiency improvements, and for redundancy. All were sited in the 200 Area, on a plateau in central Hanford. This took advantage of the higher elevation to disperse radioactive gases released from stacks, a thick layer of underlying dry sediments to absorb contaminants dumped into the ground, and distance (7 to 10 miles) from the Columbia River where contaminated groundwater would eventually enter (Parker 1959c).

(22) REDOX Plant is also known as S Plant. PUREX Plant was sometimes called the A Plant.

Reprocessing plants accounted for more than 85% of the radioactivity stored or released during the production of material for nuclear weapons.

These building are large. Four of the five plants are about 810 to 1005 feet long, 85 to 160 feet wide, and 85 to 105 feet from top to bottom. Twenty to forty percent of their height was below ground level (Freer and Conway 2002). The fifth plant (REDOX) was stockier—about 470 feet long, 160 feet wide, and 85 feet tall (except for a 130-foot-tall silo area on one side that housed 60-foot-tall gravity-feed solvent extraction columns[23]).

Each plant contained large open areas, called galleries, and chemical process areas, separated by thick (up to 7 feet) steel reinforced concrete walls. Separate galleries contained the electrical, piping, and operational floors. The chemical process areas resembled long, rectangular canyons with vertical concrete walls. The floor was called a canyon deck. Heavy cell block lids, marking the tops of about 40 enclosed chemical process cells (the number of cells varied between plants), covered the deck. Individual cell covers could be 6 feet thick and weigh 30 tons (Sanger 1995). These cells were tall, heavily shielded chemical laboratories built underground so that liquid solutions could pass between interconnected cells without exposing workers to high radiation levels. Individual process cells in T, B, and U Plants were 13 feet by 18 feet and 22 feet high (Gerber et al. 1997). Reprocessing plants were designed for remotely handling spent fuel and chemical solutions using overhead cranes and manipulators.

U.S. Department of Energy

U.S. Department of Energy 741fb.82

T and PUREX Plants. Five gray, windowless reprocessing plants were built to recover plutonium from irradiated uranium fuel. Four were used for that purpose. The first plant was T Plant (upper photograph). The last plant was PUREX (lower photograph). Most plants were 800 to 1000 feet in length and about 100 feet in height.

First, the spent fuel cladding or metal coating was dissolved in a hot solution of sodium hydroxide. This process was called decladding. Next, the bare uranium was dissolved in nitric acid. These steps turned solid uranium into a thick slurry. Plutonium was then separated from other radionuclides and chemicals using chemical precipitation or solvent extraction techniques.

Two-hundred-foot-tall concrete stacks were located at the front end of reprocessing plants, where the irradiated uranium fuel was delivered. Radioactive

(23) Solvent extraction is the separation of elements, such as uranium and plutonium, based upon their ability to be soluble in different chemical solutions. This is done by adjusting an element's valence state, that is, the combining (bonding) power of an atom for other atoms based upon the transfer of electrons.

gases generated by plutonium recovery were released into these stacks, especially when the uranium was dissolved. For the first few years, filters were not used to reduce radionuclide emissions. Spent fuel was dissolved at night (Gamertsfelder 1947) when the air was more stable, allowing radionuclides to drift offsite and therefore reduce worker exposure. However, this increased the chance for exposing the public. Because of local atmospheric conditions, fume lofting and dispersion were more effective in warmer months than during air-stagnant cooler months (DuPont 1945).

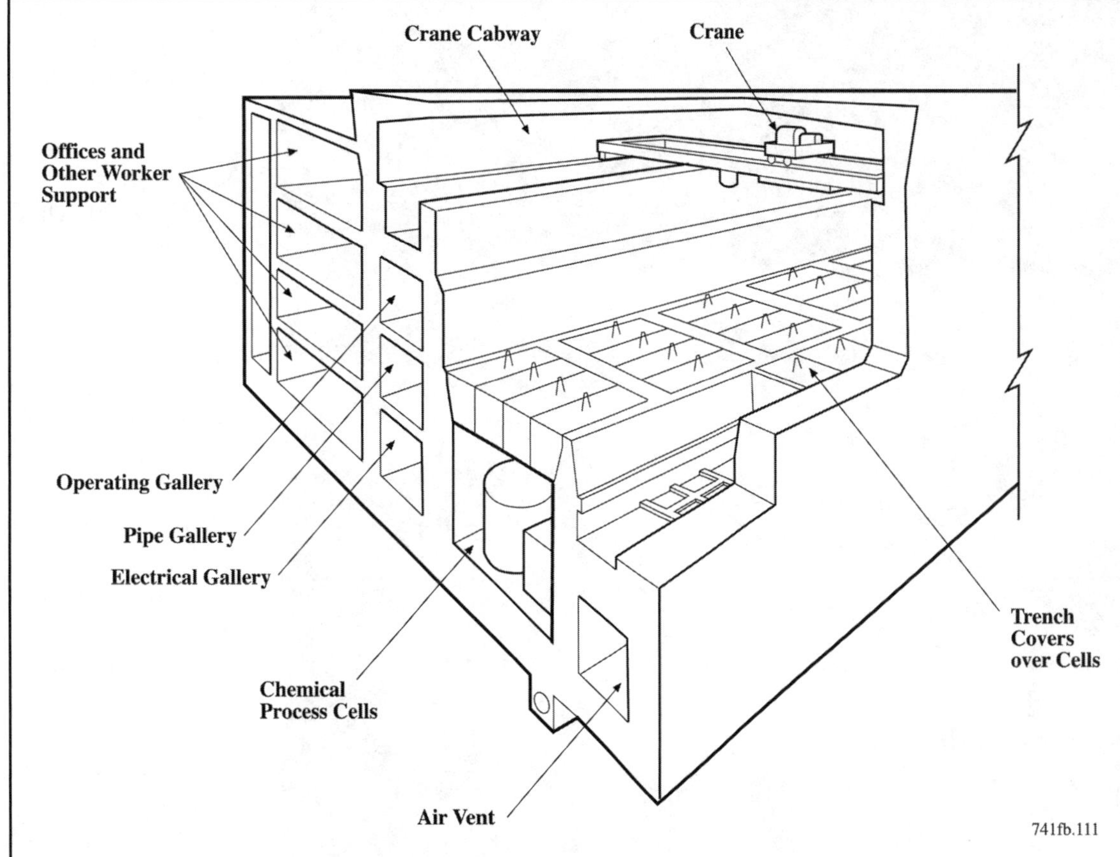

Cutaway of B Plant. This generalized cross-section of B Plant shows the internal layout for reprocessing spent fuel (RHO 1985). Everything below the level of the trench covers was underground. Thick concrete walls shielded workers from radiation.

1.6.1 Chemical Separations Processes

The technology used to chemically reprocess spent fuel at Hanford underwent significant improvements from the mid-1940s to the mid-1950s to increase safety, efficiency, and plutonium recovery.

The bismuth phosphate process had three weaknesses: it could not recover uranium, it was inefficient, and it produced large quantities of radioactive waste.

Three types of chemical separation processes were used:

- Bismuth phosphate batch processing.
- Solvent extraction using oxidation-reduction reactions.
- Solvent extraction using plutonium-uranium extraction techniques.

Bismuth Phosphate Batch Processing: The bismuth phosphate process was first used on an industrial scale in T Plant in December 1944, followed by B Plant in 1945. It worked by repeatedly alternating the chemical valence state of plutonium-239. This batch technique[24] was successful in precipitating, centrifuging, concentrating, and

(24) A single batch contained three buckets (about 3 tons) of spent fuel rods (Freer and Conway 2002).

U.S. Department of Energy 741fb.20 96040661-1d

Inside a Reprocessing Plant. The inside of a reprocessing plant resembles a long, narrow canyon with walls of cement and steel. These workers are monitoring radiation levels atop one of the heavily shielded and remotely operated reprocessing cells used to dissolve uranium fuel and extract plutonium.

purifying plutonium nitrate from liquid solutions. However, the process had three major weaknesses. First, it could not recover uranium for recycling into new fuel. Second, it was inefficient, processing only 1 to 1.5 tons of irradiated fuel a day (DOE 1997a). Third, it produced large quantities of highly radioactive waste.

Workers were required to enter the canyon decks of T and B Plants to sample chemical process solutions through special ports rising from the underlying process cells. (Remote sampling was designed into the later REDOX and PUREX Plants.) Hanford workers received some of their largest radiation exposures while working on these decks. Sometimes these areas become too radioactive to enter. In reminiscing about plant operations, one retired worker said: "[when] I stuck the meter around the corner, it went off scale, I turned around and left. I said 'forget it'" (Freer and Conway 2002, p. 2-4.28).

Following World War II, advances were made in solvent extraction techniques. These processes worked because uranium and plutonium were extractable in certain organic liquids[25] while unwanted fission products such as cesium and strontium would remain behind.

Solvent Extraction Using Oxidation-Reduction Reactions: The first solvent-extraction process used methyl isobutyl ketone (hexone) as the organic solvent with aluminum nitrate added to recover uranium and plutonium from other fission products and unwanted liquids. This continuous-flow process was called REDOX. Its use began in 1952 in the REDOX Plant. Process chemistry relied upon chemical transfer (extraction) taking place when two immiscible solutions (for example, fluids of different densities like oil and water) flowed past each other inside long chemical vessels resembling pipes.

"[When] I stuck the meter around the corner, it went off scale, I turned around and left. I said 'forget it'" (Freer and Conway 2002, p. 2-4.28).

(25) Organic compounds contain rings or chains of carbon atoms as well as hydrogen. In addition, they may include oxygen, nitrogen, and/or other elements. Common examples of organic compounds are sugar and cooking oil.

The REDOX process offered several advantages over bismuth phosphate, including reduced waste volume, improved plutonium extraction, recovery of both plutonium and uranium, and continuous plant operation. At the same time, hexone was potentially explosive and was not easily recycled. Liquid waste from REDOX complicated waste tank management because it was very radioactive, contained a host of organic chemicals, and was thermally hot. The REDOX process was designed to process 3 tons of uranium fuel a day. By 1958, technology improvements increased this capacity to 12 tons a day (DOE 1997a).

U.S. Department of Energy 38206-13

Concrete Cover Lifted Above a Processing Cell. Thick, multi-ton concrete covers were placed atop the cells in reprocessing plants to reduce background radiation levels. These workers are wearing masks, special coveralls ("whites"), and shoe covers to reduce the chance of radioactive particles lodging on their skin or being inhaled.

Solvent Extraction Using Plutonium-Uranium Extraction Techniques: A second-generation solvent-extraction process called PUREX was developed. It differed from REDOX by the use of tributyl phosphate as the organic solvent and piston-driven pulse solvent extraction columns rather than gravity flow. The solvent was diluted in a fluid similar to kerosene. After plutonium, uranium, and sometimes neptunium and thorium were extracted, they were concentrated for further purification. The solvent was recycled. As in previous processes, most acidic, unwanted solutions were neutralized with sodium hydroxide (to a pH of 9 or higher) and discharged to underground tanks.

REDOX generated waste complicated tank management because it was very radioactive, contained a host of organic chemicals, and was thermally hot.

The PUREX process was first used at Savannah River Site, South Carolina, in 1954 and in the PUREX Plant at Hanford in 1956. As much as 33 tons of uranium fuel could be processed in a single day (Anderson 1990).[26] However, less than 10 tons a day was more common. The PUREX process offered several advantages over the REDOX process, including reduced waste volume, greater reprocessing flexibility, less fire hazard, lower operating costs, and the ability to recover multiple radionuclides. In 1963, the plant's dissolver equipment was changed to accommodate the larger zirconium-clad spent fuel from N Reactor compared to the smaller aluminum-clad fuel from earlier reactors (Bailey and Gerber 1997).

Over the years, 177 underground storage tanks were built to contain the most radioactive waste from the reprocessing plants.

(26) Nearly 1000 tons of aluminum-clad spent fuel or 100 to 200 tons of zirconium-clad fuel could be processed in the PUREX Plant each month (conversation with James McClusky, former PUREX Plant manager, January 26, 2001).

As much as 33 tons of uranium fuel could be processed in PUREX each day.

As PUREX Plant was built, a project entitled the 4X Program was initiated to upgrade and increase the uranium processing capacity of Hanford (Gerber 2000). It involved renovating the old B and T Plants and adding their increased capacity to that of the REDOX and PUREX. However, after the first year of PUREX operation, the 4X Program was abandoned because the PUREX Plant demonstrated an overwhelming production capacity and economic efficiency compared to the older plants.

About 106,600 tons of uranium fuel were reprocessed at Hanford between 1944 and 1990 (see Table 1). Some plants reprocessed more uranium fuel than others (Napier 1992).

Table 1. Uranium Fuel Processed in Hanford Reprocessing Plants

Plant	Fuel Reprocessed (tons)	Operating History
T and B	8900 (8%)	1944-1956
REDOX	24,600 (23%)	1952-1967
PUREX	73,100 (69%)	1956-1972, 1983-1990
Total	106,600 (100%)	

The PUREX Plant was shut down from 1972 to 1983. By that time, only N Reactor was operating, and the U.S. government determined additional weapons-grade plutonium was no longer needed. Spent fuel was stored in the K Basins, and the PUREX Plant was upgraded. During nuclear weapons buildup in the Reagan Administration, additional spent fuel was reprocessed before the PUREX Plant was permanently shut down in 1990.

U Plant was built in 1945 though it was never used for fuel reprocessing. Rather, the plant was retrofitted with a variation of the PUREX process to recover and recycle uranium previously stored in underground tanks generated from T and B Plants. After liquids were pumped from single-shell tanks, high pressure water spraying was used to loosen uranium-bearing materials from the bottom of tanks (Freer and Conway 2002). Uranium recovery took place from 1952 to 1958, during the height of the Cold War.

1.6.2 Waste Generated by Reprocessing

As technology improved, less waste was generated for each ton of spent fuel reprocessed.

In the early years, recovering slightly more than 1 pound of plutonium produced 2000 to 25,000 gallons of radioactive liquid waste.

Reprocessing generated huge volumes of waste. The amount varied over time as reprocessing facilities became more efficient and waste management practices changed. For example, the volume of waste generated per ton of spent fuel reprocessed was reduced by a factor of 50 from the early years of using bismuth phosphate to the fuel reprocessing and low waste generation campaigns of PUREX process.

On average, the bismuth phosphate process used in T and B Plants generated 4000 gallons of waste liquid for each ton of spent fuel reprocessed (Agnew 1997). Much of this was water. Rates varied from 2000 to 25,000 gallons of liquid per ton of fuel. Initially, the REDOX Plant produced 4600 gallons of liquid per ton of fuel. Process improvements reduced this to about 1000 gallons. Depending on the reprocessing campaign, the PUREX Plant generated 50 to 1400 gallons of liquid per ton of spent fuel. The average amount of waste generated from the PUREX Plant was 500 gallons a ton.

A few hundred thousand of tons of chemicals—acids (such as nitric and sulfuric acid), solvents, nitrates, ammonia, and trichloroethylene—were used in the reprocessing plants, released to the environment, or pumped into tanks. These chemicals are common to other industries such as airplane building, electronics manufacturing, ore mining, dry cleaning, or farming. However, Hanford chemicals were unique for they were laced with a host of short- to long-lived radionuclides.

Hanford chemical waste was unique for it was laced with a host of short- to long-lived radionuclides.

During facility operations, highly radioactive waste was piped to underground storage tanks. Used water and other contaminated liquids were discharged to the ground or Columbia River. Solid waste was buried in shallow trenches or stored inside facilities. Gaseous effluents were released to the atmosphere. See Chapter 5 for a discussion on waste management.

1.6.3 Plutonium Finishing Plant

The Plutonium Finishing Plant converted plutonium-bearing solutions to plutonium metal.

In addition to reprocessing plants, other Hanford facilities were used such as the Plutonium Finishing Plant[27] that purified plutonium-bearing solutions into 4- to 5-pound hockey-puck-looking-chunks of plutonium metal known as buttons (Gerber 2002). The plant also fabricated and machined some plutonium metal pits used in nuclear weapons. The Plutonium Finishing Plant produced weapons-grade plutonium from 1949 to 1989—at a rate of 5 tons each year during the late 1950s and early 1960s. Mostly fuel-grade plutonium was produced from 1972 to 1982.

U.S. Department of Energy 94061152-1cn

Plutonium Finishing Plant. The Plutonium Finishing Plant started operating in 1949. It converted plutonium nitrate solutions received from the reprocessing plants into metallic plutonium. This photograph was taken near the time of startup.

Before the Plutonium Finishing Plant operated, plutonium-nitrate solutions were concentrated onsite by chemical treatment and heating. The nearly dry product was packaged into a "sample can," stored, and shipped offsite (DuPont 1945). When received at Los Alamos, New Mexico, it was further converted into metal and machined into the proper shapes for use in nuclear weapons. This process continued from 1945 to 1949 until the Plutonium Finishing Plant was built.

(27) The Plutonium Finishing Plant is also known as Z Plant because it housed the final steps in the production of plutonium metal.

Chapter 2

Hanford's Corporate Culture

"An overall impression of these [historical] documents is one of honesty. They were not originally intended for release to the public, so there was no incentive to 'hide' information. There was also concern for worker and public health."

—A.W. Conklin, *Summary of Preliminary Review of Hanford Historical Documents—1943 to 1957* (1986)

It is in the very nature of cultures to ensure their own survival.

The plutonium production culture at Hanford grew during a time of crisis when wars and threats of war fueled the massive arming of nations, some with first-class nuclear weapons. It also reflected a time when public trust in government and corporate institutions ran high. Later, changes in our attitudes forced re-examining previously accepted beliefs including our use of natural resources and how industrial wastes were handled. Today, people want to know why past contaminant releases at Hanford and other federal sites went unexplained, unquestioned, unresolved, and sometimes unchecked for so many years.

The U.S. nuclear weapons complex, which produced thousands of nuclear warheads for the country, consists of dozens of industrial facilities and laboratories across the country. Hanford was one of the largest of these sites.

Culture deals with the essence of a group of people—its shared identity and behavioral choice. Culture is a self-protective equilibrium that unconsciously directs the pattern of relationships and interactions between individuals as well as how they view themselves in relation to the outside world. All groups of people, whether in established institutions or ad hoc organizations, share common perspectives, language, and goals—a culture. Over time, culture is refined to serve the interests of its practitioners. Culture exists at many levels from superficial artifacts (organizational structures and work practices) to core assumptions from which shared values and actions emerge (Schein 1999).

Culture deals with the essence of a group of people—its shared identity and behavioral choice.

We think culture is what others have rather than what we hold so dear. Our beliefs lie unquestioned, taken for granted. After all, they feel so natural and immutable; we think, "if only others knew what I knew, they would believe as I do." Normally, people only confront their own beliefs when challenged by another culture, someone holding a different set of values. Name-labeling and stereotypical responses are knee-jerk reactions to protect one's belief while dismissing the legitimate concerns of others. Such reactions cloud our ability to debate crucial social and moral issues.

The more powerful a culture, the more resistant it is to questioning, let alone change. Money and ideologies are powerful cohesive forces preserving many cultures. Most cultures are neither right nor wrong; they simply exist to serve the interests of their adherents.

Money and ideologies are powerful cohesive forces preserving many cultures.

Failure to recognize the deep-seated drives of a culture protects it from adapting to new realities, becoming more effective. Over time, this can nurture extremes for controversial issues—from blind allegiance to apocalyptic pessimism.

It's easy to judge others while evading responsibility for ourselves.

One must be cautious when comparing the culture of 1940s America with today's. It is too easy to judge, to look for faults in others while evading responsibility for ourselves. No one is above pretense. No time is above pretentiousness. This is true for both the past plutonium culture as well as today's environmental culture. As in personal relationships, we tend to view others not as they are but as we wish them to be. We then paint the landscape black and white, rather than in shades of gray, to reinforce our beliefs.

2.1 Setting the Scene

"Mr. Parker, the experience that you have had there in Hanford is about one of the most complete in the storage of a great mass of this material and certainly, you have had a very fine record of no large incidents that have been a hazard to the population. You have done a careful job in handling this very deadly material. Thank you very much."

—Chet Holifield,
Chairman for the Subcommittee on Radiation
(U.S. Congress 1959a, p. 170)

The Manhattan Project was a model of technological efficiency, but social accountability was suspended (Wills 1999). General Leslie Groves, acting as the government's agent, did not have to justify before Congress or citizens the commandeering of resources and manpower. It was his job to build the world's first nuclear weapons in the shortest possible time. He worked in a command-control structure during a time of war. This inward-focused, tight-knit culture dominated Hanford operations for decades. Commonly, institutionalized cultures are molded through successes and the personality of its founders. Hanford is no exception.

During the Manhattan Project, actions were undertaken to deflect concerns about resources diverted for a project of uncertain outcome. After all, a world war involving 56 nations, destroying whole cities, devastating countries, and causing 55 million deaths demanded drastic action. Not having to justify the project meant no administrative delays in completing it. The prevailing philosophy was that the "American people were served without having to waste time explaining what was being done on their behalf" (Wills 1999, p. 312).

U.S. Department of Energy

Hanford Billboard. The urgency of Hanford workers to complete their jobs is shown on this World War II era billboard.

What was Hanford's approach to managing waste? In congressional testimony, Herbert Parker, a pioneer in radiological physics, summarized them: concentration and containment where feasible followed by dilution and dispersion when necessary. However, when "going to dilution and dispersal…it is mandatory to control the amounts so that the safety of not only the human population, but also the plant and animal resources is assured" (Parker 1959b, p. 217). Hanford waste practices received high praise before Congress (U.S. Congress 1959a). After all, they were the standards of those years.

After World War II, the Cold War began and the "grave ethical implications" of the nuclear age were suddenly thrust upon an unsuspecting world (Lilienthal 1963, p. 17). The public was informed about nuclear activities in an almost haphazard fashion, with little attempt made to help them understand the scientific, military, diplomatic, and social implications of nuclear weapons, radioactive fallout, or contaminant releases into the environment.

Hanford used a "point-of-exposure" philosophy for controlling the release of radioactive material (Parker 1959b, p. 217); that is, trying to avoid any "undue radiological exposure" along the site's boundaries rather than at the point of release. This approach relied upon the entire Hanford Site's environment to dilute, to hold until radionuclides decayed, or to re-concentrate radioisotopes in microorganisms. The environment was used to reduce radionuclide concentrations. While this philosophy was considered "more accurate and economical than the inflexible 'point-of-discharge,'" concept (Parker 1959b, p. 218), relying upon it required an accurate knowledge of environmental characteristics that influenced contaminant movement and uptake. Parker worried about an "unforeseen [radionuclide] concentration mechanism" that might cause a radiological hazard beyond that understood.

This point-of-exposure approach supported claims that radionuclides released to the ground added nothing to public's radiation exposure because the radionuclides had not yet reached the public—at least by the late 1950s. On the other hand, contaminants discharged into the air and Columbia River had reached the public for years.

However, as Abel Wolman of Johns Hopkins University pointed out in 1959, two characteristics of nuclear waste distinguish it from other more common industrial waste. First, "these wastes by and large are not detected by the human senses." Second, "their toxicity in general terms, both radioactive and chemical, is greater by far than any industrial material with which we have hitherto dealt in this or any other country" (Wolman 1959, p. 9). Radioactive waste comes with unique long-term management responsibilities that should not be "tackled from the standpoint of temporary expedience."

2.1.1 Secrecy Shrouds Contaminant Releases

Officials believed national security, public trust, and company interests were best served by not revealing too much about Hanford operations or contaminant releases.

The secrecy of Hanford plutonium production was extended to contaminant releases. While secrecy protected some vital national interests, it also controlled what people heard, squelched embarrassments, and leaked only sanctioned news. For example, in congressional testimony about waste management practices, liquid releases to underground cribs were described as being "admitted to the ground at Hanford under carefully controlled conditions" (Parker 1959a) though the practice was fraught with unknown consequences and "grave responsibilities" (Brown et al. 1958). Even the Atomic Energy Commission's own studies, captured by Sidney Williams (1948), challenged prevailing waste handling practices. "The disposal of contaminated waste in present quantities and by present methods (in tanks or burial grounds or at sea), if continued for decades, presents the gravest of problems…A tiny fraction of the money now spent for inadequate and purely temporary solutions to this problem, would provide more than enough funds for these [improved waste management] purposes" (Williams 1948, pp. 9, 74).

Secrecy concealed potential hazards from the public and health officials (Gray 1992).

"The disposal of contaminated waste in present quantities and by present methods (in tanks or burial grounds or at sea), if continued for decades, presents the gravest of problems" (Williams 1948).

More than a decade later, testimony before Congress continued to underscore concerns about the small amount of money for and lack of "pressure to get some solutions" to the problem of nuclear waste disposal (Wolman 1959). This was attributed to many insiders within the nuclear weapons complex believing that radioactive waste was a non-problem.

Secrecy also concealed potential hazards from the public and health officials (Gray 1992, p. 2). As reactors and reprocessing plants were built and plutonium production goals set, operating policies preserved nuclear material production rather than pollution mitigation. Radioactive gases were released to the atmosphere, contaminated water was discharged to the Columbia River, liquids were dumped into the soil, and highly radioactive waste was piped into tanks. Hanford created uniquely hazardous waste, though it was mostly handled by standard industrial practices. While external, independent reviews of waste management practices were not common in industry, Hanford's culture provided an additional buffer against critical assessments and a refuge for unchallenged waste practices.

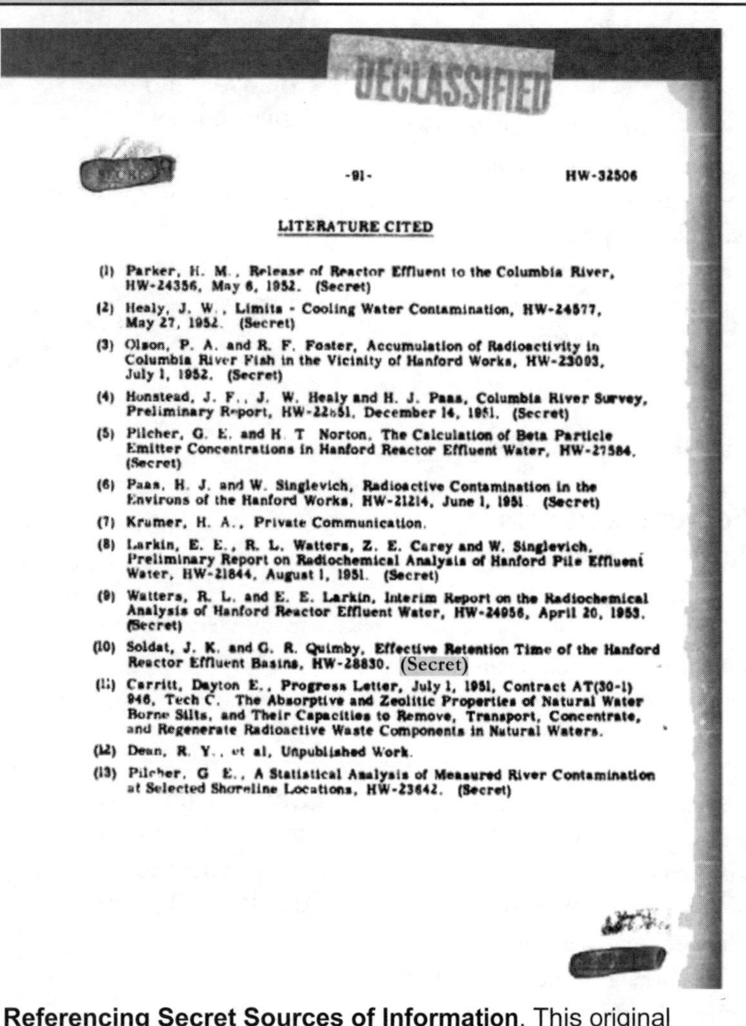

DECLASSIFIED

-91- HW-32506

LITERATURE CITED

(1) Parker, H. M., Release of Reactor Effluent to the Columbia River, HW-24356, May 6, 1952. (Secret)

(2) Healy, J. W., Limits - Cooling Water Contamination, HW-24577, May 27, 1952. (Secret)

(3) Olson, P. A. and R. F. Foster, Accumulation of Radioactivity in Columbia River Fish in the Vicinity of Hanford Works, HW-23093, July 1, 1952. (Secret)

(4) Honstead, J. F., J. W. Healy and H. J. Paas, Columbia River Survey, Preliminary Report, HW-22651, December 14, 1951. (Secret)

(5) Pilcher, G. E. and H. T. Norton, The Calculation of Beta Particle Emitter Concentrations in Hanford Reactor Effluent Water, HW-27584, (Secret)

(6) Paas, H. J. and W. Singlevich, Radioactive Contamination in the Environs of the Hanford Works, HW-21214, June 1, 1951 (Secret)

(7) Krumer, H. A., Private Communication.

(8) Larkin, E. E., R. L. Watters, Z. E. Carey and W. Singlevich, Preliminary Report on Radiochemical Analysis of Hanford Pile Effluent Water, HW-21844, August 1, 1951. (Secret)

(9) Watters, R. L. and E. E. Larkin, Interim Report on the Radiochemical Analysis of Hanford Reactor Effluent Water, HW-24956, April 20, 1953. (Secret)

(10) Soldat, J. K. and G. R. Quimby, Effective Retention Time of the Hanford Reactor Effluent Basins, HW-28830. (Secret)

(11) Carritt, Dayton E., Progress Letter, July 1, 1951, Contract AT(30-1) 946, Tech C. The Absorptive and Zeolitic Properties of Natural Water Borne Silts, and Their Capacities to Remove, Transport, Concentrate, and Regenerate Radioactive Waste Components in Natural Waters.

(12) Dean, R. Y., et al, Unpublished Work.

(13) Pilcher, G. E., A Statistical Analysis of Measured River Contamination at Selected Shoreline Locations, HW-23642. (Secret)

Referencing Secret Sources of Information. This original citation list in a Hanford report (Honstead 1954) illustrates the tight grip information control had on early publications. Of the 13 citations, 10 were labeled secret.

The compartmentalization of information, considered essential for security, also contributed to secrecy. Relatively few people knew the whole picture. Information was given only to those with a "need to know." This kept workers from independently looking into areas not under their responsibility.

Three years after the end of World War II, the Atomic Energy Commission (Williams 1948, p. 6) lamented, "a serious obstacle to progress and uniformity was and still is imposed by war-born security rules. We applaud the extraordinary security thus achieved regarding the atomic bomb but we suggest that some of these rules and the habits they engendered are still a serious and largely unnecessary handicap."

Yet, information control remained engrained across the nuclear weapons complex. Chief of Hanford's Health Physics and Biology Branch Wendell Crane noted that some scientists and managers at Hanford wanted to make reports on radiation and chemical releases available to the public. However, "it was thought that by some roundabout reasoning" the Soviets might learn exactly how much plutonium Hanford was producing, giving away valuable information on our nuclear arsenal. When the topic of making the information public came up, many of the "most eager proponents of

Water Temperature—A Security Risk

Even the amount of water pumped from the Columbia River, the temperature of the returned water, and the heating effect it had on river were classified (Johnson 1962; Wildman 1963).

Officials were concerned about someone comparing river temperatures upstream and downstream of Hanford (Johnson 1962) to estimate reactor power levels and the amount of plutonium produced. In 1962, a biologist employed by the Northwest Pulp and Paper Association used such temperatures to estimate reactor power levels to within 10%. This created quite a security stir.

This incident coupled with increasing concerns about contaminant releases led in part to representatives from state agencies in Washington and Oregon, as well as the U.S. Public Health Service, being granted security clearance to review environmental data—though for years this was considered an "unnecessary action" (Adley and Crane 1950a). By the late 1960s, after four of the nine reactors were shut down, attempts to calculate reactor power levels from river temperature data "would be so inaccurate as not to provide classified information" (Plum 1968). From then on, river temperatures were not classified.

When the topic of making information available to the public came up, proponents of more openness became very conservative. Caution prevailed.

declassification usually become very conservative...when asked to sign their names to specific declassification recommendations" (Plum 1950). Caution prevailed.

Hanford secrecy labored without the benefit of independent oversight, unshackled by preserving the operation status quo. Overall, waste management practices were carried out under the premise that there was "no immediate risk" (Silverman 2000). Hanford managers "settled on notions of acceptable risk and acceptable levels on contamination that were compatible with production goals." Hanford "displaced hazards from one [environmental] medium or location to another...By transferring environmental burdens from the ground to the water or vice versa, by shifting the deposition of airborne contaminants from lands adjacent to the facility to more distant areas, and by creating unintended and unanticipated consequences downwind, downstream, and on-site, Hanford managers reaffirmed their own risk assessment practices" (Silverman 2000). This led to exploiting "the environment's capacity to accept wastes."

Site officials were in a constant game of uncertain tradeoffs between plutonium production and environmental protection.

At the same time, as discussed in Section 4.1, national and international radiation protection standards were in their infancy. This placed site officials in a constant game of uncertain tradeoffs between plutonium production and environmental protection. For the most part, the weapon sites were on their own, guided only by site-specific knowledge and trial-and-error learning. By the mid-1950s, guidance from the International Commission on Radiological Protection and the National Council on

Radiation Protection was developed[1,2]. Much of the field data, dosimetry, and safety experience supplied to these organizations came from the same agencies operating the world's first weapon sites. Volumes 13 and 18 of the first and second International Conference on the Peaceful Uses of Atomic Energy are full of articles about the early years of radiation protection and standard development (United Nations 1956, 1958). Much was unknown. Some initial standards were qualified as perhaps reasonable. "Lacking complete information...present planning for installations or procedures must incorporate a degree of flexibility which will permit ready adaptation to future changes in maximum permissible exposure standards" (Thompson et al. 1956).

Locally derived limits to contaminant emissions were established because state and federal legislative control were nearly nonexistent (Healy et al. 1958).

Locally derived limits to contaminant emissions were established because state and federal legislative control was nearly nonexistent (Healy et al. 1958). Local standards were no less conservative than those prescribed by the Atomic Energy Commission and were "broadly responsive" to standards promulgated by either the National Council on Radiation Protection or the International Commission on Radiological Protection. In so doing, Hanford "often held a position of national leadership in radiation protection" (Parker 1959b). Hanford officials worked to link plant operations with the latest health-related research so as to "change the corresponding practices used to maintain an assured position of safety in waste disposal" (Parker 1959a).

2.1.2 State Authorities and Independent Oversight

For years, state officials hesitated to interfere with Hanford waste practices.

The pollution control authorities in the states of Washington and Oregon did not have the staff, expertise, or laboratories to independently assess Hanford's contaminant control practices. That is why the Director for the State of Washington Pollution Control Commission welcomed collaboration with the U.S. Public Health Service to survey the Columbia River because "the job is too large for State groups to handle" (Eldridge in Plum 1950). This sentiment was raised 2 years earlier when it was acknowledged that local, state, and federal public health agencies were "opposed to any large-scale transfer of responsibility" to them (Parker in Williams 1948, p. 82). Nonetheless, the Atomic Energy Commission believed that the Hanford Site operators should "consider consulting" with federal, state, and local agencies "so that responsibility in part be shared by those who might be affected" (Friedell in Williams 1948).

(1) The International Commission on Radiological Protection is an advisory body providing recommendations and guidance on radiation protection. It was founded in 1928 by the International Society of Radiology, a professional society of radiologist physicians; it was first known as the International X-ray and Radium Protection Committee. For more information, see http://www.icrp.org/.

(2) Created in 1929, the National Council on Radiation Protection and Measurements (NCRP) originally operated as an informal association of scientists seeking to make available information on radiation protection and measurements. More than 30 major reports were produced during the early period of the NCRP's history including the first recommendation specifying a maximum permissible level of exposure. Chartered by Congress in 1964, the council assembles expert guidance on the whole spectrum of radiation protection science, technology, and practice. For more information, see http://www.ncrp.com/.

During the 1950s, the general tenor of state pollution control responsibility centered on ensuring that industries followed their own self-imposed rules, rather than providing independent oversight or imposing state pollution standards. The Oregon State Board of Health echoed this sentiment, stating that the primary responsibility for preventing contaminants from reaching the environment "rests with industries concerned to carry out proper research programs and institute their own controls" (Everts in Plum 1950).

During the 1950s, the general tenor of state pollution control responsibility centered on ensuring that industries followed their own self-imposed rules, rather than providing independent oversight.

2.1.3 Early Push for Reform

Hanford officials responded forcefully to challenges and tended to dismiss dissenters.

Two themes were played out onsite for decades to protect the status quo: Hanford's assertions that everything is "under satisfactory control" (Plum 1950) and that outside authorities were not well informed on the "implications of radioactive waste disposal" (Parker 1951).

When asked by an external review group for demonstrable evidence that waste management practices were safe, Hanford officials replied that there is a wide difference between producing evidence that practices were safe versus "showing that a reasonable interpretation of existing data indicates that there appears to be no significant probability of producing an improper condition" (Parker 1951). Hanford sought to preserve existing contaminant release practices within the flexibility afforded under operation guidelines. Officials were encouraged to "avoid detailed artificial regulations which make technically improper those practices which in the light of reasonable interpretation by competent scientists appear to be safe" (Parker 1951).

Yet, officials also stressed the importance of not permitting "any laxity in protection" and rigidly following safety standards to protect workers (Cantril and Parker 1945). Though tensions existed between production and protection, the production culture dominated decisions and investments. One example was the quality of downstream drinking water. Increases in chromium and radionuclide concentrations exceeding U.S. Public Health Service maximum permissible concentrations were noted by Corley in 1960. However, instead of focusing on protection, officials focused on production and the public relations problem with workers if these contamination levels became widely known.

Instead of focusing on protection, officials focused on production and the potential public relations problem should contamination levels become widely known.

Assurances were offered to the public and review committees. Hanford's officials had "precise knowledge on the limits of [radiation] protection." Workers "never needed to be in doubt whether radiation exists in any particular location, or whether the amount is more than can be safely tolerated." The public should not be concerned about radioactive iodine releases because "the amounts are entirely innocuous." Further, "not one abnormal finding has been uncovered which could in any way be attributed to the hazards of radiation or to chemical toxicity of any of the materials used on the Plant [Hanford Site]" (Cantril and Parker 1945). Besides, "present disposal procedures may be continued, in the light of present knowledge, with the assurance of safety for a period of perhaps 50 years" (Parker 1948a). Health physicist Herbert Parker went on to say that the projection of problems, as proposed by some critics in the late 1940s, appears "to be irrelevant" for inevitable technology progress will solve them. Even major "foreseeable disasters" would not seriously jeopardize the health of communities dependent on the Columbia River. In the worst case, "radical curtailment of the use of the river water" was considered an option.

The projection of problems, as proposed by some critics in the late 1940s, appears "to be irrelevant" for inevitable technology progress will solve them (Parker 1945a).

Even major "foreseeable disasters" would not seriously jeopardize the health of communities dependent on the Columbia River because "radical curtailment of the use of the river water" was considered an option (Parker 1948a).

Some of these assurances may have laid the groundwork for a Hanford response to a report released by the Atomic Energy Commission in 1948 (Williams 1948). This report, issued by the Williams Commission, was a nationwide critique of the nuclear weapons complex. Its findings identified serious concerns about waste management at all sites.

The report stressed that the peacetime operation of facilities made it impossible to use the same operational safeguards and waste management practices that succeeded under wartime conditions. "We do not ask that safety be 'first' but that its moral, economic and catastrophe importance be fairly balanced against other considerations. We accept the principle of calculated risk, but we believe (1) that this phrase has been used to excuse risks quite uncalculated and (2) that the degree of risk justified in wartime is no longer appropriate."

Williams (1948, p. 10) noted the importance of the Atomic Energy Commission and its contractors working closely with the public as well as federal and state public health authorities because "(1) their experience will be of value on many problems and (2) no other course is consistent with American experience." The Williams Commission believed this was important because liquid and solid waste disposal practices have "not been developed with full consideration of the hazards involved." The report chastised Hanford for reducing, to a minimum, the sampling of water supply data.

Herbert Parker (1948a) wrote there was no cause for "hysteria, or for the radical expansion" of Hanford's ongoing or planned liquid disposal practices based upon the findings of an external review group. Parker, who was a member of the Williams Commission, said that Hanford's safety practices are "more thorough than is usual in industry" (Williams 1948, p. 83).

The Williams report set the tone for future exchanges between Hanford officials and agencies pushing for reform. However, information restrictions remained in place and Hanford practices remained essentially unchanged.

2.1.4 Protecting the Columbia River

"We might have legal responsibility for water pollution under certain circumstances...We are not entirely prepared, however, to clear pollution problems involving Washington and Oregon state and federal laws with the appropriate state and federal agencies"

—F.E. Adley and W.K. Crane,
Meeting of the Columbia River Advisory Group: March 6-7, 1950 (1950a)

Another example of growing external interest in Hanford operations came from the Columbia River Advisory Group. This group of federal, state, and contractor personnel established in 1949 worked with Hanford officials on radiation safety and waste management. Representatives from the U.S. Public Health Service and the states of Oregon and Washington were to cooperate with Hanford officials "to prevent possible contamination of the Columbia River by the Hanford atomic plants" (Adley and Crane 1950b). In meeting minutes for the advisory group, member H.A. Kornberg (1950) wrote about the "moral responsibility" of Hanford to protect the "environs surrounding the Hanford Works." In the same minutes, Richard Foster, a fish biologist, wrote that the federal government was responsible for protecting natural resources, "all persons utilizing the Columbia River," and "the interests of the peoples of the States of Washington and Oregon" (Kornberg 1950).

However, the first press release about the group's initial meeting set off a flurry of controversy. Arthur Gorman, Division of Engineering for the Atomic Energy Commission, took exception to a portion of the press release that read:

> "It is the consensus of the group, based upon information obtained, that so far as the Columbia River is concerned, there are no apparent water pollution hazards resulting from the operations at present. Waste disposal is a major problem. However, the operating agencies are using every means at their command to keep it under control. Future conferences are planned to review research and progress to insure that the Columbia River will continue to be protected" (Adley and Crane 1950b).

The controversy surrounding the press release from the Columbia River Advisory Group's first meeting offers early insight into the tension surrounding information openness, public trust, and preservation of scientific credibility. These same issues echo across Hanford's history.

Gorman was angry that members of the advisory group approved a press release, drafted by Hanford officials, that was unsupported by available information. He pushed the issue: "it is my opinion that insufficient data are available...to appraise what effects the wastes discharged into the river have or may have...I still doubt the wisdom of a public release based on two and one-half days sanitary engineering review of one of the most complex situations in the country, presumably developed by a professional group wholly unfamiliar with the units, the practices, the effects or the permissible limits involved in the situation" (Gorman 1950).

Gorman went on to challenge the lack of independent oversight at Hanford: "It is my sincere belief that it would be in the interest of the AEC [Atomic Energy Commission] if more basic and applied research in problems of disposal of radioactive and toxic wastes from atomic energy operations and a certain amount of off-site control monitoring were carried out by established and experienced federal and state agencies normally having jurisdiction over public health and national resources. A situation where A.E.C. or one of its contractors as a producer of wastes is the principal agency involved in decisions as to what amounts and under what conditions hazardous waste are stored, what degree of decontamination will be carried out before release of these wastes under conditions which may affect the rights of others, leaves much to be desired from the standpoint of conformance with currently accepted public health practice, to say nothing of the possible legal and public relations implication."

Some officials took exception with Gorman's statements. The manager of Hanford Operations Fred Schlemmer wrote to senior Washington, D.C. officials that members of the advisory group wondered, "why Mr. Gorman was so concerned" (Schlemmer 1950). Even the group's representative from the U.S. Public Health Service felt Gorman was unduly alarmed. Schlemmer continued: "the action taken by Mr. Gorman to be highly improper, as well as contrary to the actual facts of the situation…It would be appreciated if immediate steps could be taken to avoid recurrence of instances such as this."

However, perhaps the underlying issue was the independent role some wanted the Columbia River Advisory Group to play. As stated in the group's meeting minutes, their purpose was to "become familiar" with and to "become instructed" about Hanford waste practices as well as assist the Atomic Energy Commission in developing water pollution and safety plans (Adley and Crane 1950b). Independent oversight was not part of the plan. Perhaps this is why Fred Schlemmer (1950) stressed that the advisory group "could be an important and useful medium for transmitting to the general public and local government agencies in the Northwest the high degree of confidence which *we ourselves* [emphasis added] have that we are taking adequate steps to prevent pollution and contamination of the Columbia River."

Herbert Parker (1950) noted that the press statement attributed to the advisory board "was quite conditional…and specifically excluded reference to what may happen in the future if and when contaminated ground waters should reach the Columbia River." Overall, Parker agreed with much of what Gorman said and recommended expanding an existing aquatic biology laboratory to accommodate researchers from other agencies. Some Hanford officials were reluctant to have new staff on the site (Plum 1950).

Following initial meetings, members of the Columbia River Advisory Group wrote a letter to Schlemmer highlighting the legal responsibility of water pollution control agencies in the states of Washington and Oregon and of the U.S. Public Health Service (Eldridge et al. 1950). The group supported approaching river protection by integrating the concerns of all of the groups that used the river, not just Hanford. This work should build upon the "well planned program of research, investigations, and surveys…in effect for a number of years and competent personnel and facilities" found at Hanford. Proposed work should avoid duplication but provide independent confirmation that Hanford contractors were meeting their moral and industrial responsibilities of protecting "humans and other living things in the environs surrounding Hanford Works" (Kornberg 1950).

On the heels of the advisory group, the U.S. Public Health Service studied the Columbia River between 1951 and 1953. Results were documented the following year (U.S. Department of Health, Education, and Welfare et al. 1954). The study contained recommendations for reducing radionuclide releases, conducting new studies into radionuclide uptake by aquatic organisms and humans, and improving radiochemical analyses. Their findings are discussed in Section 3.1.

Three new reactors came on line in the 1950s. Their water discharges dramatically raised radiation levels in the river.

Nonetheless, three new reactors came on line in the 1950s. These dramatically raised radiation levels in the river and the radiation dose people received downstream (HEDR 1994). Also during the 1950s and 1960s, state and non-Atomic Energy Commission officials became increasingly informed and able to knowledgeably converse about Hanford. While congeniality prevailed, the states cautiously sought more regulatory authority as the nearby cities (especially Richland, Pasco, and Kennewick) were growing, farming expanded, hunting and fishing spread, and the public became increasingly aware of the potentially hazardous nature of Hanford contaminants.

During the 1950s and 1960s, state and non-Atomic Energy Commission officials became increasingly informed and able to knowledgeably converse about Hanford.

Twelve years after the Columbia River Advisory Group was formed, it published a report evaluating the health effects from contaminants released from Hanford reactors into the river (Tsivoglou and Lammering 1961). Most data were from Hanford contractors. Other data came from the U.S. Public Health Service and the states of Washington and Oregon. One reason for the study was changes in the generally acceptable standards of radiation protection issued by the International Commission on Radiological Protection and the National Council on Radiation Protection (see Section 4.1). Standards were being tightened, indicating an increasingly conservative approach to public protection.

The report expressed concern that while the average radiation exposure to members of the public "does not presently exceed generally accepted permissible limits," this does not mean that an undetermined number of individuals probably do receive "considerably more radiation exposure" than the averages suggest. How many people or what dose they experienced could not be determined. Bottom line? The authors believed the river was running out of capacity to absorb more contaminants without it and those living downstream suffering health and resource-use consequences. They recommended further studies and the cessation of radionuclide-containing discharges.

"Thus, in terms of the resulting human radiation exposure, the capacity of the Columbia River to receive further radioactive pollution appears to be nearly if not fully exhausted" (Tsivoglou and Lammering 1961, p. 24).

These results did not match previous assurances given by Hanford officials. The Columbia River Advisory Group and its supporting agencies were now independently examining the ecological health of the river. While their findings were not inconsistent with Hanford-funded studies, the gap between the practice of using the environment for diluting contaminants and the philosophy of minimizing releases to as low as reasonably achievable was widening.

In 1964, Hanford, Washington State, and U.S. Public Health Service officials met to discuss ways of reducing contaminant levels in the Columbia River (U.S. Public Health Service 1964). Creative solutions were sought. Consideration was given to building one or two large inland lakes, perhaps trenches, to receive reactor cooling water—piped 10 to 20 miles across the desert—rather than continuing to discharge it directly into the river. Once inland, the water would percolate into the soil, cool, and become less radioactive before seeping back into the river. Health officials wanted to reduce the radioactivity entering the river by half. This was an updated version of a mid-1950s proposal to construct three lakes in natural depressions on the north and south sides of Gable Mountain to reduce the amount of heat, radionuclides, and chemicals discharged to the river (Honstead 1955).

U.S. Department of Energy 45419-3cn

Cooling Water from the KE Reactor. Radioactive cooling water was pumped from most Hanford reactors to above-ground basins for temporary storage before release into the Columbia River. This 1967 photograph shows steam rising above one of these basins. Though designed to hold water for several hours, many basins lost water through leakage creating small streams and groundwater seeps that quickly drained into the river. Water is seen flowing from the bottom of the left basin.

"Mr. Hildebrandt [Washington State Department of Health] indicated that the state is interested in promoting atomic energy, particularly in this state, and that they wish to avoid any action that would result in shutting down any additional reactor capacity" (U.S. Public Health Service 1964, p. 18).

Hanford officials were reminded that the "primary responsibility for the administration for water pollution control rests with the states" (U.S. Public Health Service 1964, p. 18). Yet, officials from the state of Washington still hesitated to interfere with Hanford operations: "Mr. A.T. Neale of the Washington State Pollution Control, indicated that it has been the policy of his organization to proceed cautiously in establishing and enforcing [pollution] limits, believing rather that it is preferable to carefully evaluate all the economical and technical aspects of each process and plant which contribute to stream pollution. Mr. Hildebrandt [from the Washington State Department of Health] indicated that the state is interested in promoting atomic energy, particularly in this state, and that they wish to avoid any action that would result in shutting down any additional reactor capacity."

Yet, one U.S. Public Health Service official grew frustrated about delays in promised pollution control studies and measures. "Can we have any dates?...I am a little disturbed that a great deal of what we learned at the last meeting must be changed now...I am afraid that this will not satisfy our superiors at all" (U.S. Public Health Service 1964, pp. 5, 8).

In 1964, President Johnson announced cutbacks in plutonium production. In the next 7 years, eight of Hanford's nine reactors were closed.

However, where oversight failed, debates bogged down, and concepts of artificial lakes dried up, President Johnson's new administration succeeded. In early 1964, cutbacks were announced in nationwide plutonium production. Between 1964 and 1969, six Hanford reactors were shut down. Two more followed in 1970 and 1971. By then, only N Reactor operated. Radionuclide releases and the downstream radiation dose to humans dropped rapidly (HEDR 1994). The REDOX Plant was also shut down in 1967. Only the PUREX Plant remained ready to operate. Hanford plutonium production was reduced to the fewest number of facilities since late 1944—just one operating reactor and one reprocessing plant.

2.1.5 Protecting N Reactor

State and regional support for N Reactor keep it operating for another 16 years after the Nixon Administration proposed shutting it down.

In the 1970s, citizens embraced the national defense role played by Hanford, and there was wide political support for the site.

In the 1970s, the federal government began encountering public opposition to nearly three decades of nuclear dominance. Years later, it became popular to believe that the government imposed nuclear activities on an unwilling populace (Hevly and Findlay 1998). While it's true Hanford officials were not forthcoming with information about contaminant releases and their danger, it's also true that many Northwest citizens and officials welcomed the jobs, economic growth, and prosperity associated with nuclear activities. Citizens embraced the national defense role played by Hanford, and there was wide political support for the site.

Washington State's two federal senators, Warren Magnuson and Henry Jackson, supported the security role and economic growth offered by Hanford jobs. In 1971 when the Nixon Administration planned to shut down the last two Hanford reactors (KE and N), 40,000 letters from Tri-Cities residents were mailed to the White House supporting continued operations of N Reactor (Sivula 1988a). Editorials of support were written in Portland, Seattle, and Spokane newspapers. The Governors of Washington and Oregon also opposed the shutdown. The Pacific Northwest Chapter of the Sierra Club protested the shutdown in fear that the loss of electrical generation would create pressure to develop additional dams or fossil fuel plants. Labor unions were behind the reactor. N Reactor remained open for another 16 years.

Following the Chernobyl nuclear reactor explosion in Russia in 1986, calls for shutting down N Reactor grew louder. By then, both Magnuson and Jackson had passed away, and the political support for N Reactor was splintered. The need for weapons-grade plutonium was lessening. The Cold War was winding down. Besides, the Savannah River Site in South Carolina was favored to meet future plutonium demand (Sivula 1988b). The future of Hanford was being nudged towards cleanup, not new nuclear missions. Congressman Mike Lowry, soon to become Governor Mike Lowry, called for "creating something new" such as an international center for nuclear waste cleanup (Briggs 1988). Many agreed.

Following the Chernobyl nuclear reactor explosion in Russia in 1986, calls for shutting down N Reactor grew louder. The future of Hanford was being nudged towards cleanup, not new nuclear missions.

In the light of informed hindsight, the dwindling dependence on plutonium, and the growing independence from weapons-related jobs, some living in communities surrounding aging weapon sites now questioned the wisdom of past allegiances and the secrecy of the plutonium culture that dominated decision-making for so many years.

2.2 How is Nature Viewed?

"For most of us, it is difficult enough to respect those with whom we might disagree, to say nothing of those who might be different from us in culture, language, and tradition."

—Lamin Sanneh, *Translating the Message: The Missionary Impact on Culture* (1989)

Our world views are determined by personal and shared beliefs. This includes how we suspect nature responds to stress. Is nature fragile, strong, or forgiving? For decades, the Hanford environment was used as a dump for contaminants. Dilution, dispersion, decay, and technology advances were thought to eventually solve problems and return balance. Today's site operations are more protective of the environment, stressing waste minimization, containment, and restoration.

Is nature fragile, strong, or forgiving?

Unless based upon facts and experience, beliefs are underpinned by assumptions. Generally, someone is thought to be "unbiased" when they support our own beliefs and "biased" when their version of truth is different. Beliefs provide benchmarks from which we map the surrounding landscape. If benchmarks change, then the land must be remapped. Besides, change threatens to loosen the bonds we have in common with others of like mind. Change is unsettling.

Commonly, we create meaning and react as if our beliefs are undeniable truths and then are surprised if nature doesn't conform or others take exception.

Groups of people sharing world views tend to work towards common goals along a consensus of meaning. They may come together in established institutions or in informal settings. Their shared beliefs are "points of view" when open to change. They are "biased" when perspectives are fixed, unaltered by informed debate and new information.

Sometimes, opinions differ even within the same organization. For instance, the Seattle City Council once advocated breaching the four lower Snake River dams to help in salmon recovery while the city's utility division wanted to double Seattle's reliance on inexpensive power from the same dams (Lee 2000). Same city government, different attitudes driven by separate practical needs.

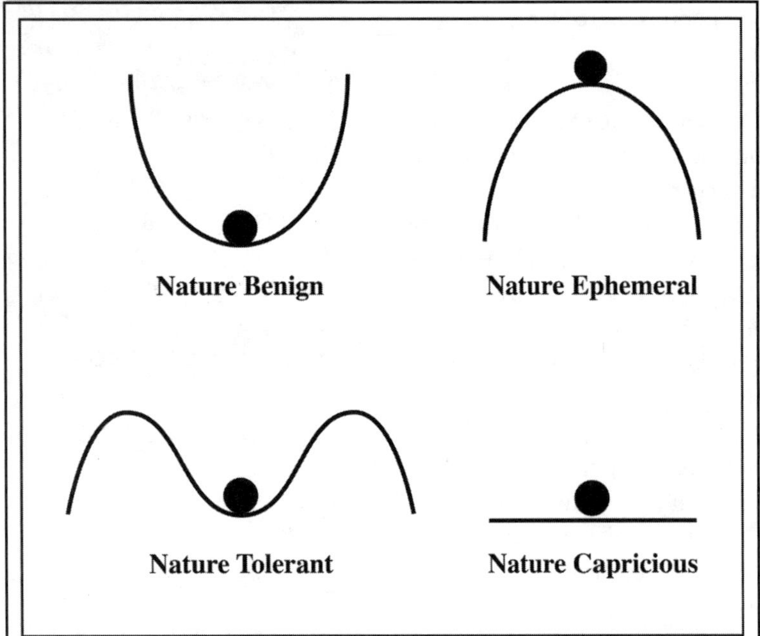

Perceptions of Nature. Each of us has unique views of how nature responds to stress such as contamination released into it. Here are four possible models (Thompson et al. 1990). In these images, movement of the dark ball represents how nature will respond to stress. In a Nature Benign world, nature is forgiving and will recover from what is done to it. It's like a ball quickly returning to the bottom of a bowl. Nature is fragile in the Nature Emphemeral world. Small errors result in extreme responses. Nature Tolerant is forgiving of most events though vulnerable to extremes. Nature responds randomly, based on blind luck, in a Nature Capricious world.

Certainly, approaches to problem solving differ between organizations. One organization might fight forest fires to protect the present scenic beauty while another lets fires burn to rejuvenate forest health over the long term. In cleanup, one organization might promote conventional technology to actively treat contaminated groundwater while another focuses upon surface barriers to buy time for more natural decay to take place before active intervention. Neither position is right or wrong; they just have different approaches, goals, costs, and time frames in mind.

The diversity of institutional behaviors and personal beliefs can be examined by "myths of nature." These myths—defined by writers Michael Thompson, Aaron Wildavsky, and Richard Ellis in their book *Cultural Theory* (Thompson et al. 1990)—are examples of habits of thinking.

These myths or models are not offered as absolutes or as an all-inclusive list but rather to illustrate how we may view nature and therefore wish to protect it.

The first four models are mutually contradictory: each provides a view considered rational at the expense of defining the others as irrational. The fifth model (not shown in illustration) accommodates a more dynamic view. Each model may be viewed as a separate world view, underpinned by unique patterns of thinking.

Approaches to problem solving differ within and between organizations.

Benign Model. Nature forgives and recovers from what's done to it—like a ball always rolling to the bottom of a bowl. The managing institution holding this view can have a laissez faire attitude because nature will always recover. This model encourages experimentation, even exploitation, for nature's dominant tendency is forgiveness. Recovery is the dominant philosophy of this model.

Ephemeral Model. Nature is unforgiving. The least jolt can trigger disaster. Nature is like a ball resting on top of a hill. Once disturbed, the ball will roll away. In this world view, the managing institution must treat nature cautiously. Tight sanctions are required to govern institutional behavior and potential threats. This model encourages hesitancy and extreme caution. Control is the dominant philosophy for nature can only tolerate modest demands.

Tolerant Model. Nature is forgiving of most events but is vulnerable to an occasional hard kick. The managing institution must regulate against extremes, so the ball never slips over the rim. Demands are placed on nature but within carefully defined limits. This model relies on mapping and managing boundary conditions. Predictability and skill are the dominant environmental management philosophies.

Capricious Model. Nature is random, like a ball resting on a flat surface. It can roll in any direction. Nature reacts like a game of chance, where raw probability controls. Institutions holding this view do not manage. They react and are taken by surprise. Learning is not possible for there are no gradients to teach us, no contours to map tendencies. Moral concerns and managing philosophies are driven by luck.

Thompson et al. (1990) go on to discuss a **nature resilient model**, in which nature is dynamic, never responding in a single manner for its response to stress depends upon time, place, and circumstances. Each stress is approached without predetermination. The prevailing management philosophy is adaptation, flexibility, and learning.

Adherents of any set of beliefs use specialized language (jargon) to define their group and who belongs within their inner sanctuary.

Within a particular cultural setting, where shared values bind people and causes, one model is commonly preferred over another. For years, Hanford waste management resembled the benign model, believing nature was tolerant and forgiving of even the most extreme stress. Though more difficult to classify, portions of both the ephemeral and tolerant models (control mixed with predictability) appear to dominate today's environmental practices. Nonetheless, proponents of both the benign and the capricious models exist.

Adherents of any set of beliefs commonly use specialized language (jargon) to define their group and who belongs within their inner sanctuary. These languages encourage mass thinking hindering cross-cultural communication by limiting exposure to new ideas. They are used to coach people to be afraid or accepting of one world view over others. Guardians of "truth" defend against change. In the extreme, adherents become cut off from new unfiltered experiences for they readily adopt, as accepted truth, only sanctioned opinions. People can become so entrenched in their own ideas that doubts are stifled. When beliefs are challenged, they retreat to the safety of traditional assumptions and generalities. When you hear the words "Hanford," "radiation," or "environmental protection" what is your reaction? Upon what is it based?

Assumptions and faith fill the gaps between facts and beliefs.

2.3 The Plutonium Culture

"No one is beyond this error of assuming a built-in advantage for culture, especially when culture is underpinned by economic and political power."

—Lamin Sanneh, *Translating the Message: The Missionary Impact on Culture* (1989)

The historical pattern of Hanford operations is often portrayed as a monolithic hierarchy (Silverman 2000)—locked inside an immense gated community living outside the mainstream of public consciousness and accepted industrial practices. This is too simplistic. Cultures are far more complex, an interwoven web of inherited and absorbed rules, values, and assumptions evolving like an organism to changing stress. Cultures that don't adapt don't survive except in a narrowly controlled and insulated setting. The closer one looks, the less coherent history becomes and the more complex culture is shown to be. Hanford is no exception.

The historical pattern of Hanford operations is often portrayed as an immense gated community living outside the mainstream of public consciousness and accepted industrial practices. This is too simplistic.

Hanford linked numerous actors in ever-changing relationships that responded to global threats and regional economic opportunities. Contractors become self-sustaining by accepting the culture of their founding fathers who created successful products.

Hanford worker and public safety were framed and then reframed to ensure achieving plutonium production goals. And, these goals were supported by an onsite culture that used the environment for contaminant disposal—like most industries and municipalities of those times. Espoused values of environmental protection were devised to preserve public acceptance of existing site operations and contaminant handling practices.

Until recently, secrecy shielded Hanford, and other nuclear weapons sites, from most independent scrutiny. Even late into the 1970s and early 1980s, "the DOE [U.S. Department of Energy] was a cold-war autocracy whose apparatchiks saw little need for consultation either with the general public or with technical advisors from outside its

Secrecy shielded Hanford and other nuclear weapons sites from most independent scrutiny until the bottom dropped out of the plutonium market.

Government Agencies Managing Hanford

U.S. Army Corps of Engineers (1942-1947)
Atomic Energy Commission (1947-1975)
Energy Research and Development Administration (1975-1977)
U.S. Department of Energy (1977-present)

Private Companies at Hanford[3]

Several private companies holding contracts with the government (known as prime contractors) built, operated, or were responsible for various Hanford plutonium production facilities and support services. Starting in the late 1980s, work shifted to waste management and cleanup.

E.I. du Pont de Nemours and Company (1943-1946)
General Electric Company (1946-1967)
Isochem, Inc. (1966-1967)[a]
Battelle—Pacific Northwest National Laboratory (1965-present)[b]
Atlantic Richfield Hanford Company (1967-1977)
United Nuclear Industries, Inc.[c] (1967-1987)
Rockwell Hanford Operations (1977-1987)
Westinghouse Hanford Company (1987-1996)
Bechtel Hanford Inc.[d] (1994-present)
CH2M Hill Hanford[e] Group Inc. (1999-present)
Fluor Hanford, Inc.[f] and support contractors (1996-present)

(a) Operated reprocessing plants
(b) U.S. Department of Energy independently operated laboratory for science and technology located in Richland, Washington
(c) Reactor operations and fuel fabrication
(d) Environmental cleanup
(e) Tank waste management and treatment
(f) Management integration of several onsite contractors

(3) Adapted from information provided by Mike Talbot, U.S. Department of Energy, Richland Operations Office.

own ranks" (*Nature* 2001). This changed during the waning years of the Cold War when the bottom dropped out of the plutonium market. Plutonium production became a non-priority, a liability. New market pressures emerged that reflected a change in America's attitude towards the environment in general and industrial waste in particular. Public oversight and reviews by scientists and other peers not paid by Hanford intensified. Public awareness and knowledge grew.

Under the new environmental culture, problems suddenly abounded, and openness became more than an option; it was a prerequisite for economic survival.

The end of the Cold War ushered in new economic, social, and political forces to clean up the waste legacy. To survive, Hanford embraced the new business of environmental protection. Under the plutonium culture, few environmental problems were openly acknowledged. Under the new culture, problems suddenly abounded, and openness became more than an option; it was a prerequisite for economic survival.

The Hanford cleanup agreement marked the most dynamic cultural shift in Hanford history.

Were concerns about waste generation and releases new? No. Within a few years of Hanford beginning operations, problems were raised about waste handling (Williams 1948). But the Hanford culture remained nearly closed to independent questioning. This set the site on an inevitable collision with an emerging environmental culture fueled by externally imposed waste management regulations, new health standards, and public outrage upon learning about formerly secret contaminant releases. Change was inevitable, though not institutionalized, with the signing of the Hanford Federal Facility Agreement and Consent Order (a.k.a. Tri-Party Agreement) in 1989. This marked the most dynamic cultural shift in Hanford history—from a single, clear-cut purpose with measurable short-term goals to a type of cultural pluralism populated with numerous actors, some playing oversight roles without responsibility for outcomes, engaged in achieving ill-defined long-term cleanup goals.

The Hanford Federal Facility Agreement and Consent Order, popularly known as the Tri-Party Agreement, requires the U.S. Department of Energy to bring Hanford into compliance with provisions of federal and state cleanup regulations. Waste management and environmental restoration activities were made into legally enforceable milestones, with penalties for not completing activities on time. For more information on this agreement, see Section 7.5.

The nationwide radiation and pesticide controversies of the 1950s and 1960s demonstrated that public agencies sometimes ignored or even developed narrow interpretations of data. This led them to discount evidence of broader ecological consequences (Andrews 1999). Risks were sometimes unknown or underestimated, or agencies were simply unwilling to openly deal with risk except as it impacted product manufacturing. As the public became more informed, frequently from non-industry and non-government sources, their confidence in institutions and traditional experts eroded.

As the public became more informed, their confidence in institutions and traditional experts eroded.

At Hanford, trust in officials and secrecy underpinned the early years. Officials believed workers could be protected without revealing information deemed sensitive about plant operations or specifics about radiation exposure.

Herbert Parker (1944) wrote, "The recommendations in the Bureau of Standards Handbook on Radium Protection specifically require that all men engaged in radium work be informed of the nature of the hazards involved. The same general principles govern operations at the Hanford Engineer Works, limited only by the requirements of Security. Security…can be maintained if hazard information is phrased in such a manner that the maximum information on the nature of the hazard is combined with the minimum information on the nature of the materials used and the processes involved."

How this was carried out is illustrated in a letter Dr. W.D. Norwood, Medical Superintendent for Hanford, wrote to Dr. Robert Stone, head of health programs for the entire Manhattan Project (Norwood 1944). Norwood asked about the risk to a worker who performed a quick emergency job in an area of high radiation. "What dose of gamma radiation may an Area Superintendent order one of his men to be exposed to with reasonable assurance that the chance of any permanent damage occurring is extremely slight? To be more specific we may say, that the exposure would not be more hazardous than that which occurs in playing football, driving an automobile, or any other occupation having hazards with which we are all in a general way familiar. It is assumed that the risk would be small enough that it would not be necessary to inform the man, other than to tell him that following the exposure, it would be necessary for him to immediately report to the Medical Department for necessary checks. Management would realize that the man would be kept away from any further exposure as long as necessary, depending upon the medical findings and that he might be kept away from any further exposure permanently."

At rare times, Hanford officials did question whether plant operations should be shut down.

At rare times, Hanford officials did question whether site operations should be shut down. In 1948, small radioactive particles were discovered released up the stacks of B and T Plants (Parker 1948c). These are hazardous if inhaled. Particles were falling to the ground outside the plants as well as 100 to 200 miles away—in Spokane, near Mt. Rainier, and in northern Idaho. In discussing this situation with an advisory panel to the Atomic Energy Commission, Parker expressed concern about a "level of contamination at which the Plant should be shut down, and whether that level was reached." The incident's seriousness was forcefully brought home when a senator asked, "If Parker had a 20 year old son, would he want him to work in the 200 Areas?" Parker replied "No." This simple response stripped away the academics of dispassionately comparing numbers and brought a personal element into the question of worker safety.

When asked during a Senate hearing if he "had a 20 year old son, would he want him to work in the 200 Areas" when radioactive particles were floating down from the sky, Parker simply replied "No" (Parker 1948c).

However, the Atomic Energy Commission's advisory panel recommended that the plants remain operating as studies of the potential beneficial use of particle-capturing sand filters were undertaken. Parker distanced his reputation and that of his colleagues and General Electric from accepting the responsibility for the panel's decision (Parker 1948c). Operation guidelines were flexed to preserve material production. This was a dominant theme in the plutonium culture.

Hanford's voluntary guideline approach to health and environmental safety reflected a tradition cautious in applying strict enforceable standards, termed "legal limits" by Parker, when flexibility was preferred. Otherwise, too strict of enforcement standards for new waste management approaches could curtail operations for long periods of time and cost tens of millions of dollars to implement (Parker 1951).

While genuine concerns for human health were never absent, maintaining plutonium production and minimizing expenses for waste management prevailed. Reactor cooling water was managed to levels "consistent with other determinants" (Parker 1945a). Health and safety regulations for controlling discharges were offered as a "working code of minimum interference with production" (Parker 1952). Radiation exposure to individuals and other significant life forms were to be kept at the "lowest practicable level" consistent with the standards of those times (Parker 1959c). Liquids were disposed to the ground to "avoid the absurd costs on tank storage, evaporation equipment, or the equivalent" (Parker 1948a, p. 3).

Health and safety regulations for controlling contaminant discharges were offered as a "working code of minimum interference with production" (Parker 1952).

Radiation protection standards and practices were established by separate groups within the same Hanford contractor whose managers were responsible for ensuring plant operations adhered to these standards (Parker 1959b). At that time, this was considered an excellent example of independent checks and balances. Though Hanford staff were urged by the Atomic Energy Commission to "make greater use of the cooperation of public health authorities, and of outside consultants" (Williams 1948, p. 1), decisions about radiological and chemical releases remained under the purview of local officials. "In our opinion, not only does the Atomic Energy Commission and the prime contractor have the responsibility to decide on the amounts and storage conditions of hazardous wastes, etc., but they are together in a better position to make such decisions than any other agency" (Parker 1950). No other agency had the knowledge, skill, expertise, or security clearances to independently critique Hanford operations.

As noted by Silverman (2000), for years Hanford was self-regulated, controlled by an internal expert-driven system rather than an external, independent rule-bound system. He writes, "The closed, expert-driven model enabled Hanford's managers to develop and maintain a consensus of environmental safety even as increasing evidence indicated weaknesses in the site's waste management practices." For that time and place, Hanford officials believed they were in the best position to know what were acceptable risks. State officials agreed.

This does not imply that Hanford health and safety research was not top notch. It was outstanding. Hanford staff pioneered world-class radiation research, engineering, and environmental monitoring, setting many global radiation standards for worker safety and public protection. Their contributions should never be underestimated nor the pioneering radiation protection work of men such as Herbert Parker. The apparent weakness was the absence of independent oversight and leadership to implement polices more protective of the environment and more informative to the public. The lack of checks and balances and the failure to acknowledge and correct mistakes—the hallmarks of credibility—invited failure in the eyes of a new generation.

Hanford staff pioneered world-class radiation research, engineering, and environmental monitoring, setting many global radiation standards for worker safety and public protection. Their contributions should never be underestimated.

To some Hanford officials, broad evaluations of water quality in the Columbia River or radiation dose received downstream of Hanford were more of an "academic interest" than other "practical problems" (Kornberg 1950) or conducting "research appropriate to the solution of local problems" (Parker 1952).

The lack of checks and balances and the failure to acknowledge and correct mistakes—the hallmarks of credibility—invited failure in the eyes of a new generation.

However, by 1959, congressional testimony from Hanford officials underscored a broader awareness of environmental pollution concerns. At those hearings, Parker (1959b) provided written testimony about Hanford staff members' extensive representation on national committees that focused on radiation protection and waste disposal issues. He also summarized Hanford contacts, exchange seminars, and other working relationships with Northwest universities dealing with the life sciences: U.S. Public Health Service, Safety and Industrial Health Advisory Board, U.S. Department of Agriculture, University of Washington, Washington State College,[4] fish and wildlife authorities, and

(4) Now Washington State University.

By the late 1950s, several events had taken place to begin dismantling the plutonium culture. These included

- A decade of interacting with members of the Columbia River Advisory Group and the U.S. Public Health Service.
- The National Council on Radiation Protection establishing maximum permissible exposure for a host of radionuclides and recommending radiation exposure to the public be 10% of that permitted for radiation workers.
- Hanford staff publishing the first radionuclide uptake and whole body and organ specific radiation doses for the public living near the site.

For more information, see Chapter 3 and Section 4.1.

pollution experts from the states of Washington and Oregon. Public exhibits, oral presentations, and unclassified reports available through the Atomic Energy Commission's Civilian Application Program were summarized to demonstrate Hanford was disseminating information to the public, external professional societies, and government agencies. Much of the site's research, waste release monitoring, and surveillance still optimized operations rather than minimized contaminant releases. With few exceptions, the plutonium culture was built upon a reliance on existing technology rather than driven by innovation to reduce waste generated and released. The same reactors, the same waste disposal, and the same contaminant problems prevailed for years.

This is why independent studies of Columbia River water quality, such as completed by the U.S. Department of Health, Education, and Welfare et al. (1954) and the Columbia River Advisory Group (Tsivoglou and Lammering 1961), were not eagerly embraced under the old culture. These studies questioned conventional practices and potential impacts to those living downstream and downwind.

People were growing increasingly opposed to involuntarily imposed hazards, especially radiation-based hazards.

Besides, why should Hanford practices be questioned? Its safety record was exemplary compared to other industries. As noted by Parker, "the insidious danger of radiation damage receives a spotlight in the popular and technical press, which is out of proportion to the hazard in comparison with injury risk in many other industries" (Williams 1948). The fact was, regardless of such safety comparisons, people were growing increasingly opposed to involuntarily imposed hazards, especially radiation-based hazards.

Hanford had lost control of the message it wanted people to hear. The public could interpret information in a manner different than the culture that created it had intended.

With the declassification of formerly secret documents in 1986, non-Hanford researchers, news reporters, and the public could now read about contaminant releases and come to their own conclusions. Hanford had lost control of the message it wanted people to hear. The public could interpret information in a manner different than the culture that created it had intended. A new self-educated and skeptical culture emerged, challenging the unflinching acceptance of past practices. Waste handling was seen from a different point of view. Declassification called the plutonium culture into question, rather than bolstering existing claims. These data became weapons for imposing greater external influence—and a new environmental compliance culture—onto Hanford. A growing desire for intervention emerged.

Embracing a new culture is neither quick nor easy. Culture is a source of identity, a reflection of beliefs that's "clung to with a vengeance" (Schein 1999, p. 12). Protection is instinctive, especially when people believe they are besieged by criticism from an uninformed outside world—as many Hanford workers and local community members felt in the mid-1980s. In 1987, the non-profit organization Hanford Family was formed in the Tri-Cities. It sought to provide alternative interpretations of Hanford history and the continuous value of onsite operations compared to a growing chorus of news reports over contaminant releases, human hazards, and violations of environmental laws.[5] However, this was a lonely battle for the political landscape was being upended as the plutonium culture was plowed underground. Outside of the Tri-Cites, few allies were found. Inside the Tri-Cities, the usually strong pro-Hanford outreach activities sponsored by the federal government and its contractors stalled out of neglect or intent (opinions differ). The voices of others, less supportive of Hanford, filled the void.

Today, Hanford is judged by new interpretations of its past. Some criticism is justified; some is not. Perhaps, the generic lesson learned is that "we cannot take warnings lightly or accept uncritically the soothing assurances of authorities" (Park 2000, p. 146). We should seek out sources of critically assessed and balanced information and make our own informed decisions.

We can move forward either trying to profess a single truth or recognize there is enough room for well-considered but differing ideas.

However, all people are vulnerable to seeing their own culture as holding a privileged and exalted position. How will future generations view our approach to cleanup, our use of limited resources, our solving of problems or just passing them along?

There is a growing need for people to work across cultural boundaries, to find common ground upon which to build credible environmental policies and cleanup approaches. We can move forward either trying to profess a single truth or recognize there is enough room for well-considered but differing ideas—as long as all is built upon high-quality knowledge, critical thinking, and enduring values. Luckily, people and institutions are not permanently locked into single mindsets. They can and do change.

Are we really sure about what we know and what we are doing?

Today, people must ask themselves the same question adherents to the past culture may have sidestepped: "Are we really sure about what we know and what we are doing?"

2.4 An Emerging Environmental Culture

"Citizens also discovered the words 'environment' and 'ecology' in the 1960s and took their meaning to heart."

—Dale Becker, *Aquatic Bioenvironmental Studies: The Hanford Experience 1944-84* (1990)

Waste-handling practices at Hanford were not carried out in isolation, and public concerns were not isolated incidents in the Pacific Northwest. Rather, they mirrored a broad growth in a citizen-driven ethic of partnering with nature, of wanting to live in a safer, cleaner environment.

(5) The Hanford Family was disbanded in 1998.

Until the late nineteenth century, the dominant environmental policies of the United States encouraged transforming raw resources into economic commodities. Resources were exploited, and the environment was used as a dumping ground for waste. Pollution and occupational hazards were accepted as necessary side effects. Slowly, conservatism and an awareness of the human-environmental interdependency grew upon America's consciousness. Public concerns about their health and the preservation of an irreplaceable environment came into play. These were novel ideas leading to revolutionary legislative and social actions in the twentieth century.

Until the late nineteenth century, the dominant environmental policies of the United States encouraged transforming raw resources into economic commodities.

U.S. Department of Energy 093363-6

Richland Housing. The Tri-Cities (Richland, Kennewick, and Pasco) grew as Hanford grew. Plutonium production was skyrocketing when this early 1950s photograph of new Richland neighborhoods was taken. Across the country, the middle class was booming.

Affluence in post-World War II America brought about travel, recreation, middle-class neighborhoods, and a growing engagement with the environment. Nature became an amenity to enjoy, a cherished respite from urban life, a resource to protect. Seeking untainted vistas was a sacred pilgrimage into time and space that transcended everyday life. The public began viewing land not just for its economic value but also for its unmanaged beauty. Enjoyment of the environment became an industry itself.

Recognizing the damage done by past exploitation, people sought environmental amelioration. Socially driven ethics and personal standards came into play.

Affluence in post-World War II America brought about travel, recreation, middle-class neighborhoods, and a growing awareness of the environment.

While the vulnerability of the environment was recognized, competing philosophies over resource use continued to clash—as they do today. Examples included the use of national forests for timber production versus preserving natural species, and the use of rivers for power production or irrigation versus fish preservation and scenic beauty. Many resource-preservation versus resource-use battles revolve around whether to exploit traditional economics or protect traditionally uneconomic, yet irreplaceable, natural resources.

Within the nuclear weapons complex, World War II and fears of the Soviet Union helped justify continuing a wartime-type secrecy surrounding the production of nuclear materials. This extended into tightly controlling records about contaminant releases and even justified the "outright deception of the public concerning the risks imposed upon them" (Andrews 1999).

During the post-war era, few officials paid attention to material consumption and waste disposal. America had evolved into a global military leader and economic powerhouse. The expansion of industry was accompanied by an enormous increase in chemical usage and disposal. The annual use of synthetic chemicals grew from 2.2 billion to 214 billion pounds in the 50 years following 1940 (Russell et al. 1992, p. 1).

However, pollution problems grew faster than the population. The likely reason rests in how consumer goods were produced and economic growth achieved. Synthetic products now substituted "for their natural counterparts—detergents for soap, plastic for rubber, metal, and glass, synthetic fibers for natural fibers, synthetic pesticides for arsenic" (Fleming 1972, p. 62). Plastics, steel, and concrete were replacing wood in home building. Many consumer products and unwanted materials did not degrade like their natural counterparts and were made from exhaustible resources, such as petroleum, using energy-intensive systems. Coupled with an increase in per capita consumption, these advances placed greater strain upon the environment.

But, bottom-line competitive economics, a lack of independent accountability, and public acceptance of industrial practices favored continued use of the air, water, and land for waste disposal. Industrialization, urbanization, and the introduction of new contaminants began dethroning the frontier mentality that nature always accommodates what's released into it, that shared and limited resources can be treated as free goods for exploitation. The unlimited opportunity to dispose clashed with the unlimited right of individuals and communities to protect.

By the last half of the twentieth century, "air and water pollution, long imposed on the public by powerful industries as the unavoidable side effects of economic progress and wartime production, were no longer automatically accepted" (Andrews 1999, p. 201). Public concerns, based upon "strong circumstantial evidence," arose over clusters of allergic reactions and serious illnesses in some neighborhoods near chemical dumps (Russell et al. 1992, p. 4). Though connecting chemical exposures to illnesses in specific people is "scientifically difficult and politically charged," Congress was forced to implement a nationwide regulatory agenda. With profits at stake, voluntary constraints were not curtailing waste dumping practices.

Starting in the late 1960s, a regulating community, empowered by law, grew in the United States as the words "ecology," "environment," and "green revolution" bolted onto the national scene. This was prompted by a groundswell of public concerns over raw sewage in rivers, air choked with industrial gases, chemicals oozing from abandoned landfills, dying animals, sick people, noise, and destroyed ecosystems. Unsightly conditions were obvious, and threats were immediate. In some fume-shrouded cities, deaths increased from respiratory illnesses. An estimated 168 people died in New York City from air pollution during the Thanksgiving season of 1966 (Whitman 2000). Thousands of others became ill. The debris and chemical-laden Cuyahoga River in Cleveland, Ohio, actually caught fire in 1969—the third time in 30 years.

Rachel Carson's book *Silent Spring,* published in 1962, took a sobering look at the ecological damage from unbounded and sometimes indiscriminate use of an emerging family of increasingly potent and ecologically toxic insecticides

The unlimited opportunities to dispose clashed with the unlimited right to protect.

With profits at stake, voluntary constraints and market forces were not curtailing waste dumping practices.

The debris and chemical-laden Cuyahoga River in Cleveland, Ohio, actually caught fire in 1969—the third time in 30 years.

U.S. Department of Energy

Nike Missiles. An air-defense system of antiaircraft guns and Nike surface-to-air missiles once surrounded Hanford. These were removed by 1958 when Russia and the United States deployed Intercontinental Ballistic Missiles. There was no defense against ICBMs.

Published at the peak of atmospheric testing of nuclear weapons, *Silent Spring* spoke to a generation worried about the impacts from contaminants released into the environment.

and pesticides.[6] She aroused public attention with specific cause-and-effect arguments, using both government-sponsored research and her own studies about how potentially toxic chemicals were transmitted through the food chain at increasingly higher and deadlier concentrations. She sought not to polarize debates or capture the moral high ground through arrogant arguments but to persuade people to be more cautions in chemical use especially at stress points in natural systems. Carson effectively intertwined her personal convictions with her passion for scientific evidence into a detective story that continues to fuel an environmental consciousness.

Carson's book was published at the very peak of atmospheric testing of nuclear weapons and the height of public and congressional concerns about radioactive fallout. It was an easy leap of logic to connect her concerns about chemicals to potential health effects from strontium-90 and other radionuclides permeating the air. Fallout was simply another dimension to the same story.

As industrial waste practices fell from popular favor, so did those at the nuclear weapons sites.

The dumping of industrial waste, global spread of radionuclides, and seemingly unbridled use of exotic chemicals brought about a re-examination of the social implications of scientific advances that largely lay shielded inside industries and government agencies. "The conservation impulse of the 1970's…was a time of rising hostility to consolidated power, walled-up pieties, and narrow specialization" (Fleming 1972, p. 22).

As industrial waste practices fell from popular favor, so did those at the nuclear weapons sites. However, the Atomic Energy Commission, and later the U.S. Department of Energy, claimed exemption from state and federal pollution control and waste-handling laws by stating its activities were covered under the Atomic Energy Act and immune from independent checks (Geiger and Kimball 1992).

This began to unravel in 1983 in the hills of eastern Tennessee when the Legal Environmental Assistance Foundation, Inc. and the Natural Resources Defense Council sued the U.S. Department of Energy for operating a uranium separations facility, known as the Y-12 Plant near the city of Oak Ridge (U.S. District Court for the Eastern District of Tennessee 1984). Previously, the U.S. Environmental Protection Agency had failed to dispute the U.S. Department of Energy's claim of immunity (Davis 1986). When the state of Tennessee attempted to assert hazardous waste regulatory authority, the U.S. Department of Energy again asserted its exemption. The state acquiesced. A lawsuit was then filed under the citizen provision of the Resource Conservation and Recovery Act,[7] a federal law to address the complete cycle of treatment, storage, and disposal of hazardous waste.

What brought the citizens of Oak Ridge to this point? As early as the late 1940s, the Atomic Energy Commission reported that radioactive and chemical releases into nearby rivers were in excess of tolerable limits and assumptions of water quality were

(6) Such chemicals as DDT, endrine, and Malathion were introduced into the environment for life-saving reasons. (DDT stands for dichloro-diphenyl-trichloroethane. It's a chlorinated hydrocarbon.) The insecticide properties of DDT were discovered in 1939 when typhus and malaria threatened to take more lives than were lost in World War II combat (Fleming 1972).

(7) The Resource Conservation and Recovery Act (RCRA), created in 1976, is a federal law governing the ongoing management of hazardous wastes. This law encompasses a formidable array of requirements designed to protect humans and the environment. For more information on environmental regulations, see Chapter 7.

substituted for actual measurements (Williams 1948, p. 69). In 1982, water sampling detected mercury contamination in a creek near the Y-12 Plant.[8] The state of Tennessee restricted fishing in the area. Subsequent inspections at the Y-12 Plant uncovered dangerous disposal practices for waste created at the site, such as unlined and leaking lagoons and trenches receiving millions of gallons of often poisonous metals and chemical-laden waste. This contributed to the contamination of local lakes, streams, and groundwater.[9] In response to a Freedom of Information Act request, the Energy Department revealed it had dumped 2.4 million pounds of mercury[10] into the creek during the 1950s and 1960s. The public was outraged. Congress got involved. When it became clear that neither the Environmental Protection Agency nor the state of Tennessee were willing to press their jurisdictional roles, the citizen lawsuit was filed.

A Tennessee federal court rejected the Energy Department's claim of immunity. The court ruled that the U.S. Department of Energy's facilities were subject to hazardous regulations under the Resource Conservation and Recovery Act and the Clean Water Act[11] except for nuclear materials, such as spent uranium fuel or plutonium metal (U.S. District Court for the Eastern District of Tennessee 1984).

In 1989, another incident occurred, this time in Colorado. Agents from the Federal Bureau of Investigation and the Environmental Protection Agency, armed with a search warrant, entered the U.S. Department of Energy's Rocky Flats nuclear weapons facility near Denver. The search exposed widespread environmental violations including "secret incineration of hazardous waste, false claims of compliance with groundwater monitoring requirements, and intentional mixing of hazardous and radioactive waste" (Cheng 1990).[12] Coming a few years after the Tennessee lawsuit, the Rocky Flats incident underscored the need for stricter environmental compliance at nuclear weapon facilities and reaffirmed public suspicions that the government continued to be a major violator of its own laws.

Coming a few years after the Tennessee lawsuit over mercury contamination, the Rocky Flats incident reaffirmed public suspicions that the government continued to be a major violator of its own laws.

The U.S. Department of Energy formed the Office of Waste Management and Environmental Restoration (now called the Office of Environmental Management) in 1989 to better manage the environmental byproducts from years of nuclear material development. Environmental budgets skyrocketed as clean-up problems were uncovered, past contaminant releases admitted, and potential health impacts studied. Across the nation, the U.S. Department of Energy entered into dozens of cleanup agreements. This is discussed in Section 8.2.

In addition, the department began sending oversight teams to its facilities to investigate environmental contamination, health and safety issues, and compliance with U.S. Department of Energy orders and regulatory laws. These groups were affectionately known as "Tiger Teams."[13]

---

(8) Between 1956 and 1963, a mercury-based chemical exchange process (COLEX) was used at Oak Ridge for enriching lithium. The resulting lithium-6 isotope was irradiated inside reactors to create tritium at other nuclear weapon sites, mainly Savannah River.

(9) Groundwater is water that exists and moves beneath the land surface.

(10) If combined, 2.4 million pounds of mercury would form a cube 14 feet on a side.

(11) Formally known as the Federal Water Pollution Control Act Amendments of 1972, this act sets the basic structure for regulating discharges of contaminants into U.S. waters.

(12) Hazardous waste contains nonradioactive metals and chemicals, known or thought to pose a risk to human health and the environment (see Chapter 5).

(13) The Energy Department initiated the Tiger Team Assessment Program in 1989 to provide non-line management oversight of site compliance with regulations and internal requirements. These teams visited all major nuclear sites within the department.

Yet, the government and its agents did not eagerly search for "alternatives to business-as-usual, nor did it begin a genuine effort to clean up old messes" (Gray 1992, p. 1). Proposals such as Complex 21 supported further warhead development and nuclear material production in spite of lowering global tensions and weapon arsenals between superpowers.

The long history of government agencies not complying with their own laws and legal maneuvering prompted Congress to pass the Federal Facilities Compliance Act in 1992. This act amended the Resource Conservation and Recovery Act by requiring all federal agencies to meet the legal requirements of federal, state, and local laws governing pollution prevention and cleanup in the same manner as private companies. For the first time, federally owned facilities, plus their agents and employees, were subject to fines and penalties for violating waste management and disposal laws. The act directed the Environmental Protection Agency to make annual inspections of facilities owned or operated by the government that treated, stored, or disposed of hazardous waste. Authorized states also conducted inspections to ensure compliance with their own hazardous waste laws.

The long history of government agencies not obeying their own laws prompted Congress to pass the Federal Facilities Compliance Act in 1992.

Today, few would disagree with the importance of Hanford moving to a more environmentally protective culture. More than a decade after signing the Tri-Party Agreement, a cultural change is still taking place. New work patterns and behaviors do not happen overnight. Change in corporate culture takes years. It requires not just new policies but a fundamental shift in work values and the underlying motivations that drive what people perceive is needed to succeed. The plutonium culture was well established, designed for manufacturing nuclear materials. However, its waste-handling practices remained primitive, even according to its own reviews.

The cleanup culture remains unproven. Nonetheless, the first steps have been taken. Work practices driving this culture are still being deciphered and will continue to evolve as more is learned and as each restoration challenge is faced.

Much remains uncertain in the cleanup culture.

What did Hanford know and when about the spread of contamination? The environmental monitoring and radiation legacy of Hanford is the focus of the next chapter.

Chapter 3

Environmental Monitoring and Surveillance

"Throughout those forty years after World War II, environmental monitoring was carried out at Hanford in perhaps the most thorough and technically excellent example in the nation. Yet, this precedent-setting endeavor faltered by classifying most of the data it collected. Early Hanford chemists…felt they were on the cutting edge of science. How much sooner might the problems of airborne, river-borne, and groundwater discharges have been solved had these unique and talented environmental-monitoring scientists shared their data?"

—Michele Gerber, *On the Home Front: The Cold War Legacy of the Hanford Nuclear Site* (1997, p. 217)

"Measurable levels of man-made radioactivity were not detected in vegetable and fruit samples collected in 1998…Radionuclide levels in Hanford-resident wildlife were similar to levels in wildlife collected at reference background locations" (Dirkes et al. 1999).

Environmental monitoring is essentially an early warning system, designed to identify symptoms of problems so that corrective actions are undertaken before large-scale damage occurs.

Today, an extensive program of environmental monitoring and surveillance takes place in and around Hanford (Poston et al. 2001). This includes sampling of the air, surface water, groundwater, vegetation, animals, and agricultural products. Results are compared to federal, state, and U.S. Department of Energy standards.

The Pacific Northwest National Laboratory[1] in Richland, Washington, provides the overall environmental monitoring and surveillance services for Hanford. The Washington State Department of Health, U.S. Geological Survey, and U.S. Environmental Protection Agency conduct independent studies. Results are reviewed by experts in health and environmental sciences and published each year. The last several annual reports are available at http://hanford-site.pnl.gov/envreport/.

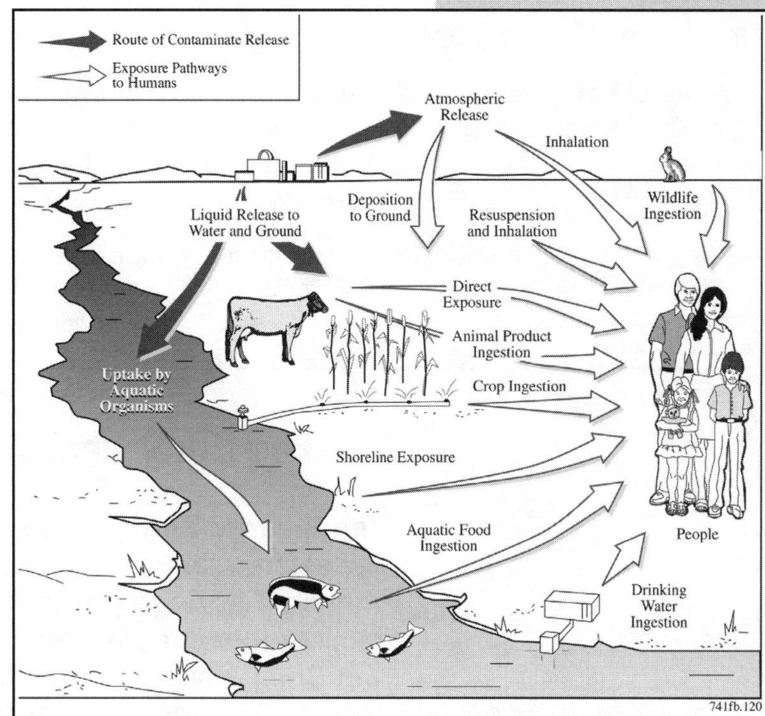

Primary Routes of Exposure to Contaminants. Contaminants are primarily released into the air, water, or the ground. They can then reach people through widely different environmental pathways as illustrated. The importance of one pathway over another is site, dietary, and contaminant specific (Poston et al. 2001).

(1) Pacific Northwest National Laboratory, operated by Battelle for the U.S. Department of Energy, performs research and development directed for Hanford, U.S. Department of Energy, other government agencies, and private businesses.

After the secret of the Manhattan Project was announced, some Hanford radiotoxicology studies started to show up in the open literature.

Private contractors are also hired by the U.S. Department of Energy to monitor waste and nuclear materials contained inside facilities and other controlled areas.

Hanford historical data constitute one of the most extensive environmental data sets for any single project in, and perhaps area of, the United States.

When World War II ended and President Truman announced the secret mission of Hanford, internal communication restrictions between Hanford groups were softened. This permitted them to more easily exchange environmental and biological data. In addition, some onsite radiologic and toxicology studies began showing up in the scientific literature as unclassified reports. Examples are found throughout an early annual report by the biology staff of Hanford (GE 1952).

However, for years, key portions of these data, and their possible biological implications, were classified as secret, restricted, or confidential. Numerous monitoring reports were published from 1947 to 1957 but were not available to the public until document declassification began in 1986 (Soldat et al. 1986). Publicly available annual reports have continued ever since.

Secret classification was applied to information that if disclosed might endanger national security, cause serious injury to the interest or prestige of the nation, or would be of great advantage to a foreign nation. (Disclosure of Top Secret information would cause exceptionally grave danger.)

Confidential classification was applied to information that might prejudice the interest or prestige of the United States, its government, or an individual. It was also applied to information that might be of some value to a foreign nation.

Restricted classification was applied to information that was "for official use only" and was not for dissemination to the public (Marceau et al. 2002).

Several intertwined events took place during Hanford's first 20 years that dramatically altered its approach to environmental monitoring. These included a growing awareness that waste management practices should change, establishment of nationally recognized radiation dose limits for the public, improved radiation detection instruments, and a better understanding of contaminant behavior and risk to humans and other life forms. This is a summary of that story.

3.1 The Early Years

"The Hanford Engineering Works was the first nuclear facility to release substantial amounts of radioactive effluents into the environment. Although government regulations did not exist at the time, responsible officials of the Manhattan Project and the Du Pont Company took the initiative to monitor those releases."

—Richard Foster, *Environmental Monitoring, Restoration, and Assessment: What Have We Learned?* (1990, p. 169)

In the early 1940s, health science was relatively young, radiation standards were poorly defined, and impacts from Hanford contaminant releases "were almost completely unpredictable, except as to gross order of magnitude" (Parker 1952). Approximations about potential health impacts were superior to available direct evidence. This prompted expanding environmental and health-related research. The first data trends in elevated radionuclide concentrations in the air, water, and vegetation collected as far away as 160 miles from Hanford were published in early 1947 (Hanford Works 1947). These findings quickly caught the attention of Atomic Energy Commission and Hanford officials.

Columbia River: Monitoring the Columbia River began before reactor operations started in late 1944.[2] The first data were from a large number of measurements of low accuracy taken over a wide area at minimal expense. Most were quick "grab samples" collected weekly. These were analyzed for total ("gross") radiation levels. The more costly and time-consuming procedures required to identify specific radioisotopes remained uncommon.

Even before Hanford began operating, General Groves, head of the Manhattan Project, was told by his supervisor (Major General T.M. Robins) about the importance of protecting the Columbia River. "Whatever you may accomplish [at Hanford], you will incur the everlasting enmity of the entire Northwest if you harm a single scale on a single salmon" (Groves 1962, p. 82). In fact, a few months after the decision was made to locate the world's first reactors along the banks of the Columbia River, the medical director for the Manhattan Project (Stafford Warren) raised questions about their potential adverse impacts on a resource that supplied food to millions of people (Stannard 1988, p. 757). This prompted research into understanding the biomedical and ecological consequences from exposure to reactor cooling waters and associated effluents. However, such research took years to complete, let alone establish scientifically defensible radiation protection guidelines. Contaminant releases far outpaced our knowledge of their potential adverse impact.

Officials recognized the "need for the perpetuation of fish life," and therefore, fish survival was considered during reactor operations and water use (Olson 1948). There were five areas of concern:

- Fish killed when sucked into large pumps supplying water to the reactors and to the rest of the site.
- Fish or their food poisoned from chemicals in the cooling water when it was returned to the river.
- High river temperatures from the discharge of heated water.
- Adverse effect of radioactivity entering the river on fish or other organisms.
- Damage to fish spawning grounds.

Radioactivity uptake by fish was measured soon after the first reactors came online (Foster 1959). This prompted the Atomic Energy Commission to establish a fish research laboratory (called the Fisheries Laboratory) in two Quonset huts near the F Reactor in 1945. Research involved exposing fish to reactor effluent water at various dilutions (Kornberg 1958). In the river itself, whitefish were used as indicators of

The first data trends in elevated radionuclide concentrations in the air, water, and vegetation collected as far away as 160 miles from Hanford were published in early 1947. These findings quickly caught the attention of Atomic Energy Commission and Hanford officials.

Contaminant releases far outpaced our knowledge of their potential adverse impact.

Beginning in the ice fields of the Canadian Rockies, the Columbia River flows through a convoluted course for more than 1200 miles until it reaches the Pacific Ocean at Astoria, Oregon, and Ilwaco, Washington. The river supports fish, hydroelectric power, transportation, irrigation, and recreation.

(2) The first contract between the U.S. Office of Scientific Research and Development and the University of Washington to study the effects of ionizing radiation on fish was signed in August 1943 (Stannard 1988, p. 757).

radionuclide uptake because they spent their entire life in the river, their eating habits caused them to accumulate the largest amounts of radioactive phosphorus, and they were easy to catch.

"Whatever you [Leslie Groves] may accomplish, you will incur the everlasting enmity of the entire Northwest if you harm a single scale on a single salmon" (Groves 1962, p. 82).

Karl Herde, Hanford wildlife biologist, wrote about the need for "more accurately determining effluent water tolerances" for fish exposed to radionuclides with half-lives greater than sodium-24 (15 hours) (Herde 1947). Chinook salmon fingerlings were discovered to concentrate radioactive sodium 10 to 30 times greater than found in water discharged from reactors (Herde 1948). Calculations, augmented by field studies, revealed that radioactive phosphorus (phosphorus-32) was concentrated in fish "hundreds of thousands of times that of water…[these numbers] may indicate levels near tolerance are sometimes approached…With higher [reactor] production rates the future level may be sufficient to produce some pathology and a noticeable increase in the mutation rate of fish" (Herde 1948).

In short order, three surprise findings were learned (Foster 1959; Stannard 1988, p. 760):

- Radioactivity in river fish was much higher than the radioactivity found in laboratory raised fish. The reason was that laboratory fish were fed uncontaminated food while river fish ate organisms containing elevated levels of radioactivity gained from Hanford.
- Radionuclides in laboratory-grown fish were different from that found in the river. Laboratory fish had high levels of sodium-24 compared to more phosphorus-32 in river fish. Why? Phosphorus was more concentrated in the natural foods eaten by river fish.
- Toxic effects of heated water and non-radioactive chromium on young fish were much greater than the effects from radioactive elements.

U.S. Department of Energy 369-49-neg8

Hanford Farming. Workers grew marketable crops on and off the Hanford Site to study the uptake of radionuclides. This farm was located downstream of F Reactor on the east side of the Columbia River. Notice the Hanford Operations (HO) number on the side of tractor. This photograph was taken in 1949.

These findings led to an early awareness of the importance of contaminant concentration within the food chain and the metabolic processes functioning inside organisms (Adley and Crane 1950a).

The Safety and Industrial Health Advisory Board for the U.S. Atomic Energy Commission released a nationwide critique of nuclear waste management practices in the late 1940s (Williams 1948, p. 74). Several problems at Hanford and other nuclear weapon sites demanded "immediate investigation." Examples included the following:

- Potential for public health impacts, especially long-term effects.

- Hazards from contaminants released into the air and surface water bodies.
- Movement of contaminants underground.
- Improved instrumentation for detecting contaminants and sampling the environment.
- Human tolerance limits for radioactive and chemical materials.
- Laboratory and field studies to improve waste disposal practices.
- Integrate studies (for example, biology, geology, meteorology, and hydrology) to understand the complete cycle of contaminant release, movement, and uptake.

Realistic contaminant release limits were desperately needed.

Hanford monitoring expanded to include crops, drinking water, air, rainwater, animals, general radiation measurements in the environment, and summaries of liquid discharges from facilities. It was increasingly recognized that realistic radionuclide and chemical release limits were needed, especially for the Columbia River (Coopey 1948).

By 1950, an animal farm and 22-acre botany field research station were established at Hanford. The field station was "to furnish data on what could be expected in the Pasco-Kennewick area from the use of the Columbia River water for irrigation purposes" (Berry and Cline 1950, p. 7). Water was pumped from the river to raise a variety of marketable crops on the old "Foster Ranch" located downstream of F Reactor.

By 1950, an animal farm and 22-acre botany research station were established at Hanford.

Radioactive phosphorus in the bones of mature ducks was found concentrated an average of 100 times that taken in river water downstream of the reactors (Hanson and Browning 1952, p. 203). Tissue samples of ducklings raised in an onsite liquid disposal ditch contained fission products 150,000 times more concentrated than that found in the water.

The U.S. Public Health Service completed a water quality study of the Columbia River in 1953 (U.S. Department of Health, Education, and Welfare et al. 1954). This was the most thorough evaluation of the river since Hanford started operating. While the study covered portions of Washington, Oregon, and Idaho, most work focused near Hanford. Recommendations included the following:

- Reducing the amount of radioactive material entering the river.
- Studying effluent treatment technologies for improving the quality of downstream water.
- Increasing research on radioactive material uptake by fish, waterfowl, and their consumption by humans.
- Increasing attention to the effects of chemical toxicity, especially on salmon.
- Improving and simplifying instrumentation for radionuclide and chemical detection, especially that used to measure low-level radioactivity.

The Columbia River has ten major tributaries. Hanford sits near the confluence of the Columbia, Snake (the Columbia's largest tributary), and Yakima Rivers.

As the above study was being written, Hanford again expanded its environmental monitoring (Honstead 1954). This was done to estimate the potential hazard to workers and people living downstream and to assist in the selection of sites for new reactors.

Aquatic studies intensified in the Columbia River as well as its tributaries (U.S. Department of Health, Education, and Welfare et al. 1954). Radioactive phosphorus was found to be more than 250,000 times concentrated in small fish (shiners) and more than 300,000 times more concentrated in caddis fly larvae than found in Columbia River

The possibility for significant radiation exposures downstream of Hanford "could no longer be considered minor" (Foster 1990).

By the late 1950s, radiation levels in the river, and the public's exposure, were reaching the highest in Hanford's history.

"Releases of radionuclides did not result in any measurable risk to the Columbia River ecosystem, as indicated by indicator species and regulatory benchmarks" (EPA 1994, pp. 3-7).

water (Foster et al. 1954). By 1955, eight reactors were operating. Wildlife monitoring became more routine and enlarged to include more waterfowl sampling (Hanson and Kornberg 1956).

Concerns were again raised about the possible impact reactors had on water quality. "The discharge of reactor effluent wastes…has posed the problem of possible effects on the use of the stream…it is apparent that operating conditions will be much closer to the permissible levels than at present and a more careful program of monitoring will be needed" (Healy 1956). The possibility for significant radiation exposures downstream of Hanford "could no longer be considered minor" (Foster 1990). The "potential [human] health significance" of these radionuclides was growing (U.S. Department of Health, Education, and Welfare et al. 1954).

Biologists Richard Foster and Jared Davis (1956) wrote about the hazard of aquatic life concentrating contaminants. "If radiophosphorus were allowed to reach the maximum level permitted for drinking water, organisms living in the water would suffer radiation damage and the fish would be unsafe for human food." They warned about biological processes introducing hazards not taken into account and about judging the acceptability of contaminant release practices by just conventional permissible drinking water standards. Davis et al. (1958) emphasized the pressing need for "radioecological research." Radiation levels in the river, and the public's exposure, were now reaching the highest in Hanford history.

Researchers were learning not only the potential impacts from radionuclides but also how heat and chemicals, such as sodium dichromate, could adversely affect aquatic life (Foster 1972). Findings guided updating permissible radiation exposure limits and controls on effluent releases.

During congressional testimony, Herbert Parker said, "as far as we are aware, there has been no damage to fish from radioactivity releases to the Columbia River" (Parker 1959a). Likewise, Parker stated these releases posed no danger to people living downstream in the cities of Pasco and Kennewick.

Years later, a review of ecological risk assessment studies for Hanford by the U.S. Environmental Protection Agency concluded that during the years of peak plutonium production, "the most sensitive life stage of fish (i.e., salmon embryo) did not appear to be at risk from radionuclide exposure in sediments or water…releases of radionuclides did not result in any measurable risk to the Columbia River ecosystem, as indicated by indicator species and regulatory benchmarks" (EPA 1994, pp. 3-7). Whether such conclusions stand the test of time will depend upon how well we can decipher indicators of biological health and what one's interpretation of risk is.

Air Monitoring: Potential hazards from air contamination received attention soon after the first reprocessing facility came on line. Of all environmental pathways, air was capable of most rapidly dispersing contaminants across Hanford and the surrounding region before radioactive decay could significantly reduce their hazard. When radioactive iodine (iodine-131) buildup was found on local edible plants, adjustments to iodine release limits were proposed (Parker 1946). By 1947, plant samples were being collected up to 100 miles from Hanford—in The Dalles, Oregon, and Davenport, Washington—as radioactive gases drifted across the landscape. The Williams Commission, appointed by the Atomic Energy Commission in 1947, questioned statements from Hanford officials that "stack emissions of radioactive iodine and xenon are insignificant" (Williams 1948, p. 64).

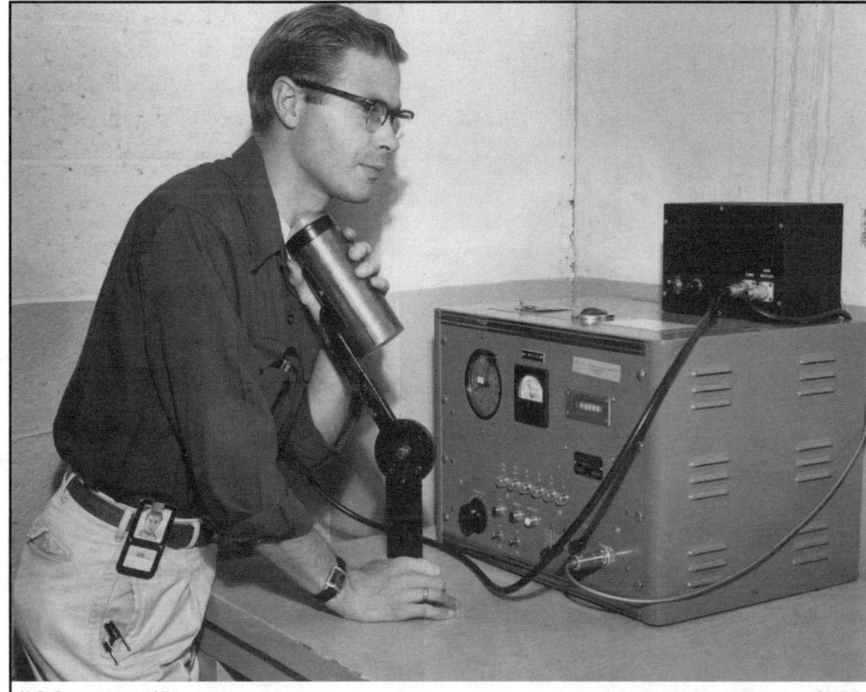

U.S. Department of Energy 9865-3

Self-Monitoring of Thyroid Glands. During Hanford's early years, workers could survey their own thyroid glands for exposure to radioactive iodine. The detector was held just below the Adam's apple.

Hanford officials soon recognized that the "revelation of a regional iodine-131 problem would have had a tremendous public relations impact" (Herde in Stannard 1988, p. 761). Therefore, any monitoring of live animals or sampling of their tissues was done in secret.

Yet, how could large animals be tested for radioactive iodine without raising suspicions? This prompted opportunistic sampling of rabbits including "clandestine attempts" to monitor radiation levels in the thyroids of livestock that grazed close to the site—especially cattle grazing downwind of Hanford (Foster 1990). Hanford volunteers with ranching experienced dressed up like cowboys and drove jeeps onto open land just east and north of the Columbia River (Stannard 1988, pp. 760-761). Hanford biologists and radiation experts accompanied them. After being chased by jeeps without being successfully lassoed, the cattle nearly collapsed from exhaustion. Biologists then held a portable radiation monitor (Geiger-Mueller survey meter) to the animal's neck where the thyroid gland[3] is found. The technique was rather primitive, and potentially hazardous to workers (angry cattle), but the results did confirm radioactive iodine uptake. On the other hand, humans were more cooperative. During the 1940s, the same types of monitors were placed alongside the necks of Hanford workers (Stannard 1988, p. 1567).

After being chased by jeeps without being successfully lassoed, the cattle nearly collapsed from exhaustion. Biologists then held a portable radiation monitor to the animal's neck where the thyroid gland is found and the uptake of radioactive iodine was measured.

These sampling results raised questions about iodine levels in the more populated Yakima Valley west of Hanford. Most of that area was upwind of Hanford. Though air, vegetation, water, and soil sampling had taken place in the Yakima Valley since the earliest days of plant operations, the monitoring of farm animals would draw too much attention from skeptical farmers. This was solved by disguising one of Hanford chief biologists: "I was no longer 'Karl Herde of Du Pont' but through that day would

(3) Iodine, from the foods we eat and the air we breathe, accumulates in the thyroid gland. Thyroxine, an iodine-containing hormone secreted by this "H"-shaped gland, regulates the body's metabolism and is essential for normal growth and development.

Hanford became a national center[4] for researching the radiation danger and chemical toxicity from inhaled radionuclides (Stannard 1988, p. 562).

This early research gave the first clear evidence of lung cancer in animals resulting from exposure to plutonium. The radiotoxicity of other radionuclides was studied.

be known and introduced as Dr. George Herd, of the U.S. Department of Agriculture. I was to simulate an Animal Husbandry specialist who had the responsibility of testing a new portable instrument (Geiger-counter) based on an unproven theory that by external readings on the surface of farm animals the 'health and vigor' of animals could be evaluated" (Herde in Stannard 1988, p. 761). Herde was accompanied by two Hanford security escorts who acted as research assistants. While measuring the "health" of the jugular vein, the radiation counter was quickly passed over the animal's thyroid. Positive but low iodine concentrations were detected. Such findings prompted building an Experimental Animal Farm on Hanford (Stannard 1988, pp. 434-450).

By 1950, radioactive iodine stood out as the most critical airborne contaminant. It was freely absorbed in the bloodstream, via ingestion or inhalation, and soon concentrated in the thyroid gland (Adley and Crane 1950a). The first guidelines for controlling iodine releases were published a few years later (NCRP 1953).

In 1951, a long-term experiment began on the hazards of iodine-131 intake by sheep (Kornberg 1958). Iodine-spiked food was fed to them and examinations carried out to directly measure biological impacts such as thyroid tumors. This was the first research to successfully define a safe limit of continuous exposure to a specific radionuclide in a large animal.

The installation of stack filters at B and T Plants in the late 1940s significantly reduced radioactive releases (see Section 5.8). Beginning in 1952, radioactive ruthenium was detected on the ground and vegetation (Tomlinson 1959b). Animal experimentation on the adverse effects of inhaling small particles of radioactive ruthenium began in 1954 after a filter failed at the REDOX Plant (Parker 1954b).[5] When large flakes of

Over the last 20 years, the ability of microorganisms living below ground level to change or transport contaminants was studied (McCullough et al. 1999). Microbial actions taking place in the shallow vadose zone (the layer of soil between the groundwater and the ground surface) or in the deeper groundwater system vary depending on the physical setting, microbes present, available nutrients, and nature of contaminants. Microbial processes can impact how contaminants move, alter their toxicity, and even chemically alter organic compounds. Understanding this microbial life will be critical for long-term predictions of contaminant movement and potential risks to higher life forms.

(4) Pulmonary radiotoxicology research was conducted in the "Pharmacology Operation" located in a Quonset hut near the 100-F Area.

(5) The reduction-oxidation solvent extraction process did not efficiently remove ruthenium, so the gases released from process solutions were routed through a filter system and out to a stack. Sometimes the filter failed.

ammonium nitrate were released from the plants, "they looked like snow" falling to the ground (Freer and Conway 2002, p. 2-4.30). Smaller flakes blew in the wind "all the way to Spokane" (Freer and Conway 2002, p. 2-4.30). As a result, sampling of fruits, vegetables, farm crops, and animals again expanded across eastern Washington. Periodic stack filter failures occurred well into the 1950s.

The first detection in North America of radioactive fallout from detonation of Russia's first nuclear weapon was reported at Hanford in 1949. By this time, the United States had tested five nuclear weapons in the atmosphere over the Pacific Ocean (UNSCEAR 2000). This prompted a rapid growth in air monitoring so that officials could distinguish between Hanford and non-Hanford radiation releases. Sampling locations expanded into Montana, Idaho, Oregon, and western Washington.

Groundwater: The first confirmation of detectable levels of contaminants in the groundwater outside of the 200 Area, moving towards the Columbia River, was reported in 1956. This promoted an expansion of the onsite groundwater-monitoring program and publishing a new series of Hanford surveillance reports. The leading edge of a fast-moving tritium plume reached the river in the early to mid-1960s, 20 years after contaminated liquids were first dumped into the ground in the 200 Area (Brown et al. 1962; Eliason 1966). Liquid releases to the ground and the movement of contamination plumes are discussed more fully in Section 5.6.

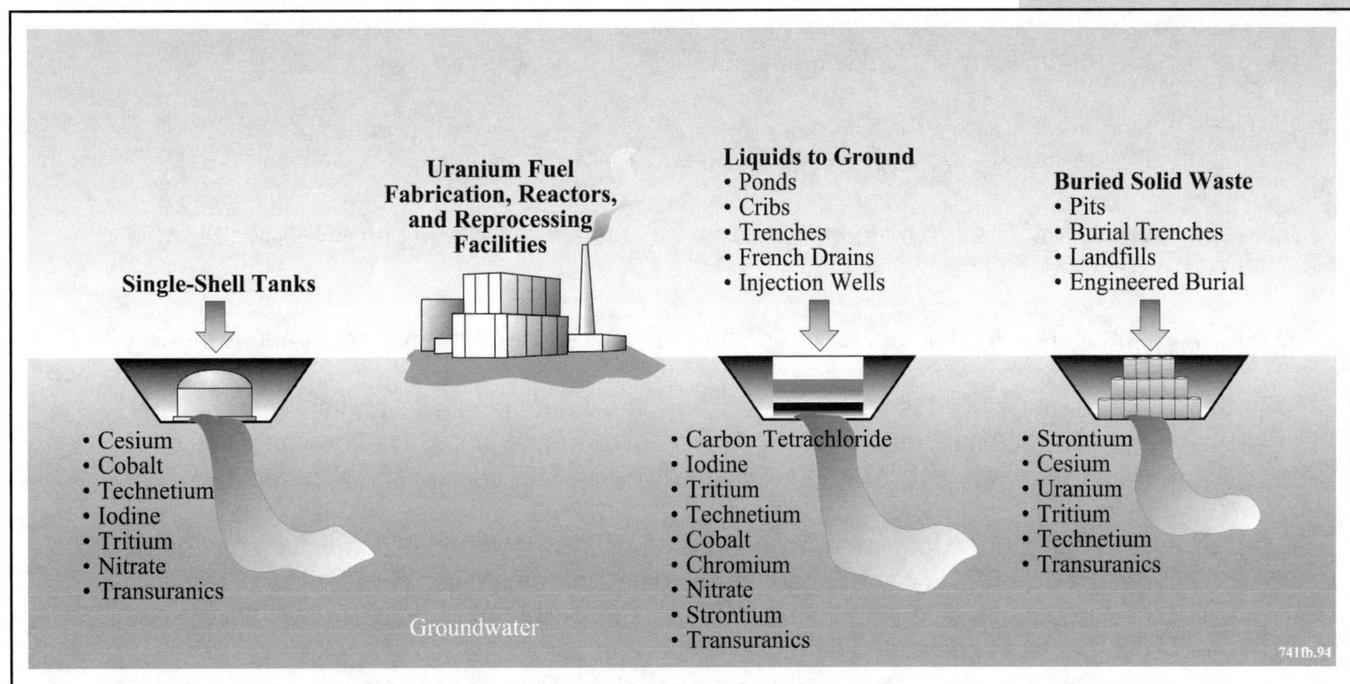

Contaminants Released Underground. These are examples of the wide variety of chemicals, metals, and radionuclides released to the ground during the years of plutonium production. Some contaminants are chemically bound to the soil. Others are mobile and have reached (or will eventually reach) the Columbia River. Understanding the long-term risk posed by these contaminants is key to making decisions regarding cleanup and site stewardship.

3.2 A Major Change in Direction[6]

"Estimates of radiation doses to individuals in the vicinity of the [Hanford] plant were not calculated until 1957 when the methods for such estimates were developed."

—J.K. Soldat, K.R. Price, and W.D. McCormack,
Offsite Radiation Doses Summarized From Hanford Environmental Monitoring Reports for the Years 1957-1984 (1986)

The time came for a change in the purpose for monitoring—from simply identifying and tracking contaminants to more sophisticated analysis that estimated the risk to people.

Hanford monitoring expanded as contamination spread and concerns arose about hazards. The time came for a change in the purpose for monitoring—from simply identifying and tracking contaminants to more sophisticated analysis that estimated the risk to people—and comparing those risks to national and international radiation protection standards. This was unfamiliar terrain; for at that time, few radiation limits were "well founded on biological experimentation" though most available limits were thought to err on the side of safety (Kornberg 1958).

Radiation Dose Reporting

At Hanford, radiation dose reporting practices for the public underwent many changes from the late 1950s to the early 1980s. This sometimes makes it difficult to compare radiation doses over time. Changes included (modified after Soldat et al. 1986):

1957-1958 Radiation doses to individuals reported as a percentage of permissible annual dose limits to whole body or specific organs.

1959-1960 Loosely defined "maximum individual" results expressed as percentage of allowable annual dose limits.

1961-1962 Annual radiation doses reported for people living in Richland, Pasco, and Kennewick.

1963-1973 Annual radiation doses for a hypothetical maximum individual and an "average" or "typical" Richland resident. Starting in 1972, the potential whole body radiation dose to members of the public living within 50 miles of Hanford was first reported.

1974-1981 Results reported for maximally exposed individual and population-wide radiation doses received during a calendar year from inhalation or ingestion plus the potential dose that could be received during the next 50 years following that single year of intake—that is, a 50-year dose committed during the year of exposure.

1982 First reporting of potential radiation doses for maximum individual and the general population in terms of a "50-year cumulative dose" from exposure to and intake of radionuclides during the reported year as well as the continued exposure to and dose from those radionuclides in the years after release. This approach was used until about 1989 when the calculation of radiation doses changed to the reporting of effective dose equivalents, which are similar in concept to committed doses.

(6) Brett Tiller, biologist, provided information about some of the future needs and trends in Hanford wildlife monitoring and health studies mentioned in this section.

The seminal publications, setting the stage for these new dose evaluations, were issued by the National Council on Radiation Protection (NCRP 1953, 1954). The council published recommendations for public exposure that were 10% of that permitted for radiation workers. Recommendations were later adopted into Hanford operating procedures (GE 1960).

Also, in 1956, a new detection instrument called a gamma ray spectrometer became available for studying environmental samples(7) (Healy et al. 1958). This was a significant advancement for it enabled the rapid identification and measurement of many specific radionuclides. Now, more meaningful assessments of radionuclide releases and better estimates of radiation doses were possible.

Driven by new public protection standards and new instrumentation, Hanford monitoring data plus dietary information were used by Healy et al. (1958) to write a landmark paper(8) estimating radionuclide intake and potential radiation doses to a "standard man rather than by any individual or group of individuals in the population." The paper scientifically and conceptually connected contaminant sources with contaminant uptake and potential hazards given as a percentage of maximum permissible limits. What was more remarkable was that the paper was written for a broad audience. Hanford became the first major nuclear facility in the United States to report potential radiation doses to members of the nearby public and specific body organs (Soldat et al. 1986).

However, monitoring and health studies remained fragmented between six different onsite organizations (Foster 1990). This made if difficult to provide a comprehensive exposure status for individuals. Parker solved this by establishing a single entity to review radioactive exposures: the Radiological Evaluation Task Force. By 1958, this task force made major revisions to the site's monitoring programs, dietary surveys, field radiation measurements (a.k.a. field dosimetry), and sample analyses. The task force evolved into the single, permanent Radiological Evaluation Group responsible for all monitoring and health research on the Hanford Site.

In 1961, the Columbia River Advisory Group reported on their evaluation of chemical and radionuclide releases into the Columbia River (Tsivoglou and Lammering 1961). Concerns expressed over the river's nearly exhausted capacity for receiving more "radioactive pollution" prompted the group to recommend new or expanded studies to evaluate the following:

- Movement and fate of individual radionuclide species.
- Radionuclides found in the Snake River that enters the Columbia River south of Hanford.
- Effects of chemicals and hot water in the Columbia River.
- Improved technologies or alternative waste management practices to reduce the amount of radionuclides entering the river.
- Eating habits of people living near Hanford and downstream to determine radionuclide uptake from local produce, livestock, and game.

"Where there was no firm numerical limit on which to base an operating decision, it was necessary to act on the best judgement of the local staff" (Parker 1948c, p. 3).

Hanford became the first major nuclear facility in the United States to report potential radiation doses to members of the public and specific body organs (Soldat et al. 1986).

(7) Before this time, the most commonly used radiation detection instrument was the Geiger-Mueller counter. It analyzed for high-energy beta emitters (electrons). It could not detect alpha radiation, low-energy beta radiation, and could not identify specific radionuclides.

(8) The Healy et al. (1958) paper was given at the second Atoms for Peace Conference in Geneva, Switzerland.

In recent decades, the public's fascination with radiation has turned to fear, and their trust of government officials has eroded. This change is demonstrated in a once-accepted program to monitor radiation levels in school children using a truck-mounted whole-body radiation counter. These children lived near Hanford as well as down the river. This excerpt is from Honstead (1967):

"In establishing a research program to investigate diets and radionuclide body burdens in local children we had to provide motivation for the schools to cooperate with us, for the children's parents to give permission for their participation, and for the children to take part in the study. We received excellent cooperation from the schools....We have been successful in eliciting cooperation from the parents through a carefully-planned public relations program coordinated with the school presentations....On the day that we approach children in their classroom with our study we attempt to have in the local news media a suitable story of the event and an explanation of its purpose....We are careful to give all of the information available to the public and to answer quite frankly any questions that arise concerning the measurements and the results....We have succeeded in obtaining participation on the part of more than 80% of the school children in the eight elementary schools so far approached...Furthermore, there has been no adverse publicity or even a single parental complaint to answer. It is one thing to gain public acceptance of this program and parental agreement for children to participate in it, but it is quite another matter to motivate the children to participate...The procedure followed is to intrigue the children with a description of the experiment and the whole-body counter, and to invite them to become 'nuclear scientists' for a few days. After they have completed their part in the experiment they are given a souvenir of the experience in the form of a Certificate of Appreciation... We provide the children with comic books to read as they wait their turn in the whole-body counter but the fascination of the instrument itself usually is greater than that of our comics."

U.S. Department of Energy 39544-2bw

Interior of Mobile Whole-Body Radiation Counter. This is a 1965 photograph of a truck-mounted whole-body radiation counter taken to schools to monitor radiation levels in children who lived near Hanford or downstream. This was a common practice in the mid-1960s and was coordinated with the school districts, parents, and the news media.

- People's natural and human-made radiation exposure from sources other than Hanford.
- Periodic and regular technical reviews of Hanford waste management programs and practices.

The groundwork was now laid for significant changes in Hanford's sampling programs and improvements in estimating radiation dose (Soldat et al. 1986). An example was the focus on dietary surveys, such as the uptake of radioactive zinc (zinc-65)—released into the Columbia River—by children and adults living along the Oregon Coast below the point where the river entered the Pacific Ocean (Essig et al. 1973). Routine surveys began in 1961 (Seymour and Nelson 1973). Before then, radionuclide samples were taken as far downstream at Vancouver, Washington, or Astoria, Oregon, occasionally (Kornberg 1950). Closer to the Pacific Coast, evidence showed radionuclide uptake by shellfish, and therefore, by people who ate shellfish. The 1964 to 1971 shutdown of all Hanford's single-pass reactors eliminated most contaminant discharge to the Columbia River, and therefore, the plume of Hanford-produced radionuclides into the downriver marine environment.

Today, some natural versus human-made radionuclides are only distinguished by their rarity.

From the mid-1940s to the late 1960s, elevated levels of Hanford-related contaminants were easily detected in the environment. Potential impacts on the public were based on measured concentrations and exposure to radioactive materials. As the number of operating facilities were reduced and effluent treatment systems improved, it became harder to distinguish Hanford contaminants from non-Hanford sources such as radioactive fallout from atmospheric nuclear weapons tests. By the early 1970s, "it was no longer possible to estimate offsite radiation doses from Hanford operations solely on the basis of samples and measurements in the environment" (Soldat et al. 1986). This prompted the coupling of monitoring and research data with environmental transport and radiation dose models to better calculate radiation doses and risk.

There is a growing importance to understand health precursors to possible injury.

Today, there is a growing importance of not just measuring contaminant concentrations and comparing them against guidelines but of understanding health precursors to possible injury at the molecular, organism, and ecosystem levels. The science of ecology is entering a new era of analysis supported by advanced bio-system technology permitting researchers to address how biological, chemical, and physical processes interact over time and space. Such capabilities can revolutionize decision-making processes (see Chapter 9), making it possible to intercept and correct pending problems before damage occurs. In a review of the science and technology needed to support Hanford restoration, the National Research Council (2001b, p. 119) reported that advanced biological monitoring capabilities should be used to define the "points at which small changes in environmental conditions can result in major changes to species habitat, abundance, or health."

This is particularly important as Hanford cleanup progresses, old waste sites are disturbed, and as more regional demands are placed on the Columbia River. To date, only one large-scale field investigation has related animal health to the amount of chemicals or radionuclides in tissues (Tiller et al. 1997).

Burrowing animals can carry buried contaminants to the surface.

Buried Waste: An Attractive Nuisance

Wildlife and plants living on Hanford and in the Columbia River are monitored because they can access contamination and, in turn, potentially expose the public to unwanted chemicals or radioactivity.

Burrowing animals such as badgers, ground squirrels, and ants carry buried contaminants to the land surface where they are spread. Plants such as tumbleweeds can grow 10-foot-deep roots penetrating buried waste sites (Klepper et al. 1985). Deep-seated root systems concentrate and help redistribute contaminants across the land.

Such events took place in the late 1950s and early 1960s (O'Farrell et al. 1973; ERDA 1975). Liquids containing strontium-90 and cesium-137 from the obsolete bismuth phosphate process at T and B Plants were disposed in long unlined trenches in a 40-acre tract of land just south of the 200 East Area. The trenches were then backfilled with soil. Between 1958 and 1960, burrowing animals dug underground and into the contaminated sediment. Rabbits, deer, and other mammals used the site as a "salt lick." Radionuclides were soon spread over a wide area. The area was surveyed, marked, and designated the B/C Crib Control Zone. In 1969, gravel was spread over the site to stabilize the trenches and isolate the remaining contamination from further intrusion.

This and similar incidents led to improved waste management practices, including the use of herbicides to control the growth of tumbleweeds and other plants, deeper burial of materials, containment of waste containing transuranic elements (that is, long-lived radioactive elements) in retrievable containers, and coverage of waste sites with plastic sheets. Tank farms were also covered with rock and gravel layers combined with herbicide applications.

Vegetation control around Hanford ponds and ditches reduced the availability of food, cover, and nesting sites for waterfowl. Some areas were covered with wire mesh to prevent access by large mammals or waterfowl.

Monitoring data collected over more than a decade are critical to demonstrate long-term plant and animal health trends.

The biological information needed to support decisions protective of the Columbia River often far exceeds that needed to satisfy routine land-management or environmental monitoring needs. Data sets gathered over more than a decade are critical to demonstrating long-term plant and animal health trends. Such trends cannot be detected using a few seasonal variations captured by short-term projects focused on immediate reporting needs.

While historical records are invaluable, they can leave many questions unanswered about contaminant hazards. For example, it took decades of monitoring followed by 15 years (1988-2002) of data reconstruction and research to assess the potential health impacts to humans from past airborne contaminant releases, primarily from one radionuclide—iodine-131 (HEDR 1994; CDC and Fred Hutchinson Cancer Research Center 2002).

Which contaminants pose the greatest future risk to humans, critical life forms, or ecosystems? Are risks of just academic interest or important enough to influence cleanup and river use plans? How will health assessments be factored into future decisions?

The biological information needed to support decisions protective of the Columbia River often far exceeds that needed to satisfy routine land management or environmental monitoring needs.

Under the title of the Columbia River Comprehensive Impact Assessment, researchers, regulators, and other representatives from the state of Washington, the state of Oregon, U.S. Environmental Protection Agency, and Native American tribes examined the potential health effects of contaminants entering the Columbia River. They studied sources on the Hanford Site; they also investigated sources outside the Hanford Site, including mining, industry, and agricultural sources. This assessment team was funded by the U.S. Department of Energy, which played a "non-negotiating role" (DOE 1998a). It's a start in looking at potential chronic and acute effects on ecosystems plus toxic and cancer-inducing impacts on humans. As expected, several metals and radionuclides were found elevated in the sediment and pore water[9] in specific areas along the Hanford shoreline. Most of the primary contaminants of concern (those which pose a risk to humans of greater than 1 in 1 million over a lifetime) along the shoreline or dissolved in the water of the Columbia, Snake, Yakima, or Walla Walla Rivers were cesium-137,[10] chromium, cobalt-60, copper, europium-154, nitrate, strontium-90, technetium-99, tritium, uranium-234/-238, and zinc. Ecological risks above thresholds were identified for copper, cyanide, lead, mercury, nickel, and zinc at select locations. Contaminants identified as above both ecological and human risk thresholds included cesium-137, chromium, cobalt-60, lead, technetium-99, and zinc.

The Integration Project Expert Panel (2000b) for the Hanford Site reported that the Columbia River Comprehensive Impact Assessment "was a major first step in both using a long history of Hanford research and in defining new research and monitoring data needs" critical to understanding potential river quality and ecological benefits from Hanford cleanup. The panel cautioned that river monitoring should not be performed

Brian Opitz, Pacific Northwest National Laboratory

Environmental Sampling. Researchers are removing water specimens from a multi-port sampler that was inserted into the sediment just offshore of the Columbia River. This sampler simultaneously collects water from several discrete zones to map the movement of contaminants into salmon spawning grounds.

(9) Pore water is groundwater moving through the small open spaces between sediment grains. Pore water was sampled in the shallow aquifer near the shore of the Columbia River. Numerous sampling locations were selected close to past contaminant release sites. Therefore, contaminant concentrations are maximized.

(10) These numbers identify the atomic weight of a radionuclide—that is, the total number of protons and neutrons found in the atom's nucleus. These protons and neutrons give the atom unique chemical and physical properties. The half-life of each radionuclide is listed in Appendix A.

The relationship between contaminants and potential biological impacts requires closer investigation in the Hanford environs.

State Monitoring of the Hanford Shoreline

Contaminants in groundwater beneath most of Hanford are below the depth of roots typical of a shrub-steppe environment.(11) However, the water table is shallower near the Columbia River. There, tree roots and shoreline plants can access contaminated groundwater before it seeps into the river. This increases the chance for plant-eating animals, and the public, to access contamination. For this reason, vegetation sampling takes place along the shoreline of the Columbia River.

In 1999, the Washington State Department of Health carried out routine sampling and radiological analysis of terrestrial vegetation growing along Hanford's shoreline (Washington State Department of Health 2000). Sample sites were selected in the 100 Area, 300 Area, and Old Hanford Townsite where groundwater contains radionuclide concentrations above background levels. Samples were also collected from mulberry trees growing along the shoreline in the 100-H Area known for elevated radioactive strontium (strontium-90) levels. Analyses were reported and potential impacts to humans and biota estimated. "These doses are far below levels for which there is a known human health impact, and also below limits set by federal regulations. An individual must consume twenty five tons of [mulberry] fruit in one year to reach the maximum dose of 100 mrem/y [millirem per year] allowed to a member of the public from facility operations" (Washington State Department of Health 2000). The estimated dose to a typical deer consuming 100-H Area fruit and leaves is "several hundred times lower than the biota dose rate limit of 0.1 rad/day set forth in DOE orders. As reported by the State, the radionuclide concentrations found in shoreline vegetation near 100-H Area and the Old Hanford Townsite are not indicative of environmental impacts to organisms consuming the vegetation." Nonetheless, the relationship among radioactivity, other metals, and potential biological impacts require closer investigation in the Hanford environs, especially for radionuclides upwelling in the riverbed.

just as a service activity but as an essential, integrated part of all Hanford work to continuously assess contaminant movement and risk. This distinguishes between proactive and reactive monitoring programs.

Agencies and groups affiliated with the Hanford Natural Resource Trustee Council are now acting as an independent advisory board for resource protection (Hanford Natural Resource Trustee Council 2001). The council agrees there is a need to identify potential contaminants of concern and the plants and animals that could be impacted, to assess river-related human health and ecological risk, and to evaluate the sustainability of the river ecosystem to future stresses (Findings 97-01 and 99-04).

(11) A shrub-steppe environment is one where short, woody shrubs and grasses dominate. This is typical of an arid to semi-arid landscape.

U.S. Department of Energy

Columbia River Shoreline. The Columbia River and its adjoining shoreline environs are an irreplaceble natural resource, especially that portion adjoining Hanford. Biological information needed to support wise protection of this resource often exceeds that collected for routine environmental monitoring and land management.

The risk of contamination to people working on the Hanford Site today is low because the old plutonium production facilities are shut down and access to waste sites is restricted. Environmental monitoring demonstrates that risks downstream of Hanford from site releases are also extremely small (Poston et al. 2001). "However, potentially increased risk is possible if people were to move onto the Hanford Site and derive large percentages of their daily food uptake from crops and animals" grown in certain spots along the river's bank where contamination is higher (DOE 1998a). For this reason, large portions of Hanford will remain under restricted use for a very long time.

The risk of contamination to most Hanford workers is now low.

The Hanford Natural Resource Trustee Council has the authority to seek legal damages for injury to natural resources. The council ensures natural resource values are factored into Hanford cleanup decisions, integrates natural resource restoration into cleanup actions, and encourages natural resource planning throughout the site. Members include representatives from the U.S. Environmental Protection Agency, U.S. Department of Interior, U.S. Department of Energy, Bureau of Land Management, Washington State Department of Ecology, state of Oregon, U.S. Fish and Wildlife Service, plus the Confederated Tribes of the Umatilla, Nez Perce Tribe, and Yakama Indian Nation.

Environmental monitoring demonstrates that risks downstream of Hanford from site releases are extremely small (Poston et al. 2001).

Chapter 4

Radiation: What Was Known and When

"U.S. regulatory standards to protect the public from the potential health risks of [low-level] nuclear radiation lack a conclusively verified scientific basis....Regulating at these levels, well below the range where radiation effects have been conclusively verified, is essentially a policy of judgment."

—General Accounting Office,
Radiation Standards: Scientific Basis Inconclusive, and EPA and NRC Disagreement Continues (2000f, pp. 4, 29)

"There is a growing feeling that it is unwise to continue the assignment of authority over the public health aspects of atomic energy to the same agency that has a prime interest in the promotional aspects in the field" (Science 1959).

Concerns over the potential health effects from radiation have fueled debate about the adequacy of Hanford's past waste practices. These same concerns will significantly influence how cleanup is undertaken.

In the early 1940s, naturally occurring radioactive elements,[1] such as uranium, were concentrated and radioactive elements long decayed away since the earth's formation were re-created, on an industrial scale, at Hanford. Nuclear weapons testing introduced significant quantities of anthropogenic radionuclides into the global environment, while material development operations released radionuclides into the environment near nuclear facilities.

All the while, cutting-edge radiotoxicology and environmental research moved forward, though at a pace out of step with the knowledge needed to understand the potential impacts from contaminant releases. Nearly 15 years separated the beginning of nuclear waste disposal practices from our grasping the health consequences of those practices (see Chapter 3).

Forty percent of the 1 billion curies of radionuclides remaining at all U.S. nuclear material and weapons sites rests at Hanford. And, when Hanford plants operated, more than 140 million curies of

741fb.56

Radiation Sign at Tank Farm. Radiation signs are common at Hanford. These are fastened to a chainlink fence surrounding the B Tank Farm, 200 East Area.

(1) Radiation is part of the environment. Naturally occurring radionuclides come from two primary sources: primordial (terrestrial) and extraterrestrial (cosmic). Both contribute to our background radiation dose.

Broadly speaking, the potential danger from ionizing radiation depends upon its energy and the interaction of that energy with the matter penetrated.

Alpha particles don't penetrate matter very far, but over a long time, they are biologically hazardous.

mostly short-lived radionuclides were released into the air and the Columbia River. Chemicals were also discharged. Workers and members of the public were exposed to these releases. As a result, some people suffered illnesses. The extent? Still being studied and debated.

Broadly, radiation is of two types—ionizing and non-ionizing (see Appendix A).[(2)] Ionizing radiation can break chemical bonds in matter and damage or kill living cells. This produces potentially damaging electrically charged particles (called ions) inside the matter struck. It does this by removing tightly held electrons orbiting an atom's nucleus. Broadly speaking, the potential danger from ionizing radiation depends upon its energy and the interaction of that energy with the matter penetrated.

Several types of particles and rays can create ionizing radiation. Common types include the following:

- Alpha particle—a positively charged particle containing two protons and two neutrons. It resembles the nucleus of a helium atom. Alpha particles don't penetrate matter very far, so they release their energy within the small amount of tissue penetrated.
- Beta particle—a negatively charged electron (negatron) or positively charge (positron) particle emitted from the nucleus of a radioactive atom. Beta particles are smaller and more penetrating than alpha particles.
- Gamma rays—a bundle of high energy released from the nucleus of a radioactive atom. Gamma rays are very penetrating.

Neutrons, x-rays, and various other subatomic particles such as fission fragments can also cause ionizing radiation.[(3)]

Non-ionizing radiation does not create ions. Generally, it releases energy in the form of heat. Non-ionizing radiation is common in our everyday lives. Examples include ultraviolet light (causes sunburns), visible light, microwaves (cooking), and radio waves. This book addresses only ionizing radiation.

---

(2) A set of easy-to-read fact sheets explaining radiation as well as describing the properties of many radioactive and hazardous elements at Hanford and other U.S. Department of Energy sites is available online at http://riskcenter.doe.gov/techinfo/factsheets.cfm.

(3) Not all types of radiation have the same biological effects. Alpha particles are very ionizing. They deposit a lot of energy over a short distance in the tissue penetrated. High-energy beta particles are also ionizing. X-rays and gamma rays, being electrically neutral, are not themselves ionizing. However, they transfer their energy to electrons that do ionize. Neutrons are not ionizing particles, but they produce ionizing particles when they strike the nuclei of atoms. Such different types of radiation are factored into calculating the dose equivalent (potential for producing biological damage—expressed in rems). The dose equivalent from exposure to beta, x-ray, and gamma sources is generally equal to the radiation energy received (absorbed energy). Alpha particles have a dose equivalent that is much higher than the absorbed energy, owing to a higher weighting ("quality") factor in the dose calculations (Bodansky 1996, p. 31).

Radiation Jargon: A Quick Guide for Most of Us

The potential health impact of radiation is measured in rem. Rem stands for *roentgen equivalent man*; this is a unit of radiation dose that indicates the potential for damage to humans. The unit of a millirem (one thousandth of one rem) is used to describe a low radiation dose such as the public routinely receives from their environment or during certain routine medical procedures such as x-rays.

Potential health impacts depend on the radiation absorbed (measured in radiation absorbed dose or rad), the type of radiation received (for example, alpha particles, beta particles, gamma rays, or neutrons), the radionuclide delivering the dose, the tissues exposed, and how the dose is received.

Words used to describe radiation dose are sometimes unclear and inconsistent. One team of health physicists writes, "there are many ill-chosen names and phrases and much jargon that permeate our professional speech and writing…we even have institutionalized confusion" (Strom and Watson 2002). Some terms were developed by researchers having technical, rather than linguistic, skills. This can make comparing radiation doses reported in different publications very challenging. At the same time, terms used in the popular literature are even worse and frequently wrong. Words such as dose, risk, radiation, radioactive, contamination, exposure, or irradiation are interchangeably wielded across text and headlines.

For this book, let's remember two things. One, "dose" refers to the energy deposited to tissue, an organ, or the whole body. It reveals nothing about the potential biological hazard from that exposure. Second, the general phrase "radiation dose" is used to describe the potential health impacts (risk) from exposure to radiation. This term is simpler than repeatedly using a phrase like "effective dose equivalent" which is more technically correct, but awkward. For the sake of discussion, we'll ignore subtleties and focus on broad comparisons.

For those who like particulars—many radiation dose values reported in the Hanford literature actually refer to the "effective dose equivalent"[(4)]—that is, the radiation dose received by the whole body of an individual over 50 years. For example, when Poston et al. (2001) report a given radiation dose for the year 2000, that value is actually an integrated radiation dose spread over the years 2000 to 2050. Therefore, the value reported is a conservative number—attributed to a single year.

The glossary defines terms frequently used in the literature to describe radiation doses such as dose equivalent, committed dose equivalent, and effective dose equivalent.

Commonly, terms used in the popular literature to describe radiation or radiation doses are wrong.

While the nuclear weapons sites are synonymous with radiation, the potential hazards from chemicals and metals must be factored into estimating human-health risk.

(4) "Effective dose equivalent" means the same thing as the U.S. Nuclear Regulatory Commission's term "committed effective dose equivalent."

Some say Hanford is safe. Others say it's dangerous. Can they both be right? Responses depend on context. Some facilities contain materials emitting lethal levels of radiation. Others are less dangerous or not hazardous at all. Physical barriers and other safeguards control access into the most dangerous sites. However, the public is most concerned about radionuclides that were or could be released into the environment.

> During the earliest years of Hanford, officials of the Atomic Energy Commission wrote about unchallenged tolerance limits for radiation and toxic chemicals. Wolman and Gorman (Williams 1948, p. 68) stated: "the maximum allowable amounts...were established during the war period...There is little indication that these limits were reviewed, even during the war period, by public health officers normally concerned with and responsible for such problems in civilian life."
>
> Regarding radionuclide releases to the Columbia River, Tsivoglou and Lammering (1961) reported that studies "should be cognizant of the importance of protecting individuals as well as groups of people, and should not neglect special dietary groups within larger population groups." For more information on releases to Columbia River, see Section 5.7.

Some say Hanford is safe. Others say it's dangerous. Can they both be right?

Radiological and biological programs were initiated to protect the Columbia River to the extent possible during the "stress of wartime conditions" (Parker 1952). Many of these practices continued unchanged.

Contaminant release policies were molded around three concerns: plutonium production, worker safety, and offsite safety. Worker safety was immediate and most easily managed (Silverman 2000). Besides, they were the first line of defense to preserve plutonium production. Rules for assessing the adequacy of controlling contaminant releases were offered as a "working code of minimum interference with production" (Parker 1952). Radiological and biological programs were initiated to protect the Columbia River to the extent possible during the "stress of wartime conditions" (Parker 1952). However, many waste management practices continued unchanged for decades.

Just after remediation of the nuclear weapons complex began, the Office of Technology Assessment (1991, p. 8) reported: "despite some DOE [U.S. Department of Energy] statements about the lack of immediate health threats, public health concerns have still not been investigated adequately by DOE or by other government agencies... Unless and until the contamination-related health issues of most concern to the public are recognized and addressed, the most ambitious, sophisticated, and well-meaning cleanup plans and activities will likely meet with skepticism, suspicion, and legal challenges."

"Existing records reflect only the immediate health effects, not the undetermined remote ones" (Williams 1948, p. 67).

Preliminary findings suggest that some workers become ill from exposure to contaminants released. These conclusions were based on a review of dozens of studies and medical data covering an estimated 600,000 employees at 14 sites, including Hanford (Associated Press 2000). An unprecedented national program was implemented to compensate thousands of ailing workers, or their survivors, for medical costs and lost wages. Before this, the government's policy was to litigate occupational illness claims (H.R. 4205, 2000). This deterred many workers from filing workers' compensation claims. Contractors were held harmless, and the employees were denied compensation for occupational diseases. One-third of the employees who worked in the weapons complex were at Hanford. In 1999 and 2000, legislation was proposed and Congress passed a compensation package for workers harmed from exposure to contaminants.

Questions remain about the impact past releases had on the public. The issue of health impacts will continue to be examined. Decades after Hanford first released radioactive and chemical contaminants, lawsuits are still sought by plaintiffs. In the year 2000, a class action suit for $100 billion was filed in Spokane, Washington, against former and present Hanford contractors (U.S. District Court Eastern District of Washington 2000). Two years later, the Ninth District Court of Appeals in California overturned a 1998 U.S. District Court ruling that had dismissed more than 3000 plaintiff claims from a consolidated suit against Hanford contractors (Cary 2002a). The initial suit had been filed in 1990—4 years after the first declassification of reports began about downwind radiation releases.

For years, the ethics driving human protection focused upon society—if society is protected, then individuals are protected (Clarke 2001). Yet, this can lead to inequities in the distribution of risk and benefit.

Each of us has strongly held opinions about exposure to nearly any level of contaminants. Reactions are very negative to involuntary exposure. If voluntary, and benefits exist (for example, jobs), then exposure tends to be more acceptable. The following sections examine issues of radiation protection and radiation dose. This discussion is not intended to justify any single belief or negate any concerns people have; rather, it's written to encourage readers to think about what radiation protection means.

4.1 The Debate: What is Safe?

"Just as there are some people who have been found unusually sensitive to sunlight, so there must be some who are more sensitive than others to radioactivity. For such reasons it has been considered desirable to set the tolerance dose for the entire population lower than that for medically controlled radiation workers."

—Consumers Report, *Hearings* (1959)

The need for radiation protection was recognized by the late nineteenth century with devastating accounts of health impacts to scientists and health care workers who pioneered the use of x-rays and other ionization sources. The first reports of eye damage and skin burns were in 1896, one year after W.K. Roentgen[5] discovered x-rays. It's reported that Henri Becquerel, discoverer of radioactivity in the late nineteenth century, carried a small vial of radium in his vest pocket to demonstrate its unique radioactive properties in front of an audience. Unfortunately, this also caused radiation burns to his skin (Stannard 1988, p. 23). Protective heavy glass plates, metal filters, lead-lined housings, and special clothing started to be used in the late 1890s. During the twentieth century, there was increasing caution about radiation exposure, and greater care was taken to lower the dose to workers. Public protection came decades later, particularly awareness of the long-term latency hazards associated with internally deposited radionuclides compared to the sometimes more immediate effects of external radiation sources such as x-rays.

During the twentieth century, there was increasing caution about radiation exposure, and greater care was taken to lower the dose to workers. Public protection came decades later.

There are many stories of early radiologists and x-ray technicians who had radiation-damaged fingers or arms amputated. Physicist Pierre Curie's hands became swollen

(5) W.K. Roentgen died of bone cancer in 1923.

One science historian reported Marie Curie received as much as 1 rem (1000 millirem) of radiation dose each week during her pioneering experiments (Nies 2001)! That was very high.

from radiation burns, and he suffered from radiation-induced illnesses. Neither he nor his wife Marie took precautions during their pioneering experiments, as the importance of radiation safety was not understood at that time. Marie Curie died of radiation-induced leukemia in 1934.[6] One science historian reported Marie Curie received as much as 1 rem (1000 millirem) of radiation dose each week during her pioneering experiments (Nies 2001)! That was very high.

Little effort was expended to protect the public from radiation hazards until after World War I when the diagnostic use of x-rays became widespread (World Health Organization 1956). By then, the hazards of artificially produced radioactive elements and radiation exposure were widely recognized.

During the 1920s, occupational hazards associated with radium began showing up. Though a number of chemists, physicists, and physicians received fatal radium exposures during this time, the most famous case involved radium poisoning of female workers who painted small dials in watch factories during World War I and the years following. These women had ingested radium when they licked thin brushes containing luminescent radium-laced paint.[7]

In 1924, a New York dentist and some perceptive physicians and pathologists began reporting in the scientific literature cases of jaw and mouth disorders and cancers in about 800 women who had worked in a New Jersey radium dial-painting plant (Stannard 1988, pp. 23-24). Symptoms began showing up 1 to 7 years after the workers terminated their employment. The company protested any claim of responsibility saying the health impacts were from other causes such as syphilis. Factory officials contracted for their own medical studies. The result was the same: radium was the culprit.

In 1924, a New York dentist and some perceptive physicians and pathologists began reporting in the scientific literature cases of jaw and mouth disorders and cancers in women who had worked in a radium dial-painting plant (Stannard 1988, pp. 23-24).

Medical evidence gained from living workers and autopsies increasingly confirmed the health hazards of ingested radium. "With terrible, almost mathematical regularity, I [Dr. Martland] have been crossing off these names as the girls have died or developed symptoms of radium poisoning" (Martland 1929 in Stannard 1988). In the 1930s, the U.S. Public Health Service performed their own investigation.

The International Commission on Radiological Protection was formally established in 1928. This was a non-government organization whose members were experts in radiation and related health sciences. Over the years, the commission provided leadership in developing protection guidelines based upon top-notch scientific knowledge and empirical evidence. Their findings had no legal standing. Recommendations incorporated into regulations and codes of practice remained the purview of individual countries, industries, and sites. The commission's first recommendations focused on the medical profession, luminous dial painters, and radiographers. Broader health considerations grew after fission was discovered in 1938 and industrial-scale uses of that technology became more commonplace.

(6) Pierre Curie died 30 years earlier in a street accident.

(7) The first glow-in-the-dark watch dials used luminescent paint containing radium. Radium decays by releasing an alpha particle. The dial's light was produced when radium was mixed with a chemical compound, such as zinc sulfide, that interacted with these alpha particles. These interactions created a small amount of green light.

In the United States, the Advisory Committee on X-Ray and Radium Protection[8] was established in 1928 under the National Bureau of Standards (Stannard 1988, p. 1387). The committee's name was changed to the National Committee on Radiation Protection in 1946 as its research expanded to cover the basic principles and philosophy of radiation protection supporting the disposal of radioactive waste, handling of radioactive materials, radiation monitoring, and the development of maximum permissible radiation exposures. These were the first years after uranium refinement and plutonium production began in the United States. By 1964, the committee was reorganized and chartered by Congress as the National Council on Radiation Protection. In this book, the term "National Council on Radiation Protection" is used to avoid tripping over the time-dependent words "council" and "committee" used during the organization's history.

In 1946, the nation's radiation protection organization expanded its research to cover the basic principles of radiation protection for disposing of radioactive waste.

There was a growing awareness that the force of law was needed to exert greater enforcement of worker and public protection. Voluntary compliance was not working well. By the early 1950s, the courts had begun settling radiation damage cases based upon protective measures issued by the National Council on Radiation Protection. A few municipalities and one state (California) had radiation control regulations in force. Greater control through legislation was required. However, with the exception of the Food and Drug Administration and revision of the Atomic Energy Act in 1954, the federal government's role remained quite limited (Taylor 1956a).

There was a growing awareness that the force of law was needed to exert greater enforcement of worker and public protection.

In 1953, the National Council on Radiation Protection published maximum permissible exposure limits for workers for 70 radionuclides. While recognizing there was considerable uncertainty in these values, the council believed it "desirable to agree upon what are considered as safe working levels for these radioisotopes now rather than wait until more complete information is available" (U.S. Department of Commerce 1953).

Taylor of the U.S. National Bureau of Standards wrote about the early history of permissible radiation exposures (1956b, pp. 196-197) (see Table 2). He stressed that while the prime consideration in the past was the exposure of individuals based upon scientific principles alone, "future decisions may have to include political and economic factors." However, Taylor cautioned, "We should start to condition our thinking for a change in philosophy with regard to radiation exposure." Specifically, the radiation dose received by members of the public may need to be reduced to a "very small fraction" of that received by radiation workers. Public protection had finally reached the forefront of the radiation debate.

Table 2. Trend in Reducing the Maximum Permissible Radiation Dose for Workers from 1931 to 1959 (Morgan 1959)

Period	Maximum Permissible Radiation Dose (Annual)
1931-1936	60,000 millirem (60 rem)
1936-1948	30,000 millirem (30 rem)
1948-1958	15,000 millirem (15 rem)
1958-1959	5,000 millirem (5 rem)

Radioactivity and Health: A History by J. Newell Stannard (1988) is a comprehensive story of the research and early case histories underpinning our present knowledge of the behavior and effects of radioactive material on people and the environment.

(8) The committee served as the national analog to the International Commission on Radiological Protection.

The first maximum permissible exposure limits for workers were published in 1953.

Future decisions about radiation protection may have to include "political and economic factors" (Taylor 1956b).

In 1957, congressional hearings were held about health risks from radioactive fallout from the atmospheric testing of nuclear weapons.

A Potentially Dangerous Business When Safety Is Sidestepped

Los Alamos, New Mexico (1945)—A worker was killed during the final stages of the Manhattan Project from a radiation burst when a critical mass of fissile plutonium was accidentally brought together. A similar incident occurred 9 months later; eight people were exposed, one of whom died days later.

Tokaimura, Japan (1999)—Officials at the Tokaimura uranium processing plant sidestepped work procedures resulting in a nuclear criticality. Two workers died from receiving a high radiation dose and 667 others (workers and members of the public) were exposed to increased radiation levels (Feder 2000).

In 1954, unwanted radioactive ruthenium particles were being released from the newly built REDOX Plant at Hanford. These created a ground contamination problem—found in local communities, orchards, and crop fields—more severe than experienced in previous years though the total risk was believed "low" (Parker 1954b). Particles were also a skin contamination problem potentially delivering an effective radiation dose of 200 to 2000 rad over a small area of skin in 1 day. If the low value was correct, there would be no visible effect on a person's skin. If the high value was correct, there would be "tanning, persistent erythema, and perhaps some degree of desquamation." [9] However, "this point can be answered only by human experimentation. This appears to be an entirely safe procedure, and four volunteers are available." The report does not mention whether these human experiments were carried out.

Radiation protection standards dealing with the permissible amounts of radioactive elements in the human body, air, and water were extended to the public at levels more restrictive than those governing occupational exposure. In 1957, congressional hearings were held about health risks from radioactive fallout from the atmospheric testing of nuclear weapons as well as from a nuclear war, should one occur (Holifield 1958). Discussions highlighted a growing political and social concern over exposure to radiation above background levels.

At low radiation doses received over a period of time, human cells normally repair damage done to them. The low-level range includes exposure up to about 10 rem (10,000 millirem) though the term is commonly used to refer to exposures of less than a few hundred millirem (GAO 2000f). In 1990, the National Research Council (1990a)

(9) Desquamation means to peel, shed, or come off in scales.

concluded that an instantaneous (acute) whole-body exposure of 10,000 millirem (10 rem) increases an average person's chance of suffering from a fatal cancer by 0.8%—that is 1 chance out of 125.

Debates continue about the potential health effects from low levels of radiation.

At acute doses (around 100 rem), mild clinical symptoms occur because cells may not repair themselves, are permanently altered, or die.[10] While the body can replace dead cells, damaged cells may reproduce abnormal cells that could lead to cancer and/or other abnormalities. At a higher acute radiation dose (around 300 rem), cells can't replace themselves fast enough, causing tissues to fail. Radiation sickness symptoms such as nausea, diarrhea, and weakness occur when water and nutrients are no longer taken up by the intestinal lining. The body's immune system can become so damaged that it can't fight infections. Statistically, we have a 50% chance of surviving an acute exposure of 300 rem. At a higher dose (1000 rem), vascular damage threatens blood flow, the nervous system, and brain functions. This is the dose received in leukemia therapy, by some firefighters following the Chernobyl nuclear power plant explosion, or atomic bomb blasts. Thus, high levels of ionizing radiation, especially high levels received over a short period of time, are very dangerous, and exposure must be controlled. Cause and effect are well documented. However, debates continue about the potential health effects from low levels of radiation.

Public acceptance of any change in protection standards, especially if standards became less stringent, is a social issue. Understanding the potential biological impact from radiation is a scientific issue.

The establishment of defensible radiation protection standards always fuels debate. Public acceptance of any change in protection standards, especially if standards became less stringent, is a social issue. Understanding the potential biological impact from radiation is a scientific issue. The scientific defensibility and social acceptability of radiation standards are tightly entwined when radiation protection policies are discussed.

The average resident of the United States is routinely exposed to about 300 millirem of radiation every year from the soil beneath their feet and the sun shining overhead—without concern. Yet there is considerable debate about the potential risk posed by low levels of radiation such as common in our everyday environment. Some claim low radiation doses adversely impact human health. Others believe available data demonstrate low doses are not harmful. When risk is plotted against dose, respected authorities disagree if the lines drawn should be straight (no risk-free dose), curved downward (less risk at low dose), or curved

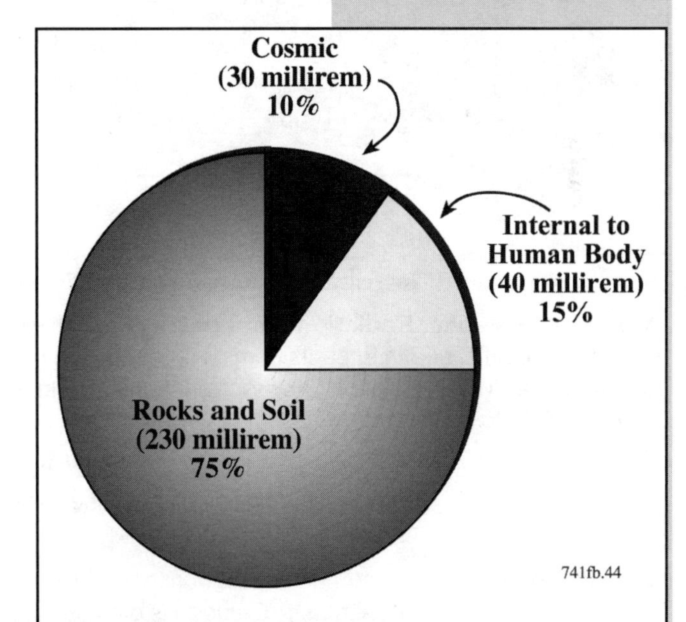

Natural Radiation Dose. The national average radiation dose people receive from natural sources is about 300 millirem a year—or about 1 millirem a day (NCRP 1987b). This is close to the natural background radiation dose for people living near Hanford.

(10) All life forms spend a considerable amount of their energy making and repairing cell-altering events. Damaged DNA that is not repaired properly can lead to mutations. In humans, each cell repairs anywhere from tens of thousands to one million damaged sites in its DNA every day. The human body has an active repair process to counter cell mutations, though these processes are not risk-free and their relevance to human cancer is unclear. Exposure to low-level radiation increases the number of these events by some small amount. Evidence suggests that damage done by ionizing radiation-induced events has a higher chance of resulting in DNA damage than that done by other normally occurring cell-altering events (Ward 1994; Sutherland et al. 2001).

"The possible deposition of unknown amounts of product [plutonium] in the body is a source of continual worry to men handling this material" (Parker 1945c).

upward (increased risk at low dose). Low-dose radiation protection standards have been set by extrapolating from adverse biological effects observed in high-radiation dose studies.

Lacking conclusive evidence, a "linear, no-threshold" model is used widely to estimate potential health impacts. According to this model, even the smallest radiation exposure carries a quantifiable cancer risk; health effects are proportional to exposure. The National Council on Radiation Protection and the International Commission on Radiological Protection endorse the linear model. It's been used for years as the preferred model for regulating radiation exposure.

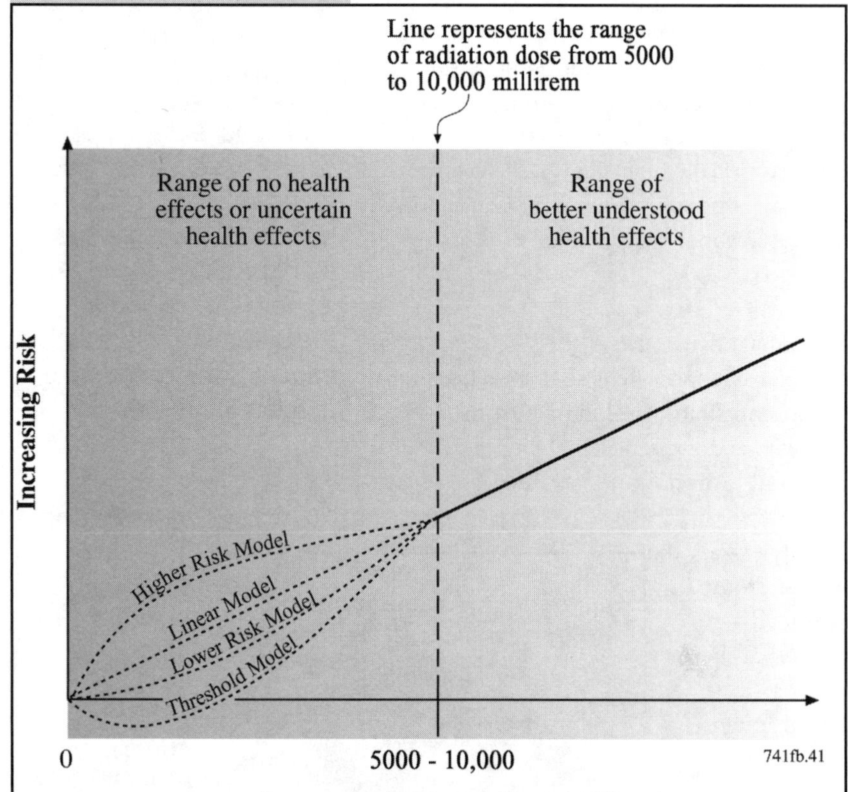

Models of Potential Radiation Damage. High radiation doses are harmful. Potential health impacts from lower doses are more controversial. This illustration shows four potential health impacts from low radiation doses (after GAO 2000f).

Higher-Risk Model	Assumes low levels of radiation become more harmful than a one-to-one (linear) dose would predict.
Linear Model	Assumes any amount of radiation increases health risk. Damage is directly proportional to dose. This effect is shown as a straight line.
Lower-Risk Model	Assumes low levels of radiation cause less damage below a certain dose.
Threshold Model	Assumes that below a certain threshold value, low-level radiation doses have no adverse health effect or possibly beneficial effects (hormesis theory).

The linear model is relatively simple to apply and conservative; that is, it likely does not underestimate risk. Regulators use the model in doing risk assessments and cost-benefit analyses to estimate risk reductions and hypothetical lives saved from regulating at one radiation level versus another.

Yet, many researchers question the correctness of this model. "The standards administered by EPA [U.S. Environmental Protection Agency] and NRC [U.S. Nuclear Regulatory Commission] to protect the public from low-level radiation exposure do not have a conclusive scientific basis…According to a consensus of scientists, there is a lack of conclusive evidence of low-level radiation effects below total exposures of about 5,000 to 10,000 millirem" (GAO 2000f, p. 10). The Washington State Department of Health (Wells 1994) reports, "The uncertainty of linear extrapolation is sufficiently great that the risk of very low levels of radiation may be zero." Existing data do support the possibility of no adverse health effects at low radiation levels.

However, a single dose-response model, even the linear model, does not fit all available information. Some people and cultural groups may be more sensitive to radiation than others.[11] Some organs

(11) Advances in human genetics over the last 20 years now enable us to understand that certain people and cultures may be genetically pre-disposed to illness or disease from exposure to select contaminants and environmental conditions.

weaken or become more cancer-prone when exposed to radiation; others do not. For example, thyroid[(12)] cancer appears to correspond to a linear model of dose-health impacts; that is, the more radiation the thyroid is exposed to the greater the risk of cancer. Yet, bone cancer appears controlled by threshold radiation levels; that is, the person must be exposed to a certain level of radiation; below this level, cancer doesn't develop. Are multiple models needed?

"The standards administered by EPA [U.S. Environmental Protection Agency] and NRC [U.S. Nuclear Regulatory Commission] to protect the public from low-level radiation exposure do not have a conclusive scientific basis" (GAO 2000f, p. 10).

The linear model is so widely accepted that it could only be superseded if evidence clearly demonstrated the need, if new scientific evidence was provided, and perhaps if Congress required federal agencies to re-examine radiation protection standards. Updated human epidemiological studies[(13)] and other research are under way to examine new evidence shedding light on possible low-level radiation effects. Since 1999, the largest epidemiological study in the United States, funded by the U.S. Department of Energy (DOE), is investigating cell and tissue responses to low-level radiation, and discerning between radiation-induced cell damage and other naturally occurring cell mutations.

Why new studies? Our understanding of the potential carcinogenic effects from exposure to contaminants is often based on conservative and outdated assumptions. Russell et al. (1992, p. 2) noted that while "there has been an explosion of data and understanding regarding the extent of human and environmental exposure to toxic substances, a precise scientific consensus as to the health effects of exposures" remains years away. The potential harm from exposure to single contaminants as well as mixed chemical and radioactive contaminants, especially at low radiation doses, remains largely unknown

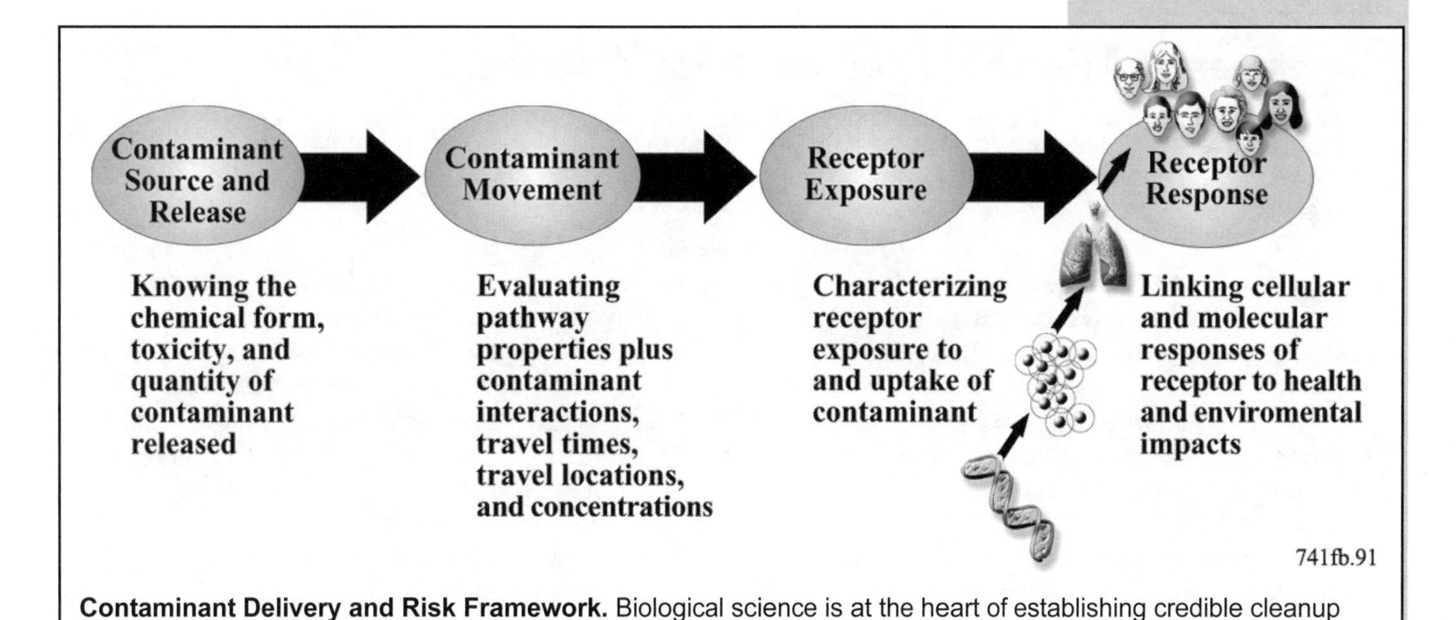

Contaminant Delivery and Risk Framework. Biological science is at the heart of establishing credible cleanup standards. Cleanup decisions are based on making informed choices about protection. This requires understanding how contaminants are released into, move through, and interact with the environment.

(12) The thyroid gland is located in the front of the neck, just below the Adam's apple. It secretes hormones necessary for proper body growth and metabolism. (Metabolism is the chemical activity that takes place in cells enabling the body to release energy from nutrients or to use energy to create other needed substances.)

(13) Epidemiology is the branch of medicine that studies the causes, distribution, and control of diseases in a population.

and is a prime research target (GAO 2000a, 2000f). In recent years, a scientific revolution has taken place in understanding the molecular-level basis for diseases and the prevention, repair, and removal of damaged DNA in the human body. This new capability and the knowledge it provides are only now being used to re-examine molecular, cellular, tissue, organ, and whole body responses to environmental stresses such as radiation.

As reported by a Presidential and congressional commission on risk, this situation is complicated because "scientists and regulators dealing with chemical hazards or with radiation hazards have been so independent of each other that there has been little combined analysis or combined risk management for medical, industrial, nuclear power, nuclear weapons production, and waste disposal settings where radiation and chemical contamination coexist" (Presidential/Congressional Commission on Risk Assessment and Risk Management 1997).

Current weaknesses in our knowledge about biological responses to stress can lead to environmental policy and remediation decisions that are at best overly stringent and at worst ineffective.

Upcoming molecular-level studies will help identify the pathway characteristics of cancer as well as identify populations that may be more susceptible or resistant to radiation exposure.[14] This will aid in performing more accurate risk assessments and developing environmental protection standards ensuring cleanup actions are truly protective. Otherwise, current weaknesses in our knowledge can lead to environmental policy and remediation decisions that are at best overly stringent and at worst ineffective. Some report that "cleanup beyond a threshold needed to provide protection to human and ecological health wastes money" (Integration Project Expert Panel 2000b). Time and research will tell.

Lifestyles and Regional Cancer Mortality Rates

If a correlation exists between low radiation dose exposure and adverse human health impacts, it might be obscured by cancers caused by human habits, such as smoking, drinking, high-cholesterol diets, and lack of exercise.

For example, heavy smokers expose their lungs to radioactive lead and polonium found in tobacco—up to 20,000 millirem (20 rem) a year for a heavy smoker (Presidential/Congressional Commission on Risk Assessment and Risk Management 1997).[15] This is the greatest radiation dose received by the public from any consumer product (Moeller 1990). Two-pack-a-day cigarette smokers expose themselves to a radiation dose of 1300 millirem a year. Yearly, the average person in the United States receives about 60 millirem of radiation above background levels from consumer products and medical procedures (NCRP 1987b). Airline crewmembers receive about 1000 millirem a year because more cosmic radiation is found at higher elevations. Passengers flying in the Concorde receive 1 millirem of cosmic radiation for each hour of flight (UNSCEAR 2000). The radiation dose received from a single chest x-ray is 5 to 20 millirem. Such exposures are a matter of personal choice—and normally unquestioned.

---

(14) According to the Washington State's Model Toxics Control Act (WAC 173-340-702), "cleanup standards and cleanup actions selected...shall be established that provide conservative estimates of human health and environmental risks that protect susceptible individuals as well as the general population."

(15) These radionuclides are deposited on tobacco leaves from airborne materials originating from the decay of natural radon. Radon comes from the radioactive decay of uranium. When a cigarette is lit, radionuclides enter the smoke and are inhaled deep inside the lungs.

Over the years, differences between the U.S. Environmental Protection Agency and the U.S. Nuclear Regulatory Commission over radiation protection standards and acceptable risks have contributed to higher oversight and cleanup costs while at the same time raising public concerns about what cleanup levels are truly protective (GAO 2000f).

We know that people receiving the highest radiation dose are not necessarily those at greatest risk (Oughton 1996). For a similar dose, the chance of contracting cancer or another disease is higher for children or the unborn. We also know certain radionuclides target certain organs. Strontium is chemically similar to calcium; therefore, it accumulates in hard tissues such as bones, eggshells, and antlers. Cesium-137 chemically resembles potassium and accumulates in the muscle of humans and wildlife.

Preferential protection of the most sensitive members of our population can be justified on the grounds of fairness and equity. This is why, particularly in our democratic society, equal protection of individuals is favored over equal treatment of all people.

A study by National Institutes of Health (Devesa et al. 1999) displays a county-by-county comparison of cancer mortality for every state. The Washington State Cancer Registry (see http://www3.doh.wa.gov/WSCR/) reports on the incidence and mortality of all forms of cancer for each county in the state.

4.2 Radiation Protection Goals

"Public health is at less immediate risk than is worker safety, largely because most waste and contamination is being managed and contained at present. There is still cause for real concern in this area, though, particularly over the medium to long-term. The current mechanisms for managing and containing potential public health risks (such as controlling access to the site) will degrade long before the contamination becomes benign... Because of a lack of agreed-upon cleanup standards, commitments made before problems were understood, and limited risk studies, in many cases it is not possible to separate truly hazardous conditions from those where contamination is measurable, but not a serious health risk. Nevertheless, this uncertainty must not be used as a rationale for not moving ahead aggressively with cleanup programs."

—National Research Council,
Improving the Environment: An Evaluation of DOE's Environmental Management Program (1995, pp. 5-6)

Through hard work, enforcement of environmental laws, and the application of the best available technology, industrial waste management has improved since the 1960s. The benefit of pollution control was immediate and obvious in our air, land, rivers, and streams. Nonetheless, problems persist, and the public must remain diligent.

We know that people receiving the highest radiation dose are not necessarily those at greatest risk.

Today, some of the toughest cleanup issues have shifted from avoiding immediate, overt threats to protecting against potential harm from contaminants based upon important though low-probability "what if" scenarios. This is a speculative, value-laden arena, where making decisions based upon "obvious" facts is difficult. This challenge is superimposed on our living in a time when people are unwilling to voluntarily accept any risk imposed by others, especially risk created by institutions believed untrustworthy.

Some of the toughest cleanup problems have shifted from avoiding immediate, overt threats to protecting against potential harm from contaminants based upon important though low-probability "what if" scenarios.

"We are loath to let others do unto us what we happily do to ourselves" (Lamarre 1992).

Hanford's boundary is the point of compliance for protecting the public to ensure the yearly dose from routine radiological releases is less than 100 millirem (DOE Order 5400.5 [DOE 1990a]) and that not more than 10 millirem are received from airborne sources (40 CFR 61) or 4 millirem from groundwater sources (40 CFR 141). A temporary annual dose limit to the public, not to exceed 500 millirem a year, is allowed by regulations in the event of an accident or other unusual circumstance.(16)

The current nuclear worker exposure limit for the whole body is 5 rem (5000 millirem) a year. The reason for this large difference between permissible worker and public exposure is because workers voluntarily accept employment where they may be exposed to higher-than-normal radiation. Benefit and protection are traded. The public does not have that choice.

The general principle that radiation doses should be kept "as low as reasonably achievable" is incorporated into most radiation protection standards. It was introduced by the National Council on Radiation Protection in 1954 and incorporated into DOE (then the Energy Research and Development Administration) site operating manuals in 1975 (Becker 1990).

How do these numbers compare to the past? During the 1950s, the recommended limits for yearly radiation exposure to members of the public outside of Hanford was 500 millirem to the whole body and 1500 millirem to most soft tissues from internally deposited radionuclides (Parker 1959b, p. 226). Parker reported that the representative annual radiation dose to members of the public was probably "not less than 50 mrem and not more than 200 mrem." However, he cautioned that if all radiation exposure factors were maximized, an annual dose of 600 to 700 millirem might be reached. "It may thus be supposed that the safety margin [between the recommended radiation limit and the actual dose received] for this case is not a comfortable one" (Parker 1959b, p. 229).

Not operating plants to protect maximum exposed individuals was one way that Hanford officials could report that radiation exposures from the atmosphere pathway were "well within the maximum permissible limits recommended by national authorities" for people living near the site (Tomlinson 1959b, p. 281).

For a waste site in Washington State to be considered clean and available for unrestricted use by the public, the Washington State Department of Health (1997) specifies a radiation dose limit of 15 millirem a year above background(17) from all environmental pathways. This is to a maximally exposed individual. This limit applies for 1000 years after cleanup is complete. It corresponds to a lifetime excess cancer risk of less than 3 in 10,000 over a 30-year exposure period (EPA 1994). If a waste site can't be cleaned to such levels, then the site owner is required to implement measures, such as land-use limitations or engineered barriers, to control exposure.

---

(16) Today, at least 18 federal agencies in the United States have responsibility and implement policies related to regulating radiation exposure or nuclear research (Young 2001).

(17) "Local area background is the external radiation and environmental radionuclide concentrations in the area near Hanford but not contaminated by past Hanford activities" (Washington State Department of Health 1997). Fifteen millirem a year is 5% above the average background radiation dose of 300 millirem a year for someone living in the United States.

The Not-So-Average Person

When calculating the amount of radiation received by a person that did not work at Hanford, health physicists and researchers develop scenarios to determine the "maximally exposed person." This person is a hypothetical member of the public who could receive the highest possible radiation dose. This person's lifestyle routinely brings them into contact with high levels of contamination. In fact, the person's exposure pathways are selected to maximize the combined radiation dose from Hanford. For example, this person cooks, showers, and otherwise uses water from a local water treatment system that pumps water from the Columbia River just downstream of Hanford; spends time outside where they are exposed to radionuclides deposited on the ground; eats food irrigated with river water; spends time in the river swimming and fishing; spends time along the river shoreline; and eats locally caught fish or game. Because no one person is believed to have this lifestyle, this person represents the highest, or maximum, dose someone not working at Hanford could receive. During the year 2000, this person received a radiation dose that was 0.005% higher than other people in the area received from the natural background (Antonio et al. 2001).

According to the Washington State Department of Health (1997), the process of determining if this cleanup objective is met "involves field and laboratory measurements of radioactivity at the site and modeling of expected doses based on proposed land uses and the site-specific physical parameters."

The Model Toxic Control Act for the state of Washington notes: "In some cases, cleanup levels...are less than natural background levels or levels that can be readily measured. In those situations, the cleanup level shall be established at a concentration equal to the practical quantitation limit or natural background concentration, whichever is higher" (Washington State Department of Ecology 2001b).

When federal standards are in place, they may be higher than 15 millirem a year but likely never lower (Washington State Department of Health 1997).

When radiation protection standards are examined, especially those specifying cleanup levels close to background, it's a challenge to reconcile why the potential risk of 1 out of 10,000 or 1 out of 100,000 a year to a maximally exposed individual in the future appears more important than the much greater risk the average person accepts today from both artificial and natural sources (Bodansky 1996).

Does a *de minimus* threshold exist below which health impacts are not a concern? Natural radiation variations are common across the country (Oakley 1972) and have contributed to radiation protection debates brewing since the 1950s (Terrill 1958; Morgan 1959).

Does a *de minimus* threshold exist below which health impacts from radiation are not a concern?

For example, should anyone treat areas of high natural radiation, such as around Spokane Washington, as a radiation hazard? After all, according to the Washington State Department of Health (Hughes 1994), the yearly radiation dose received by the public in northeastern Washington is 500% higher than the national average.[18] This is also 500% higher than background levels for communities near Hanford. Yes, this is natural radiation versus human-made radiation, and yes, it is a semi-capricious comparison. However, cells don't discriminate. An alpha particle is an alpha particle.

(18) A review of health data for the Spokane area shows an average annual incidence and mortality by cancer compared to other Washington State counties (Washington State Department of Health 1998).

The yearly natural background radiation dose received by the public in northeastern Washington is 500% higher than the national average or the levels for communities near Hanford.

This highlights a fundamental problem—how do we reconcile spending huge sums of money to clean some waste sites to local background levels compared to much higher radiation dose received by residents within a short drive of Hanford? Even at Hanford, natural radiation levels vary. The answer likely rests in our despising involuntary radiation exposure, our tendency to conservatively protect when facing uncertainty, and our invoking personal values.

Cleanup goals are commonly set within 5% (15 millirem a year) of natural background radiation, though little data justify that requirement or link it to meaningful health-based dangers (Bodansky 1996; Cary 1999b). In fact, during any year and at any location, background radiation exposures can vary by a few tens of percent (NCRP 1987a).

Each year, people receive about 4 millirem more cosmic radiation dose for every 1000 feet above sea level they live compared to those residing at lower elevations (NCRP 1987b). This is why someone living in Denver, Colorado, receives a cosmic radiation dose of 50 millirem a year compared to 26 millirem a year for those living at sea level.

Cleanup goals are commonly set within 5% of natural background radiation, though little data link it to meaningful health-based dangers.

How might this relate to identifying a radiation dose from cosmic sources for a future resident at Hanford? That person would receive about 10 millirem more radiation each year if residing atop Rattlesnake Mountain (elevation 3500 feet) compared to living in the middle of Hanford (elevation 700 feet). People don't live at either location, but this illustrates how dose from just one source—outer space—differs across the site. Couple this with natural radiation variations that take place at any location throughout the year and one begins to understand why it will be a challenge to prove that a contaminated site is cleaned to within 15 millirem of pre-Hanford conditions.

Key cleanup decisions should be based upon more than a single number.

However, key cleanup decisions should be based upon more than a single number. Reliance "should be supplemented by the use of other figures of merit, including comparisons with other hazards faced by mankind and an overall judgment criterion of the quality of scientific understanding" (NRC 2001a).

"Regulators have a tough job. But they make it even more difficult, by trying to regulate using a 'speed limit' concept. 'Thou shalt have no more than so many ppms, or curies, etc'. They then suffer and agonize over where to set this speed limit. They try to set it at such a level that the decision maker is forced to make the best decision for the community as a whole. Difficult to do. Worse, they attempt to set the limit without explicitly, and quantitatively, doing the decision analysis. That's next to impossible. No wonder they suffer. We are asking the wrong question. The question is not, 'How much risk is acceptable?' The question is, 'What is the best decision option?'"

—Stan Kaplan, *Risk Analysis* (1997)

Oughton (1996) writes: "It seems this debate might give important insight into the public's different attitudes to natural and technological sources of radiation risk. For example, one might argue that cancers brought about by the practice of the nuclear industry (acts of commission) are more morally reprehensible than cancers allowed by non-intervention in the accidental or natural radiation situation (acts of omission).

Hence, on these premises, the distinctions might be used to support the stringent dose limits on the nuclear industry. However, it does not necessarily follow that one can justify a higher level of spending to reduce doses from existing practices than from natural or accidental sources."

Future generations don't choose what they inherit. We are their caretakers.

On the other hand, future generations don't choose what they inherit. We are their caretakers and should not pass along hazards greater than those we face. It is our responsibility to invest in solutions and knowledge for ourselves and for those who follow. Perhaps it is also our responsibility to reduce potential radiation exposure, even when near background levels, to as low as possible. These are challenging issues especially when we are faced with limited resources to do the most good.

For the high-level waste geologic repository, researchers are being asked to predict the behavior of a waste site for thousands of years; "this is scientifically unsound and will lead to bad engineering practice...the U.S. [repository] program is bound by requirements that may be impossible to meet" (NRC 1990b). Hanford faces similar challenges.

Our imprecise understanding of the natural environment, contaminant behavior, and potential health impacts raises a cautionary flag not to overstate predictions or to impose clean-up goals that can't be confirmed. As noted by the National Research Council (NRC 1990b, p. vii) in their rethinking of the high-level waste geologic repository program, researchers are being asked to predict the behavior of a waste site for thousands of years, "this is scientifically unsound and will lead to bad engineering practice...the U.S. [repository] program is bound by requirements that may be impossible to meet." Though waste sources differ, Hanford faces similar challenges.

Some issues of knowing what is safe now lie along the fringe of science where personal values and ideology prevail. Distinguishing between science-driven evidence and value-based perceptions is important. If not wisely done, the inability to discern will impede making the right investments to improve our knowledge and cleanup capabilities (EPA 1994). Of course, the choice of what an acceptable risk is remains personal and socially defined.

4.3 Radiation Doses of Yesterday

"It is my personal feeling that the [radiation] levels that have been used have been safe during wartime, but now that the operations embark upon a peacetime era, the levels should be reduced as low as is practicable."

—H.L. Friedell, *Report of the Safety and Industrial Health Advisory Board* (Williams 1948)

More than 140 million curies of radioactivity were released to the atmosphere and Columbia River when Hanford reactors and reprocessing plants operated.

Thirty-two million curies were released into the air—12 million came from reactors and 20 million originated in reprocessing plants. Some radionuclides accumulated in the human body, had a long enough half-life, and were in high enough quantities to increase radiation exposures for people living downwind.

In 1986, DOE began declassifying formerly secret documents revealing that large amounts of radioactive material were once dumped into the environment. Representatives from the states of Washington, Oregon, and regional Native American tribes

Year	Iodine-131	Ruthenium-103 and -106	Cerium-144	Strontium-90	Plutonium-239
1944 - 1949	697,000	290	1740	30	2
1950 - 1959	43,000	1130	630	10	<1
1960 - 1969	460	130	1350	25	<1
1970 - 1972	<1*	1	50	1	<1

99% of dose from iodine-131

1% of dose from these radionuclides

Source	Released to Atmosphere
Reactors	12 million curies
Reprocessing Plants	20 million curies
Total	32 million curies

*< symbol means "less than."

741fb.11

Radionuclide Releases to Atmosphere. Thirty-two million curies of radioactivity were released into the atmosphere from Hanford plants between 1944 and 1972. Most of the resulting radiation dose came from six radionuclides. Iodine-131 accounted for 99% of that dose.

Those living adjacent to and directly downwind of Hanford received the highest radiation dose.

worked with the Centers for Disease Control and Prevention to form the Hanford Health Effects Review Panel to evaluate this information. The panel recommended undertaking two analyses: (1) a dose reconstruction study for all radionuclides people who lived near Hanford were exposed to and (2) a thyroid disease study focused on one radionuclide—iodine-131. The first study evolved into the Hanford Environmental Dose Reconstruction (HEDR) Project. The second ended up as the congressionally mandated Hanford Thyroid Disease Study.

The Pacific Northwest National Laboratory[19] was directed by DOE to conduct the HEDR Project. Starting in 1988, and to further distance the study from DOE, the project was headed by an independent technical steering panel whose members were selected by the deans of research departments at major universities in Washington and Oregon. In 1989, the Fred Hutchinson Cancer Research Center in Seattle, Washington, was contracted to support research on thyroid disease.

The HEDR Project was completed in 1994. It contained information later used in the thyroid study. Shortly thereafter, the Centers for Disease Control and Prevention issued a pilot report recommending a full-blown thyroid epidemiological investigation. The draft findings were issued in 1999 (Fred Hutchinson Cancer Research Center 1999). The final report was published in 2002 (CDC and Fred Hutchinson Cancer Research Center 2002).

The HEDR Project along with referenced studies reported that gaseous releases from reactors produced "very small doses" to humans—between 2 and 4 millirem a year (Napier 1992; HEDR 1994; Heeb 1994). This was about a 1% increase compared to local natural background sources. Years earlier, Parker (1959b) had reported that "no distinguishable contamination levels in the air or ground" beyond the immediate vicinity of the reactors had resulted from normal operations.

(19) Then entitled the Pacific Northwest Laboratory.

Iodine-131 released from reprocessing plants dominated the radiation dose received by adults and children. Those living adjacent to and directly downwind of Hanford received the highest dose. The 27-year cumulative dose to an adult that received the highest possible dose (known as the maximum exposed adult) between 1945 and 1972 ranged from 6 to 1000 millirem. The highest possible doses for surrounding areas were 6 millirem in Wenatchee, 31 millirem in Spokane, 470 millirem in Richland, and 1000 millirem in Ringold, Washington, immediately east of Hanford (HEDR 1994). The cumulative dose over the same period from non-iodine-131 radionuclides was not greater than 50 millirem to any organ or 20 millirem to the entire body (Napier 1992).

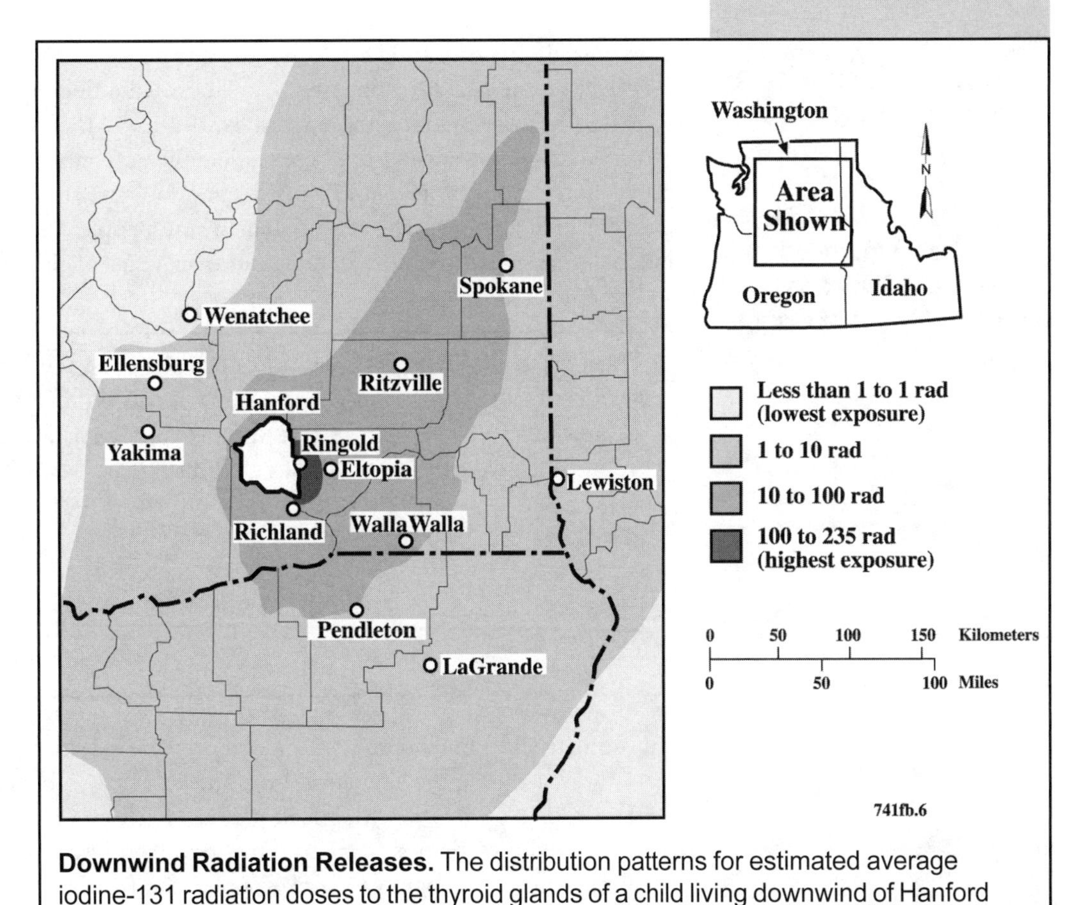

Downwind Radiation Releases. The distribution patterns for estimated average iodine-131 radiation doses to the thyroid glands of a child living downwind of Hanford between 1944 and 1951 are shown. Children living east and northeast received the highest dose (HEDR 1994, p. 31).

The average dose to the thyroid of an adult living downwind of Hanford and exposed to the highest possible dose was 36 rad with a range of 10 to 150 rad (HEDR 1994). This range equates to the thyroid receiving 10 to 150 rad (10,000 to 150,000 millirem), with most of the dose received before iodine-capturing silver iodide filters were added to the reprocessing plants in 1950 and 1951. The dose to the thyroid of an adult living in the area but receiving a smaller amount of radiation was 0.01 rad (10 millirem) with a range of 0.001 to 0.06 rad (1 to 60 millirem). The average dose to the thyroid for a child exposed to the highest possible dose was greater than what adults received—235 rad (235,000 millirem) with a range of 54 to 870 rad (54,000 to 870,000 millirem). At the lowest exposed locations, a child's estimated dose was 0.07 rad (70 millirem) with a range of 0.01 to 0.3 rad (10 to 300 millirem).

The average dose to the thyroid for a child exposed to the highest possible dose from iodine-131 was much greater than what adults received.

The final report for the Hanford Thyroid Disease Study stated: "Data show that the risks of thyroid disease were about the same regardless of the radiation doses people received. In other words, no association between Hanford's iodine-131 and thyroid disease were observed" compared to worldwide occurrences (CDC and Fred Hutchinson Cancer Research Center 2002, p. 9).[20]

(20) Researchers studied a wide range of thyroid diseases and abnormalities including thyroid cancer, benign nodules, and underactive thyroids.

"The findings show that if there is an increased risk of thyroid disease from exposure to Hanford's iodine-131, it is probably too small to observe" (CDC and Fred Hutchinson Cancer Research Center 2002).

However, the report cautioned: "the findings do not prove that Hanford radiation had no effect on the health of the area population…the findings show that if there is an increased risk of thyroid disease from exposure to Hanford's iodine-131, it is probably too small to observe using the best epidemiologic methods available." The regional population's exposure to Hanford-generated iodine-131 was unique to human experience for little is known about situations where people are exposed to low levels of internally deposited iodine-131 radiation over years of time.[21]

The failure of a single fuel slug in the KE Reactor in May 1963 resulted in one of the largest single releases of radionuclides in the first 20 years of Hanford (Hall 1963). This failure added 4000 to 5000 curies to the river. Most was from radioactive neptunium[22] along with strontium, iodine, barium, and zirconium. The maximum dose to the thyroid of an infant living downstream in Pasco, Washington, was about 8 millirem. In those days, this was considered small compared to the average annual radiation limit of 500 millirem to adults set forth by the Federal Radiation Council (1961).

An estimated 110 million curies of radiation were released to the Columbia River from 1944 to 1971 during the operation of Hanford's first eight reactors (Heeb and Bates 1994). From 1956 to 1965 (the peak period of release), a typical nearby resident may have received a yearly dose of 1 to 5 millirem from the river (HEDR 1994). This is a small percentage compared to background dose from natural sources and atmospheric fallout from weapons tests. On the other hand, a person who constantly went swimming in the river and ate lots of fish caught from it may have received a higher annual dose of 25 to 135 millirem—a 15 to 45% increase compared to other residents (Farris et al. 1994). Dosage decreased down river because of dilution and radioactive decay.

Native Americans who used the river were exposed to 10 times more radiation than the maximally exposed non-Native American.

In a review of these past radiation dose estimates from releases to the Columbia River, health physicist John Till confirmed that fish ingestion was the dominant exposure pathway (Till et al. 2002). Till also reported, "However, the significance of this [fish] pathway for Native American users of the river was greater than that for non-Native Americans by a factor of ten because fish consumption rates reported for Native Americans tended to be higher than the value assumed for the maximum representative individual in the HEDR Project." Native American lifestyles more closely linked them with the health of the river.

---

(21) The Centers for Disease Control and Prevention reported that thyroid diseases found in adults and children exposed to iodine-131 from nuclear testing in the Marshall Islands (1954), the Nevada Test Site (1950s), or the Chernobyl nuclear reactor explosion (1986) are not directly applicable to the Hanford situation (CDC and Fred Hutchinson Cancer Research Center 2002, pp. 13-15). Key differences include how the radiation was received (internal vs. external exposure), the period over which exposure took place, and the extent to which iodine deficiency pre-existed in the population exposed.

(22) The half-life of neptunium-239 is 2.4 days. It decays to plutonium-239.

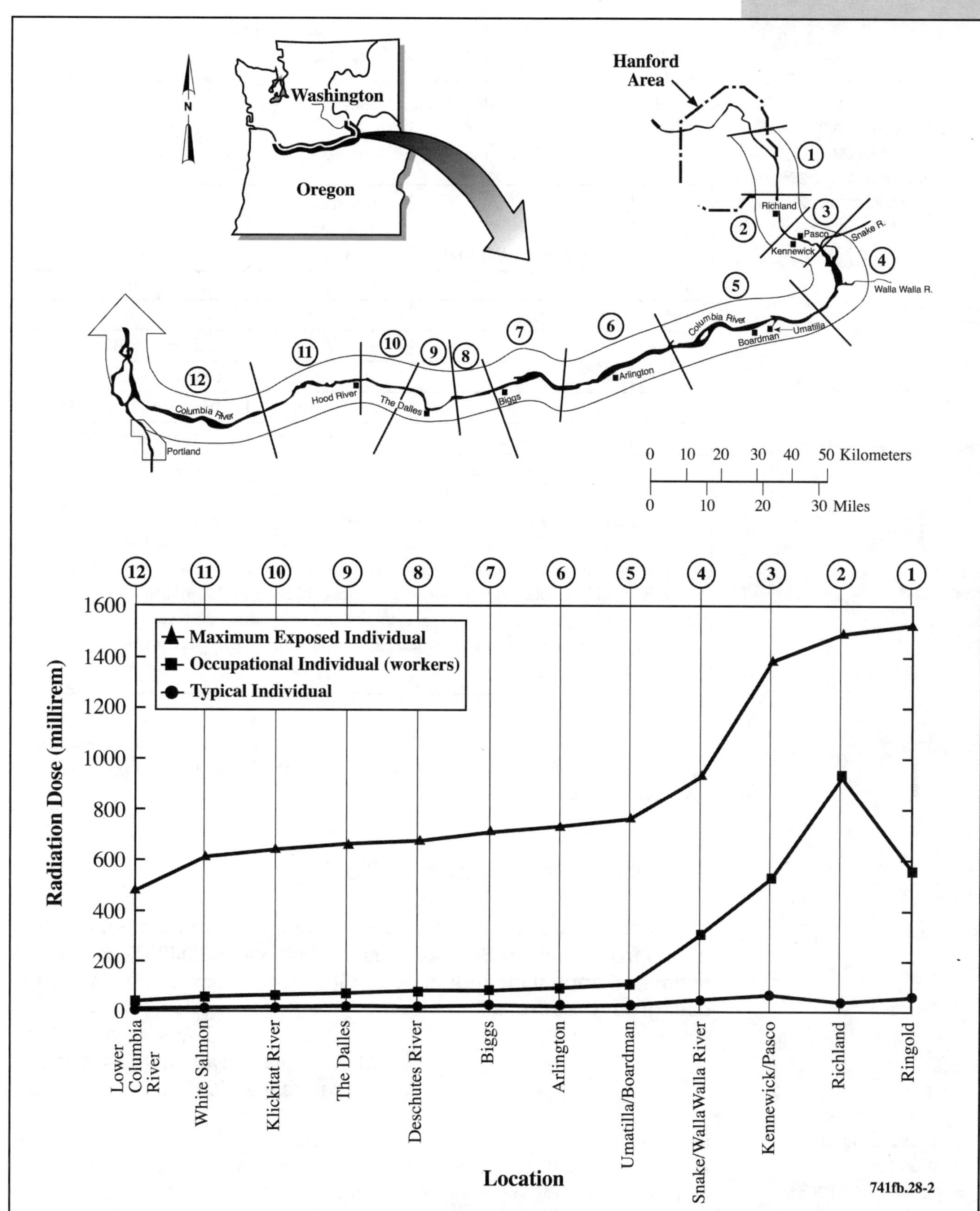

Past Radiation Doses Downstream from Hanford. The estimated cumulative radiation dose to Hanford workers and members of the public living downstream from Hanford varied based on river water drunk, fish eaten, and time spent recreating on the Columbia River. These cumulative doses are for the years 1944 through 1992. Those living closest to Hanford received the highest dose (Farris et al. 1994).

As the reactors and reprocessing plants shut down, so did radionuclide releases and the radiation dose received by people.

As the reactors and reprocessing plants shut down, so did radionuclide releases and the radiation dose received by people. Table 3 summarizes the estimated average and hypothesized maximum whole-body radiation doses received each year by the public from Hanford. It covers, in decade increments, the last half of the twentieth century.

Table 3. Average and Maximum Dose Received by the Public (1950-2000)

Year	Average Individual[a]	Maximum Exposed Individual[b]	Reference
1950	0.2 millirem	35 millirem	HEDR (1994)
1960	3 or 5 millirem[c]	80 millirem	Nelson (1961)
1970	2 millirem	12 millirem	Corley (1973b)
1980	0.0024 millirem	0.01 millirem	Sula and Blumer (1981)
1990	0.004 millirem	0.03 millirem	Woodruff et al. (1991)
2000	0.0008 millirem	0.014 millirem	Antonio et al. (2001)

(a) Over these years, the point of reference for an "average individual" changed from a person living in Richland to someone living within 50 miles of Hanford. Radiation doses to a typical Richland resident sharply increased in 1963 when the city changed its drinking water source to the Columbia River (HEDR 1994).
(b) The models, data, and assumptions used for estimating the maximum exposed individual have been refined over the years (Soldat et al. 1986). See Chapter 3.
(c) The wording in Nelson (1961) describing the radiation dose received by an average Richland resident is unclear. The intended value appears to be either 3 or 5 millirem.

All but two reactors were shut down by 1970 and only the PUREX facility continued reprocessing spent fuel. This was the primary reason for the significant drop in radiation dose compared to earlier years.

4.4 Radiation Doses of Today

"The estimated annual average individual dose to members of the public from Hanford Site sources in 2000 was ~0.0003% of the estimated annual individual dose (300 mrem) received from natural background sources."

—E.J. Antonio and others,
Hanford Site Environmental Report for Calendar Year 2000 (Poston et al. 2001)

The average radiation dose from the natural environment received by the public each year depends on where they live, especially the rocks and soil on which they live.

Some people living near the Hanford Site were exposed to high levels of human-made radiation when Hanford facilities were operating. Today, the radiation dose received by the public is extremely low (Poston et al. 2001).

The average radiation dose from the natural environment received by the public depends on where they live, especially the rocks and soil on which they live. For example, according to the Washington State Department of Health (Hughes 1994), the natural radiation dose in southeastern Washington State (Walla Walla area) is 564 millirem a year; 175 millirem a year in Puget Sound (Seattle area); and 1618 millirem a year in northeastern Washington State (Spokane area). Hughes reports, "In the Spokane Valley…the residential radon levels are among the highest in the state." Radiation dose varies by nearly 1500 millirem across the state.[23]

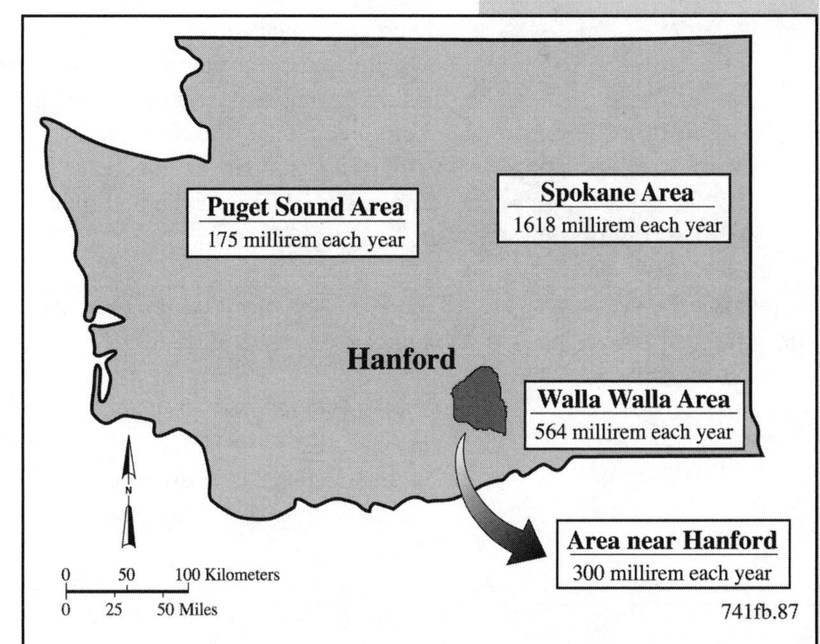

Radiation Doses Across Washington State. The public's average exposure to natural background radiation varies across the state. Most exposure comes from radon and that dose depends upon the type of rock underlying an area. According to the Washington State Department of Health, people living in the Spokane area receive some of the highest radiation doses in the state (Hughes 1994; Antonio et al. 2001).

The background radiation dose for an average person living in the geographic area surrounding Hanford is 300 millirem a year (Antonio et al. 2001). This is also the national average reported by the National Council on Radiation Protection and Measurements (NCRP 1987b).[24]

Each year, an average person living near Spokane, Washington, receives five times the radiation dose of an average person living near Hanford. Why? It's because of the greater abundance of natural radiation sources in granite-like rocks underlying the Spokane area. Radiation doses in the Puget Sound area are lower because that area's rock contains fewer radionuclides contributing to dose.

Today, groundwater containing radioactive elements from Hanford seeps into the Columbia River. This plus exposure to airborne radiation and consumption of locally grown food are the main contributors to a yearly increase above background of 0.0008 millirem for an average person living within 50 miles of Hanford (Antonio et al. 2001).[25] How much radiation is this? It's 0.0003% of the estimated radiation dose an average person living near Hanford receives each year from natural sources. Another way of looking at this radiation dose is that it is equal to receiving one additional minute of local background exposure or living 10 seconds in the Spokane area. Using yet another comparison, this Hanford contribution is about 25 times smaller than the 0.02 millirem dose increase in cosmic radiation received on the head of a 6-foot-tall person compared to his or her feet.[26]

The background radiation dose for an average person living in the geographic area surrounding Hanford is 300 millirem a year. This is also the national average.

(23) According to Hughes (1994) naturally occurring radon gas, seeping into homes and buildings, contributes 33% to 89% of the radiation dose received by the general population across Washington State. Radon is created during one of the several decay steps that natural uranium-238 undergoes.

(24) The worldwide average radiation dose from natural sources is 200 millirem a year (UNSCEAR 2000). Doses ranged from less than 100 millirem to more than 700 millirem a year.

(25) The regional collective dose (called the effective dose equivalent—see glossary) from Hanford operations is estimated from calculating the radiological dose to the entire population (380,000 people) living within a 50-mile radius of the site.

(26) This comparison is based upon differences in cosmic radiation dose as a function of elevation as reported in National Council on Radiation Protection (1987b). Doses from cosmic radiation increase about 4 millirem per 1000-foot rise in elevation.

"The only claim that can be made with certainty is that the risk of cancer at radiation doses below about 10,000 mrem (millirem) is so small that no one has been able to unambiguously measure it" (Wells 1994).

The Washington State Department of Health reports that radioactivity in Columbia River sediment is now dominated by natural sources.

Each year, calculations are also done to estimate the maximally exposed (hypothetical) person living in the general area of Hanford. This person has a lifestyle that "makes it unlikely that any other member of the public would receive a higher radiological dose" (Antonio et al. 2001). During the year 2000, this person received a radiation dose of 0.014 millirem[27] and lived in an area between Richland and Pasco, Washington. This exposure is 0.005% of the 300 millirem a year an average person receives from background sources.

These average or maximum radiation doses vary slightly from year to year depending upon the relative contributions from human-made sources.

How can human-made radiation doses be so low in communities near Hanford? The answer is that most Hanford radionuclides now lie inside underground tanks, basins, and buildings. In addition, most radionuclides once released into the air and the Columbia River were short lived or are no longer available for biological uptake.

Some long-lived radionuclides released into the Columbia River are buried in sediment, especially in slack waters behind islands or behind McNary Dam, located 30 miles downstream of Hanford. These include isotopes of cobalt, strontium, cesium, uranium, and plutonium (Patton 1999). According to the Washington State Department of Health (Wells 1994), if these sediments were dredged and placed on top of the ground, the maximum average radiation dose to a person over a 75-year lifetime would be 1.6 millirem a year. This is about 0.5% of the yearly dose a Hanford area resident or a U.S. citizen receives from natural radiation sources. Wells (1994) goes on to report that radioactivity in Columbia River sediment "is dominated by natural sources" such as potassium, uranium, and thorium (and their decay products) found in rocks. Wells notes that artificial radioactivity in the Columbia River now is from worldwide fallout from past atmospheric testing of nuclear weapons and Hanford operations. Most contributions are from tritium.

Wu (1994), a graduate student at Oregon State University, also estimated the radiation dose received by members of the public from background and Hanford-originated radionuclides in the "Lower Columbia River sediments." The maximum exposed individual was calculated to receive 10.4 millirem each year while the average person might receive 4.9 millirem each year. Approximately one-third comes from Hanford. Wu reported, "that the dose contributions of the radionuclides derived from Hanford operations during 1944-1971 are only a small component compared with the dose contributions from radon, natural external and internal exposure, medical and consumer product radiation."

Stewart (1996), also from Oregon State University, modeled some "highly conjectural" maximum exposure scenarios for a farmer living near and farming atop contaminated sediment dredged downriver of Hanford from behind McNary Dam. Estimated

(27) Voluntary activities comparable to receiving an increased risk from 0.014 millirem of radiation include riding in a car for 0.6 mile, eating 6 ounces of charcoal-broiled steak, swallowing 1.8 ounces of beer, or drinking 1 quart of chlorinated tap water (Antonio et al. 2001). Yes, such comparisons between radiation and voluntary activities drive some people crazy, but they do encourage discussions—like reading life-expectancy actuary tables published by the insurance industry.

radiation doses ranged from 28 to 106 millirem each year with the bulk coming from naturally occurring radionuclides.

Certain locations along the Columbia River shoreline adjoining Hanford have radiation levels higher than background. The potential radiation dose from continuous occupancy at such a site containing the maximum concentrations of radionuclides measured at any of these sites would be 14 millirem a year—a 5% increase over local background radiation (Cooper and Woodruff 1993). More recent reports suggest the highest radiation dose rate along the Hanford-Columbia River shoreline during the year 2000 was in the N Reactor area (Antonio et al. 2001). For every hour someone spent there, they would receive a dose of about 0.005 millirem above the average shoreline rate. If a person lived on the N Reactor shores for 1 year (this is not possible, but assuming so), they would receive a radiation dose of 44 millirem from this one Hanford source.

Certain locations along the Columbia River shoreline adjoining Hanford have radiation levels higher than background.

Though potential exposure to long-lived radionuclides in river sediment is now below the level at which routine surveillance is required, periodic sampling continues to ensure no significant changes take place. Someday, the extent and amount of radionuclides in river sediments and the actual amount transported to the river may require updating as waste site cleanup is carried out (Danielson and Jaquish 1996; DOE 1999c).

4.5 Radiation Dose from Global Fallout

"Senator Clinton P. Anderson, chairman of the Joint Atomic Energy Committee, took sharp issue at the morning session with Merrill Eisenbud, manager of the AEC's [Atomic Energy Commission] New York operations office. Eisenbud had just stated that the faster rate of fallout now taking place had the advantage of bringing the atomic debris out of the atmosphere sooner.

Anderson interrupted: "When you were here at the fallout hearing 2 years ago, you told us how good it was that the atomic debris was not coming down so fast, since this gave it time to lose some of its radioactivity. Now you tell us, how good it is that it's coming down so fast, since this means there is less left in the stratosphere. No matter what happens, you keep telling us how good it is. Now which is it?"'

Eisenbud had no reply. There was an embarrassing pause and the hearing moved on."

—Edward Gamarekian,
Washington Post and Times Herald (1959)

"No matter what happens, you keep telling us how good it is. Now which is it?" (Gamarekian 1959)

The testing of nuclear weapons in the atmosphere from 1945 to 1980 caused the largest release of artificial radiation ever experienced by the world—nearly 70 billion curies (UNSCEAR 2000). Most of this radioactivity was contributed by radionuclides with half-lives of less than a month; these have not persisted in the environment. Longer-lived radionuclides remain.

The testing of nuclear weapons in the atmosphere caused the largest release of artificial radiation ever experienced by the world—nearly 70 billion curies (UNSCEAR 2000).

After the Limited Test Ban Treaty was signed in 1963, the United States, Soviet Union, and Great Britain moved testing underground.

The kiloton-yield fission bombs of World War II and the early Cold War deposited most of their radioactive debris near the detonation sites. Later megaton-yield thermonuclear weapons dispersed fallout on a global scale.

The United Nations Scientific Committee on the Effects of Atomic Radiation reported 543 weapons were exploded in the atmosphere —219 by the United States,[28] 219 by the Soviet Union, 50 by France, 22 by China, and 33 by the United Kingdom (UNSCEAR 2000). Eight of these were underwater. The U.S. tests were about evenly split between Nevada and islands or atolls in the Pacific Ocean. In addition, three weapons were exploded over the Atlantic Ocean.

Following World War II, the United States began testing nuclear weapons in the Pacific Ocean. This became logistically complex and expensive, so some testing was moved to Nevada beginning in 1951 (NRC 1999f). After the Limited Test Ban Treaty was signed in 1963, the United States, Soviet Union, and Great Britain moved testing underground.[29] France continued atmospheric testing until 1974 and China until 1980. No nation has conducted an atmospheric test since 1980 (this includes nuclear tests conducted by India and Pakistan).

Most weapon tests were carried out in the Northern Hemisphere. The kiloton-yield fission bombs of World War II and the early Cold War deposited most of their radioactive debris near the detonation sites. Later megaton-yield thermonuclear weapons dispersed fallout on a global scale. Between 1945 and 1980, 440 megatons of yield were exploded in the atmosphere (UNSCEAR 2000). Forty percent of this (170 megatons) took place in just one year—1962.

Three main sources of radioactivity contributed to fallout (NCRP 1987a). The primary source was from byproducts of fission within the weapons' materials, such as uranium. The radionuclides produced were those in the middle range of atomic weights such as cesium-137 and strontium-90. According to the United Nations, 26 million curies of

Radiation Exposure of Servicemen

Approximately 382,000 U.S. servicemen, their families, and nearby civilians participated in and witnessed nuclear weapon detonations on islands or atolls in the Pacific Ocean and Nevada. On occasion, these personnel performed maneuvers, surveillance, and cleanup in and around the detonation site without adequate monitoring, protective clothing, respiratory devices, or health-care follow-up. Some were exposed to high levels of radiation. Some became ill. The National Association of Atomic Veterans (http://www.naav.com/), founded in 1979, helps these people obtain health care and financial assistance. President Bush signed the Radiation Exposure Compensation Act in 1990 (GAO 2001a). It established a trust fund and criteria for compensation of people exposed to radiation from these test sites. The act has been amended and broadened.

(28) The 219 U.S. explosions included 22 safety tests and 2 combat uses.

(29) There were many motivations for implementing the Limited Test Ban Treaty, not the least of which was public concerns, mounting in the mid-1950s, over escalating fallout (U.S. Congress 1959). Litigation was another. By the 1970s, hundreds of claims had been filed against the federal government (NRC 1999f).

cesium-137 and 17 million curies of strontium-90 were globally dispersed from all atmospheric tests (UNSCEAR 2000).[30]

The second source was neutron activation of metals in the weapon or in the environment surrounding the explosion. When a nuclear bomb detonated, neutrons were released. These were captured by nearby nonradioactive elements to produce large quantities of radioactive products such as tritium and carbon-14 or smaller amounts of radioactive iron, magnesium, or other metals.[31] How much was produced depended upon whether the bomb was detonated on the ground (surrounded by elements making up rocks), in the air (surrounded by air molecules), or in the ocean (surrounded by water molecules).

The third source was radionuclides in the weapon. Examples include uranium, plutonium, and tritium. An estimated 3.7 tons of plutonium were released into the atmosphere (UNSCEAR 2000). Eighty percent of this was plutonium-239.

Winds carried radionuclides aloft from the Nevada Test Site. Rainstorms deposited locally high concentrations in far-away communities in the Midwest and New England.

Different locations were used for aboveground testing, each contributing to unique local or global radionuclide distributions: land surface, towers, ocean barges, balloons, airplanes, and high-altitude rockets. Test locations, bomb designs, and meteorological conditions influenced whether radioactive debris entered the local, regional, or global environment.

Test designs and weather patterns could create radiation hot spots. Winds carried radionuclides aloft from the Nevada Test Site, and rainstorms deposited locally high concentrations in far-away communities of the Midwest and New England—in just a few days after detonation. Counties receiving some of the largest doses were in Colorado, Montana, Idaho, Utah, and South Dakota (NCI 1997). With the prevailing west-to-east winds, the West Coast and the Pacific Northwest generally experienced the lowest fallout from Nevada tests.

On average, certain radioactive metals, for example, cerium-144 or zinc-95, were deposited locally while others (for example, strontium-90, cesium-137, and iodine-131) were widely distributed. Highly mobile radionuclides, such as tritium and carbon-14, easily disperse globally.

The average drift time for radioactive debris lofted into the lower atmosphere was 1 month to 1 year (NCRP 1987a; UNSCEAR 2000). Materials lifted into the upper atmosphere had an average drift time of an additional 1 or 2 years. (Tritium and carbon-14 remained aloft and in the global hydrologic cycle for longer times.) These delays allowed some of the shorter-lived radionuclides to decay before coming into contact with humans.

---

(30) The 26 million curies of cesium-137 equals about 650 pounds. The 17 million curies of strontium-90 equals about 270 pounds (based upon Table 9 in UNSCEAR 2000).

(31) Most tritium was produced during the fusion process in thermonuclear weapons. It was released into the atmosphere as tritiated water. During the peak of atmospheric testing, the global concentration of tritium increased hundreds of times above natural levels (NCRP 1987a).

Radioactive fallout is spread around the globe and over generations.

During peak atmospheric testing in the early 1960s, the average radiation dose from fallout to a person living in the Northern Hemisphere was 12 millirem a year.

The present annual radiation dose from global fallout is 0.6 millirem per person in the Northern Hemisphere.

Today, radionuclides from these tests are found in rainfall, surface water, groundwater, plants, and people throughout the world. Higher concentrations are left in the environment near test sites. The small amount remaining, particularly carbon-14 with a half-life of 5715 years, will continue to recycle in the environment for several thousand years.

Fallout delivered an external and internal dose to humans. A large portion of the external dose came from the gamma emission of cesium-137. Beta emissions from both radioactive cesium and strontium were key contributors to internal dose. Over the years, health organizations have tracked the concentrations of tritium, carbon-14, plutonium-239, iodine-131, krypton-85, and iron-55 for estimating potential hazards.

The radiation dose received from fallout, plus artificial radionuclides present in the body, will continue for years. This includes exposure, especially internal exposure, to "individuals not yet born at the time of the explosions" (UNSCEAR 1964).

Thus, fallout is spread over generations. The United Nations Scientific Committee on the Effects of Atomic Radiation (2000) estimated a total average radiation dose of 358 millirem over many generations to individuals living in the Northern Hemisphere.[32] The cumulative radiation dose for a person living there between the years of 1945 and 1999 was 108 millirem. (This averages to 2.5 millirem per person per year though most exposure took place in the 1950s and 1960s). The remaining 250 millirem are spread across the future.

During peak atmospheric testing in the early 1960s, the average radiation dose from fallout to a person living in the Northern Hemisphere was 12 millirem a year (UNSCEAR 2000). According to a 1964 study (Wilson 1964), a typical Richland resident received about 14 millirem of whole-body radiation from fallout during 1963. Most was from strontium-90. However, the study's author, R.H. Wilson, considered this value as "unrealistically high." A year later, another report dropped the whole-body radiation exposure to between 2 and 3 millirem (Soldat and Essig 1966).

Fallout is no longer a significant source of radiation exposure. Public and scientific interests have waned. Radionuclides that still remain "have tended to become less available for biological uptake" (NCRP 1987a). The present annual radiation dose from global fallout is 0.6 millirem per person in the Northern Hemisphere. This is 0.2% of the 300-millirem dose an average person in the United States receives each year from natural sources.

The Hanford monitoring programs demonstrate another example of this lessening of fallout because of radioactive decay. In 1970, the average concentration of tritium in the Columbia River *upstream* of Hanford was 840 picocuries per liter (Corley 1973a). That tritium came from fallout—not from Hanford. By the year 2000, the concentration upstream dropped to about 35 picocuries per liter (Patton 2001).[33] The downward trend will continue, provided there are no new atmospheric explosions or use of nuclear weapons, until a natural background level of about 15 picocuries per liter is reached.

(32) This and other numbers are average values. Specific radiation doses will vary from location to location, between young and old, between cultures, and over time.

(33) These same two reports note that the average concentration of fallout-derived strontium-90 (half-life of 29 years) in the river upstream of Hanford has dropped from 0.4 picocurie per liter to 0.07 picocurie per liter over the same time period.

4.6 Health Studies

"In the more technical fields, chemists, engineers, and technicians in many fields had to learn to work with new technics [sic] and precautions for the good of themselves and others. Other workers who did not need to fully understand the complexities of the hazards were told sufficiently about them however, to permit them to work safely...At no time have we seen signs of group fear or alarm in contemplating or doing an assigned job. To be sure there have been a few individuals, as there are in any plant or army, who cannot assimilate the true picture of hazard conditions, and who have needed reassurances and further guidance in understanding the conditions as they are and not as they have read elsewhere or imagined them to be."

—S.T. Cantril and H.M. Parker,
The Status of Health and Protection at Hanford Engineer Works (1945, p. 4)

Hanford was a large industrial site producing metals unique from many other industries. Some employees worked in potentially hazardous conditions, conditions not acceptable today. Radioactive and chemical contaminants were also released into the environment exposing people who lived downwind and downstream. As such, it's not surprising that some people were harmed. For example, studies suggest an increase in certain diseases, such as beryllium-related illnesses, multiple myeloma, and possibly leukemia, among certain workers in the nuclear weapons complex.

Yet, the cause of certain illnesses may be verifiable only in exceptional cases, given the current limited understanding about how cancer develops[34] (GAO 2000f) and the sometimes poor record keeping during the early years of the weapons complex.

Health data collected through 1977 suggested cancer of the pancreas was correlated with radiation dose—higher risk with higher dose (Gilbert and Marks 1980). In later analyses, no meaningful correlation was found (Gilbert 1989). Similarly, a correlation between radiation dose and Hodgkin's disease was also called into question (NRC 1990a). The continued appearance of multiple myeloma[35] deaths in workers who received high doses of radiation cannot be dismissed (Gilbert et al. 1993b). In that study, Gilbert et al. wrote that the positive correlation between radiation values and an increase in cancer of the pancreas and Hodgkin's disease were "probably spurious." However, when results of this same study were summarized by the U.S. Department of Health and Human Services (NIOSH 2002), this qualification went unmentioned.[36] Linking any cause and effect for cancers will take years, if ever, to unambiguously explain.

Studies suggest an increase in certain diseases, such as beryllium-related illnesses, multiple myeloma, and possibly leukemia, among certain workers in the nuclear weapons complex.

The cause of some illnesses may be verifiable only in exceptional cases.

(34) Cancer is not the only health problem potentially associated with unacceptably high levels of radioactive or chemical contaminants.
(35) Multiple myeloma is a cancer of the blood-forming tissues in the bone marrow.
(36) It's little wonder the public is sometimes confused about the potential health effects from exposure to contaminants. Counter claims between groups are common even when referencing the same health study.

Gilbert et al. (1993b) also reported that Hanford workers continue to "show a strong healthy worker effect with death rates from most causes substantially below those of the general US population." This is attributed to workers undergoing preemployment examinations, having good jobs, better diets, and more access to medical services than the general population. This effect was also noticed after the first years of Hanford construction and operation (Cantril and Parker 1945).

Hanford workers continue to "show a strong healthy worker effect with death rates from most causes substantially below those of the general US population" (Gilbert et al. 1993b).

In some circles, controversy still surrounds the termination of an epidemiological study of radiation workers headed by University of Pittsburgh radiation researcher Thomas Mancuso in the mid-1960s (D'Antonio 1993). Mortality rates were assessed for employees principally from Hanford and Oak Ridge. Mancuso was reported to have uncovered excess cancer risk for Hanford workers at a time when acknowledging such possibilities was not popular.

In some circles, controversy still surrounds the termination of an epidemiological study of radiation workers headed by University of Pittsburgh radiation researcher Thomas Mancuso in the mid-1960s (D'Antonio 1993).

According to radiation toxicologist J. Newell Stannard (1988, p. 1472), it was difficult to ascertain whether or not internal radiation doses were factored into the Mancuso study results because most measurements of radiation exposure used in the study were made using badges and dosimeters detecting only external radiation sources. Charges of "bias and suppression of information" were made by the study's sponsors. In 1977, Mancuso terminated his involvement. The potential benefit from an objective analysis of the study's results was lost in the resulting political foray.

Electronic access to Hanford health studies is found at several Web sites. Health agencies in the states of Washington, Oregon, and Idaho, along with Native American nations, once maintained the Hanford Health Information Network. Updates stopped in May 2000. Their work is still accessible at http://www.doh.wa.gov/hanford/. The Hanford Health Studies Information (http://www.hanford.gov/safety/healthstudies/studies.html) Web site is maintained by DOE. The U.S. Department of Health and Human Services maintains the Energy Related Health Research Program (http://www.cdc.gov/niosh/2001-133.html). It also accesses health studies about Hanford and other DOE sites. Health-related radiation research findings for major DOE sites are also found at the Centers for Disease Control and Prevention site (http://www.cdc.gov/nceh/radiation/). Readers are encouraged to search these and other information sources.

Examples of seven health studies are summarized in this section. Most involve Hanford workers.

Hanford Thyroid Disease Study: This study investigated whether thyroid disease increased among people, particularly children, exposed to atmospheric releases of radioactive iodine from Hanford reprocessing plants. Congress mandated the study after DOE declassified documents revealing that a large quantity of radioactive material, including iodine, was released into the soil, air, and Columbia River. Preliminary results for the Hanford Thyroid Disease Study were published in 1999 (Fred Hutchinson Cancer Research Center 1999). Final results were published in 2002 (CDC and Fred Hutchinson Cancer Research Center 2002).

According to the study, children were especially at risk because, on average, radiation exposures to their thyroids were much higher when compared to adults. Children's thyroids are smaller, and they consume more milk. Milk and other dairy products were of concern because they came from cows that grazed where Hanford-originated radioactive iodine had fallen.

Preliminary health findings, discussed in Section 4.3, suggested there was no evidence linking radiation exposure to the rate of thyroid disease found in the study population. Nonetheless, concerns were expressed about those findings (Nussbaum and Grossman 1999; DOE 2000d). One was that the study only examined diseases of the thyroid rather than other cancers or aliments. An informal study conducted by the Oregon Chapter of Physicians for Social Responsibility suggested greater-than-normal cases of hypothyroidism[37] than reported by the Hanford Thyroid Disease Study (Cary 1999a). Another concern was that the damage claimed by some organizations and members of the public was unsupported by health studies reported by the state of Washington, county health departments, the American Cancer Society, and the Fred Hutchinson Cancer Research Center (Woodcock 2000).

The National Research Council believed that investigators initially overstated the certainty of their thyroid disease study.

After reviewing the thyroid study, the National Research Council noted that investigators completed a carefully designed and executed study (NRC 2000d). However, the council believed that investigators overstated the certainty of their results. "Given the inherent imprecision in exposure estimates and the effect of other statistical issues, the absence of any observable radiation effects is not proof that there is none. It means that the iodine-131 exposure did not have large effects" (NRC 2000d). In other words, the National Research Council stated that one could not rule out individual tragedies caused by Hanford releases though widespread health problems were not evident.

Three years later, final results of the Hanford Thyroid Disease Study were published. As revealed in the preliminary study, the Centers for Disease Control and Prevention did not find any increases in thyroid diseases in the populations living downwind of Hanford compared to other populations (CDC and Fred Hutchinson Cancer Center 2002). If there were any effects, they were too small to be detected.

Exposure of People to Iodine-131 from Atmospheric Tests in Nevada: Between 1951 and 1962, 86 aboveground nuclear weapons tests were conducted in Nevada (UNSCEAR 2000). An estimated 150 million curies of iodine-131 were released (NCI 1997). Radiation dose was highest for those living immediately downwind of the test site. Rain and wind deposited fallout across the country. In 1983, Public Law 97-414, in part, directed the Secretary of Human Health and Human Services to research and develop estimates of the iodine-131 doses to people living in the United States during the 1950s.

The National Cancer Institute released its report 14 years later (NCI 1997). Publication followed news articles that the report was unpublished though completed 6 years earlier. The best estimate for the average cumulative radiation dose from iodine-131 fallout to an individual's thyroid was 2 rad (2000 millirem). Uncertainty was estimated to be a factor of 2. Thus, the per person dose may have ranged between 1 and 4 rad (1000 to 4000 millirem). Some residents living in states east and northeast of Nevada received the highest dose to their thyroids: 9 to 16 rad (9000 to 16,000 millirem). Children, ages 3 months to 5 years, "exceeded the average per capita thyroid dose…by a factor of about 3 to 7" (NCI 1997). An estimated cumulative dose of 10 rad (10,000 millirem) to the thyroid was received by a large part of the population that was under the age of 20. Perhaps "tens of thousands" of people received cumulative radiation doses of iodine greater than the exposure limit of "5 rem" (5000 millirem) (NRC 1999f, p. x).

(37) Hypothyroidism is a condition in which the thyroid gland fails to produce enough thyroid hormone to regulate body metabolism.

The correlation between radioactive iodine concentrations in animal thyroids and atmospheric tests was accidentally discovered in the mid-1950s through studies of cattle thyroids (Van Middlesworth 1954).

According to the National Cancer Institute, exposure to these iodine releases may cause 11,300 to 212,000 excess lifetime cases of thyroid cancer. The average is 49,000 cases. In an independent review of the study, the National Research Council (1999f) commended the researchers for performing a "careful, detailed, and responsible effort" while taking on a "very difficult task that depended on limited data of uncertain reliability and validity."[38] Questions were raised including the validity of the linear dose-response model used to estimate cancer risks. The council concluded that given the limitations "the analyses suggest that the excess of cancer cases is far below the highest value in the estimated range provided by NCI [National Cancer Institute] and is probably in the lower part of the range" (NRC 1999f). Additional analysis suggested that about "45 percent of iodine-131 related thyroid cancers have already appeared…the chance of a significant exposure is highest for those who were young children at the time and who routinely drank milk from backyard cows and in particular, goats."[39] Regretfully, available data do not allow identifying individuals at risk, detecting increases in thyroid cancer risk related to fallout patterns across specific counties, or distinguishing between thyroid cancer induced by iodine-131 and naturally occurring thyroid cancer.

According to the National Cancer Institute, the public's exposure to iodine releases from the atmospheric tests of nuclear weapons may cause 11,300 to 212,000 excess lifetime cases of thyroid cancer. The National Research Council questioned the validity of the linear dose-response model used to estimate those risks.

Health Study of Female Nuclear Workers: A study of women who worked at Hanford and 11 other DOE sites found fewer cancer-linked deaths than expected. However, it also uncovered more cancer among women who experienced the highest levels of radiation exposure (Wilkinson et al. 2000).

The study was sponsored by the National Institute for Occupational Safety and Health and headed by the University of Buffalo, State University of New York. It examined the deaths among 67,976 female nuclear weapons employees who worked at 12 DOE sites before January 1980. The number of deaths that occurred was compared with the deaths expected in the U.S. female population. The study attempted to determine if there was a relationship between exposure to radiation and deaths from certain diseases.

For most causes of death, including cancers related to ionizing radiation, fewer female workers died than expected based on the U.S. female population. At all sites, the number of deaths was either similar to or lower than expected.

Additional analyses of 21,440 female workers monitored for external radiation were conducted to explore the relationship between radiation and deaths from specific causes. These found

An increase in all cancer-related deaths was found among female workers who were exposed to the highest levels of external radiation.

- An increase in leukemia-related deaths among female workers who were exposed to higher levels of external radiation.
- A possible increase for all cancers among female workers who were exposed to the highest levels of external radiation.

The study had limitations. The fact that fewer deaths than expected were found might suggest problems in tracking deaths. Also, some DOE sites did a better job than others at monitoring radiation exposures, especially in the early years of plant operations. Other potentially important factors like lifestyle (such as smoking), radiation doses from medical procedures, age, and additional workplace exposures could not be evaluated.

(38) Radiation measurements were collected at fewer than 100 sites across the United States.
(39) Goat milk concentrates iodine-131 more than cow milk.

Study on Multiple Myeloma Mortality in Nuclear Industry Workers: Multiple myeloma (cancer of the blood-forming tissue in the bone marrow) primarily affects older people and is often fatal. A study of multiple myeloma mortality for workers at four DOE sites was released in 1998 (Wing 1998). The sites studied were Hanford, Oak Ridge (Tennessee), Los Alamos National Laboratory (New Mexico), and Savannah River Site (South Carolina). The study was conducted by researchers at the University of North Carolina and was funded by the National Institute of Occupational Safety and Health under a grant from DOE. It attempted to determine whether workers who had been exposed to a variety of chemical and physical agents (solvents, metals, welding fumes, asbestos, and radiation) might have the disease. The study concluded that workers 45 years of age or older were associated with an average 7% per 1 rem (1000 millirem) increased risk of multiple myeloma compared to non-exposed workers. The odds for developing the disease increased for those whom had a cumulative radiation exposure of 5 rem (5000 millirem) or greater compared to workers exposed to less than 1 rem.

Among the 31 other types of cancer studied, a positive association was observed only for multiple myeloma.

Studies of Cancer Mortality from Low Radiation Exposure Among Nuclear Industry Workers: Studies of the mortality among nuclear industry workers in the United States, United Kingdom, and Canada were undertaken to assess the potential carcinogenic effects of low-level exposure to external, predominantly gamma, radiation (Cardis et al. 1995). The studies examined mortality data on 95,673 workers—85% of them male. No direct link was found between radiation dose and mortality from all causes of cancer. Death from leukemia, excluding chronic lymphocytic leukemia,[40] was associated with higher cumulative radiation doses. Leukemia is the cause of death most consistently related to radiation dose in studies of atomic bomb survivors and others exposed to high radiation doses. Among the 31 other types of cancer studied, a positive association was observed only for multiple myeloma.

Mortality data on workers were analyzed to examine health risks resulting from protracted low-dose radiation exposure. The occurrence of cancers seemed to be a matter of chance.

Study of Worker Mortality Data for Three DOE Nuclear Weapons Sites: Mortality data on workers at Hanford, Oak Ridge, and the Rocky Flats weapons plant (Colorado) were analyzed to examine health risks resulting from protracted low-dose radiation exposure (Gilbert et al. 1993a). For leukemia, no increased risk was found. Of 24 cancer types tested, 12 appeared correlated with increased radiation dose, and 12 appeared to be uncorrelated. This is what would be expected by chance. Cancer of the esophagus, cancer of the larynx, and Hodgkin's disease showed statistically significant correlations with increased radiation dose; however, further research suggested cancer was not related to nuclear weapons work. Evidence of increased risk for all cancers at both Hanford and Oak Ridge was found as workers aged (75 years and older).

Chronic Beryllium Disease Prevention Program: Beryllium—used primarily in the production of nuclear fuel rods—was recognized as a health hazard within the nuclear weapons complex in the 1940s (Williams 1948, p. 9). In 1998, DOE established the Chronic Beryllium Disease Prevention Program, involving exposure screening and medical surveillance for employees who worked around beryllium metal. Beryllium was used at Hanford from about 1960 until 1986. Research at Rocky Flats, Colorado, indicated the potential for some workers to develop acute or chronic beryllium disease.

(40) Chronic lymphocytic leukemia is a disease where too many infection-fighting white blood cells called lymphocytes are found in the body. Lymphocytes are made in the bone marrow and by other organs of the lymph system. This form of leukemia can cause suppression of the immune system, failure of the bone marrow, and infiltration of malignant cells into organs.

U.S. Department of Energy 26989-23

Forming Uranium Fuel. Billets of uranium metal were preheated and extruded into cylindrical-shaped fuel elements used in Hanford reactors. This 1961 photograph shows a plunger pushing uranium into an extrusion press. Flames (left of front worker) shoot from the press as lubricants are ignited from the heat of metal compression. Workers wore protective clothing appropriate to the industrial standards of those days.

Acute beryllium disease occurs from exposure to very high levels of airborne beryllium dust particles. This disease is a form of chemical pneumonia that occurs soon after exposure. Chronic beryllium disease is an allergic condition to beryllium dust and fumes in which tissues of the lungs become inflamed after exposure.

In 1999, the Clinton Administration proposed legislation to establish a program to compensate current and former nuclear weapons site workers who became ill from beryllium exposure. The bill also directed DOE to conduct a pilot program at Oak Ridge, Tennessee, to examine the relationship between workplace exposure to radiation and hazardous materials to occupational illness. Secretary of Energy Bill Richardson noted that this legislation signaled DOE's "changing from an agency that opposed worker health claims to one that is actually trying to help resolve those claims" (Wicker and Sherwood 1999). Proof rests in long-term agency conduct.

In 2000, Congress enacted the Energy Employees Occupational Illness Compensation Program Act (Public Law 106-398). The law became effective in July 2001. The act provides compensation for workers, or their survivors, who have occupational illnesses from exposure to the unique hazards associated with building and operating the nation's nuclear weapons factories. This includes radiation-induced cancers, asbestosis (exposure to fire-retardant and insulation materials), chronic beryllium disease, and silicosis (exposure to rock dust from underground tunneling at bomb test sites). An estimated 3000 to 4000 former workers are expected to be eligible for compensation (Rizzo 2000). Depending upon the cause of illness, lump sum payments of between $50,000 and $150,000 plus payment of medical expenses from the date of claim are available. The compensation program is expected to cost $1.4 billion over the first 10 years. Another $460 million would be spent on medical care and other benefits to sick uranium miners. As of 2002, more than 24,000 workers nationwide had filed claims (Stiffler 2002). Some 2700 of these had been approved and $172 million in compensations paid.

Beryllium was recognized as health hazard within the nuclear weapons complex in the 1940s. It was used at Hanford from about 1960 until 1986.

Chapter 5

Waste and Nuclear Materials

"The experiences at Savannah River and Hanford point out means by which disposal may be accomplished, but not without the assumption of an extensive and exhaustive research program, and the gravest responsibilities associated with that disposal."

—Randy Brown, Hanford geologist
(Brown et al. 1958, p. 100)

Plutonium production created large amounts of radioactive and chemical waste. Some was stored. Some was released. Why? Because, the entire Hanford Site and nearby environs were used for waste disposal or dilution.

Hanford's approach to managing contaminant releases was to control radiation exposures within "permissible limits" and to keep waste-handling expenses to a minimum (Pearce 1959).

"The site has accordingly been used for nearly ten years as a full-scale 'test' site for the disposal of plant process wastes" (Parker 1954a).

Site operations focused on controlling chemical and radiation exposures where citizens of nearby communities would encounter the contaminants—miles from the point of release. This allowed for the "economic use" of the entire Hanford Site—and sometimes the surrounding region—for contaminant absorption, filtration, and dilution rather than "basing waste disposal practices on restrictions at the immediate point of discharge" (Pearce 1959). In fact, even the ability of algae and other plants, animals, and microorganisms to accumulate radionuclides in their bodies was known to dilute radioactive elements, so concentrations downstream of Hanford would be lowered (Berry and Cline 1950).

The central plateau of Hanford was considered ideal (a "fortunate situation") for waste disposal because it was underlain by 200 feet of nearly dry sediment that acted like a "sponge" to physically retain contaminants (Parker 1959a). While the deliberate contamination of the groundwater was "normally inadvisable" (Parker 1954a), it was permissible because a large chunk of real estate was available to meet self-imposed limits on waste disposal (see Chapter 2).[1]

Waste or Nuclear Material Source

100 Area:
Reactor Operations

200 Area:
Nuclear Fuel Reprocessing and Plutonium Recovery

300 Area:
Nuclear Fuel Fabrication and Research

Type of Storage or Disposal

Highly Radioactive Waste → Stored in Tanks (Some Leakage to Soil)

Solid Waste → Buried in Landfills and Stored in Surface Facilities

Water Containing Low to Intermediate Contaminated Liquids → Released into the Columbia River, Ponds, Trenches, and Cribs

Reactor Cooling Water → Released to the Columbia River

Gaseous Effluent → Released to the Atmosphere

Nuclear Materials → Stored Above Ground

741fb.40

Waste Source and Disposal. Liquid and solid wastes were stored or released based on their sources, potential hazards, and physical properties.

(1) During the early years of the nuclear weapons complex, externally imposed criteria did not exist. Waste management was in its infancy.

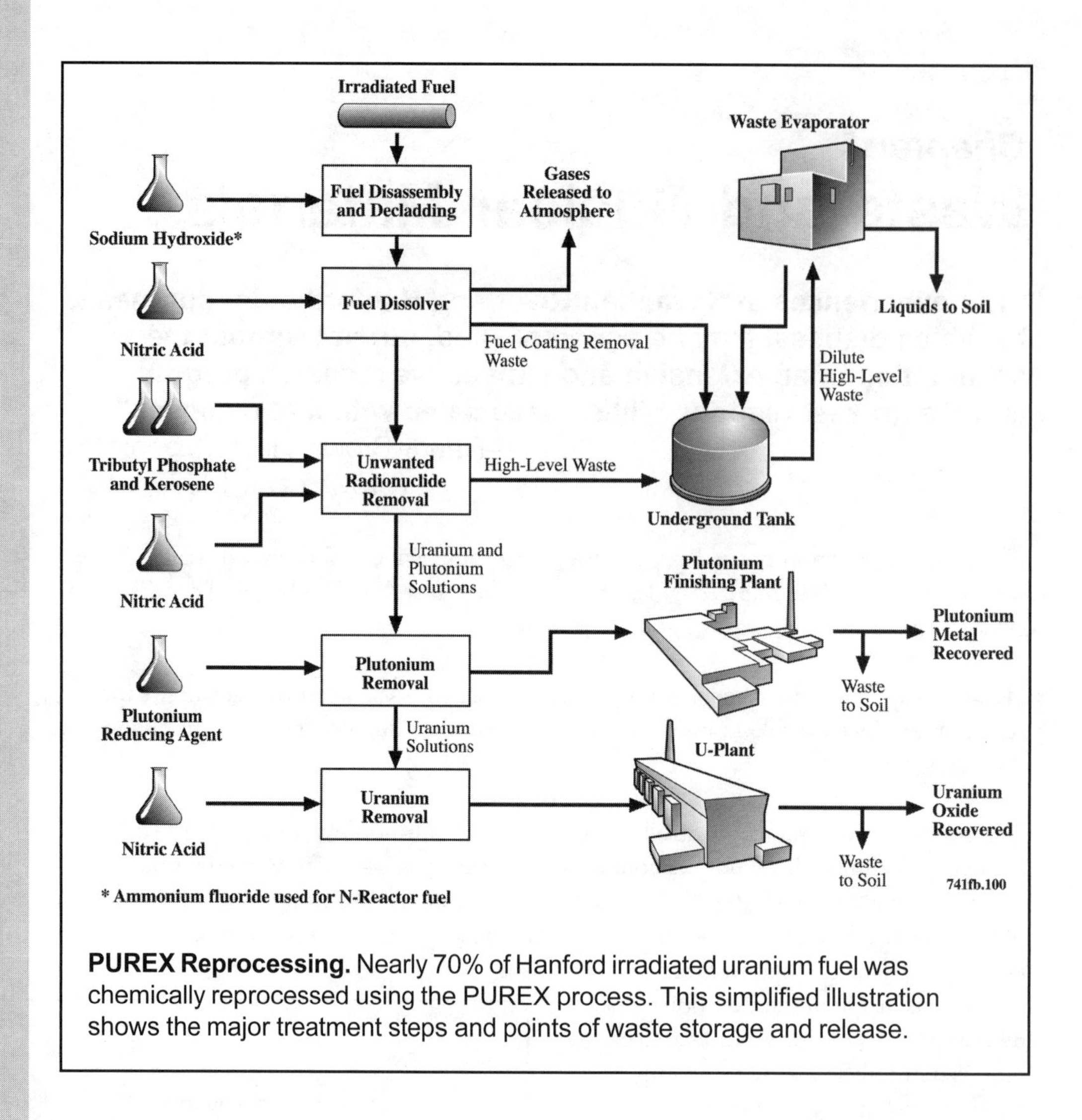

PUREX Reprocessing. Nearly 70% of Hanford irradiated uranium fuel was chemically reprocessed using the PUREX process. This simplified illustration shows the major treatment steps and points of waste storage and release.

Highly radioactive liquids first went into hastily built underground storage tanks.[2] Later constructed tanks were of higher quality. Less-contaminated liquids were discharged to the soil. Gases were released to the atmosphere. Most solid waste was buried in shallow trenches. Nuclear materials, such as cesium- and strontium-filled capsules and spent fuel, remained inside buildings. A record of problems encountered from 1946 to 1968 in operating the reprocessing plants and disposing of waste is summarized by McCullugh and Cartmell (1968).

Plutonium production dominated Hanford decisions.

While conscientious efforts were made to protect workers, the Columbia River, and the public, plutonium production dominated Hanford decisions. In the early years, contaminant releases were controlled to the extent it did not interfere with plutonium production. For example, though longer cooling times for spent fuel were known to dramatically reduce radionuclide releases to the air, times were not increased even after

(2) Underground tanks are used to store the most radioactive and dangerous waste generated from the chemical removal of radionuclides, such as plutonium, from the nuclear reactor fuel rods. Today, 177 tanks store about 53 million gallons of waste (see Section 5.3).

rainwater was found to contain radionuclide concentrations "up to 3 times the tolerable value" and ground contamination was discovered in Walla Walla, Washington—50 miles downwind of Hanford (Parker 1945a).

However, Hanford's top health physicist Herbert Parker[3] noted, "there is no hazard in these areas at this time, and that generally adequate observations are being made to follow the pattern of such contamination" (Parker 1945a). In the hierarchy of allocating funds, reactor operations dominated and reprocessing came second. Waste management was a distant third. In the late 1940s, radiation protection programs at the nuclear weapons sites, overseen by the Atomic Energy Commission, were consuming only 3% of operating costs (Williams 1948, p. 82).

In the late 1940s, radiation protection programs at nuclear weapons sites were consuming only 3% of operating costs (Williams 1948).

Was the approach to managing Hanford waste unusual in its day? No. Industries and cities routinely used the air and the closest river as an environmental sink for anything unwanted. Until the mid-twentieth century, waste treatment was uncommon. So, even when contaminated groundwater reached the Columbia River or radionuclide levels substantially increased in the river, dilution was relied upon to return water quality to acceptable levels. The philosophy of environmental protection guidelines, independent of facility operations, was years away for American industry.

"At the Hanford Engineer Works, the ground disposal of low- and mid-level atomic wastes and the underground tank storage of high-level wastes were first conceived as temporary, wartime expedients. However, increasing volumes of liquid process wastes generated during the ensuing twenty years presented new and unexpected difficulties to site scientists and officials. Beginning in the later 1950s, public concern over nuclear waste increased and became vocal, providing one of the first challenges to AEC [Atomic Energy Commission] hegemony" (Gerber 1997, p. 169).

There are many unknowns about the volumes and characteristics of the contaminants generated and released at Hanford as well as other nuclear sites. Two key reasons are that this type of information was not originally collected, and much of the original data—not required for business records or reprocessing operations—was destroyed (Mercier et al. 1981).

The philosophy of independent environmental protection guidelines was years away for American industry.

Waste information in the following sections relies upon many formal and informal sources having a range of reliabilities. Inventory records for some sites are limited and contradictory (DOE 2001). There are many unknowns. Numbers are best estimates and will change as an improved understanding is gained of Hanford's contaminant inventory. However, deciphering this entire inventory is less important than pinpointing, or at least bounding, those portions posing the greatest potential health risk.

(3) For more than 50 years, Herbert Parker was a leading force in radiological physics, particularly at the Hanford Site. In addition to his pioneering work on developing radiological units with clear physical and biological bases, he made important contributions to the development of scientifically based radiation protection standards.

5.1 Types of Waste

Most Hanford waste and nuclear material exist inside buildings, water-filled basins, or underground tanks. The remainder lies uncontained in the soil and groundwater.

Today, Hanford is home to some 1200 sites where waste or nuclear material are stored or were released (Ecology et al. 1998). Approximately 60% (700 sites) of these sites are in the 200 Areas (DOE 2000g). Most of the remaining are near the reactors (400 sites in the 100 Area) and near the former fuel rod manufacturing and research facilities (100 sites in the 300 Area). Another document suggests there are 1700 waste sites at Hanford (DOE 2002d).

All sites contain about 390 million curies of radioactive materials remaining from the production of plutonium. Most radioactivity is emitted from cesium-137 and strontium-90. Each year, radioactive decay decreases the number of curies on the Hanford Site by about 10 million curies. As a result and as of the year 2002, Hanford contains 130 million fewer curies than when the cleanup agreement[4] for the site was signed in 1989.

Waste Source	Radioactivity	Chemicals	Volume	Volume as Depth in Football Field
Tank Waste	195 million curies	240,000 tons	53 million gallons	150 feet
Solid Waste	6 million curies	70,000 tons	25 million cubic feet	500 feet
Soil and Groundwater	2 million curies	100,000 to 300,000 tons	35 billion cubic feet	100 miles
Facilities	1 million curies	--	200 million cubic feet	4000 feet
Nuclear Material	185 million curies	--	25,000 cubic feet	1/2 inch

741fb.12

Waste and Nuclear Material Inventory. The waste and nuclear material inventory remaining from the plutonium production mission contains 390 million curies of radioactivity and 400,000 to 600,000 tons of chemicals. There are significant unknowns in this inventory, especially for specific radionuclides and their chemical forms.

(4) This agreement is the *Hanford Federal Facility Agreement and Consent Order*, known informally as the Tri-Party Agreement (see Section 7.5).

Hanford also contains another 25 million curies not originating from the production of plutonium at the site (see Section 5.9).

The radioactivity at Hanford makes up about 40% of the 1 billion curies of radioactivity existing at all of the U.S. Department of Energy's nuclear sites. A general description of waste and materials at these other sites is in *Linking Legacies: Connecting the Cold War Nuclear Weapons Production Processes to Their Environmental Consequences* (DOE 1997a).

Each year, radioactive decay decreases the number of curies on the Hanford Site by about 10 million curies.

Hanford waste also contains large volumes of chemicals (such as sodium nitrate and carbon tetrachloride) and nonradioactive metals such as chromium. About half (240,000 tons) of all this chemical waste lies in underground storage tanks (Gephart and Lundgren 1998). Another 100,000 to 300,000 tons were released into the soil and groundwater—most of this beneath the 200 Area (Waite 1991; Wodrich 1991). Chemicals and metals are also contained in buried and stored solid waste.

5.2 Facilities

The nine nuclear reactors and five reprocessing plants at Hanford are the largest engineered structures remaining from the days of plutonium production.

One-third of Hanford's 1500 facilities are contaminated. The reactors and reprocessing plants contain the bulk of these residues.

The total radioactivity remaining inside all Hanford's facilities is uncertain, though it's likely about 1 million curies.[5] Most is found inside air filters, especially at B Plant, which recovered radioactive strontium and cesium from tank waste. B Plant filters contain about 455,000 curies (Campbell 1999). On the other hand, filters at the REDOX Plant hold 8300 curies while those at U Plant have 7600 curies (H&R Technical Associates and Kerr 2000; Kerr et al. 2000). An additional 150,000 curies reside in the canyons and galleries of B Plant, REDOX Plant, U Plant, and PUREX Plant; of this, 85% is found inside B Plant (Campbell 1999; Dodd 1999; H&R Technical Associates and Kerr 2000; Kerr et al. 2000).

Nearly half of the 1 million curies inside Hanford facilities is found in B Plant (Campbell 1999).

Documents suggest 250,000 curies of radioactivity remain inside Hanford's nine reactors as of the late 1990s. Most of this is tritium and carbon-14. Each of the older reactors (B, C, D, DR, H, and F Reactors) contains 7000 to 12,000 curies (Rodovsky et al. 1996; Bond and Rodovsky 1998; Kerr 1998; Koster 1998; Rodovsky and Bond 1998). The jumbo KE and KW Reactors each hold 25,000 to 27,000 curies (Kerr 1998). N Reactor has the highest amount, 131,000 curies, because it most recently operated and less time has elapsed for radioactive decay. N Reactor was shut down in 1987 compared to 1964 to 1971 for the other eight reactors.

(5) This number may change as more is learned about residual contamination, especially that found in emission control filters and waste evaporators.

Hanford has 177 underground tanks arranged in 18 tank farms.

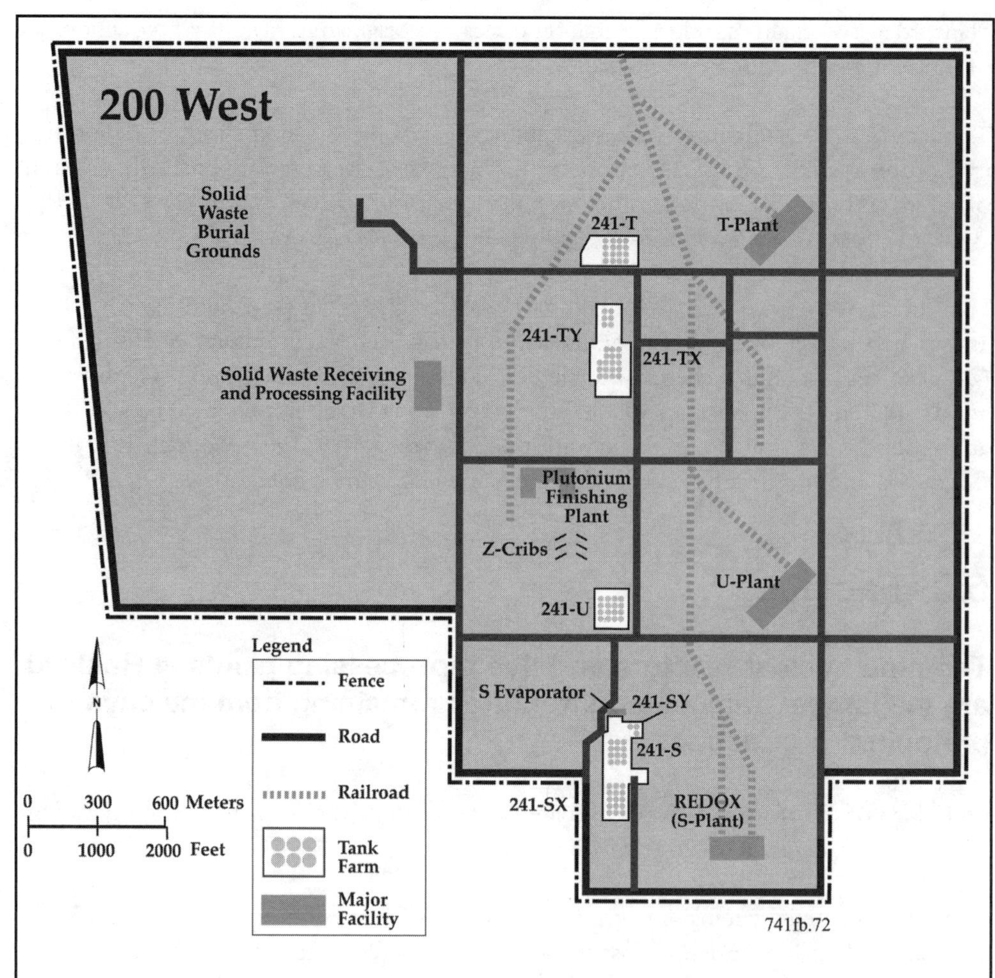

200 West Area. The 200 West Area contains seven tank farms, four major facilities, solid waste burial grounds, and numerous liquid waste disposal sites. Some key facilities are shown.

5.3 Tanks

"*Representative Price*. What do you believe the life of the tank to be? *Mr. Parker*. I will answer the question by saying that for a longer time than any operation heretofore contemplated by man, these wastes will have to remain isolated from the environment."

—Herbert Parker, congressional testimony before the Special Committee on Radiation (1959a, p. 165)

Tanks were an interim method for waste disposal pending more permanent solutions (Pilkey et al. 1958).

The storage of waste in underground tanks was considered an "interim method of waste disposal" pending development of "more permanent methods" (Pilkey et al. 1958). Tanks were considered a "cheap" alternative.

The "sensible management of these wastes" amounts to waiting until such time that a method of "ultimate disposal" is developed, a reliable method of immobilizing the waste in place is created, or a "profitable use" of these materials is developed (Parker 1959b).

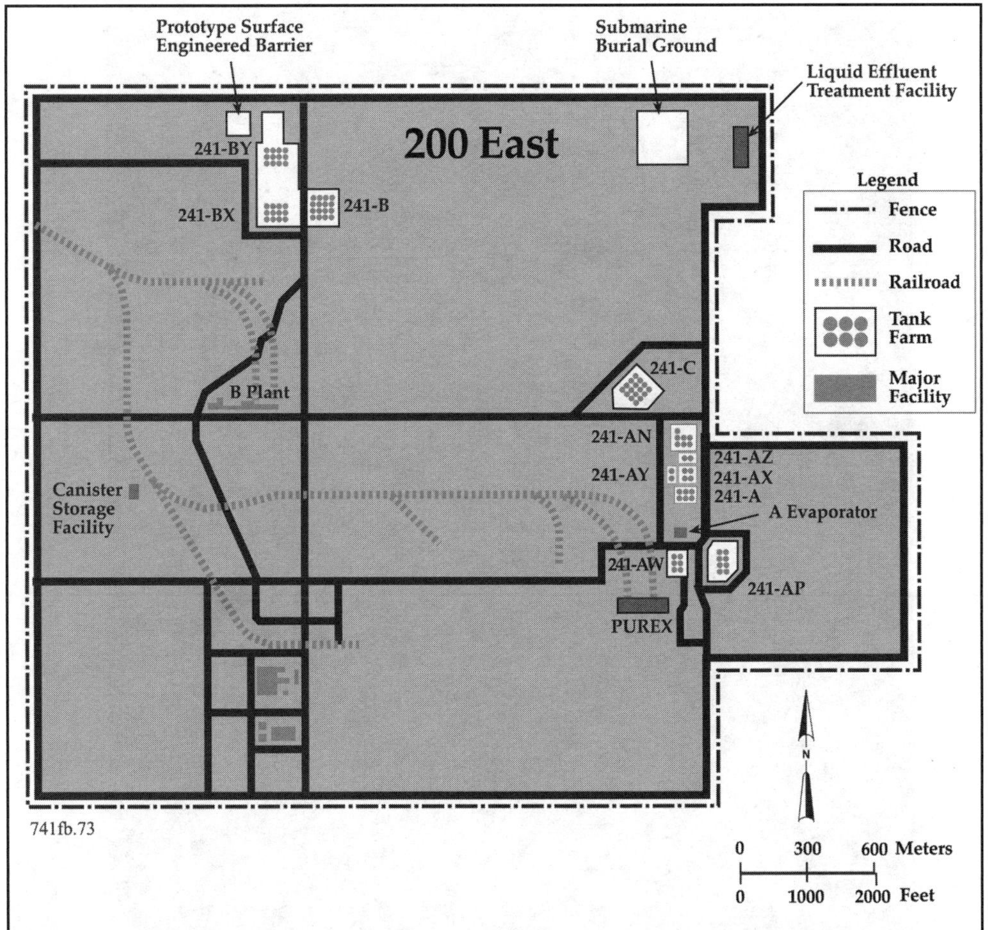

200 East Area. The 200 East Area contains 11 tank farms, 2 reprocessing facilities, solid waste burial grounds, liquid waste disposal sites, and a liquid effluent treatment plant. Some key facilities are shown.

Hanford generated 525 million gallons of tank waste.

Between 1944 and 1988, 525 million gallons of radioactive waste were generated in the reprocessing plants and piped into underground tanks (Agnew 1997). By the year 2002, this volume was reduced to 53 to 54 million gallons by liquid evaporation, pouring waste into the ground, treatment, and leakage (Hanlon 2001, 2002). Today, there is enough waste in the tanks to fill 2600 railroad tanker cars—creating a train 26 miles long.

This waste is stored in 177 single- and double-shell carbon-steel tanks. These are cylindrical shaped concrete structures with one or two inner carbon steel liners. They range in volume from 55,000 to 1.16 million gallons. Their tops are buried about 10 feet underground. More than 99% of the 195 million curies of radioactivity in this tank waste comes from two radionuclides, cesium-137 and strontium-90,

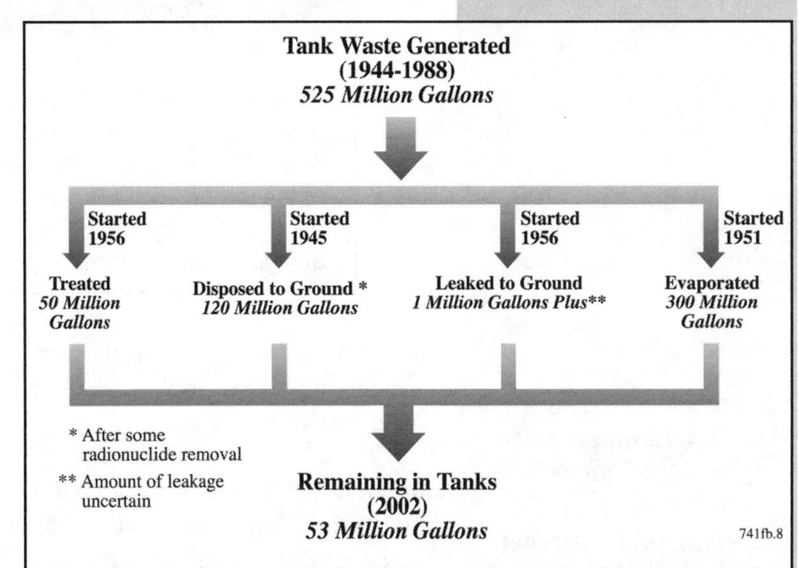

Tank Waste History. Hanford tanks contain 53 million gallons of chemical and radioactive waste. This is 10% of the tank waste generated onsite. The other 90% was treated, discharged to the ground, leaked to the soil, or evaporated.

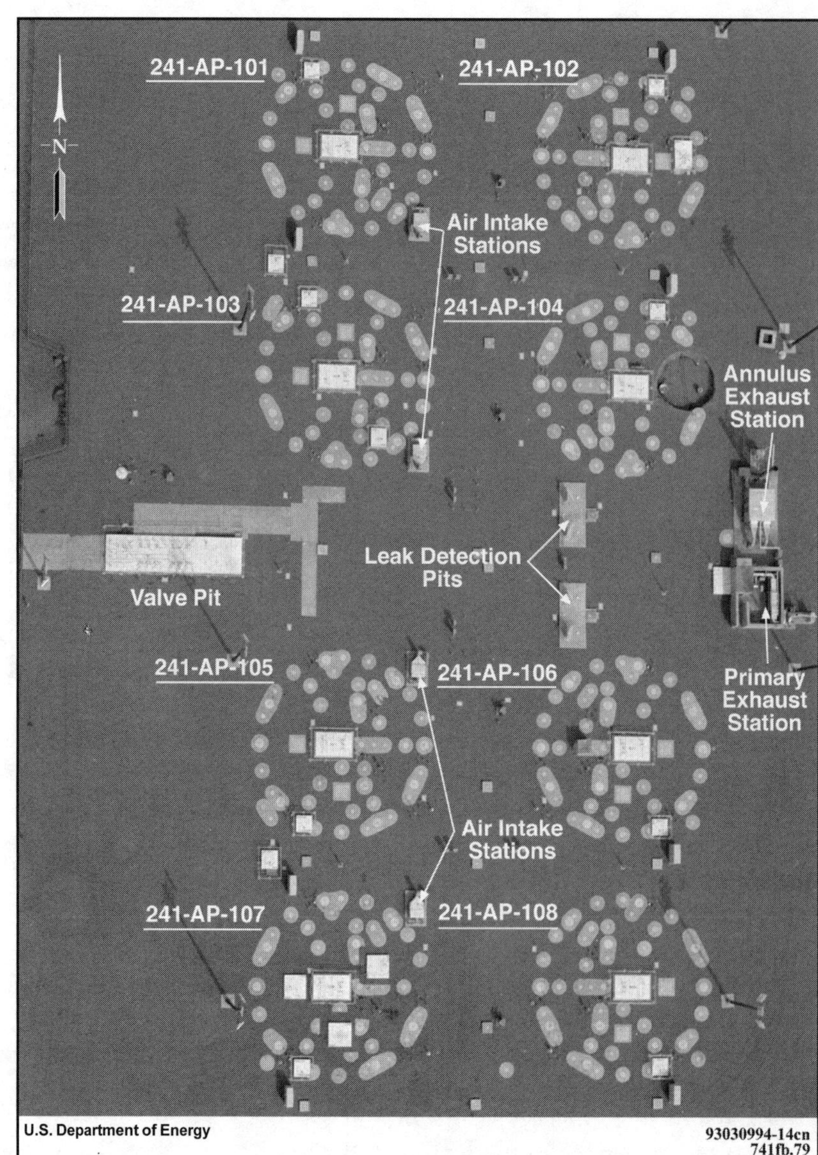

AP Tank Farm. This is an aerial photograph of the concrete pads overlying eight double-shell tanks in the AP Tank Farm, 200 East Area. Each circular pattern is slightly less than the 75-foot width of the underlying tank. One to three access ports are located in each pad. Entry points into the older single-shell tanks are fewer and generally smaller.

To reduce costs, the U.S. Government built carbon steel tanks (rather than stainless steel tanks) for storing high-level radioactive waste (IEER/IPPNR 1992, p. 96).

and their radioactive decay products. Some 240,000 tons of chemicals, such as sodium nitrates and nitrites and various metal hydroxides, oxides, and carbonates, are also in the tanks.

Single-Shell Tanks: Hanford has 149 single-shell tanks, built from 1943 to 1964, located in 12 groupings, called tank farms. They were constructed during four periods as the need for waste storage grew (Brevick 1994, 1995a, 1995b, 1995c):

1943-1944....... 64 tanks (Tank Farms B, T, C, U)
1946-1949....... 42 tanks (Tank Farms BX, TX, BY)
1950-1955....... 39 tanks (Tank Farms S, TY, SX, A)
1963-1964....... 4 tanks (Tank Farm AX)

Waste Types

The following terms are used to describe the types of waste at Hanford and elsewhere in the nuclear weapons complex. The terms are often based on the processes that created the waste, not the actual level of radioactivity in or the biological risk posed by that waste.

High-level waste is chemical solutions created during the initial reprocessing of used (irradiated) uranium fuel. This waste contains most of the unwanted and concentrated amounts of radionuclides; solids created from these solutions. Liquid high-level waste, along with other reprocessing solutions, was pumped into underground tanks and sometimes into the ground.

Hazardous waste contains nonradioactive metals and chemicals known or thought to pose a risk to human health and the environment.

Transuranic waste is radioactive waste that contains alpha-emitting transuranic elements, elements beyond uranium on the periodic chart, such as plutonium, having half-lives greater than 20 years and present in concentrations of more than 100 billionths of a curie (100 nanocuries) per gram[6] of waste.

Low-level waste is a general term describing any radioactive waste that is not irradiated fuel or high-level waste or does not contain large amounts of transuranic materials. It can include liquids or contaminated solids such as clothing, tools, and equipment. Low-level waste is found in solid waste burial grounds and inside facilities. It contains a broad range of radionuclides. The principal radionuclides include activation products (such as cobalt-60 and iron-55), fission products (such as cesium-137 and strontium-90), and other generally short-lived radionuclides.

Class A low-level waste has the lowest concentrations of radionuclides that can be disposed with the least stringent requirements on waste forms and disposal packaging.

Class B low-level waste contains higher concentrations of short-lived radionuclides. This waste is packaged more stringently than Class A waste.

Class C low-level waste must meet the form and stability requirements for Class B plus actions are required at the disposal site to protect against inadvertent human intrusion for 500 years.

Mixed waste contains both radioactive (either low-level waste or transuranic waste) and hazardous waste components.

Half of the chemicals and radioactivity onsite rests inside tanks.

(6) Twenty-eight grams equals 1 ounce.

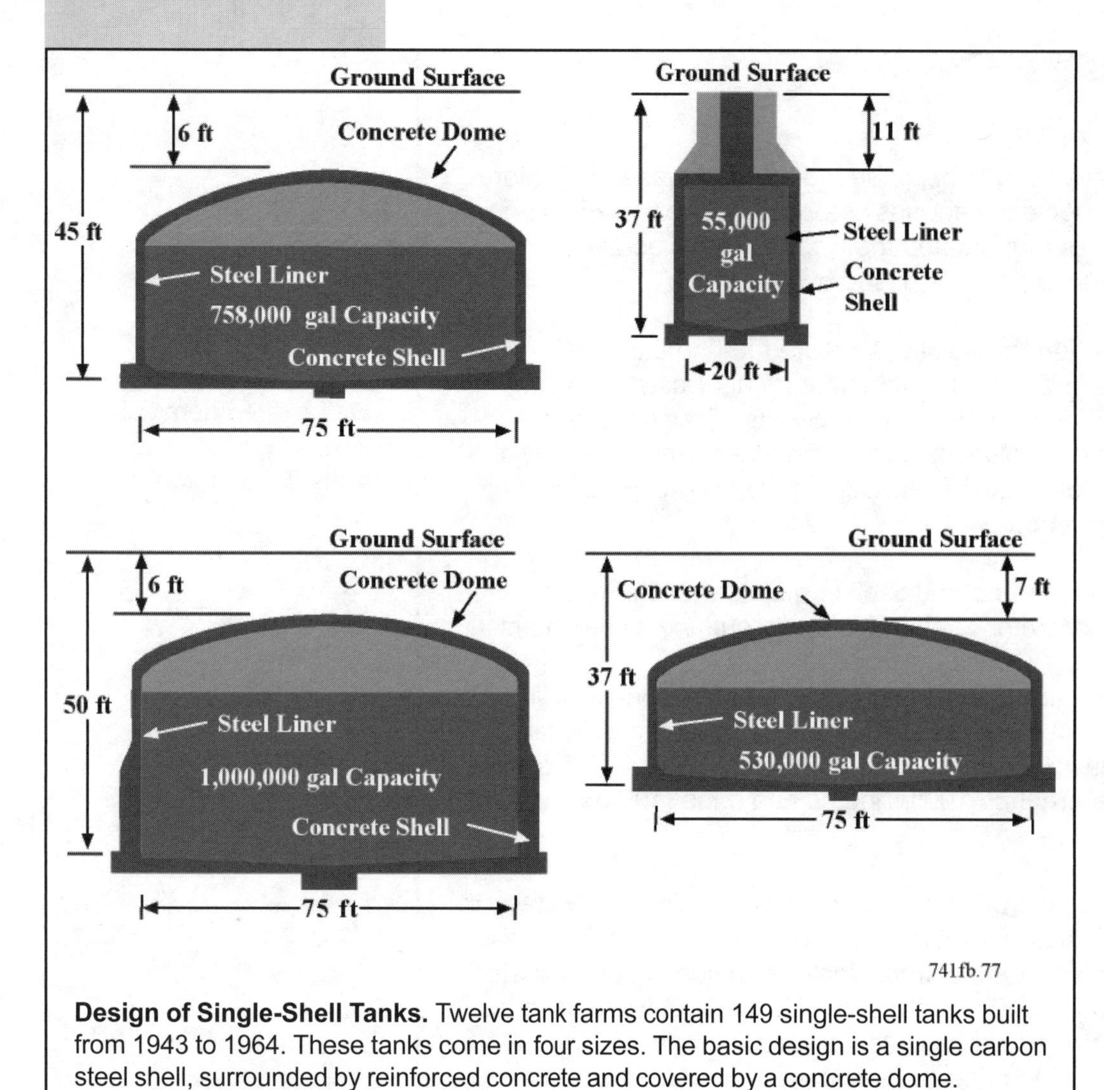

Design of Single-Shell Tanks. Twelve tank farms contain 149 single-shell tanks built from 1943 to 1964. These tanks come in four sizes. The basic design is a single carbon steel shell, surrounded by reinforced concrete and covered by a concrete dome.

Single-shell tanks come in four sizes:

- 16 have a capacity of 55,000 gallons.
- 60 have a capacity of 530,000 gallons.
- 48 have a capacity of 758,000 gallons.
- 25 have a capacity of 1 million gallons.

Their design life ended up being 10 to 20 years; originally, it was expected to be much longer. Laboratory tests conducted in the late 1950s using simulated waste showed that a tank's carbon steel liner would corrode at a rate between 0.00001 and 0.0001 inch each month. Based upon these data, a 3/8-inch-thick steel liner should last for "at least 500 years." The bottom plate and upper liner exposed to vapors might corrode more rapidly. Nonetheless, "a life expectancy measured in decades appears to be very conservative" (Tomlinson 1959a, p. 307). Hanford's health physicist Herbert Parker echoed this multi-decade life though he acknowledged "the life of the tanks is not yet known" (Parker 1959a, p. 165).

At that time, Parker and others believed that if the tanks began to fail, new tanks would be constructed and waste pumped into them. Tank failure was considered "inevitable" though people managing and handling waste expected to "demonstrate that there is a superior method of packaging materials in dry form" before failures were experienced (Parker 1959a, p. 166). The failures came first, prompted by corroded welds and buckled steel liners.

During the mid-1940s, a problem posed 20 years in the future seemed a long time away and of no great importance considering other post-war urgencies. Nonetheless, generating and storing large volumes of waste in underground tanks was considered a "short-term solution—and a poor one" (Lilienthal 1980, p. 80).

Of these 149 single-shell tanks, 67 are suspected to have leaked less than 750,000 gallons to as much as 1.5 million gallons of waste into the underlying sediment or soil (Hanlon 2002; NRC 2001b). Nearly 32 million gallons of waste remain in the single-shell tanks.

No new waste is added. Over the years, most liquids were pumped into the newer double-shell tanks. Thick sludge, moist to dry layers of precipitated or crystallized waste (saltcake), and 4.3 million gallons of liquids exist in the single-shell tanks as of July 2002 (Table 4).

Sixty-seven tanks are suspected to have leaked as much as 1.5 million gallons of waste to the soil (NRC 2001b).

Table 4. Tank Waste Volumes and Percentages by Type (Hanlon 2002)

	Total Waste Volume	Supernatant Liquid	Saltcake	Sludge
Single-Shell Tanks	31.7 million gallons	—[(a)]	68%	32%
Double-Shell Tanks	22.7 million gallons	80%	15%	5%

(a) Inside the single-shell tank waste, 2.5 million gallons of drainable liquid plus another 1.8 million gallons of pumpable liquid remain. These are not listed as free-floating supernate liquids.

Double-Shell Tanks: Twenty-eight double-shell tanks were built in six tank farms from 1968 to 1986. These provided improved liquid containment, greater waste storage capacity, and better access for monitoring and sampling. The two steel liners are separated by about 3 feet of space called the annulus. It provides a margin of safety if the inner liner leaks. Four tanks have a capacity of 1 million gallons, and 24 have a capacity of 1.16 million gallons.

Nearly 23 million gallons of waste rest inside the double-shell tanks. Their expected design lives range from 25 to 50 years. To date, no double-shell tank has leaked, though the oldest ones have reached their life expectancy.

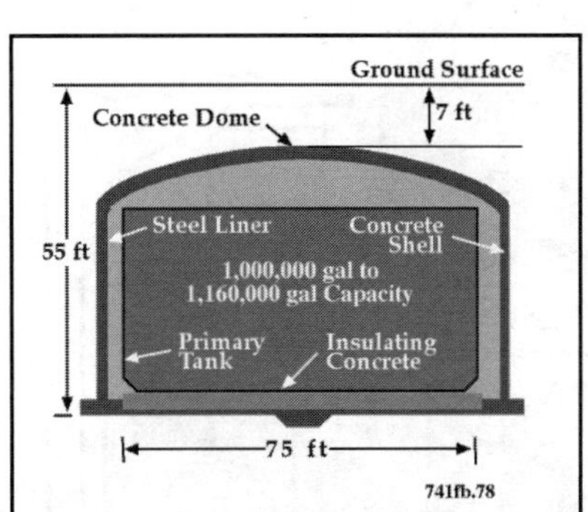

Design of Double-Shell Tanks. Six tank farms contain 28 double-shell tanks built between 1968 and 1986. Double-shell tanks provided more storage volume and better containment than the older single-shell tanks.

In addition to these large tanks, there are about 40 smaller underground tanks that are not currently used and 20 small miscellaneous storage tanks that are still relied upon. Numerous pipes, concrete diversion boxes,[7] and support structures interconnect reprocessing plants, tank farms, and individual tanks (DOE 1998d). The amount of waste remaining in these structures and the amount of waste leaked from them are not well known. Leaks from broken pipes were suspected as early as the 1940s (Parker and Piper 1949, p. 85).

An unknown amount of waste has leaked from smaller underground tanks, pipes, and diversion boxes.

Tank wastes are chemically complex mixtures. They were generated from three separate reprocessing techniques (bismuth phosphate and solvent extraction using either hexone or tributyl phosphate). Numerous other processes were used to recover select radionuclides. In addition, acidic waste streams were made caustic[8] by adding

(7) Diversion boxes are concrete-lined spaces, located underground between the tanks, that contain switching pipes to route liquids between tanks.
(8) A caustic solution is one with a high pH compared to acids that have low pHs. Bleach and ammonia are examples of household caustic solutions.

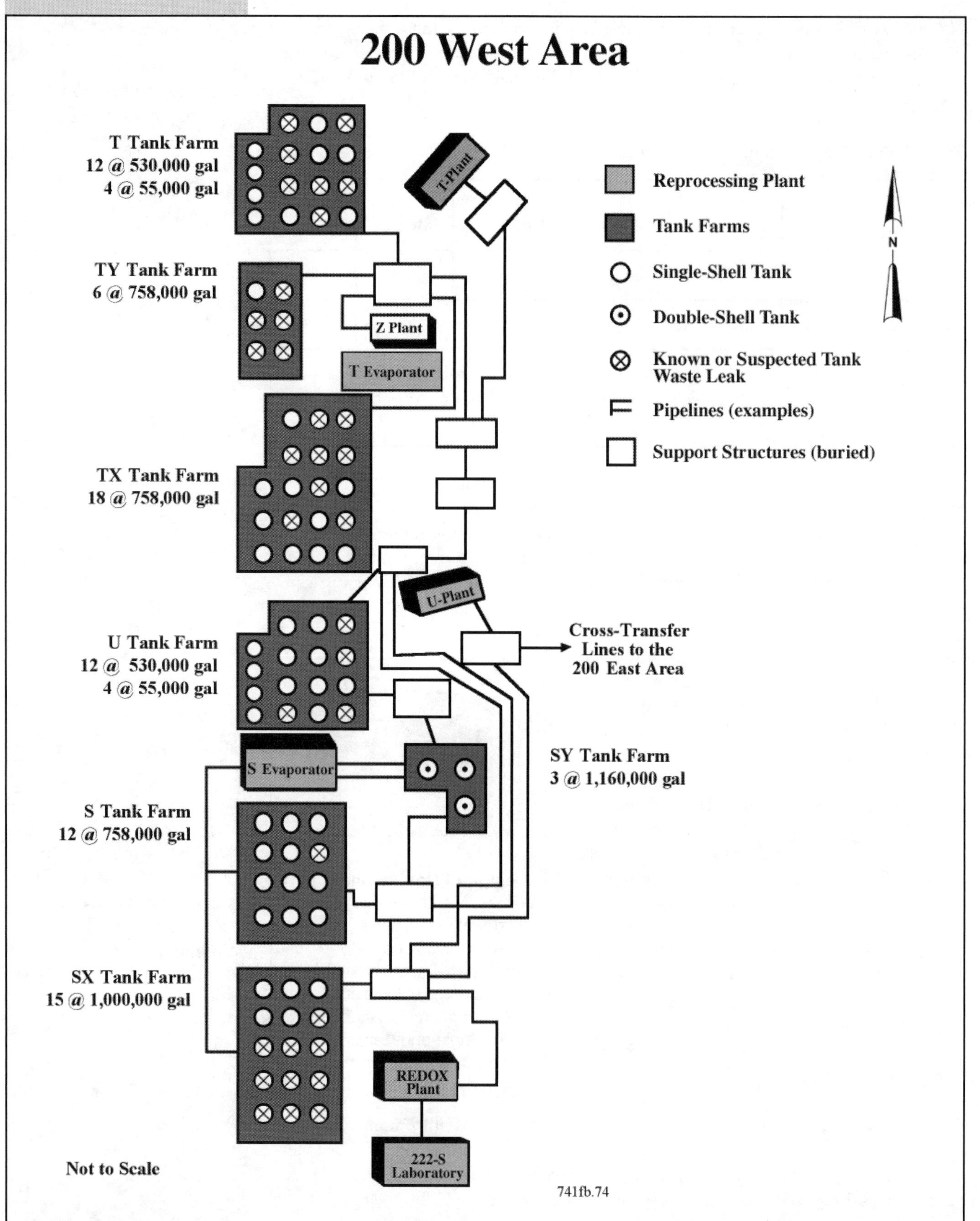

200 West Area Tank Farms. Tanks are grouped into tank farms interconnected by pipes and formerly operating facilities. The 200 West Area contains 86 underground tanks holding radioactive waste—83 single-shell tanks and 3 double-shell tanks. Thirty-five of these single-shell tanks are known or suspected to have leaked waste to the soil.

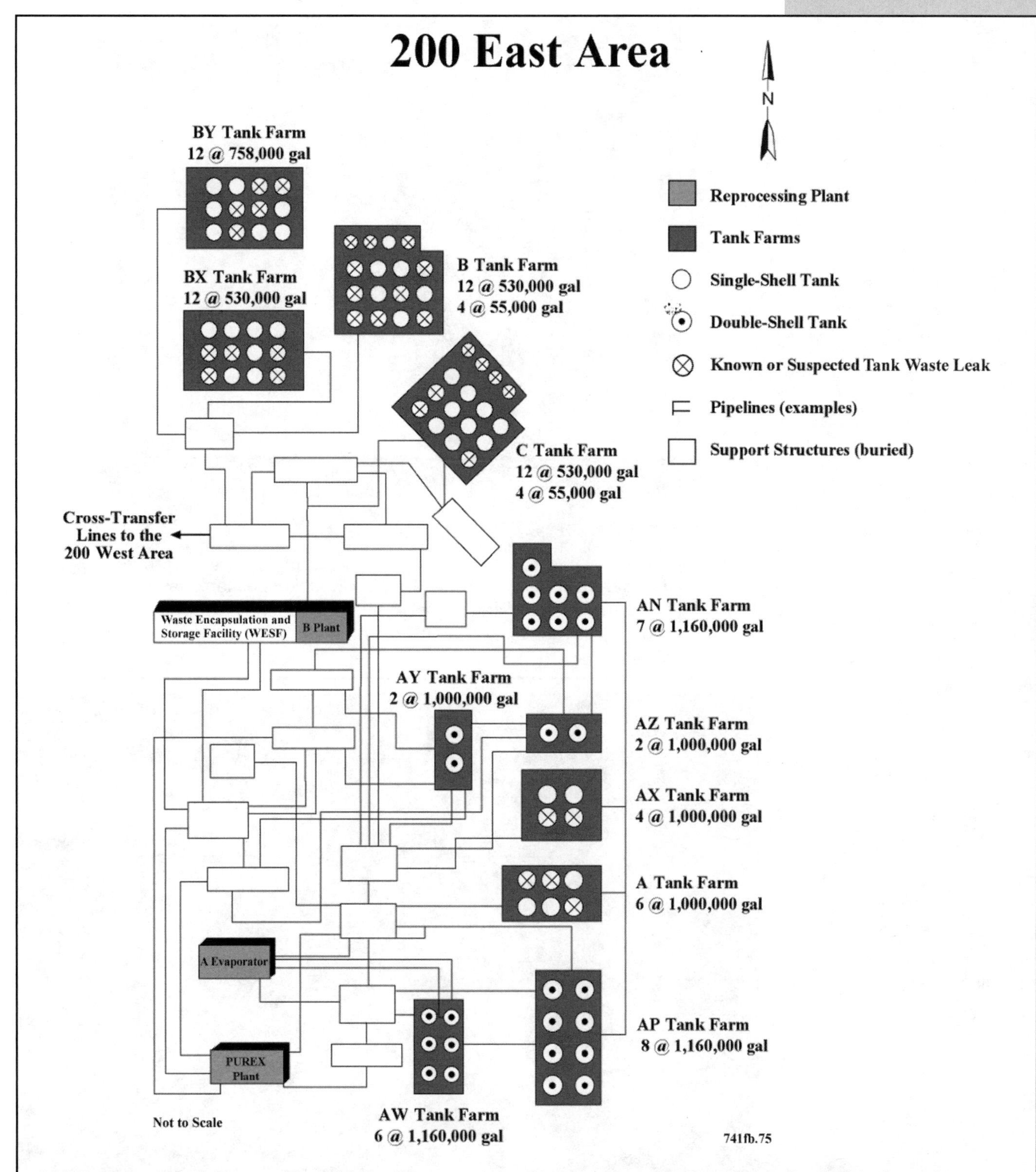

200 East Area Tank Farms. The 200 East Area contains 91 underground tanks—66 single-shell tanks and 25 double-shell tanks. One-third of the single-shell tanks are known or suspected to have leaked. Waste transfer lines connect the tanks.

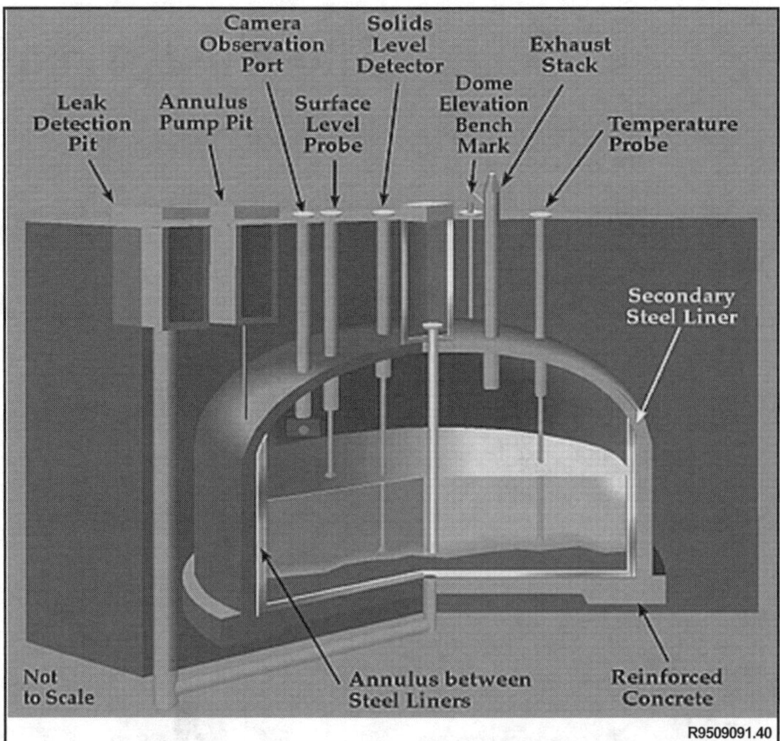

Typical Double-Shell Tank. A typical double-shell tank has several entry points and risers extending to ground level. These allow waste monitoring, sampling, treatment, and gas venting to the atmosphere.

Tank waste can resemble peanut butter, small broken icebergs, foam, or whitish crystals.

Tank Waste. The waste inside the 75-foot-diameter Tank SX-105 has the consistency of soft peanut butter. A temperature probe is hanging down at the right. Halfway up one pipe is a coating of saltcake remaining from when the waste level was several feet higher.

Tank waste is composed of mixtures of liquids and solid particles. A general description is[9]

Supernatant liquid (or supernate): a liquid that floats above a layer of settled solids or under a floating layer of crust. Chemicals found in liquids include sodium nitrates, nitrites, phosphates, aluminates, and carbonates. Radioactive cesium and technetium are common.

Sludge: a layer containing fine particles made of water-insoluble chemicals that settled to the bottom of a tank when sodium hydroxide was added to a reprocessing plant's acidic waste stream, making the waste suitable for storage in the carbon-steel tanks. Most particles are made of metals, such as aluminum, silicon, iron, uranium, and zirconium combined with oxygen, hydroxides, and phosphates. Strontium and transuranic elements tend to concentrate in sludge.

Saltcake: a moist material, sometimes resembling a wet or semi-dried beach sand, created from the crystallization and precipitation of chemicals after liquids were evaporated. Saltcake is usually made of water-soluble chemicals such as sodium nitrate.

Drainable or interstitial liquid: a liquid found within the pore spaces of saltcake and sludge. Some interstitial liquids are drainable; others are not. Interstitial liquids are often thought of as the water trapped in a kitchen sponge. The water is trapped in small spaces, or pores, inside the sponge. With effort, much of the water can be wrung out—or in the case of a tank, pumped out.

Vapor: gases such as hydrogen, nitrous oxide, and ammonia produced from chemical reactions and radioactive breakdown of water and organic compounds. Gases rise from the waste and are released into the space above the waste inside the tank (this area is called the head space).

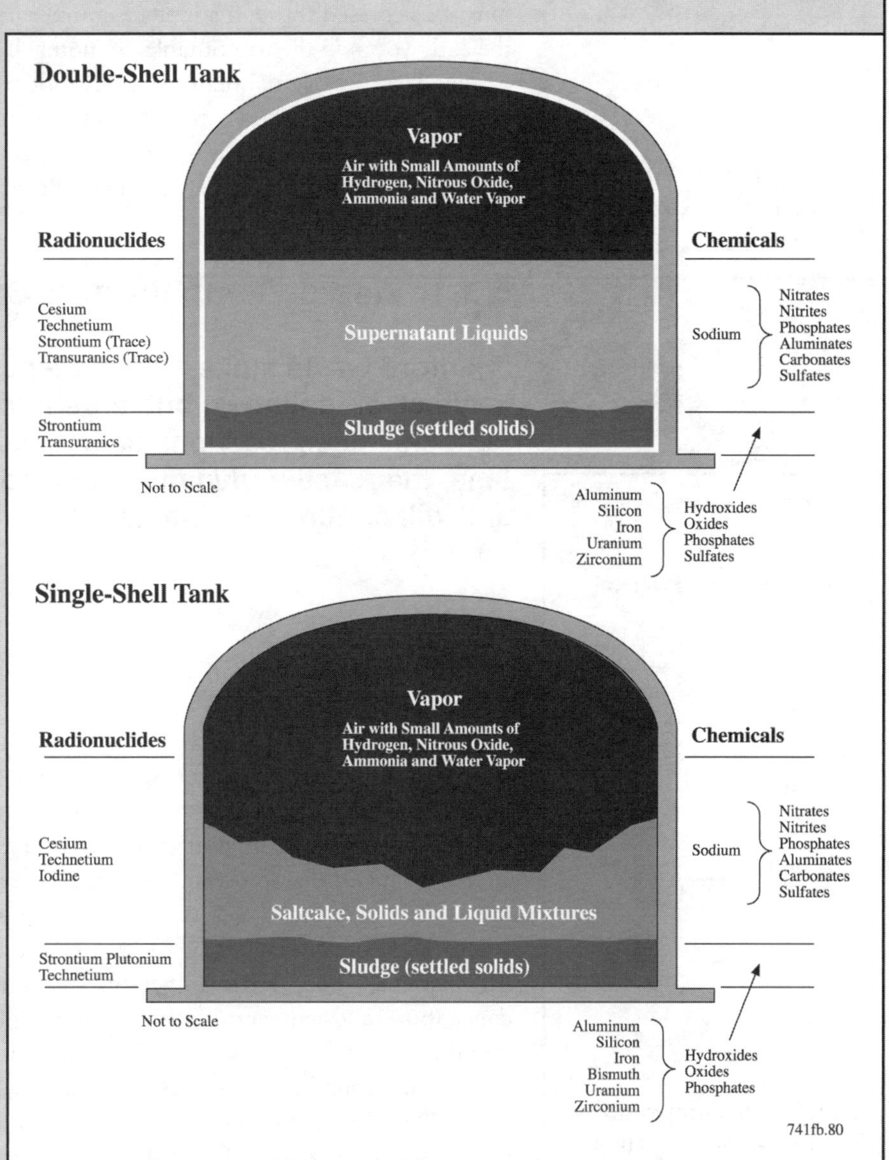

Tank Waste Contents. These are idealized illustrations of the radionuclide and chemical content of Hanford tanks. Waste characteristics can vary significantly within individual tanks and between tanks.

(9) Tank waste is described in many ways. This description relies upon some traditional terms without getting lost in a lot of terminology.

The highest temperature found in a tank was 594° F in 1963 (Mercier et al. 1981).

sodium hydroxide.[10] This substantially increased the amount of chemicals in the tanks plus it chemically segregated materials that could not remain in solution in a caustic environment. Waste was mixed between tanks and additional chemicals were added to some. This included ferrocyanide to scavenge cesium (that is, cause it to precipitate), diatomaceous earth and cement to soak up liquids, casks of experimental fuel elements, plastic bottles containing plutonium and uranium, and organic ion-exchange resins (Gephart 1998). More than 40 types of waste have been identified (Bunker et al. 1995).

It is difficult to measure the physical and chemical properties of tank waste. Samples must be accessed through a limited number of openings (called risers—or "nozzles" in the early years) in the top of tanks, remotely handled, and then examined inside specially designed, sealed mini-laboratories called hot cells. Because the characteristics of a waste sample can change when exposed to the heat, moisture, and other conditions outside the tank environment, what is measured may not represent what exists in the tank. This is why some of the most meaningful characterization is conducted in situ—right in the tank.

5.3.1 Waste Bumping and Burping

"Hanford waste tanks are, in effect, slow chemical reactors in which an unknown but large number of chemical (and radiochemical) reactions are running simultaneously. Over time, the reaction dynamics and compositions have changed and will continue to change..."

—Art Janata et al.,
Report of the Ab Initio Team for the Hanford Tank Characterization and Safety Issue Resolution Team
(Colson et al. 1997 p. B-11)

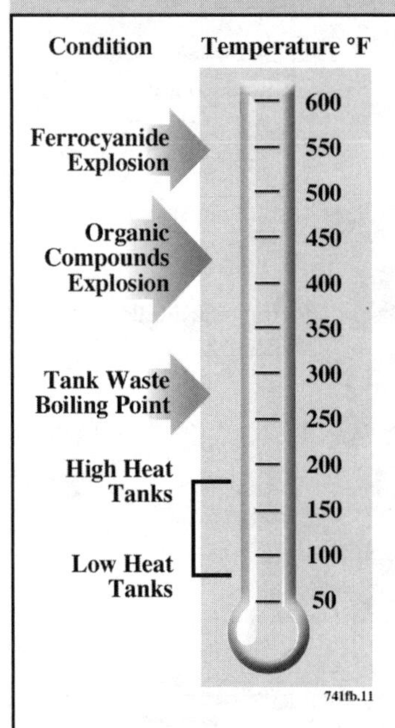

Tank Waste Temperatures. The temperatures of tank waste played a key role in understanding the risk of unwanted reactions. For example, waste can be ignited under certain conditions of dryness, composition, and high heat. Today, waste temperatures are much lower than they were in the past.

When uranium fuel was reprocessed only a month or two after irradiation, its waste liquids contained a large amount of heat-generating radionuclides—especially in the sludge layer found in the bottom of tanks. This "young" or "green" fuel created waste so thermally hot that it would "periodically burst into a violent surging boil" that pressurized the tanks by releasing gases more than 20 times the normal rate (Pilkey et al. 1958). The average temperature of waste entering many tanks was 100° to 130°F (Mercier et al. 1981). High temperatures caused tank liquids to evaporate, chemicals to concentrate, and temperatures to rise. Boiling temperatures were reached after another month. Water was added to dilute the waste and lower the temperature. The highest temperature found in a tank was 594°F in the A Tank Farm in 1963 (Mercier et al. 1981). Old photographs show snow preferentially melting atop single-shell tanks compared to other areas during the winter months (Pilkey et al. 1958).

Elevated temperatures caused tank expansion and contraction. This degraded the tank's shell of concrete. The steel bottoms in some tank bottoms may have buckled 4 to 8 feet (Brownell 1958; Hanford Works 1965). After about 6 years, short-lived radionuclide heat sources decayed away leaving the less radioactive cesium-137 and strontium-90 to more slowly and less violently warm the waste.

(10) Acidic waste could not be put in the tanks as it would quickly corrode the carbon-steel container.

The Watch List

In 1990, the U.S. Congress passed Public Law 101-510, Section 3137, creating the "Watch List" of Hanford tanks. Tanks were considered dangerous or potentially dangerous based on their waste contents. The waste issues that put tanks on this list were

Ferrocyanide: Between 1954 and 1957, 150 tons of ferrocyanide-bearing chemicals were added to several tanks to precipitate radioactive cesium to the tank bottoms. Later, it was discovered that under certain conditions, the ferrocyanide could ignite. Waste characterization, further studies, or waste transfers closed this issue in 1998.

High Heat: The waste in one tank produced high levels of heat. Most of its liquids were transferred to nearby tanks, and the waste temperatures dropped from over 150° to 57°F (Hanlon 1999a, 2001), closing this issue in 1999.

Nuclear Criticality: Concerns were raised that plutonium remaining in the tanks could collect and lead to a nuclear chain reaction. After researching the issue, it was determined that no one tank contained enough plutonium for a reaction and that neutron absorbers in the waste further reduced that possibility. This safety issue was closed in 1999 (DOE 2000b).

Organic Materials: If temperatures rose above 360°F, the organic materials in tanks could explosively react with nitrate and nitrite (DOE and Ecology 1996). Waste characterization, further studies, or waste transfers closed this issue in 2000.

Flammable Gas: Large volumes of potentially flammable gases were created in tank waste and released as vapor. This issue was closed in 2001.

Initially, 54 of Hanford's 177 tanks were on the Watch List. As many as 60 tanks made the list at one time or another.

Single-shell tanks in the S, SX, A, and AX Tank Farms received this self-boiling waste (Agnew 1997). After accepting REDOX Plant waste in 1951, tanks in the S Tank Farm started to boil (Brevick 1995a). Four years later, nine tanks in the nearby SX Tank Farm also received self-boiling waste from the same plant. All nine tanks subsequently leaked (Agnew and Corbin 1998). In fact, 10 of the 15 tanks in the SX Farm are now suspected or confirmed of leaking. Some of these tanks were subjected to temperatures of 280° to 320°F for 10 or more years. One tank reached a peak temperature of 390°F in 1958 (Mercier et al. 1981).

Tank waste from the first solvent extraction plant (REDOX) was more concentrated than that produced using bismuth phosphate extraction in T and B Plants.

The ten single-shell tanks in the A and AX Tank Farms received self-boiling waste from the PUREX Plant. (Some B Plant waste was also received in the AX Tank Farm.) Wastes boiled for several years at a temperature of 250°F (Brevick 1994). The self-boiling of PUREX waste accelerated to the point in 1957 that 10 gallons a minute of liquid was lost from evaporation within individual tanks (DOE 1999d). It became necessary to add water to keep waste from drying and increasing the chance for igniting (Anderson 1990).

Monitoring liquid levels was important to early tank operations for it revealed leaks. Temperature measurements were secondary until the idea of using self-boiling waste as an in-tank waste volume reduction process became accepted in the early 1950s. Since 1974, tank temperatures have been routinely collected. Water additions, air circulation, and other measures were used to keep temperatures at acceptable values for "calculations indicate that if the sludge were allowed to dry without supplementary cooling, peak temperatures of about 1600°F would be reached in about 18 months" (Isochem 1967).

Temperatures were highest in the tank sludge.

Temperatures were highest in a tank's lower layer of sludge (DOE Grand Junction Office 1998). A cooler upper layer of liquid covered the sludge. Experiments demonstrated that when the temperature difference between the two layers reached 20° to 25°F, heat and steam were released from the sludge in a sudden burst that Hanford workers called "bumping" (Gerber 1993a). This mixed the waste layers. Years later, pumps, additional vents, and overflow lines were installed to avoid waste bumping and to better regulate in-tank temperatures.

Hydrogen and other flammable gases are created from the chemical and radiolytic breakdown of water and organic compounds in tanks.

The double-shell Tank SY-101, located in the 200 West Area, was a top safety issue for nearly all of the 1990s. Every 3 or 4 months, 5000 to 10,000 cubic feet of potentially flammable gas was suddenly released from beneath the upper waste layer (Stewart 2000). This caused a buoyancy-induced lift called a waste rollover, nicknamed a "burp." Hydrogen and other gases are created from the chemical and radiolytic[11] breakdown of water and organic compounds in the waste plus the corrosion of steel tank walls (Stock 2000). Except for ammonia, most of these gases come out of solution as bubbles. A burp took place when bubbles, trapped in a thick waste layer, made the layer so light that it rapidly rose to the waste surface, releasing the trapped gas. The gas was vented into the tank's head space (Stewart et al. 1996).

In 1993, a seven-story-tall mixer pump was installed in Tank SY-101. The pump took liquids from above the sludge and forced them out at the bottom of the tank. This gently stirred the sludge allowing gases to release at a steady rate instead of in sudden events. Extra safety measures for this tank cost $30 million dollars a year (Tyree 2001).

However, over time, this mixing acted like an eggbeater, entrapping small bubbles in the waste. Slowly, the waste level rose 10 feet (Stang 2000e). This brought it within 2 feet of overflowing into the annular space separating the two steel walls of the double-shell tank. From late 1999 to early 2000, 400,000 gallons of liquid were pumped into an adjoining tank. During this period, additional water was added to dilute the waste. The waste no longer burps or rises. The flammable gas safety issue in SY-101 was closed, and the tank was removed from the Watch List in January 2001 (Hanlon 2001).

5.3.2 Waste Evaporators

"The business of constructing more and more containers [tanks] for more and more objectionable material has already reached the point both of extravagance and of concern..."

—Abel Wolman and Arthur Gorman,
Report of the Safety and Industrial Health Advisory Board (Williams 1948, p. 70)

To reduce the amount of tank waste stored, some liquids were evaporated. In 1951, the first waste concentrators began operating. They were steam-heated, pot-like evaporators located on the ground. Waste was piped from single-shell tanks into these concentrators to partially boil down the liquids. An 80% reduction in volume was achieved (DOE 1999d). A more concentrated waste mixture was returned to the tanks where solids

(11) Radiolysis is the breakdown of molecules by radiation. For example, a molecule of water is dissociated into hydrogen and oxygen.

precipitated as the liquids cooled. The less-contaminated and evaporated liquids, called condensate, were disposed to the soil.

Another early technique used in-tank evaporation. One approach resorted to inserting an electric heater directly into the waste. The heated mixture was then circulated into other tanks. A second approach involved circulating hot air into tanks through a perforated pipe.

741fb.60

Waste Evaporator. Evaporators, such as the 242-A Evaporator located in the 200 East Area, boil off water from tank waste. This reduces the amount of liquid stored.

Large-scale waste evaporation began in the 1970s with the operation of the 242-S and 242-A Evaporator-Crystallizers. These are located in the 200 West and 200 East Areas, respectively. Their life expectancy was 20 years though upgrades lengthened that time (Burris 1999). Tank liquids were pumped into them and evaporated until a thicker solution, containing about 30% by weight of solids, was created. The slightly hot, concentrated slurry was returned to a tank where it cooled and crystallized. The principal product was a large volume of soft sodium nitrate saltcake and thick slurries rich in chemical compounds such as sodium hydroxide and sodium aluminate.

One early technique of evaporating tank waste was to insert an electric heater directly into the waste.

Between 1951 and the mid-1990s, some 300 million gallons of liquids—55% of the tank waste created at Hanford—was evaporated (Agnew 1997). Another 120 million gallons of liquid condensates from evaporators and tanks were discharged to the ground via single-shell cascade tanks (DOE 1991b; Waite 1991). These interconnected tanks were arranged in cascades of three. When the first tank was filled, liquids overflowed into the next tank, and that one overflowed into a third (Mercier et al. 1981). Excess liquids from the third tank were diverted to a crib, which resembles an underground tile field. Cribs[12] were the most common way of disposing of low-level radioactive liquids.

The monitoring and control systems in the 242-A Evaporator-Crystallizer were upgraded in 1994. Two to three evaporation campaigns are carried out each year. During the year 2000, 2.4 million gallons of liquids were added to the double-shell tanks while 682,000 gallons were evaporated (Diediker and Dyekman in Poston et al. 2001). Nearly all of this additional liquid was water-diluted single-shell tank waste. Residual cooling waters and steam condensate are now piped to the state and federally regulated 200 Area Treated Effluent Disposal Facility. Such waste additions and deletions cause the volume of tank waste to vary from year to year.

(12) Box-like cribs built underground received contaminated liquids. The waste drained through the cribs; the soil absorbed certain radionuclides while allowing the most mobile contaminants to pass through (see Section 5.6.1).

U.S. Department of Energy 96020684-2cn 741fb.24

Waste Encapsulation and Storage Facility. Water-filled pools at this facility hold 1936 stainless steel capsules, containing 130 million curies of cesium and strontium plus decay products. This is 30% of the radioactivity at Hanford. The water surrounding each capsule shines blue from Cherenkov glow, formed when high-energy beta particles released from the capsules interact with water.

5.4 Nuclear Materials

Nuclear materials include cesium and strontium capsules, spent nuclear fuel, and plutonium in various forms.

In 1968, B Plant was refurbished and began separating cesium and strontium from tank waste, allowing less-radioactive liquids to be poured into the soil. This freed additional tank space for holding newly generated high-level waste. Separation activities continued until 1985 (WHC 1995). A new facility, the Waste Encapsulation and Storage Facility, was added to B Plant in 1974 to encapsulate these radionuclides inside stainless steel cylinders. Of the 2217 capsules manufactured, 1936 remain onsite in concrete water-filled basins (DOE 2000i). This includes 1335 capsules containing cesium chloride and 601 capsules holding strontium fluoride. Each capsule is 2.6 inches wide by 20 inches long and emits 165 to 225 watts of heat. Combined, the onsite capsules contain 130 million curies of radioactive cesium and strontium plus their decay products.[13] This is about one-third of the total radioactivity at Hanford.

The radiation dose adjacent to each capsule is extremely high, more than 1 million rad an hour. A person would receive a lethal dose of radiation (about 300 rad) in less than 10 seconds standing 1 foot from an unshielded capsule or in about 1 minute at a distance of 3 feet.[14] The capsules are kept under water; this shields workers and visitors from all but extremely low radiation levels.

Some 2310 tons of irradiated uranium fuel from N Reactor were stored under water in two aging, leak-prone concrete basins at the KE and KW Reactors. These basins are in the 100 Area within 1400 feet of the Columbia River. They were built in 1951 and are well beyond their design life of 20 years (GAO 1999e). Spent fuel was first stored in the K-East Basin in 1975 and in the K-West Basin in 1981 (DOE 2000j). The last fuel was received in 1989. Spent fuel removal, drying, and shipment to the 200 East Area for storage began in the year 2000.

As of 2001, this fuel contained 55 million curies of radioactivity.[15] Some fuel has corroded, releasing strontium, cesium, plutonium, tritium, and other fission products and actinides. These elements are now found in the basins' water, in 15,000 gallons of sludge-like material[16] (Stang 2000g), and in concrete walls. Before 1994, 15 million gallons of water leaked through floor joints and cracks of the basins, releasing radionuclides into the local soil and groundwater (GAO 1998e). Those cracks were later sealed.

Hanford also has nearly 12 tons of plutonium onsite, in various chemical and physical forms (DOE 1996d). Most rests in two locations—spent fuel in the K Basins contains 4.5 tons of plutonium (Reilly 1998), and the Plutonium Finishing Plant holds 4.7 tons of

(13) Conversation with Brian Oldfield, health physicist with Fluor Hanford, Inc., July 27, 2001.
(14) Communication from Daniel Strom, health physicist, May 3, 1999. Calculation assumed each capsule contains about 50,000 curies.
(15) Based upon radioactivity inventory data calculated for the Hanford Groundwater/Vadose Zone Project. Information received from Fluor Hanford, Inc. staff on August 9, 2001. The number of curies in these basins decreases as spent fuel is removed and the radioactive materials decay.
(16) This sludge is created from particles of corroded spent fuel mixing with windblown dust, dirt, and water.

Spent Fuel in K Basins. Some 2310 tons of irradiated uranium fuel were once stored in two water-filled concrete basins near the Columbia River. This photograph is looking down into the basin water at two dozen stainless steel canisters holding several 2.4-inch-diameter, 2-foot-long, double-cylinder-shaped fuel rods. Some rods are intact; others are heavily corroded. Basin cleanup is now under way.

U.S. Department of Energy

92021955-2cn 741fb.23

scrap plutonium (DOE 1999) in 20 tons of plutonium-bearing materials and solutions. The volume of these solutions decreases with time as a more stable dry plutonium-oxide material is formed by chemical precipitation and heating.

5.5 Solid Waste

"All solid contaminated material that is not readily decontaminated to usual levels is discarded into trenches, and suitably covered... [This] does represent relatively haphazard control and so merits the Committee's attention."

—Herbert Parker, *Summary of Hanford Works Radiation Hazards for the Reactor Safeguard Committee* (1948b)

Few records were kept documenting solid waste burial activities before 1960 (DOE 1993). From 1960 to 1967, records generally listed the waste shipper, radioactivity level, volume, burial site, and type of disposal container. They did not identify all burial locations, dates of waste shipments, or the chemical nature of material dumped. Little care was taken to enable later waste retrieval, repackaging, or shipment offsite because radionuclide leaching from solid waste was considered "not a problem in arid regions" (Healy et al. 1958).

Before the 1960s, little care was taken to enable later retrieval of buried solid waste.

Solid waste buried since 1967 is better characterized. According to a directive issued by the Atomic Energy Commission in 1970, transuranic-contaminated solid waste was to be segregated from low-level radioactive waste and placed in retrievable storage pending shipment to a geologic repository (DOE 2000f). At Hanford, this separation began in 1972. Hanford stopped burying transuranic solid waste in 1988 and started storing it in buildings in the 200 West Area (Stang 2000c). The first shipment of transuranic solid waste to a permanent underground geologic repository built in salt (at the Waste Isolation Pilot Plant in New Mexico) occurred in the year 2000.[17]

(17) Congress authorized the Waste Isolation Pilot Plant in 1979. It began receiving waste 20 years later.

U.S. Department of Energy

741fb.21 13101-1

Old Solid Waste Burial. This photograph shows the haphazard solid waste burial practices of years ago. Sixty percent of the solid waste at Hanford was buried before 1970—when it contained a mixture of low-level radioactivity, long-lived transuranic elements, chemicals, and various undocumented materials.

Over time, corrosion weakens containers. Most likely, the thousands of steel barrels buried before 1970 have corroded, releasing all or portions of their content. Drums buried in the 1970s have or will soon reach their design life (Stang 2000c). The amount of corrosion and type of waste mixtures determine how this waste might be exhumed, if ever. Solid waste that can be moved and handled by workers is called contact-handled waste. More dangerous or highly radioactive waste that must be retrieved by robotic devices is remote-handled waste.

Hanford contains 25 million to 26 million cubic feet of solid waste; 97% of it buried, and the remainder is stored in facilities.[18] This includes low-level, transuranic, and mixed chemical and radioactive waste. This is enough waste to fill a cube 300 feet on a side. Solid waste generated from Hanford operations contains about 6 million curies of radioactivity (decayed as of 1998) and 70,000 tons of chemicals. An estimated 0.4 tons of plutonium and 650 tons of uranium exist in solid waste in the 200 Areas (Wodrich 1991). About 1.7 tons of plutonium is buried or stored as solid waste across Hanford (DOE 1996d).

Hanford's burial grounds contain 2.7 million cubic feet of transuranic-contaminated solid waste having 67,800 curies (decayed to 1997) of radioactivity (DOE 2000f).[19] This is about 10% of Hanford's solid waste volume and is the largest inventory of transuranic-contaminated buried waste for the nuclear weapons sites studied.

Some old burial grounds contain fragments of spent fuel.

Sixty percent of Hanford's solid waste was buried before 1970. In some of this, radiation readings ranged from hardly detectable to so intense that remote-handling equipment (sometimes a truck) dragged contaminated solid material out of facilities to dump it in a burial ground. Occasionally, burial trenches were backfilled to "prevent spontaneous combustion and spread of contamination" (Parker 1954a). Some of these old burial grounds contain dangerous chemical mixtures including possible munitions and materials capable of igniting upon exposure to air.

---

(18) One U.S. Department of Energy (1997a) report listed 24 million cubic feet of low-level and transuranic solid waste on the Hanford Site. Another documented nearly 46 million cubic feet of low-level and "previously disposed" transuranic contaminated waste onsite (DOE 2001b, pp. 6.7, 7.52). A conversation with Michael J. Turner of Fluor Daniel Hanford Inc., February 26, 1999, noted 26 million cubic feet. Other sources suggest 19 to 31 million cubic feet. This range of numbers reveals the uncertainty in waste inventories. The author used the lower end of these numbers.

(19) A volume of 2.7 million cubic feet forms a cube 140 feet on a side.

There are 75 solid waste burial grounds at Hanford (DOE 1998h):

- 28 in the 100 Area
- 34 in the 200 Areas
- 13 in the 300 Area and other locations.

Some reports identify additional burial sites. As cleanup continues, more sites will be uncovered. In the early days, there was little control over where burial grounds were placed, who could start a new one, and what documentation was required.

The first shipment of transuranic contaminated solid waste to an underground repository built in salt took place in the year 2000.

Two examples of solid waste burial sites, receiving considerable attention in recent years, are the 618-10 and 618-11 sites. Each received high-activity, low-activity, and transuranic contaminants from nuclear fuel experiments and other research conducted in the 300 Area.

Four miles north of the 300 Area, the 618-10 Burial Ground was used from 1954 to 1963 (DOE 1993). The site measures 485 by 570 feet—about 6 acres in size. It contains 12 trenches and 94 long buried pipes made of 55-gallon bottomless barrels welded one atop another. A total of 500,000 cubic feet of material[20] is estimated to rest at the site (Kincaid et al. 2001). The trenches received low-level waste in cardboard boxes, cement barrels containing radioactive waste, liquids, and miscellaneous items such as laboratory hoods and filters. Remote-handled high-level radioactive waste removed from a laboratory was placed in the pipes. The site may hold 2.2 to 4.4 pounds of plutonium and other transuranic elements (Petersen et al. 2001).

Between 1962 and 1967, waste from nuclear fuel experiments and other research conducted in the 300 Area were buried in the 618-11 Burial Ground located 8 miles north of the 300 Area (DOE 1993). The site measures 375 feet by 1000 feet, and contains 750,000 cubic feet of material (Kincaid et al. 2001).[21] The U.S. Department of Energy (2000f) reports the site contains 680 curies of radioactivity—decayed as of 1995. Kincaid et al. (2001) suggests a higher value of 42,100 curies of cesium-137 and strontium-90 as well as 430 curies of plutonium-239. Records reveal waste was disposed inside 3 trenches, 50 vertical pipes (steel barrels welded together) and 5 galvanized metal caissons. The bottoms

U.S. Department of Energy G999090118.15 741fb.61

Solid Waste Burial. Today, solid waste is well characterized, packaged, and stored. This is one of the open trenches receiving solid waste in the 200 West Area burial grounds.

(20) A volume of 500,000 cubic feet would fill a cube 80 feet on a side.

(21) A volume of 750,000 cubic feet would fill a cube 90 feet on a side.

of the trenches and pipes are 20 to 30 feet below ground level (Dresel et al. 2000). Some waste contains slivers of irradiated nuclear fuel including 11 to 22 pounds of plutonium and other potentially dispersible radionuclides (Petersen et al. 2001). When buried, the radiation level measured at the surface of a container was up to 500 rad an hour. This is dangerously high. Groundwater samples collected in 1999 and 2000 revealed elevated levels of tritium—as much as 400 times above drinking water standards (Hartman et al. 2001). This is the highest tritium concentration found onsite in recent years. "Some, if not all" of the tritium came from the disposal of materials from a mid-1960s onsite tritium production demonstration project (Dresel et al. 2000).

A unique type of solid waste storage exists in two tunnels next to the PUREX Plant. Some radioactively contaminated processing equipment used in PUREX and other facilities could not be safely moved (ERDA 1975). Instead, the equipment was loaded onto flatbed railroad cars and pushed by a remote-controlled locomotive into a tunnel built beneath a mound of soil. The first tunnel is 360 feet long, 22 feet high, and 19 feet wide. It's appended to the southeast corner of the PUREX Plant and received contaminated equipment from 1960 to 1965. An adjoining 1700-foot-long tunnel was built in 1964. Both tracks were constructed with a slight downward slant to ensure the railcars would not roll out.

Some radioactively contaminated equipment couldn't be safely moved. So, it was loaded on flatbed railroad cars and pushed by a remote-controlled locomotive into tunnels built beneath mounds of soil.

As of 1960, the first tunnel contained 40,000 curies of radioactivity stored in containers atop three railroad cars (DOE 2000h). By the early 1970s, the second tunnel contained about 85,000 curies of radioactivity plus mixed waste. In 1995, the second tunnel held 21 railroad cars of waste, including lead, mercury, chromium, cadmium, barium, plutonium, and miscellaneous radionuclides. Two concrete boxes holding 2 million curies of waste from the decontamination of facilities in the 300 Area were placed inside the second tunnel in 1996. Accounting for radioactive decay, about 2 million curies of radioactivity likely exist in both tunnels.

In addition to the old burial grounds, there are eight active low-level waste burial grounds; two are in the 200 East Area and six in the 200 West Area. They cover 556 acres (about 1 square mile) (DOE 2000d). The waste in these burial grounds was first placed inside drums, boxes, or caissons made of steel and concrete depending upon their potential hazard. Approximately 10 million cubic feet[22] of low-level waste was buried at these sites between 1960 and 1999 (DOE 2002a). A small portion of it was never covered with soil to facilitate retrieval. One trench in the 200 West Area accepts mixed chemical and radioactive waste.

Most solid waste that is not buried is stored in the Central Waste Complex, an aboveground set of facilities on the western site of the 200 West Area. Solid waste created at Hanford and at other nuclear sites is stored here. The Waste Receiving and Processing Facility is also located there. It began full operations in 1998 and can process transuranic and low-level radioactive waste as well as provide limited treatment for some mixed chemical and radioactive waste. In addition, T Plant and a nearby facility are used for remotely handled solid waste characterization, equipment decontamination, repackaging, and storage of spent reactor fuel from an offsite commercial reactor client (see Section 5.9).

(22) A volume of 10 million cubic feet would fill a box 215 feet on a side.

5.6 Soil and Groundwater

"The disposal of liquid waste to the ground in the process areas...was considered originally a temporary measure to be discontinued as soon as process development permitted. Then, as now, waste disposal techniques were an attempt to secure maximum economic benefit from waste disposal to the ground consistent with the welfare of human and other life."

—J. McHenry, *Adsorption and Retention of Cesium by Soils of the Hanford Project* (1954, p. 4)

The soil and groundwater beneath Hanford are estimated to contain 1.8 million curies of radioactivity and 100,000 to 300,000 tons of chemicals. Intentional liquid releases into cribs, trenches, and other underground structures account for 1.1 million curies of this total.[23] The remaining 700,000 curies came from waste leaked from tanks (Simpson et al. 2001).[24]

Contamination was released from hundreds of sites. Studies are under way to better understand the nature and behavior of this material though this effort "lacks a systematic framework for identifying and addressing" critical unknowns (NRC 2001b, p. 3).

Hanford is reported to contain 1.1 million cubic feet of soil contaminated with long-lived transuranic radionuclides contained in liquids released into 23 cribs, ditches, trenches, and reverse wells in the 200 Area (DOE 2000f).[25] This contains 25,400 curies of radioactivity (decayed to 1997). This may well be the largest volume of transuranic-contaminated soil of any U.S. nuclear weapons production site.

The soil and groundwater beneath Hanford contain 1.8 million curies of radioactivity and 100,000 to 300,000 tons of chemicals.

Wind and surface water are dynamic forces capable of quickly moving contaminants from waste sources to expose humans and other parts of the environment. Plants and animals can also accumulate and disperse contaminants. On the other hand, chemicals or radionuclides released into the ground move more slowly, taking years or centuries to travel appreciable distances.

With the exception of some long-lived radionuclides found in the Columbia River, most of the 140 million curies of radionuclides released into the air and river have decayed away. However, the underground sediment contains wastes that are sources of present and future contaminant releases. Below ground, most contaminants move where groundwater moves.

Mobile contaminants move underground when groundwater moves.

The potential risk from radionuclides below the ground will change over time. Short-lived radionuclides, with half-lives of minutes to months, have all decayed away. Most radionuclides having half-lives of less than a few years have also decayed into stable

(23) Based upon a summary of radioactivity deposited beneath the 200 Area calculated by the Hanford Groundwater/Vadose Zone Integration Project. Information received from Fluor Hanford, Inc. staff on August 9, 2001.
(24) The author took Simpson et al.'s (2001) numbers and decayed them to the year 2000.
(25) A volume of 1.1 million cubic feet forms a cube 105 feet on a side.

elements, with concentrations indistinguishable from background levels. It will take 300 years for all but about 2000 curies of the 1.8 million curies now found underground to decay away.[26] Long-lived radionuclides, such as technetium-99 and iodine-129, exist for hundreds of thousands to millions of years.

Extensive well drilling began onsite in 1947 to determine the extent of contaminant migration underground (Parker and Piper 1949, p. 88). Geologist Randy Brown and his colleagues reported their underground studies disclosed that limited onsite waste disposal was "adequately safe" (Brown et al. 1958, p. 95). Thereafter, even larger volumes of untreated liquids were released. Disposal stopped at a site when "biologically significant isotopes" were detected in the groundwater in concentrations up to one-tenth the limits set by the National Committee on Radiation Protection.

Perhaps the prevailing disposal philosophy was best expressed by Parker (1948b) when he stated: "There has been much criticism of this method of disposal, largely based on inappreciation of the amount involved. We have been following the fate of this material underground for about a year, and know that the plutonium is rather firmly fixed to the soil, and the fission products at least not freely mobile. If all this material escaped to the river we might have a poor condition, but hardly a disastrous one." A few years later, Parker (1952) wrote, "What the water does at Hanford is irrelevant, if the toxic matter is retained."

Early studies emphasized the ability of the soil to chemically bind many contaminants near the release point. For those not bound, dispersion was relied upon to lower concentrations.

Early studies emphasized the ability of shallow soil and deeper sediments to chemically bind many contaminants near where they were released. For the more mobile contaminants, dispersion was relied upon to lower concentrations. Parker (1948a) wrote, "This gives [the 200 Area] a capacity for hundreds of years of plutonium disposal, and for perhaps 50 years of fission product disposal." However, Art Piper of the U.S. Geological Survey commented that the factual basis for understanding radionuclide movement in the ground was "fragmentary" and "premature" (Parker and Piper 1949, p. 93)

5.6.1 Water and Contaminant Release

"The Hanford Works area, originally believed to be suitable for the temporary or interim underground disposal of certain radioactive [materials]...has served as a test area for the Atomic Energy Commission for such disposal practices...[however] certain factors and conditions limited the waste disposal practices and strongly indicated the need for considerable additional attention."

—R.E. Brown and H.G. Ruppert,
The Underground Disposal of Liquid Wastes at the Hanford Works Washington, Interim Report Covering the Period up to January 1, 1950 (1950)

During the first months of plutonium production and waste generation, mildly contaminated liquid was simply discharged through pipes to a natural low spot on the

(26) This assumes most curies below the ground surface come from cesium-137, tritium, and strontium-90. Longer-lived radionuclides also exist underground. While they contribute little to the curie load, they do control long-term health risks.

Cribs (used 1944-1990s)

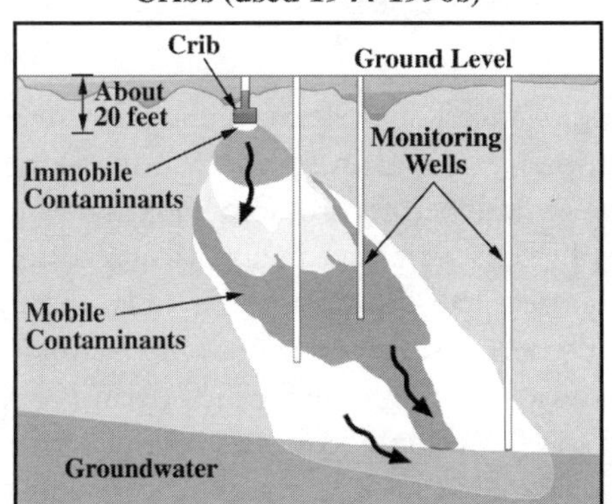

Injection Wells (used 1945-1955; one to 1980)

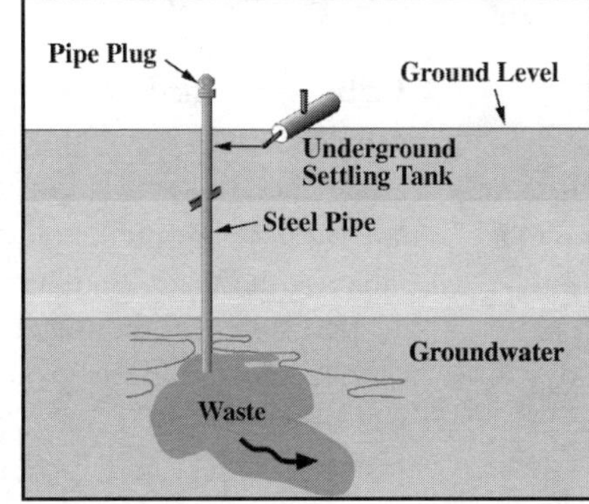

French Drains (used 1944-1980s)

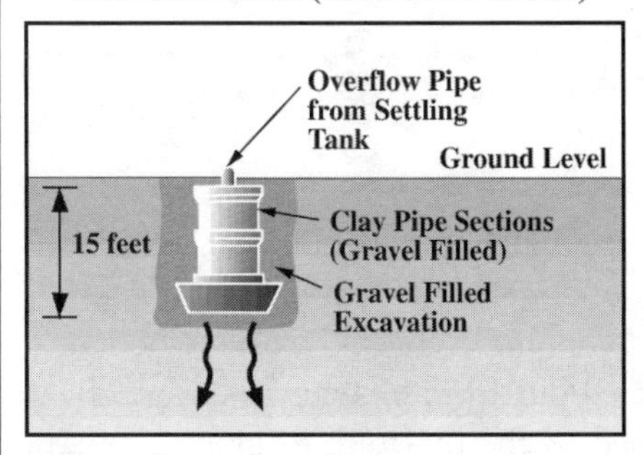

Specific Retention Trenches (used 1944-1973)

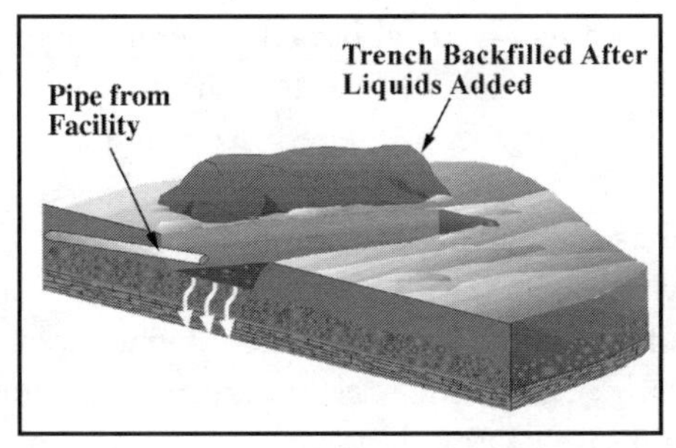

Ponds and Ditches (used 1944-1990s)

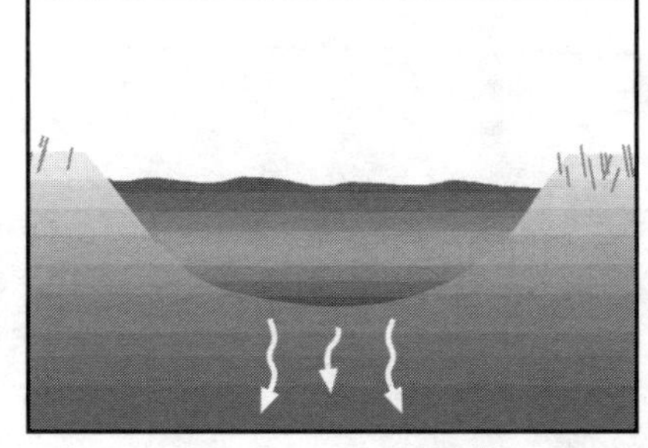

741fb.10

Liquid Releases to Ground. Clean water and contaminated liquids were released to the ground using different techniques including cribs, trenches, French drains, wells, ditches, and surface ponds. Most water was released to three large ponds that no longer exist.

ground. Liquids seeped into the sandy soil, adding contaminants to the shallow sediment and groundwater. Some liquids evaporated, leaving surface residues for plant and animal uptake as well as being spread by the wind. This disposal practice created contaminated wetlands identified by such names as B-Swamp, T-Swamp, U-Swamp, and the Redox Swamp (Healy 1953). A few years later, Parker wrote: "It is impracticable to create surface lakes or swamps of radioactive liquids. These sources contaminate waterfowl and are prone to create particle hazards" (Parker 1956).

Some liquids evaporated, leaving contaminated surface residues for plant and animal uptake.

Such practices quickly became unacceptable. Liquids were then pumped down wells (called reverse wells or injection wells) that ranged in depth from 150 to 300 feet (Brown and Ruppert 1950). While this got contaminants underground, it also inserted them closer to and sometimes directly into the groundwater aquifer.[27] Such disposal bypassed the sediment column relied upon to absorb contaminants. Reverse wells frequently plugged. Within a few months, wells were regarded as another mistake,

Within a few months, injection wells were regarded as another mistake (DuPont 1945).

(27) An aquifer is a permeable underground layer of sediments or fractured rock that can hold and transmit large quantities of groundwater.

Today, isotopes of uranium, strontium, cesium, and plutonium are found in the groundwater near injection wells (Hartman et al. 2001).

The U.S. Geological Survey labeled many of Hanford's early assumptions about contaminant retention in the soil as "unjustified," "speculative," and "unduly optimistic" (Parker and Piper 1949, pp. 89, 92, 93).

except for the disposal of small waste quantities such as from laboratories (DuPont 1945). Depending upon the contaminant nature of the liquids, they were then pumped into:

Cribs—intermediate-level radioactive liquids[28] were pumped underground into either box-like structures that had open bottoms or 20- to 1600-foot-long, gravel-filled tile fields (Pearce 1959). Cribs were designed to promote liquid seepage as the soil absorbed certain radionuclides. Liquids were repeatedly discharged to cribs until "radionuclides with greater than a three year half-life are detected in concentrations exceeding 10 percent[29] of the maximum permissible concentration in drinking water for occupational exposure in the ground water beneath the facility" (Irish 1961). The first cribs were built of wood boxes 5 to 10 feet on a side with an open bottom and were installed 12 feet underground (Parker 1954a). Later, they were constructed as an underground system of interconnected drainage pipes covered by polyethylene sheets and backfill. At Hanford, 131 cribs were used (DOE 1991b).

French drains—vertical buried concrete pipes filled with gravel (DOE 1991b). The 67 drains were each 10 feet deep and received low volumes of cooling water and steam condensate from the reprocessing plants.

Trenches—single-use open trenches receiving select batches of liquids that would otherwise interfere with the chemical absorption of contaminants in the soil if discharged to a crib. In some cases, trenches were temporarily covered with a wooden roof, canvas, or plastic sheets to keep contaminated dirt from blowing in the wind (Parker 1954a). Disposal volumes were determined by the moisture-retaining properties of the soil rather than its affinity to absorb contaminants. Near the reactors, trenches sometimes received liquids discharged when a fuel element ruptured, releasing fission products such as strontium.

Ponds—a naturally low spot or broad depression dug into the ground. Ponds received the largest amounts of liquids disposed. These were mostly uncontaminated cooling waters. In addition, steam condensates from the reprocessing plants and support laboratories were sent to ponds. Such liquids were considered low-level radioactive waste because they were sometimes contaminated.

Economics played a major role in Hanford's waste practices. Discharge to swamps or ponds was the cheapest approach as only simple piping was needed, and the disposal system required little maintenance. The cost per gallon of liquid disposed to ponds was about one-hundredth of one cent (Pearce 1959). Crib disposal cost one-tenth to three-tenths of a cent per gallon disposed. Disposal to trenches was more expensive—three to five cents per gallon. However, the short- and long-term capital and worker costs of well drilling, environmental monitoring, analysis, and reporting were not factored into these expenses. Storage of these liquids in tanks was considered uneconomical unless a major campaign of evaporation was first undertaken.

---

(28) Intermediate-level radioactive waste was defined as liquids requiring additional decontamination (such as filtering and ion exchange through the soil) before release into the environment outside of Hanford but not needing long-term storage in tanks (Parker 1959b).

(29) Ten percent was the cut-off (safety factor) because it was recognized that nearby monitoring wells might not detect the maximum concentration of contaminants.

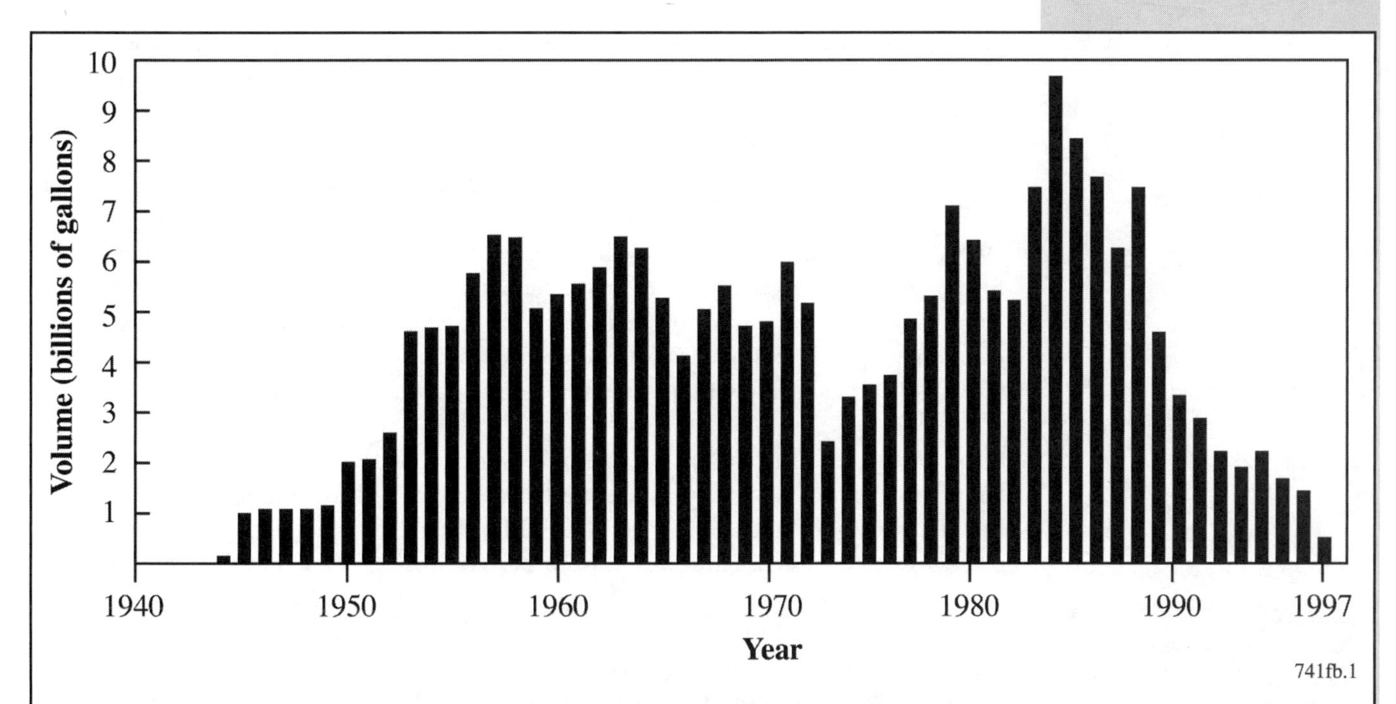

Water Disposal. Some 450 billion gallons of clean and contaminated water were pumped into ponds and ditches at Hanford. This volume equals 5 days continuous flow of the Columbia River. Water releases rapidly decreased as plutonium production was ending in the late 1980s.

U.S. Department of Energy

RG98060057.20
741fb.39

U Pond. Ponds and ditches once received water discharged from Hanford plants. Sometimes they were located near tank farms and other waste storage or release sites. The 14-acre U Pond shown in this 1962 photograph, located in the 200 West Area, was one of the largest. Pond water helped move contaminants deeper and faster through the subsurface. Water discharges to U Pond ended in 1984. The area was covered with soil.

Promises were made to improve liquid disposal practices. However, most untreated discharges continued to the early 1990s.

Water seeping into the soil above the tank farms can spread contaminants deeper underground.

Nonetheless, releasing mildly contaminated wastes into the ground remained "most troublesome" (Parker 1956). This is why substantial areas of land were reserved for waste disposal.

During the first 50 years at Hanford, 400 to 450 billion gallons of liquids were pumped to 30 ponds and ditches (ERDA 1975; DOE 1992a, 1992b, 1998c; Hartman and Dresel 1998). This volume is the same as the amount of water flowing down the Columbia River over 5 days. These estimates are uncertain, especially for liquids released before 1979. Ninety-five percent of these discharges occurred in and around the 200 Area where most ponds and ditches were located. While most liquid was uncontaminated, about 10% contained low to intermediate levels of radionuclides and chemicals.

Ground disposal of liquids was recognized as a temporary practice tolerable only as a temporary "expedient, until practical means can be devised for making such fluids inert" before discharge (Parker and Piper 1949, p. 93). Promises were made to improve these practices (Parker 1951). However, most untreated discharges continued to the early 1990s.

Water also naturally enters the ground as precipitation from rain and melting snow. The amount of this water, called recharge, varies across Hanford from near zero to more than 4 inches a year depending on precipitation amounts, vegetation type, and soil cover (Gee et al. 1992).

This recharge causes an interesting phenomenon to occur in areas where the vegetation cover was removed such as above the tank farms. After the tanks were built, they were covered by coarse gravel and sand. Vegetation is not permitted to grow atop the tank farms to prevent fires and for general safety reasons. These coarse, plant-free areas allow water to easily enter the ground rather than evaporating or transpiring. Today, we know tank domes act like umbrellas that channel infiltrating water down the sides of tanks (GAO 1998a). Sometimes water ponded across tank farms after a rapid temperature increase caused snow to melt quickly. This happened in 1979 atop the T Tank Farm, and "there is little doubt that this is not the only event of this type that has occurred" (Hodges 1998). Open fire hydrants have also flooded the ground above tank farms. Where past waste leaks exist, such water recharge can spread contamination deeper underground.

5.6.2 Subsurface Environment and Contamination

"Historically, scientists, regulators, managers, and decision makers concerned with subsurface contamination have focused on groundwater and contaminant movement below the water table. This focus seemed warranted because groundwater is the principal system for moving contaminants away from a disposal site...Today, the vadose zone is recognized as a key player in determining the long-term impacts of contamination..."

—Marilyn Quadrel and Regina Lundgren,
Vadose Zone: Science and Technology Solutions
(Looney and Falta 2000, p. 62)

Eventually, all contaminants not irreversibly attached on sediments, chemically precipitated, lodged between sediment grains, or decayed away will reach the Columbia River at some location, concentration, and time.

When released, liquids drain though layers of sediments sandwiched between the surface and the water table. This region is called the vadose zone.[30] The vadose zone acts like a time-release capsule, holding some contaminants permanently, slowly releasing others, and quickly allowing mobile contaminants to move deeper and outward. It provides the future source for contamination to enter the groundwater.

The vadose zone acts like a time-release capsule for contaminants.

Except for shallow wells located around Hanford tank farms, the vadose zone is not extensively monitored. Very few wells are drilled near waste sites for sampling sediment and studying contamination movement in the vadose zone. This is why the National Research Council (2001b, p. 63) concludes that contaminant distributions, chemical forms, and future movement beyond the immediate "footprints" of waste sites are poorly known.

Beneath the 200 Area, the vadose zone is 200 to 350 feet thick (DOE 1999d). In the 100 and 300 Areas, near the Columbia River, it's generally less than 100 feet thick (Bechtel Hanford 1998; Newcomer and Hartman 1999).

At some distance below the vadose zone is the water table. It marks the top of the groundwater aquifer. A hard crystalline rock called basalt marks the bottom of this aquifer. Mobile contaminants reach the aquifer. Once there, they travel down the hydraulic gradient (from high to low water table elevation) towards the Columbia River, enter the river, and move downstream. Beneath the 200 Area, the water table is 60 to 100 feet higher than along the Columbia River. The aquifer thickness beneath the 200 Area varies between 30 and 400 feet. South of the 200 Area and just north of Gable Mountain, the aquifer is

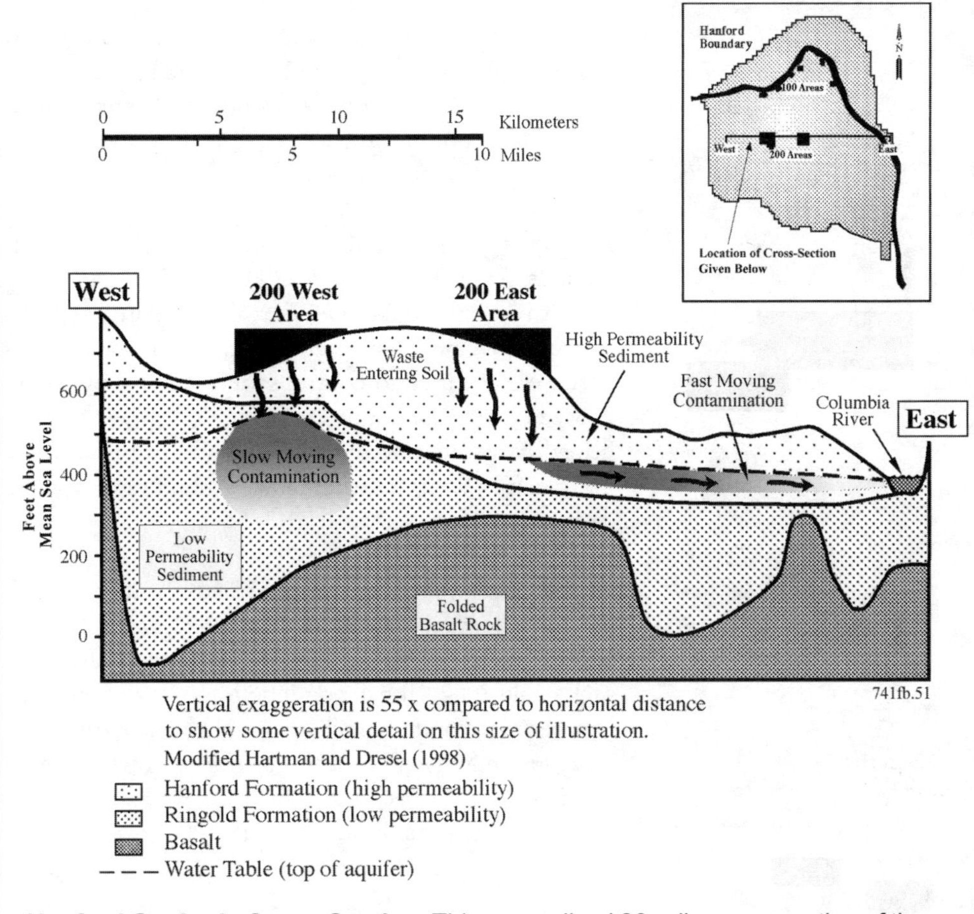

Hanford Geologic Cross-Section. This generalized 20-mile cross-section of the Hanford subsurface shows contaminated groundwater entering and moving in the underlying aquifer. Mobile contaminants travel rapidly toward the Columbia River through permeable sediments making up the upper aquifer beneath the 200 East Area. Contaminants move more slowly through less permeable sediment under the 200 West Area.

(30) The vadose zone contains most of the chemical and radioactive contaminants discharged or leaked into the environment. Underground tanks, solid waste burial grounds, cribs, and trenches lie in the upper vadose zone. This zone is built from a porous framework of sediment grains made of different minerals containing microbes and plant life plus liquids and gases filling the interconnected open spaces between sediment grains. All of these can influence contaminant movement. Continued liquid infiltration promotes more contaminant movement and may exacerbate cleanup efforts.

thickest, reaching 600 feet. This is where the underlying basalt rock is folded into broad sediment-filled valleys. Generally, the aquifer is 200 to 300 feet thick near the Columbia River (Poston et al. 2000).

> The ease and direction of contaminant movement in the subsurface are affected by many factors. These include how sediments are layered, what minerals make up those sediments, chemistry of the contaminants, chemistry of the groundwater, types of microbes living underground, and the hydrology of the aquifer.

Disposed liquids provided a hydraulic force to drive some contamination deeper into and faster through the ground than would have otherwise taken place. This problem was recognized early in the site's history. Parker (1948a, p. 5) reported, "Disposal of large volumes of inactive water adjacent to the waste disposal points might also affect retentivity [of contaminants]."

Past liquid releases significantly raised the water table. Large rises, called groundwater mounds, formed beneath the 200 Area. From 1945 to 1984, the water table rose 79 feet beneath the 200 West Area (Hartman and McDonald 2002). By 1969, it peaked 17 feet above pre-Hanford conditions near the 200 East Area. These mounds were unsuccessfully used to control groundwater flow patterns by cycling water discharges between individual ponds (Brown and Ruppert 1950). Such control might have helped delay the movement of contaminants. However, the site geohydrology was too complex to accurately predict groundwater flow in response to changing water disposal practices.

As discharges were reduced and ponds no longer used, groundwater levels began to drop, and groundwater mounds shrank in size. The water table has dropped 36 feet beneath the 200 West Area and 8 feet beneath the 200 East Area (Hartman and McDonald 2002). It will continue to go down until a new equilibrium is established between water inflow (recharge) and water outflow (discharge). This cessation redirects some groundwater flow patterns. For example, halting discharges to Gable Mountain Pond (located between the 200 East Area and Gable Mountain) beginning in 1984 now permits more groundwater from the 200 East Area to move southeast than northwest as it did when the water table was higher (Lindberg et al. 2002).

"The [Energy] Department's understanding of how wastes move through the vadose zone to the groundwater is inadequate to make key technical decisions on how to clean up the wastes at the Hanford Site in an environmentally sound and cost-effective manner" (GAO 1998a, p. 1).

Informal estimates suggest 30 to 50 cubic feet (225 to 375 gallons) of Hanford groundwater enter the Columbia River every second. This compares to an average river flow of 115,000 cubic feet (860,000 gallons) every second (Patton 1999).[31]

During fiscal year 2001, groundwater samples were collected from 706 wells distributed across Hanford (Hartman and McDonald 2002). Most monitoring uses water samples collected at or near the water table. This is where most mobile contamination is believed to advance outward.

(31) A river flow of 115,000 cubic feet per second forms a cube of water 50 feet on a side moving past an observer every second.

Groundwater monitoring reveals that each contaminant behaves differently depending upon a host of geologic, hydrologic, and geochemical influences (DOE 1997b; Newcomer and Hartman 2001). Some of the fastest-moving contaminants are tritium, technetium-99, iodine-129, and nitrate. These travel at or close to groundwater speed. Various forms of plutonium, uranium, neptunium, strontium, and cesium move slower because they are less soluble or they chemically react with minerals in the surrounding sediment to retard their movement. However, it's important to anticipate their long-term chemical sorption and potential alterations. Over time, contaminants once strongly sorbed might become more mobile. Some chemicals such as carbon tetrachloride and trichloroethylene slowly dissolve. Carbon tetrachloride, only slightly soluble in groundwater, naturally degrades to chloroform, which is more soluble. Small colloid particles, containing chemicals or radionuclides, may also move underground.[32] Fifty years ago, Parker (1954a) said colloidal movement was of interest because these materials may not follow the "absorption laws of solutions." The current state of knowledge about contaminant mobility and the conditions controlling that mobility are discussed in the U.S. Department of Energy report *Groundwater/Vadose Integration Project* (DOE 1999d).

The movement of radionuclides and chemicals is intricately linked to the geology, hydrology, and chemistry of the waste, soil, and groundwater.

Contaminant mobility can be altered intentionally or accidentally. For example, the chemistry of groundwater is adjusted by inserting certain chemical agents. The mobile form of chromium (hexavalent chromium) is of concern in the groundwater near D and DR Reactors. A chemical agent (sodium dithionite) was injected into the groundwater to create a zone of electron donation. This chemically reduces the mobile form of chromium to immobile (trivalent) chromium (Fruchter et al. 2000). Initial results are encouraging and demonstrate a 60% cost savings compared to conventional groundwater pumping and treating (Apley 2001).

On the other hand, chemicals found in leaked tank waste cause some radionuclides to be retarded while others move deeper into the ground than otherwise expected.

The 300 Area contained a "waste disposal system which is a model of what one should not do" (Parker 1948b). Because many of materials handled in 300 Area were considered less dangerous than their 100 or 200 Area counterparts, less care was taken in their disposal. Acids (nitric, sulfuric, and hydrofluoric) containing uranium, strontium-90, zirconium, copper, and beryllium were released to the ground (DOE 1997a; Hartman et al. 2001). Chemicals such as nitrate, trichloroethylene, and dichloroethene are found in the local groundwater at concentrations unacceptable for drinking water (Lindberg and Dresel 2002).

The 300 Area contained a "waste disposal system which is a model of what one should not do" (Parker 1948b, p. 8).

Liquids from 300 Area were once piped into an underground tank and discharged to ponds and trenches built next to the Columbia River. Drinking water wells were 600 feet from these ponds and during times of high river water, contaminated water was flushed into those wells (Parker 1948b). In 1948, a dike broke, spilling 14.5 million gallons of chemical solutions into the river (Adley and Crane 1950a). New waste ponds, process trenches, a waste treatment system, and solar evaporation basins were later used. Today, no untreated wastewater is released in the 300 Area. Contaminated sediments beneath these old ponds and trenches are being dug up and moved to a landfill in the 200 Area.

(32) Colloids have the potential to travel in groundwater. At this time, there have been no credible reports of substantial colloid-assisted contaminant transport taking place (NRC 2001b, p. 70).

Radionuclide Movement in Soil

After investigating factors that control the chemical sorption of cesium-137 (^{137}Cs) on sediments sampled from beneath Hanford, geochemist John Zachara reported "our work has shown that approximately 60% of the in-ground $^{137}Cs^+$ pool in the S-SX tank farm is immobile and is unlikely to desorb from the sediments under any reasonable geochemical scenario" This finding "has reduced by more than 50% the effective in-ground inventory of $^{137}Cs^+$ that needs be considered from the [biological] risk perspective" (Zachara 2002, pp. 3.2-3.3).

Geochemist Jeff Serne stated that his studies of technetium-99 contaminated sediment collected from beneath the SX and BX Tank Farms "have not revealed any evidence that the technetium leaked or spilled from these tank farms significantly interacts with the sediment to inhibit movement. Given the current environmental conditions below the tanks, if more water enters the sediment (including normal rainfall and snow melt), the technetium is available for deeper transport, including to the water table."(33)

Such research is two examples of the growing scientific basis for better understanding the behavior of radionuclides as they may interact with minerals (such as mica, quartz, and carbonate coatings) and geochemical conditions found in Hanford sediments. The more predictions are based upon knowledge rather than assumptions, the more accurately we can discern long-term biological risks. The role of science in supporting cleanup policies and decisions is discussed in Chapter 10.

5.6.3 Groundwater Plumes and Travel Times

"Theoretically, restoration of contaminated ground water to drinking water standards is possible. However, cleanup of contaminated ground water is inherently complex and will require large expenditures and long time periods, in some cases centuries."

—National Research Council,
Alternatives for Ground Water Cleanup (1994a, p. 2)

Hanford groundwater discharges to the Columbia River.

About 150 square miles of groundwater beneath Hanford contain high to diluted levels of contamination. Monitoring reveals a number of contaminated plumes(34) emanating from the 100, 200, and 300 Areas. Groundwater monitoring reveals that two-thirds of the area covered by these plumes, or 100 square miles, is contaminated at levels unacceptable for drinking water (that is, above the regulated standards set for safe drinking water) (Hartman et al. 2002).

---

(33) Conversation with Jeff Serne, Pacific Northwest National Laboratory, October 14, 2002. Examples of these geochemical data and interpretations are reported in Serne et al. (2002a, b, c).

(34) A plume defines the three-dimensional extent of contaminants present in an aquifer in amounts above an acceptable level such as drinking water standards. Contaminant plumes also exist in the vadose zone.

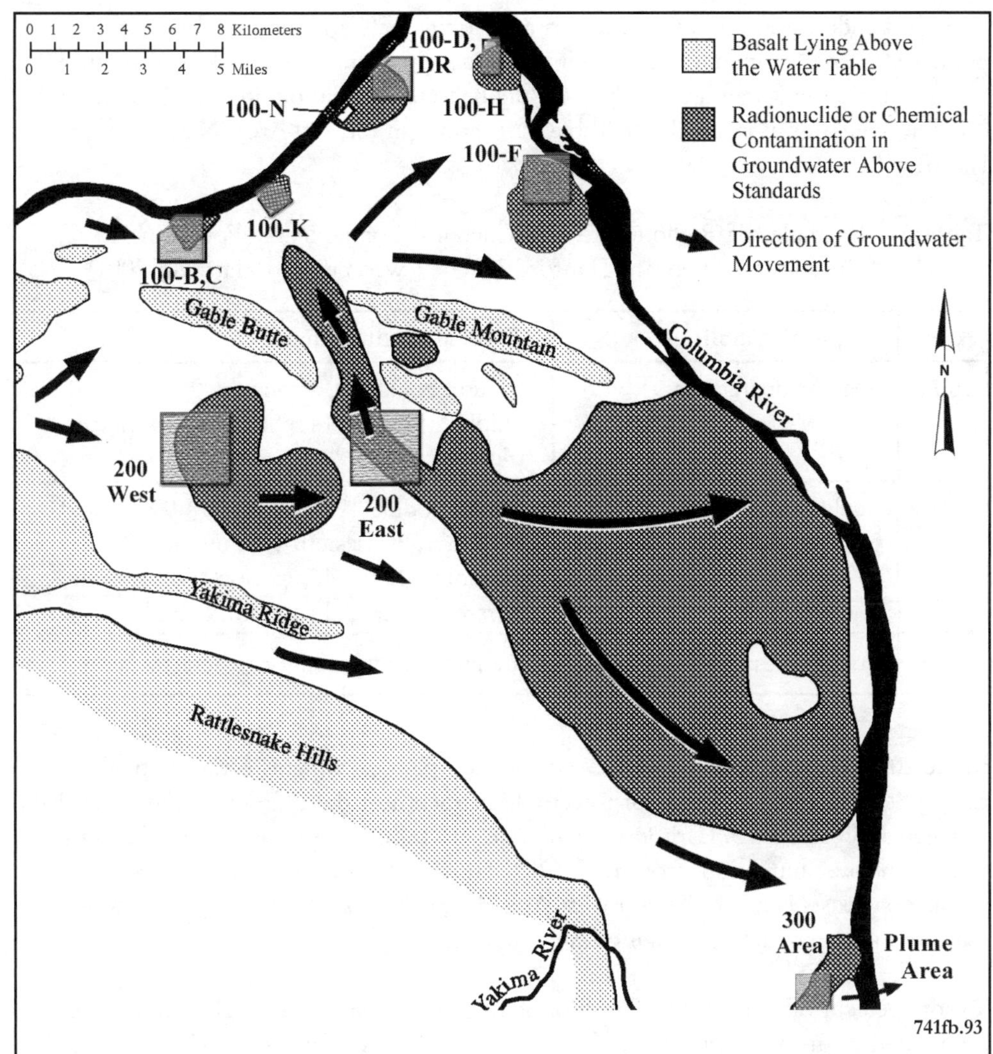

Contaminated Groundwater Plumes. Nearly 100 square miles of Hanford groundwater contain chemical and radionuclide contamination above drinking water standards. The largest plume extends from the 200 East Area. Smaller plumes exist elsewhere. Generally, groundwater moves west to east beneath Hanford.

See the Agency for Toxic Substances and Disease Registry's fact sheets about the potential health hazards for contaminants at waste sites (http://www.atsdr.cdc.gov/toxfaq.html).

For more information about health-related impacts of contaminants, see the fact sheets written for DOE (http://riskcenter.doe.gov/techinfo/factsheets.cfm).

Groundwater plumes are thickest near their sources and thinner near the Columbia River.

The thickness of contaminated plumes extending outward from the 200 Area is estimated at about 50 feet. In the 100 and 300 Areas, closer to the Columbia River, plume thickness is less—10 to 15 feet. Plumes are thicker near the center of Hanford because more vertical mixing of groundwater takes place in areas of water recharge before outward movement dominates. Brown et al. (1962) wrote, "The control exerted by geological factors (structural and stratigraphic) in addition to the gravity effects operating on high salt wastes may greatly increase the tendency toward downward movement." Plumes are thinner near the river as groundwater discharges upward into the river.

The most widespread groundwater plumes contain tritium, iodine-129, and nitrate.

The most widespread plumes contain tritium, iodine-129, and nitrate. Other smaller plumes also exist containing carbon tetrachloride, chromium, carbon-14, strontium-90, technetium-99, and uranium. Cesium-137, cobalt-60, and plutonium-239/240 exceeded drinking standards at isolated (see Table 5) locations in the 200 Area (Newcomer and Hartman 2001).

Table 5. Examples of Radionuclide and Chemical Contaminants Found in Groundwater Beneath Hanford (after Newcomer and Hartman 2001)

Area	Type of Facilities	Typical Contaminants
100	reactor operations	tritium, cobalt-60, strontium-90, uranium, carbon-14, chromium, nitrate, trichloroethene
200	fuel reprocessing and plutonium purification	tritium, iodine-129, technetium-99, strontium-90, cesium-137, uranium, plutonium-239/240, carbon tetrachloride, nitrate, fluoride, chromium, cyanide
300	fuel fabrication	uranium, nitrate, trichloroethene

In the 200 West Area, one of the largest chemical plumes contains carbon tetrachloride.[35] Some 630 to 1015 tons of it were discharged near the Plutonium Finishing Plant between 1955 and 1973. Carbon tetrachloride and tributyl phosphate were used in the plant to recover plutonium from recycled waste. A 4-square-mile carbon tetrachloride plume now exists beneath the 200 West Area that, in places, contains chemical concentrations 1400 times above drinking water standards.

Twenty percent of the carbon tetrachloride originally discharged is believed to have evaporated to the atmosphere (Rohay 1993). Vapor extraction, from wells drilled into the vadose zone, started in 1992. Groundwater extraction began 2 years later. As of late 2001, 91 tons—perhaps 10% of the total carbon tetrachloride released—were recovered (Dresel et al. 2002; Rohay 2002).

Tritium was created at Hanford as a fission product when uranium fuel was irradiated (Phillips and Easterly 1980). In addition, two limited tritium production programs took place.[36] The largest tritium plume covers 100 square miles, extending from the 200 Area to the Columbia River. It may contain 100,000 to 200,000 curies of tritium.[37] A study of tritium released from the PUREX Plant found that 68% of the site's tritium was released in liquids discharged to underground cribs (Jeppson 1973). Of the remainder, 5% went to tanks, 22% was pumped to surface ponds, and 5% escaped as gaseous effluents up the stacks of reprocessing plants.

(35) People are exposed to carbon tetrachloride mostly from breathing air, drinking water, or coming in contact with contaminated soil. This generally occurs around hazardous waste sites or in certain businesses, such as dry cleaning. Exposure to very high amounts can damage the liver, kidneys, and nervous system (ATSDR 2002).

(36) Most tritium production in the nuclear weapons complex took place at Savannah River, South Carolina.

(37) Tritium has a specific activity of 10,000 curies per gram. Therefore, 100,000 to 200,000 curies of tritium contain 10 to 20 grams (one-third to two-thirds of one ounce) of tritium.

The first tritium production mission at Hanford was the P-10 Project, operating from 1949 to 1952, when lithium-containing targets were irradiated inside B Reactor. The second was the CoProduct Program that ran from 1965 to 1967. Aluminum-lithium targets were irradiated in the K and N Reactors (Dresel et al. 2000). The fate of waste generated remains unknown though material from the second project is likely buried north of the 300 Area.

The Columbia River is 10 miles east of the 200 East Area. Between 1962 and 1965, the leading edge of a fast-moving tritium plume reached the Columbia River (Brown et al. 1962; Eliason 1966), 20 or so years after liquid waste disposal began in the 200 Area. Tritium concentrations in groundwater above drinking water standards reached the river about 30 years after disposal began (Newcomer and Hartman 1999).

Contaminated groundwater reached the Columbia River 20 years after liquid disposal began in the 200 Area.

Years earlier, Parker (1954a) wrote, "Although some estimates of the time of [groundwater] movement to the river range from 1000 to 1500 years for locations presently blocked by ground water mounds, it is believed that a minimum time will be on the order of 50 to 100 years."

Groundwater travel time to the river from the 200 West Area may approach a century or more. Sediments surrounding the local aquifer have lower permeability than sediments beneath the 200 East Area. This is why contaminated groundwater from the 200 East Area has spread across 100 square miles, while that from the 200 West Area covers only a few square miles.

Mobile contaminants discharged to the sediments in the 100 and 300 Areas moved rapidly to the Columbia River because these areas lie adjacent to the river and generally rest atop highly permeable sediments. In some cases, travel times to the river were hours to days.

A long groundwater travel time is good when it provides extra time for cleanup, radioactive decay, or chemical breakdown. However, over time, some contaminants will spread farther and become increasingly difficult to contain or remove.

One challenge facing Hanford is accounting for contamination released into the ground, particularly contaminants of future concern. What was released? How much? What is its chemical form? Where will it move? Have the greatest biological risks passed or are they yet to come?

Has the greatest biological risk from groundwater contamination passed or is it yet to come?

5.6.4 Wastewater Treatment

Since 1995, untreated wastewater is no longer discharged into the ground at Hanford.

Two state- and federal-licensed liquid treatment plants were built in the early 1990s to remove low-level radioactive and chemical contamination from Hanford wastewater. This was done to better monitor and control the discharge of untreated water. Materials removed from the wastewater are packaged and shipped to solid waste burial grounds in the 200 West Area or to the Environmental Restoration Disposal Facility located near the 200 West Area. That facility is a large open trench, compliant with landfill

Two liquid treatment plants were built in the early 1990s... 50 years after liquid discharges began onsite.

regulations, built to receive Hanford-generated low-level radioactive, chemical, or mixed solid waste (Oestreich et al. 1998). It is one of the largest engineered radioactive and mixed waste disposal sites in the world (Hughes 2001).

One treatment plant located just north of the 300 Area receives wastewater from approximately 45 office buildings, research laboratories, and support facilities in the 300 Area (Danielson 1999). The plant removes metals, radionuclides (except tritium), and organic contaminants. Purified water is discharged to the Columbia River.[38] A second plant, sited in the 200 East Area, is called the 200 Area Liquid Effluent Treatment Facility. It began operations in 1995. All releases are regulated under Washington State or federal discharge permits. Wastewater fed to the 200 Area facility is received from several sources, including a tank waste evaporator, the Plutonium Finishing Plant, a chemical laboratory, T Plant, and the Waste Encapsulation and Storage Facility. It disposes of treated water, containing some tritium, to a State-approved site located east of the 200 East Area. During fiscal year 2001, 122 million gallons of water were pumped to the facility (Lindberg et al. 2002).

5.6.5 Waste Leaked from Tanks

"Considering the depth and extent of some of the relatively immobile contaminants...a great concern has been raised as to the distribution of the more mobile radionuclides such as ^{99}Tc [technetium], tritium, and ^{129}I [iodine]."

—U.S. Department of Energy,
Groundwater/Vadose Integration Project (1999d)

The first tank leak was suspected in 1956 and confirmed 3 years later.

In Hanford's early years, tank leaks were a concern, but the "safety factor" of having 200 feet of dry sediment beneath a tank was believed capable of retaining "several tens of thousand gallons of waste" before liquids would reach the water table and move outward (Tomlinson 1959a). In addition, solids present in the leaked waste would reduce the permeability of the soil—making it "self-sealing." Thus, leaked waste was believed to remain confined to the soil immediately beneath a tank.

Sixty-seven single-shell tanks have leaked or are suspected to have leaked as much as 1.5 million gallons of mostly sodium nitrate and cesium-contaminated liquids into the soil (NRC 2001b, p. 22). This volume could be larger. Long-lived radionuclides, such as iodine-129 and technetium-99, are present in this waste (DOE and Ecology 1996).

Los Alamos National Laboratory researchers Simpson and others (2001) estimate this leaked waste contains about 870 tons of chemicals and 780,000 curies of radioactivity (decayed as of 1994). This curie count drops to 660,000 when decayed to the year 2000. When this curie load is added to the 1.1 million curies estimated to be in the soil and groundwater from intentional liquid discharges,[39] then a total of about 1.8 million curies could exist in the subsurface as of the year 2000.

---

(38) The treatment plant uses chemical precipitation to remove heavy metals (metals are precipitated when the wastewater is made slightly alkaline by adding ferric chloride and sodium hydroxide), ion exchange to remove metals and most radionuclides, and ultra-violet light oxidation and hydrogen peroxide destruction of organic compounds. Filters and clarifiers remove suspended solids from the process stream.

(39) Based upon a summary of radioactivity deposited beneath the 200 Area calculated by the Hanford Groundwater/Vadose Zone Project. Information is from Fluor Hanford, Inc. staff on August 9, 2001.

In January 1956, unusual waste level changes were noted in one tank (Tomlinson 1959a). This was attributed to the buckling of a bottom steel plate. The tank was later filled with water. No leakage was detected. In 1957, fluid losses brought the integrity of a second tank into question. This time the loss was attributed to evaporation. In May 1959, the bottom steel plate of a third tank that had received hot waste from the REDOX Plant was found to have bowed 4 feet upward. Liquid waste was removed, and its integrity remained under investigation. Hard-to-explain fluid level changes in these tanks were often described as "suspicious occurrences" (Parker 1959a).

The first waste leaks were suspected in 1956 and confirmed in 1959. One was in the U Tank Farm in a tank used for 13 years, and another was in the younger TY Tank Farm in use for just 6 years (Brevick 1995a, 1995b). From 1959 to 1968, 12 tanks were confirmed to be leaking (labeled "leakers"). Liquids continued to be added to some single-shell tanks until 1980 (GAO 1989). The last new tank leak was declared in 1988.

Some leak estimates, such as for four tanks in the SX Tank Farm in the 200 West Area, might be much higher than reported. The commonly quoted 1-million-gallon value for all Hanford tanks (Hanlon 2002) does not include potentially higher volumes estimated by using newer modeling techniques or waste lost from underground pipelines and diversion boxes (Agnew 1997; GAO 1998a).

The U.S. Environmental Protection Agency (2000) is concerned that adequate tank leak and contaminant detection is not in place.

Based upon leak volumes and radioactivity levels (Waite 1991; Wodrich 1991; Pajunen et al. 1994; Agnew 1997), plus discussions with Hanford staff,[40] the estimated amount of radioactivity in these liquids varies from less than 0.2 curies to 2.4 curies a gallon.[41] An average value of 1 curie a gallon is used in this book.

The largest leak began in April 1973 in Tank T-106 located in the T Tank Farm, 200 West Area. About 115,000 gallons, containing 40,000 curies (16 ounces) of cesium-137, 14,000 curies (4 ounces) of strontium-90, 4 curies (2 ounces) of plutonium, and various other radionuclides, was released (Atomic Energy Commission 1973). So much waste was lost because procedural checks of the tank's liquid level were not properly followed by workers. This is the largest tank leak reported for the U.S. nuclear weapons complex.

Many single-shell tanks contained thermally hot waste, contributing to metal corrosion, high stresses, and metal buckling (Agnew and Corbin 1998).

Uncertainty also exists about the chemical and radiological characteristics of the leaked waste as well as its distribution underground. Contamination spread is receiving considerable attention because some radionuclides have moved deeper than originally thought possible (GAO 1998d; DOE 1999d). On the other hand, others, such as cesium, move less than originally expected (Zachara 2002). The U.S. Environmental Protection Agency (2000) expressed concern that adequate detection systems are not installed in all Hanford tanks to detect future leaks. That Inspector General report (EPA 2000) goes on to say that an effective leak investigation process is needed to ensure that tank leaks are expeditiously identified and remediated. This includes monitoring waste movement in the sediment beneath tanks.

In addition to leaked waste, an additional 120 to 130 million gallons of tank liquids were intentionally discharged to cribs and trenches in the 200 Area (Waite 1991; Agnew 1997). This was done to "economize"; that, is lessen the volume of liquids kept inside

(40) Communication with Al Pajunen, Fluor Hanford, Inc. (now retired), October 9, 2000.
(41) Values decayed as of several dates ranging between 1989 and 1996.

tanks to free up space for newly generated high-level waste (Parker 1954a). Most was discharged between 1946 and 1958 during the early years of plutonium or uranium recovery at T, B, and U Plants. Some waste was pretreated[42] to remove select chemicals and radionuclides (Parker 1954a; DOE 1999d).

Many tanks and liquid disposal sites are located near each other. For example, within 500 feet of the 40 single-shell tanks in the B, BX, and BY Tank Farms (200 East Area), approximately 74 million gallons of liquids were released through 28 cribs, trenches, reverse wells, and French drains (Narbutovskih 1998). This took place from the mid-1950s to the early 1970s. Between 1971 and 1984, 20 of these tanks were confirmed or assumed to have leaked about 120,000 gallons of high-level waste to the underlying soil (Hanlon 2001).

One pretreatment approach was to simply cascade waste between tanks before it overflowed into a crib. Waite (1991) wrote that some wastes "were discharged to the ground after they had settled for a period of time in tanks." As of 1989, this amounted to 120 million gallons of waste that contained an estimated 300,000 tons of chemicals and 60,000 to 65,000 curies of radioactivity (Waite 1991; Wodrich 1991). As of the year 2000, this radioactivity would have decayed to 45,000 to 50,000 curies. Not included in these release estimates is about 1 million curies of short-lived ruthenium-106 that has since decayed away. Agnew (1997) estimated that 4.7 million curies of radioactivity was contained in 130 million gallons of tank liquids discharged to cribs. Those numbers are higher than reported by Waite and Wodrich.

The hazard posed by leaked waste has been disputed since Hanford's earliest days.

As noted earlier, the Waite, Wodrich, and Agnew numbers reasonably bound the amount of radioactivity released, then each gallon of intentionally released liquid contained less than 1 curie of radioactivity.

The hazard posed by leaked waste has been disputed since Hanford's earliest days. Parker (1948b) wrote, "that all of this [tank] material could be released without its constituting a major disaster. It is proper to advise the Committee that Mr Gorman, A.E.C [Atomic Energy Commission] and Mr. Piper, U.S.G.S. [United States Geologic Survey] believe that there are areas of doubt in these speculations." At the time of this quote, Hanford tanks contained 50 million curies of radioactivity, 500 tons of uranium, and 110 pounds of plutonium.

Understanding contaminant behavior beneath tanks will influence many cleanup decisions.

In 1997, the federal government acknowledged that tank waste from several single-shell tank farms had reached groundwater. Cesium, technetium, and strontium had migrated deeper than previously acknowledged. Other tank-originated metals such as chromium, sodium, and nitrate are also likely in the groundwater. Iodine and tritium, from the waste evaporators, had reached groundwater.

Understanding contaminant behavior beneath tanks will influence many cleanup decisions, such as how tank waste is removed, the possible need for engineered barriers, and whether leaked waste should remain in place, be treated, or be recovered.

(42) Pretreatment means waste evaporation, waste cascading between tanks (to settle out solid particles before liquids are released to the soil), or removal of select radionuclides.

5.7 Contaminants Released to the Columbia River

"Efforts should be continued toward the reduction of the amounts of radioactive materials that enter the river from atomic energy installations, especially for those radionuclides which are concentrated to a high degree by aquatic organisms."

—U.S. Department of Health, Education, and Welfare et al., *Water Quality Studies on the Columbia River* (1954, p. 6)

An estimated 113 million curies of radiation were released to the Columbia River from 1944 to 1971 during operation of Hanford's first eight reactors (Heeb and Bates 1994). Many of these radionuclides were created when neutrons, emitted during the fission process inside a reactor, were captured by natural elements dissolved in the cooling water as well as by chemicals added for water treatment (Haushild et al. 1973). Such radionuclides are called activation products.[43] Radionuclides released from broken fuel slugs were fission products; that is, they were created by the fission of uranium making up the slug. In addition, hazardous chemicals were dumped into the river.

"The more carefully one looks at it, the more isotopes does one find" (Parker 1951).

In most reactors, cooling water was temporarily stored in concrete or steel basins before being returned 200 to 1600 feet into the river through large underwater pipes.[44] The plan was to hold the water for about 6 hours before dilution in the river (Parker 1945b). This would allow time for short-lived radionuclides to decay and total radioactivity to decrease by a factor of 20. By the late 1950s, holding times were reduced to between 1 and 3 hours (Parker 1959a). As radionuclides moved downstream, radiation levels further decreased. By the time water had flowed 35 miles south of the reactors (near Richland), radiation levels were 10% of that found in the storage basins (Foster 1959).

Minimum river travel times to the nearest downstream town (Richland) were 11 hours at times of high-river flow and about 22 hours at low-river flow (Honstead et al. 1960). The flow time between the reactors and the downstream water intake systems in Pasco and Kennewick ranged between 8 and 30 hours (Davis et al. 1958). In other words, it took less than 1 day for contaminated river water to flow to the nearest downstream municipality.

Was chemical toxicity exceeding radiation toxicity in the Columbia River?

When released, reactor water was 120°F higher than ambient river temperatures (Foster et al. 1954). When river flows were low during the summer months and all reactors were operating, the river temperature for the first 56 miles downstream was increased by 5°F (Becker 1990). Arrangements with upstream dam operators allowed higher releases of cool water to protect fish.

(43) During 1958, 95% of the radiation exposure received by people living downstream came from these activation products (Foster 1959). The remaining 5% was from fission products released when the metal jacket enclosing a uranium fuel slug ruptured. A trace amount of activation products was also created when uranium, naturally dissolved in river water, captured neutrons while passing through a reactor. This was informally called "tramp uranium" (Stannard 1988, p. 759).

(44) The first cooling water basins were made of concrete (Foster 1959). However, the concrete soon began to crack from the thermal shock of receiving and discharging hot water. Steel basins replaced the concrete.

Increased river temperatures were thought to "constitute the greatest potential threat to the local salmon" (Foster et al. 1954). Concerns also centered around the chemical toxicity of sodium dichromate (Parker 1948b). Parker wondered if, in some instances, chemical toxicity was exceeding radiation toxicity. For example, years later during just 12 months (July 1966 to June 1967), nearly 40,000 tons of chemicals (sodium dichromate, sulfuric acid, aluminum sulfate, bauxite, and chlorine)[45] were released into the Columbia River from reactor effluent in the 100 Area (Wells 1967). During the same period, 165 tons of chemicals (for example, sodium hydroxide, nitric acid, and aluminum sulfate) used in fuel fabrication were dumped into the 300 Area process ponds. Contaminated liquids from these ponds seeped into the river.

As a reactor operated, a film of oxides and other debris built up in the process tubes and on the fuel elements. This interfered with cooling water flow and slug removal. A fine abrasive slurry of diatomaceous earth was periodically injected into the water stream to scour out these materials in a process known as purging. Purges began in 1945 and continued until the last of the single-pass reactors was shut down in 1971. The radioactivity contained in this purged material increased radionuclide releases by about 3% compared to releases from routine operations and fuel ruptures (Heeb and Bates 1994).

Fuel slugs ruptured when cracks broke through their thin metal coating.

The Columbia River also received radionuclides from the failure of irradiated fuel slugs. Slugs ruptured when cracks broke through their thin metal coating (cladding). This allowed river water to directly contact the uranium and newly formed fission products. These fission products were more biologically hazardous than many of the shorter-lived activation products (Parker 1952).

Approximately 20 million fuel slugs were irradiated in Hanford reactors. Two thousand slug ruptures occurred between 1951 and 1965 (DeNeal 1965; HEDR 1994). Marceau et al. (2002) lists 2092 failures. Between 1955 and 1964, 69 to 200 slugs ruptured each year (Jerman et al. 1965). The amount of irradiated uranium lost to the river water depended on the severity of the rupture. As much as 1 pound could be released during a severe break. Average uranium losses during 1964 were less than an ounce per rupture. The 97 fuel failures that took place in 1964 contributed 0.7% to the total radioactivity discharged into the Columbia River from all of the reactors.

Radioactive fission products were more biologically hazardous than many shorter-lived activation products (Parker 1952).

"Every attempt was made to remove the ruptured slug as soon as possible" (Heeb and Bates 1994). When a slug ruptured, the reactor was shut down and water from the affected process tube was pumped into a crib (Parker 1952). For example, the strontium-90 now found in trenches near N Reactor came from cooling water diverted to liquid disposal sites after slugs ruptured. Most of the radioactivity released remained "in solid form as a sludge on the bottom of the [reactor] basin" (Irish 1959).

Most radionuclides released to the river had short half-lives and have decayed away (Wells 1994). For example, manganese-56 contributed nearly 65% (80 million curies) of the curies discharged but its half-life is only 2.6 hours. It decayed away in about 1 day and therefore did not exist long enough to significantly contribute to radiation doses to people living outside the Hanford Site boundary. About 1% of the radioactivity came from long-lived radionuclides.

---

(45) Three percent or 1200 tons of this chemical load released was sodium dichromate (Wells 1967). About 70% (27,500 tons) was sulfuric acid, 15% (6000 tons) aluminum sulfate, 10% (3800 tons) bauxite, and 2% (1060 tons) chlorine.

The five radionuclides contributing most (94%) of the estimated radiation dose to humans via the Columbia River are shown in Table 6 (Heeb and Bates 1994).

An average of 10,000 to 12,000 curies a day was discharged to the Columbia River from 1956 to 1965 (HEDR 1994).

Table 6. Radionuclides Released to the Columbia River Contributing Most Radiation Dose (sorted by half-life)

Radionuclide	Curies	Half-Life
Sodium-24	12,600,000	15 hours
Phosphorous-32	230,000	14 days
Neptunium-239	6,300,000	2.4 days
Zinc-65	490,000	244 days
Arsenic-76	2,500,000	26 hours

The largest releases occurred between 1956 and 1965 when as many as eight reactors were operating simultaneously. An average of 10,000 to 12,000 curies a day was discharged during that time (HEDR 1994). In 1963, an average of 14,500 curies were released each day (Gerber 1997). Larger peak releases took place during any single day.

Contaminants can be dissolved in river water or attach to sediment. Sediments would settle to the river bottom when in slow-moving or quiet waters such as behind islands or dams. Downstream from Hanford, as much as 1 foot of sediment accumulates each year behind McNary Dam located along the Washington and Oregon border. Much of this sediment originates from soil erosion from agriculture and enters the Columbia River from the Snake, Yakima, and Walla Walla Rivers. Radionuclides are found in this sediment.

Sediment samples collected along the shoreline of the Columbia River adjacent to Hanford contain elevated levels of radioactive cobalt, strontium, cesium, europium, and plutonium compared to

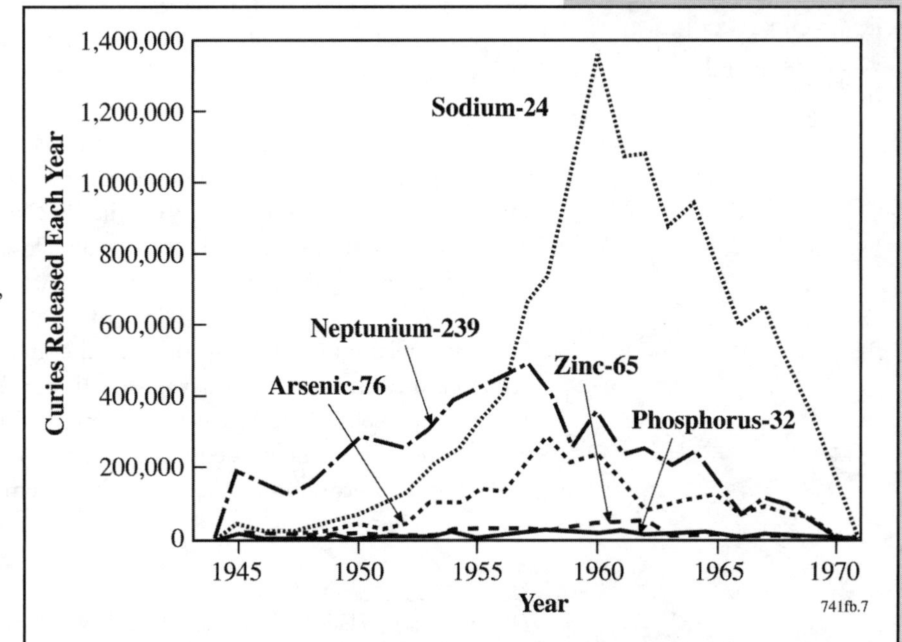

Radionuclide Releases to Columbia River. Nuclear reactors at Hanford released nearly 110 million curies of radioactivity into the Columbia River between 1944 and 1971. Ninety-four percent of the radiation dose received each year by the public came from five radionuclides. Radiation doses ranged from less than 1 millirem to 130 millirem a year (HEDR 1994).

Sediment behind McNary Dam contains radionuclides released from Hanford. Examples include cobalt-60, strontium-90, cesium-137, uranium-238, and plutonium-238, -239, and -240 (Patton 1999). Except for cobalt-60, the other radionuclides listed have half-lives longer than 29 years.

sediments upstream from Hanford (Cooper and Woodruff 1993). The concentrations of most nonradioactive metals are generally the same upstream and downstream of Hanford (Patton 1999).

Nitrate, sulfate, and chloride concentrations dissolved in the water are elevated on the side of the Columbia River opposite Hanford. These come from groundwater seepage and irrigation runoff from farms and orchards lying north and east of Hanford.

When groundwater seeps into the Columbia River, it's mixed with other natural and human-made elements already present. Various industrial-, fertilizer-, and pesticide-[46] contaminated irrigation runoff and groundwater seepage enter the river. For example, of the 132 potentially hazardous chemicals studied in a fish contaminant study of the Columbia River Basin, the U.S. Environmental Protection Agency detected 92, some at levels of concern[47] (EPA 2002). Though the concentrations of many contaminants found in fish were generally lower than reported in the literature in the 1970s, the health of some people, particularly Native Americans, may be at risk. The environmental agency noted, "Many of the chemical residues in fish identified in this study are not unlike levels found in fish from other studies in comparable aquatic environments in North America…The results of this study, therefore, have implications not only for tribal members but also the general public" (EPA 2002, p. E-8).

Pesticide-contaminated runoff is a growing concern across the Pacific Northwest. Seven pesticides in the central Columbia River drainage basin were identified at concentrations "above criteria set to protect aquatic life" (Lind 2002).

Also, metals such as arsenic, cadmium, chromium, lead, silver, mercury, and zinc are leached from natural rock and mine tailings in northern Washington, Idaho, and Canada (DOE 1998a; McLean 1998). These also enter the Columbia River. The Hanford contribution to such metals is "generally factors of two to three or less" than upstream sources, making the identification of sources difficult (DOE 1998a). Some tributaries emptying into the Columbia River, such as the Spokane River, contain especially high levels of metals and other industrial contaminants. This underscores the need to assess the water quality and health of the Columbia River as an integrated whole regardless of contamination source.

A complex food web links algae, insect larvae, crayfish, water fowl, and higher forms of plants and animals living in and along the Columbia River. Contaminants introduced into the water are passed along to higher life forms. In addition, organic matter created by photosynthesis consumed by plants and algae can assimilate dissolved chemicals into small particles for movement in river currents.

The accumulation of Hanford-originated contamination in the Columbia River was recognized from the earliest days—and thoughts were given to its removal. According to Parker (1948a), "One has also not considered the accumulation of river pollution in algae or mud, with the subsequent scouring of the river beds at times of early summer high flow. Heroic, but feasible, methods of removing such contamination exist."

(46) Pesticides include herbicides, insecticides, fungicides, and rodenticides. It's a general word referring to a chemical agent used for managing a pest by destroying, repelling, or mitigating its potential damage.

(47) Contaminants contributing to potential hazards are persistent bioaccumulative chemicals (PCBs, DDE, chlorinated dioxins, and furans) as well as naturally occurring arsenic and mercury (EPA 2002, p. E-8).

Years later, the amount of radioactivity (curie load) flowing down the Columbia River is significantly lower. Water analyses detected about 10 curies of radioactivity dissolved in the water each day as it entered the Hanford Site (Patton and Poston 2000).[48] Over 98% of this radioactivity comes from tritium found in rainfall, and therefore in river water.[49] According to the Washington State Department of Health, radioactivity now in the river comes predominantly from natural sources (Wells 1994). Most is from uranium, potassium, and thorium (and their decay products).

As the river passes Hanford, groundwater from the site and other nearby sources enters. Each day downstream of Hanford, about 20 curies of natural and human-made radioactivity are detected (Patton and Poston 2000). Compared to up-river water samples, over 99% of this 10 curie-per-day increase comes from tritium-containing groundwater seeping off Hanford.

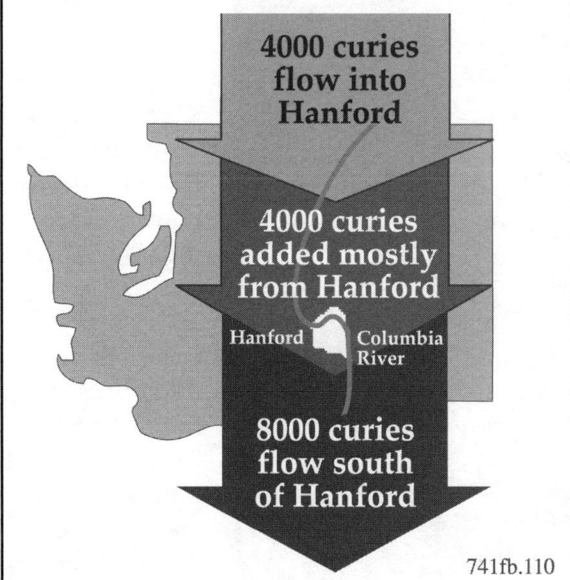

Tritium in the Columbia River. The yearly flow of the Columbia River contains about 4000 curies (0.01 ounce) of tritium as it enters the Hanford Site. This tritium comes from the global environment. Groundwater from beneath Hanford adds another 4000 curies. Therefore, downstream of Hanford, the Columbia River contains about 8000 curies (0.02 ounce) of tritium each year (Patton and Poston 2000).

The amount of tritium seeping into the Columbia River is decreasing over time. The average concentration sampled between 1989 and 1993 just south of Hanford was about 109 picocuries per liter[50] (Dirkes and Hanf 1995). The average for 1995 to 1999 had dropped about one-third—to 70 picocuries per liter (Patton 2001). Tritium values will continue to decrease until a natural background level is achieved (see Section 4.5).

Iodine-129 concentrations are also slightly higher downstream. During some past years, an increase in strontium was detected. Radionuclide levels are well below federal and Washington State standards (Poston et al. 2001).

The U.S. Geological Survey routinely monitors water quality in the Columbia River (Wiggins et al. 1996). These analyses include chromium, nitrate, dissolved oxygen, dissolved solids, and phosphorus; they do not monitor radionuclides. Slight increases in chromium and nitrate were detected down-river of Hanford.

The river has mostly recovered from past Hanford contaminant releases. The Hanford Reach portion "is one of the most valued and ecologically important stretches of the Columbia River" (NRC 2001b, p. 123).

Today, the Columbia River water is very safe to drink, following normal chlorination. From Grand Coulee Dam to the Washington-Oregon border, which includes the

(48) The amount of natural and human-made radioactivity detected in Columbia River water varies from year to year.
(49) Tritium was formed during past atmospheric testing of nuclear weapons. Small amounts are also created naturally in the atmosphere.
(50) A picocurie is one trillionth of one curie. One liter is 0.26 gallons—almost 1 quart.

The Columbia River water is very safe to drink, following normal chlorination. From Grand Coulee Dam to the Washington-Oregon border, Washington State has classified the river as "Class A, Excellent" (Patton 1999).

Hanford Reach, Washington State has classified the river as "Class A, Excellent" (Patton 1999). This means it meets or exceeds the state requirements for essentially all uses such as human consumption, salmon migration, wildlife habitat, and recreation.

5.8 Contaminants Released to the Air

"The following measurements on thyroid activity of sheep in regions adjacent to the Plant [Hanford] were obtained under conditions which avoided the excitement of public curiosity."

—K.E. Herde,
I^{131} Accumulation in the Thyroid of Sheep Grazing Near H.E.W (1946)

Thirty-two million curies of radionuclides were discharged to the atmosphere—12 million from the reactors and 20 million from the reprocessing plants (Napier 1992; Heeb 1994). Nearly 740,000 curies of iodine-131 were released from the reprocessing plants.(51) This accounted for 99% of the radiation dose people received from the air pathway (HEDR 1994).

Today, radioactive airborne releases are small and well within allowable limits—all radionuclides monitored are 0.03% to 0.000001% of concentration guidelines (Gillespie 2001).(52) Air sampling results are routinely monitored and reported (Poston et al. 2001). Releases are from old plant stacks, tank farm vents, operating laboratories, or wind-blown sediment. During the year 2000, the total radiation dose received from the air pathway(53) by a hypothetical person whose lifestyle exposed them to the highest possible dose in the Hanford area(54) was 0.0022 millirem (Antonio et al. 2001). This person is exposed to 0.02% of the 10-millirem-a-year air standard set by the U.S. Environmental Protection Agency.

Today, radioactive airborne releases are small and well within allowable limits (Gillespie 2001).

Some additional contaminant releases took place after a range fire burned 300 square miles of sagebrush and grasses on Hanford in June 2000. That fire raised the estimated maximum radiation dose to a member of the public from inhalation to 0.18 millirem for a 30-day period (Antonio et al. 2001).

The first onsite experiments studying the potential hazard of atmospheric dispersion from gaseous emissions began in 1943 (DuPont 1945). Smoke generators burned petroleum oils to produce a visible and persistent "oil fog"; the oily clouds were traced for miles across the landscape. Naturally, most burns were carried out where the reprocessing plants and reactors were being built. In 1944, temporary stacks were constructed in an attempt to use sulfur dioxide gas for monitoring air currents

(51) In comparison, 15 curies of iodine-131 were released during the Three-Mile Island nuclear reactor accident in Pennsylvania in 1979 (Gerber 1999).

(52) National Emission Standards for Hazardous Air Pollutants (40 CFR 61, Subpart H) specify that no member of the public should receive a radiation dose greater than 10 millirem a year from airborne radioactive effluents, other than naturally occurring radon.

(53) Exposure from the air came from three pathways: inhalation, external exposure, and airborne radionuclides deposited on foods.

(54) This person is known as the maximally exposed individual.

(Grossman 1994). This didn't work because the gas was hard to detect and threatened to asphyxiate nearby workers. A meteorological tower and auxiliary buildings were built between the 200 East and 200 West Areas in 1944 to provide weather forecasts for scheduling spent fuel reprocessing—and the release of radioactive gas.[55]

U.S. Department of Energy 92081233-15cn 741fb.53

Wind Testing at T Plant. Oil was burned and a thick exhaust released into the unfinished stack of the first reprocessing facility—T Plant. These plumes were part of early meteorological studies of wind dispersion that would later carry radioactive gases off the Hanford Site. Photograph taken in 1944.

No serious air-mixing problems were thought to occur during normal summer daytime conditions when a steady breeze accompanied releases. However, under calmer conditions, such as prevails during winter months, it wasn't possible to predict confidently where effluents would go (Cooper 1943). After a year of plant operations, Parker (1945a) cautioned: "it has been tacitly assumed that atmospheric concentrations of iodine in the residence areas will be very low. That may not be the case during a prolonged period of atmospheric stagnation such as was observed by the Meteorological Group in the winter of 1943-1944."

Restricting spent fuel dissolving to times when conditions favored dispersion gave mixed results. It lowered radiation exposure to Hanford workers while increasing exposure to the public. For example, Parker wrote that the ground contamination in areas such as Walla Walla, Washington (50 miles southeast of Hanford) was as high as in the 300 Area (Parker 1945a). "Again it should be emphasized that there is no hazard in these areas at this time, and that generally adequate observations are being made to follow the pattern of such contamination" (Parker 1945a).

In 1944, sulfur dioxide gas was used to monitor air currents. This didn't work because it was hard to detect and threatened to asphyxiate workers.

In 1945, Parker established a routine program for checking iodine-131 accumulation in thyroid glands. "Despite the belief that there is no critical hazard due to the emission of iodine from the stack…in the interest of morale it may prove more desirable to restrict the evolution of fumes under certain atmospheric conditions" (Parker 1945c). Three years later, John Healy[56] (1948) noted that the iodine-131 "has been long recognized as a possible hazard…The presence of other fission products with long half-lives in the stack gases has presented the possibility of a similar hazard with these materials."

"It has been tacitly assumed that atmospheric concentrations of iodine in the residence areas will be very low" (Parker 1945a).

(55) That meteorological station continues to monitor Hanford and forecast weather conditions used by site cleanup contractors.

(56) John (Jack) Healy, a chemical engineer, was part of Hanford's Special Studies Group in the mid-1940s and 1950s. This group was responsible for creating new means of radiation measurement and environmental monitoring (DOE 2002c).

In 1945, Walla Walla, Washington, had as high of levels of radionuclides on the soil as the fuel manufacturing area at Hanford.

Increasing the cooling time for spent fuel was known as the simplest and most expedient way to reduce hazardous gases, and their potential ill effects, to "negligible proportions" (Parker 1945a). Carl Gamertsfelder[57] (1947) reported that extending the cooling period from the "present level of approximately 60 to 90 days would reduce the levels [of iodine-131] by a factor of about 16, and the contamination in surrounding territory would drop to undetectable levels." Indeed, this happened a few months after the end of World War II when the immediate pressure for plutonium and thus reprocessing young fuel (that is, about 1 month out of the reactor) lessened. About 75% (560,000 curies) of all Hanford iodine-131 releases had taken place by December 1945 (HEDR 1994; Heeb 1994).

Parker (1946) and Healy (1946) recognized the importance of iodine-131 accumulation in local farm animals, leafy vegetables, and farm products within the first year of reprocessing plant operations. It was detected in the thyroids of animals collected onsite—and on vegetation collected 200 miles from Hanford (Parker 1948b). Some areas, within a radius of 50 miles of Hanford—including the cities of Pasco, Richland, Kennewick, Benton City, and the Wahluke Slope north of Hanford—had iodine levels "above the tolerance level" (Gamertsfelder 1947). The effect on grazing animals was reported as "not injurious when short periods of time are involved. Effects of prolonged exposure are not known…The effect on human inhabitants would be no greater than for communities of Richland and Pasco" (Gamertsfelder 1947).

There was great uncertainty about what levels of iodine were safe.

- "More information is clearly required" (Parker 1946).
- Human tolerance to iodine dose now "depends upon opinion" rather than available data (Gamertsfelder 1947).
- The Safety and Industrial Health Advisory Board for the Atomic Energy Commission "saw little evidence that AEC has profited from the vast experience of industry (particularly the smelting industry) in coping with stack pollution problems" (Williams 1948, p. 64).

Two types of stack emission problems were occurring: release of radioactive particulates and radioactive gases. Particulates caused "spot contamination." In late 1947, this was noticed coming from reprocessing plant stacks and falling near the facilities (Hanford Works 1948). That same year, the necessity to install filters in the reprocessing plants was recognized (Tomlinson 1959b).

Flakes of radioactive paint blew in the wind like gray snow.

Sand filters were added to T and B Plants in 1948 to reduce the emission of small particles (Roberts 1958). The origin of the particles? The original air blowers between a reprocessing plant and its 200-foot-tall stack were made of steel painted with an asphalt-based paint (Tomlinson 1959b). Rust formed beneath the paint causing flakes to be torn off and blown out the stack like gray snow. Parker (1948b) wrote that these fission-product-containing particles were a serious hazard to workers and the environment unless they became too large to be retained in the lungs. The largest particles fell near the plant stacks, while lighter and less radioactive particles may have traveled as far

(57) Carl Gamertsfelder served as a physicist on the Manhattan Project. He was a pioneer in radiological sciences and health physics (DOE 2002b).

as 100 miles. When General Electric Company management was informed about these particles, and before sand filters were installed, workers outside of T and B Plants were required to wear respirators (Parker 1948c). Sand filters had an efficiency of greater than 99.5%. Fiberglass filters were added in 1950.

Water scrubbers[58] were also installed in 1948 to remove an estimated 85% of the iodine-131 from emissions (Roberts 1958). Once both sand filters and water scrubbers were online, combined with an increase in fuel cooling times, the amount of iodine released rapidly dropped. Water scrubbers were replaced in 1950 by silver "reactor" filters,[59] which had a removal efficiency of greater than 99% for iodine-131 when functioning at peak performance (Work 1951). These began continuous operation in early 1951 (Heeb 1994). However, periodic higher-than-normal radionuclide releases still took place during the 1950s when filters temporarily failed. Two failures occurred in 1954 at the REDOX Plant, each releasing 300 to 400 curies of ruthenium (Freer and Conway 1999). Thereafter, iodine-131 releases averaged 1 curie a day or less (Healy et al. 1958).

As noted, radioactive ruthenium was also in the stack gases. After one unplanned release of ruthenium-laden particles from the REDOX Plant in 1952, Hanford officials wanted to analyze the kidneys and livers of cattle from local stock "but could not do it without risk of exciting too much comment [from the public]" (Parker 1954b). Some particles, found in Richland and other neighboring communities, could give a radiation dose to the skin well above safe limits. Hanford officials agreed that if a "demonstrable hazard to the public exists," the public and state officials should be notified (Parker 1954b). Concerns over notification appeared motivated out of a concern for public health as well as to stave off criticism when "that information, or rather misinformation, on the off-site condition will leak to the public" (Parker 1954b).

Originally, 60 to 65 days were considered adequate for irradiated uranium fuel to cool down before reprocessing (Cooper 1943; Way 1944). This was needed so short-lived radionuclides could decay away before the spent fuel was dissolved. Some less than 40-day-old fuel was reprocessed. This proved troublesome, so times were lengthened to about 100 days. By 1951, improvements in air treatment enabled plant operators to reprocess irradiated fuel within 46 days of irradiation (Heeb 1994).

When the stack filters worked, "cooling times for the fuel elements were again shortened and meteorological controls relaxed" to support increased plutonium production (Tomlinson 1959b).

Later, operating procedures called for a 90- to 120-day holding period. When pressure mounted for plutonium production, young uranium fuel was again reprocessed (Heeb 1994; Goldberg in Hevly and Findlay 1998). By the late 1950s and early 1960s, the average storage time for irradiated fuel had increased to between 100 and 250 days. A plant-by-plant history of the uranium processing and average spent fuel cooling times is summarized in *Radionuclide Releases to the Atmosphere from Hanford Operations* (Heeb 1994).

Another event in the release of radionuclides to the atmosphere began with the Soviet Union's first nuclear bomb explosion on August 29, 1949. This gave rise to Hanford's "Green Run." Green refers to fresh irradiated fuel reprocessed after a short cooling

(58) Water scrubbers were installed in the dissolver lines that received radioactive gases released when the uranium fuel was dissolved.

(59) Silver reactor filters contained a heated column of silver nitrate-coated packing material through which the gas stream passed. The iodine rapidly combined with the silver and was retained on the filter material. A silver iodide (capturing iodine-131) coating formed on the filter surfaces.

In September 1949, U.S. officials were startled by the Soviet's first nuclear bomb explosion. It was detected in air samples from near Kamchatka Peninsula (Rhodes 1995, p. 371).

Hanford officials disabled the emission control filters at T Plant to conduct the Green Run experiment.

time. The Green Run was an experiment to estimate Soviet reprocessing operations and plutonium production that began in 1948 at the Mayak site, southeast of the Ural Mountains. Officials in the United States thought the Soviets were reprocessing spent fuel less than 3 weeks old.

The Green Run was planned for T Plant. It permitted a comparison of air samples, and therefore plutonium production, between Hanford and Mayak. It took place on December 2 and 3, 1949, when 1 ton of 16-day old spent fuel was reprocessed and the weather looked favorable (IEER/IPPNR 1992). About 11,000 curies of iodine-131 and xenon-133 were released into the air.

To conduct the Green Run, Hanford officials "deliberately disabled" emission control filters at T Plant (Silverman 2000). Some contractor personnel "may have initially had reservations about the experiment but acquiesced at the urging of the air force" (Schwartz 1998, p. 452). Then, weather conditions deteriorated. Rain, snow, and wind deposited radionuclides in concentrated patterns northwest and southwest of Hanford. Radionuclide concentrations in the thyroids of waterfowl "increased abruptly" to about eight times above government-allowed levels (Herde et al. 1951). Some thyroid tissue samples collected 15 to 20 miles off the Hanford Site contained iodine-131 levels that "significantly exceeded" the chronic maximum permissible concentration for humans (Herde et al. 1951).

The amount of radionuclides released during the Green Run was small compared to previous routine plant releases. However, when this information was declassified and released, it raised the ire of many because it was an intentional experiment that exposed an unsuspecting public to increased radiation levels, where temporary operational needs overrode public health concerns and safety practices (Pritikin 1994/1995; Silverman 2000).

Heeb (1994) reported that between 1944 and 1970, approximately 10 million curies of radioactivity were discharged into the atmosphere from Hanford's first eight reactors. Between 1963 and 1987, an additional 2 million curies were released when N Reactor operated. This raised the cumulative release from all reactors to 12 million curies. More than 99.9% of these releases were argon-41 formed when argon-40, found in air that leaked into the reactor stack-gas system, captured an extra neutron. This radionuclide decayed away in less than a day because its half-life was just 1.8 hours. Small amounts of carbon-14 and tritium were also released.

Between 1944 and 1972, about 20 million curies of radionuclides were released into the atmosphere from Hanford's reprocessing plants (Napier 1992; HEDR 1994; Heeb 1994). These are listed in Table 7.

Between 1944 and 1970, about 10 million curies of radioactivity were discharged into the air from Hanford's first eight reactors. Later, N Reactor added another 2 million curies. Between 1944 and 1972, about 20 million curies of radioactivity were released into the air from Hanford's reprocessing plants.

Table 7. Radionuclides Released into the Atmosphere From Hanford Reprocessing Plants Between 1944 and 1972

Radionuclide	Curies	Half-Life
Krypton-85	18,500,000	10.7 years
Iodine-131	739,000	8 days
Xenon-133	418,000	5.2 days
Cerium-144	3770	285 days
Ruthenium-103	1160	39 days
Ruthenium-106	388	1 year
Strontium-90	64	29 years
Plutonium-239	2	24,000 years

After the U.S. atomic bomb test on Bikini Island in July 1946, air currents were monitored at "points around the world to determine if an atomic bomb has been detonated in the air" as well as when and where it had exploded (Rhodes 1995, p. 262). The experiment worked.

Six radionuclides released into the atmosphere during reprocessing plant operations contributed 99% of the potential radiation dose to humans received from the air pathway: iodine-131, ruthenium-103, ruthenium-106, strontium-90, plutonium-239, and cerium-144.

Though krypton-85[60] and xenon-133 make up 95% of the curies released during reprocessing, they contributed little to radiation dose because they do not interact with the environment, do not remain in the body, and have low energy levels—10,000 to 400,000 times lower than the less plentiful but far more dangerous iodine-131. Napier (1992) reported, "The doses, either to adults or infants and from any pathway, are four to seven orders of magnitude [that is, ten thousand to 10 million times] smaller from any other radionuclide than they are from iodine-131."

5.9 Other Onsite Radionuclides and Chemicals

Solid waste and nuclear materials, not originating from Hanford's plutonium production, are stored or disposed onsite.

Twenty-five million curies of radioactivity and some chemicals not produced by the plutonium mission are stored at Hanford.

Private companies and federal agencies other than the U.S. Department of Energy use portions of Hanford for the storage or disposal of about 25 million curies radionuclides and some chemicals.[61] In summary, these include

- 4.5 million curies of commercial low-level waste in the U.S. Ecology landfill located between the 200 East and 200 West Areas.

(60) Seventy-five percent of the krypton-85 releases occurred over a single decade: 1958 to 1968 (Napier 1992).

(61) Radioactivity inside the Fast Flux Test Facility and the one commercial nuclear power plant located on Hanford is not included. The Fast Flux Test Facility, located in the southern part of Hanford, was a pilot plant for the nation's breeder reactor program. It is a sodium-cooled reactor built between 1969 and 1978. In 1992, the reactor was placed on standby status, awaiting another long-term mission or shut down. The reactor contains 12 tons of heavy metal having 15 to 20 million curies of radioactivity (decayed as of 2000). In late 2001, the Secretary of Energy issued an order to permanently shut down this reactor.

- 5 million curies inside the Navy reactors stored in a burial ground in the 200 East Area.
- 2 million curies in spent fuel from the decommissioned Shippingport, Pennsylvania, commercial nuclear reactor.
- 13 million curies of "special waste," including 7 to 8 million curies in glassified waste from Germany removed from the 300 Area and now stored in the 200 West Area.

Commercial Low-Level Waste: In 1964, a 1000-acre (1.5-square-mile) tract of land was leased to the state of Washington to promote nuclear-related development (DOE 1998c). It was located between the 200 East and 200 West Areas. A commercial low-level radioactive waste disposal site, run by U.S. Ecology, Inc., currently operates on 100 acres of this land. In the 1990s, the remaining 900 acres were returned to the U.S. Department of Energy. This site is one of three locations in the United States used for disposing commercially generated low-level or mixed radioactive and chemical waste. The other two are at Barnwell, South Carolina, operated by Barnwell Waste Management, and a site 80 miles west of Salt Lake City, Utah, managed by Envirocare of Utah, Inc.

Commercial Low-Level Waste Sites

Beginning in the 1960s, radioactive waste in the United States was disposed at six commercial facilities (DOE 2000f). These were Barnwell, South Carolina; Hanford, Washington; Beatty, Nevada; Maxey Flats, Kentucky; Sheffield, Illinois; and West Valley, New York. Some waste at these sites was classified as transuranic contaminated because disposal began prior to transuranic waste being identified by the Atomic Energy Commission in 1970 as a separate waste category. By the late 1970s, three of these sites (Maxey Flats, Sheffield, and West Valley) had closed.

This U.S. Ecology site operates under licenses issued by the Washington State Department of Health and the U.S. Nuclear Regulatory Commission. The first waste shipment arrived in 1965. Today, it receives solid waste from 11 western states.

The site consists of 20 unlined trenches (four in use) and three underground tanks (none in use). Waste can include contaminated equipment and protective clothing packaged in steel boxes or drums. Containers are placed in trenches typically 45 feet deep, 1000 feet long, and 150 feet wide. After a trench is filled, it's covered with soil and gravel. From 1965 to 1998, about 13 million cubic feet of waste, containing 3 million curies of radioactivity, was buried (GAO 1999d).

Little public concern was voiced about the radioactive Trojan commercial reactor core coming onsite.

In 1999, the core from the Trojan Nuclear Power Plant, built northwest of Portland, Oregon, was buried at the U.S. Ecology site. (All spent fuel was removed before shipment.) This reactor core added 1.54 million curies of short-lived radionuclides to the burial ground.[62] Ninety-five percent of this radioactivity comes from three activation products—cobalt-60, nickel-63, and iron-55. These are metals in the reactor that became radioactive when they captured neutrons or other types of radiation during

(62) Communication from Michael Murdock, Portland General Electric Company, August 24, 1999.

reactor operation. This single shipment raised the radioactivity inventory at the U.S. Ecology site from 3 million to 4.5 million curies. Little public concern was voiced about the addition of this non-Hanford curie load to the onsite radionuclide inventory.

At the U.S. Ecology site, slightly radioactive Class A solid waste is placed inside trenches and backfilled. Higher-contaminated Class B and C wastes are placed in concrete containers and then lowered into a trench.

The burial ground is scheduled for closure in 2056 (GAO 1999d), at which time a final protective ground cover is to be installed. The state and federal government will monitor the site through at least the mid-twenty-second century.

DynCorp Tri-Cities Services 741fb.42

Core of Trojan Nuclear Power Plant. The 1020-ton nuclear reactor core from decommissioned Trojan power plant in Oregon arrived at Hanford in 1999 for burial at the U.S. Ecology commercial landfill. Though the highly radioactive uranium fuel had been removed before shipment, 1.5 million curies of short-lived radioactive activation products remained in the vessel. This added to the onsite radionuclide inventory from non-Hanford activities.

Navy Nuclear Reactors: More than 40% of the U.S. Navy's vessels are nuclear powered. When a vessel is retired, that portion containing the reactor is cut away, sealed, shipped to Hanford, and buried in a open landfill in the 200 East Area. The Navy plans to retire the nuclear reactor compartments of perhaps 200 submarines and surface cruisers by the year 2010. As of early 2002, 100 compartments had arrived. On average, six to ten reactor shipments are received each year.

The Navy began disposing of reactor compartments at Hanford in 1986. The site's security played a key role in selecting it for this mission because some submarine technology remains classified.

Shipments are prepared at the Puget Sound Naval Shipyard in Bremerton, Washington. There, the reactor and piping systems are cut from the vessel, and steel plates are welded on each end. These compartments resemble large, stocky cylinders—33 to 42 feet high, 40 to 55 feet long, and weighing 1130 to 2750 tons (Oregon Office of Energy 1999a). The compartments are shipped down the Washington coast, up the Columbia River, unloaded, and hauled to the 200 East Area. The spent fuel is sent to the Idaho National Engineering and Environmental Laboratory near Idaho Falls, Idaho.

Each Naval reactor contains 16,000 to 86,000 curies of radioactivity (U.S. Department of Navy 1996).

These reactors remain a source of radiation even after the uranium fuel is removed because metal was "activated" when exposed to neutrons during reactor operations. The amount of residual radioactivity depends upon the reactor type and when it ceased operating. One year after shutdown, each reactor contains 16,000 to 86,000 curies (U.S. Department of Navy 1996). If each reactor is assumed to contain an average of 50,000 curies, then the 100 reactor compartments now onsite would contain 5 million curies of radioactivity. The compartment's outer shell is not radioactive.

These compartments are regulated as dangerous waste by the state of Washington because they contain lead once used for shielding. They are also regulated as mixed waste because of the radioactivity and certain chemicals (polychlorinated biphenyls in insulation and cable coverings) present.

Commercial Power, Test Reactor, and Special Waste: Two million curies of radioactivity are contained in commercial spent nuclear fuel stored inside T Plant.[63] This 16 tons of fuel are contained in 72 assemblies shipped to Hanford in 1978 and 1979 from the decommissioned Shippingport Nuclear Reactor in Pennsylvania (DOE 1997). Each narrow square-shaped assembly is about 12 feet long, 8 inches on a side, and weighs nearly 1200 pounds. In contrast to Hanford's short uranium metal fuel rods, the Shippingport fuel consists of uranium oxide wafers sandwiched between zirconium metal plates. These plates are separated and stacked atop each another. Starting in 1957, Shippingport was the first reactor to provide commercial electricity in the United States.

741fb.58

Naval Nuclear Reactor Solid Waste. After nuclear-powered submarines and cruisers are decommissioned, their spent fuel is removed, and the reactor section of the vessel is cut away and sealed. These 33- to 42-foot-tall steel hulls are now a disposal container. Each package contains residual radionuclides plus lead. They are barged to Hanford from Bremerton, Washington.

Since 1974, 0.4 tons of irradiated nuclear material from experimental reactors have been placed in retrievable storage in the 200 West Area low-level burial grounds. These materials originated at Hanford as well as from the Idaho National Engineering and Environmental Laboratory, a research reactor at Oregon State University (located in Corvallis, Oregon), and other experimental reactors. They are stored as remote-handled transuranic waste.[64]

(63) Communication with Fluor Hanford, Inc. staff, March 13, 2001. This fuel is being dried, repackaged inside stainless steel canisters, and moved to the Canister Storage Facility in the 200 East Area. There it is stored below ground level in the same facility as the repackaged spent fuel from the K Basins.

(64) Remote-handled waste, contaminated with high-penetrating gamma-emitting transuranic radionuclides, has a measured radiation dose at the surface of its container of 200 millirem per hour or greater. It's handled robotically and transported in lead and steel-shielded containers. On the other hand, contact-handled waste mostly holds less penetrating alpha and beta radiation. The radiation dose at the container's surface is 200 millirem per hour or less. Contact-handled waste does not require special shielding and handling.

Chapter 6
The End of Secrecy

"An overall impression of these [historical] documents is one of honesty. They were not originally intended for release to the public, so there was no incentive to 'hide' information. There was also concern for worker and public health."

—A.W. Conklin, *Summary of Preliminary Review of Hanford Historical Documents—1943 to 1957* (1986)

It took several decades to chip away the wall of secrecy surrounding Hanford contaminant releases.

National security concerns once surrounded plutonium production. Some information remains classified. For years, Hanford was rimmed with antiaircraft guns and bunkers housing Nike missiles. Security guards touched worker badges, searched lunchboxes, and patrolled the Columbia River. Airplanes were prohibited from Hanford air space. Those of us old enough still remember when the Premier of the former Soviet Union, Nikita Khrushchev, threatened "We will bury you" or the face of a young President informing the American people about Soviet missiles 90 miles off the Florida coast. Uneasy times. Yet, site secrecy did not necessitate nor legitimize the contaminant legacy of Hanford. We never faced an either-or situation. As mentioned in Chapters 2 and 3, concerns about waste management practices date back to the 1940s—those concerns were even published by the government's own Atomic Energy Commission. Eventually, the public and various agencies became more conversant and aware of Hanford affairs. Today, people still seek to understand why contaminant releases went unexplained, unquestioned, unresolved, and sometimes unchecked for such a long time. This led many to become involved in Hanford issues, particularly issues dealing with creating a new future for the site.

Site secrecy did not necessitate nor legitimize the contaminant legacy of Hanford. We never faced an either-or situation.

Many events over the last 30 years led us to this point. Some of these are discussed here.

6.1 Public Awareness

"These [public health] assurances, combined with the veil of secrecy that shielded operations from outside scrutiny, sufficed for decades to calm whatever public fears may have existed and to inhibit any truly independent review of the plant's emissions."

—Daniel Grossman, *Pacific Northwest Quarterly* (1994)

For years, major Hanford decisions were made without public awareness or involvement. If a new reactor was needed, it was built. If more contaminants were released, they were released. Government officials and private contractors were trusted to protect the public's interest.

Hanford history is steeped with contradictions in how officials dealt with potential hazards. On one hand, radiation protection was strictly monitored, routine medical exams conducted, and exposure limits set for workers. On the other hand, the public was kept largely uninformed and, at times, lied to about emissions. Hanford scientists performed world-class research to understand the potential biological effects of radiation exposure. However, public resentment and news media attention slowly grew in the 1970s and then blew across the landscape like a wildfire in the 1980s as they grasped what some officials knew years earlier.

Hanford history is steeped with contradictions in how officials dealt with potential hazards.

One Hanford researcher wrote: "The waste management policy reflects a desire to conserve a natural resource through the effective and prolonged use of ground disposal facilities. However, due consideration must be and is given to economic factors when evaluating ground disposal against alternate or supplemental treatment and disposal methods" (Irish 1961). Fifty years passed until pressure from regulatory agencies and the public forced the building of the first liquid effluent treatment plants at Hanford, so untreated liquids were no longer discharged to the soil.

From the early 1970s to 1990, a series of events captured the public's attention. A few examples are summarized here.

1973 Largest Tank Leak in the History of U.S. Weapons Production: The leak of 115,000 gallons of waste from an underground storage tank was announced in the spring of 1973. This was from Tank T-106 in the 200 West Area. Aging tanks and human error were blamed. About 40,000 curies of radioactivity were contained in this waste. As noted in Section 5.6.5, this was and continues to be the largest tank leak in the history of the nuclear weapons complex.

The accidental release of 115,000 gallons of tank waste continues to be the largest leak in the history of the nuclear weapons complex.

1975 First Environmental Impact Statement: Hanford's first environmental impact statement[(1)] was published in 1975 (ERDA 1975). It described how waste was managed and in some cases gave amounts for past contaminant releases. Publication took place amid the first newspaper and popular magazine articles about tank leaks, billions of gallons of low-level radioactive liquid dumped to the soil, and unplanned radioactive ruthenium releases to the air 20 years earlier.

1976 Chemical Explosion Contaminated Worker: Hanford worker Harold McCluskey was contaminated with radioactive americium during a chemical explosion in the Plutonium Finishing Plant, used to convert plutonium-bearing solutions into plutonium metal. McCluskey was working in front of a sealed glove box housing a vessel filled with nitric acid, an ion-exchange resin, and about 4 ounces (100 grams) of americium (McMurray 1983). The vessel exploded during reactivation of a recovery process shut down 5 months earlier. External decontamination and multi-year chemical chelation therapy reduced the radiation burden on and in McCluskey's body. This accident focused attention on the potentially dangerous nature of Hanford work and upon work protocols followed.

(1) An environmental impact statement is a document prepared to describe the effects of proposed activities on the land, water, air, and living organisms. In addition, it addresses environmental values and associated social, cultural, and economic implications. Laws and regulations require the federal government to prepare an impact statement when contemplating certain types of major activities.

1979 Three-Mile Island Reactor Accident: In March, the partial meltdown of a commercial nuclear core in the Three-Mile Island reactor near Harrisburg, Pennsylvania, released about 15 curies of iodine-131 into the atmosphere. The maximum dose a person outside of the reactor site might have received was 70 millirem (U.S. Nuclear Regulatory Commission 1980). Though radiation release and exposure were low, the accident focused national attention on reactor safety and the public's involuntary exposure to radiation.

The perceived insulation of the nuclear power industry from public scrutiny prompted anti-nuclear activism.

> The government's and industry's unwillingness to acknowledge and realistically deal with the potential environmental and economic risks of nuclear power "exposed it to increasing public attacks, leading to the virtual demise of the U.S. nuclear power program in the 1970s. It also kindled a broader public loss of trust in scientists' and engineers' authority as public officials, a distrust that came to pervade U.S. environmental policy debates" (Andrews 1999, p. 185).

1976 to 1981 Statewide Initiatives: Escalating construction costs of nuclear power plants in the Pacific Northwest and the perceived insulation of the nuclear power industry from public scrutiny prompted anti-nuclear activism. This led to several voter initiatives: the Nuclear Safeguards Initiative of 1976 opposed building two nuclear power plants northeast of Seattle, the Don't Waste Washington Initiative of 1980 prohibited importing out-of-state radioactive waste, and the Don't Bankrupt Washington Initiative of 1981 required public approval for issuing bonds to finance building commercial nuclear reactors.[2] These initiatives focused public attention on nuclear issues and stirred more regional activism on a growing number of nuclear-related concerns.

Karl Fecht, Basalt Waste Isolation Project

Basalt Repository. The deep basalt underlying Hanford was studied as a potentially suitable rock for building a geologic repository for high-level radioactive waste. Meetings held by the Basalt Waste Isolation Project opened the door for onsite news media attention and public involvement more than any other project in the previous 40-year history of Hanford.

1983 Geologic Repository Studies: The U.S. Department of Energy announced that Hanford was one of nine sites being studied for possible construction of a permanent underground repository for highly radioactive waste. In May 1986, the basalt rock

(2) Voters defeated the Nuclear Safeguards Initiative (I-325) by a two-thirds vote (Pope in Hevly and Findlay 1998). The Don't Waste Washington Initiative (I-383) received 75% voter approval. It was later ruled invalid by a Federal District Court judge. Voters approved the Don't Bankrupt Washington Initiative (I-394) by 58%.

beneath Hanford was selected as one of three finalists for the repository (DOE 1988a). The prospect of more high-level waste coming to Hanford and its potential release into the environment galvanized public concerns. In 1986, voters in Washington State approved Referendum 40 by 84%. This directed the state legislature to veto any designation of Hanford as a geologic repository. In late 1987, Congress revised the Nuclear Waste Policy Act to focus repository studies on a volcanic rock, called tuff, in Nevada. Within a few days, Hanford basalt studies were stopped.

Commercial Nuclear Reactor Bond Default in 1983

In 1983, the largest bond default in U.S. history occurred when construction was stopped on all but one of five commercial nuclear reactors being built in the Pacific Northwest. At Hanford, construction on two of three reactors was terminated. The only commercial nuclear reactor in Washington, located at Hanford, began operating at Hanford in 1984.

Atomic Harvest by Michael D'Antonio (1993) summarizes many of the news events surrounding the publication of the first articles about Hanford contaminant releases.

1984 Hanford Education Action League: A diverse group of Spokane, Washington, citizens founded the Hanford Education Action League (popularly known as HEAL). Membership grew to between 300 and 400. The league's initial purpose was to promote public involvement and education about environmental issues associated with the potential use of Hanford as a geologic repository and the operation of the PUREX Plant, a reprocessing facility (D'Antonio 1993). The league was a leading public advocate for bringing regulatory oversight to Hanford. Over the years, their educational, political, and investigative activities expanded across a spectrum of issues including contaminant releases to the air and Columbia River, tank safety, N Reactor operation, waste transportation, and environmental cleanup. Their publications, issued between 1985 and 1999, are available at the Foley Center Library at Gonzaga University in Spokane, and are online at http://inlan.gonzaga.edu:1080/img/0550/HEAL.html. The league was disbanded in 1999. Several past members remain involved in Hanford issues.

According to former Secretary of Energy John Herrington "'Chernobyl broke the log jam' to overcome the secrecy and the production-oriented mindset within the department [of Energy]" (D'Antonio 1993, p. 241).

1985 Downwinder Articles: In the summer of 1985, the Spokane *Spokesman-Review* newspaper published its first article about the Hanford "downwinders," people living downwind of the Hanford Site who were exposed to past airborne radioactive elements. Follow-up articles, written by Karen Dorn Steele, helped galvanize public concerns and brought attention to the problem of contaminant releases from Hanford and other nuclear weapons sites.

1986 Chernobyl Reactor Explosion: In April, human error caused an explosion at one of the Chernobyl nuclear reactors, near Kiev, in the former Soviet Union. Fifty million to 80 million curies were released during the explosion or from the subsequent fire in the nuclear core (Bradley 1997). Approximately 116,000 people living in communities near the power- and plutonium-producing reactor were evacuated (Vargo et al. 2000). Thirty-one deaths occurred as an immediate result of the accident. Radioactive fallout in the United States from the explosion was first detected in rain samples collected near Hanford. Thyroid diseases in the downwind population have risen. This accident heightened public concerns about nuclear issues, specifically the safety of N Reactor, the last operating defense reactor at Hanford. N Reactor was shut down in 1987.

1986 Previously Classified Documents Made Public: In September 1985, Hanford Site manager Mike Lawrence committed to releasing formerly secret documents about past contaminant releases. This was an unprecedented move by a senior U.S. Department of Energy official. The first 19,000 pages of documents were made public in February 1986.[3] Though spearheaded by growing public pressure, the release of these reports was a first for the nuclear weapons complex. The Energy Department continues to declassify documents. An estimated 3 to 4 million pages of Hanford information remains classified (Hanford Openness Workshops Fact Sheet 1999a).

Hanford Site manager Mike Lawrence committed to releasing formerly secret documents about past contaminant releases. This was an unprecedented move.

Some of this information related to the airborne release of radioactive elements such as iodine-131. Current and former residents who lived downwind of Hanford organized into advocacy groups to share their stories of illnesses, seek medical monitoring and treatment, and pursue legal redress from past contractors (Pritikin 1994/1995). In 1988, Congress mandated a study of potential health impacts. This grew into the Hanford Environmental Dose Reconstruction Project and the Hanford Thyroid Disease Study. In January 2000, the federal government acknowledged that some workers in the nuclear weapons complex would likely become ill from past radiation and chemical releases (see Sections 4.3 and 4.6).

Current and former residents who lived downwind of Hanford organized into advocacy groups.

1987 Heart of America Northwest: A regional, non-profit organization, Heart of America Northwest, was established around concerns of potentially using Hanford as a geologic repository and the 1986 declassification of contaminant release reports. Today, it works on a range of environmental, public health, and economic issues including Hanford cleanup and transportation of nuclear waste to the Northwest. Their Web site (http://www.heartofamericanorthwest.org/) provides news releases, reports, upcoming events, and discusses how to become involved in Hanford issues. Their headquarters is in Seattle, Washington.

1987 Defense Waste Impact Statement: The U.S. Department of Energy published a draft version of the environmental impact statement on Hanford defense waste in 1986. The final version was issued the following year (DOE 1987). This further described the quantity of waste at Hanford. When coupled with the news that Hanford was a finalist for further study as a potential national repository, public and political concerns escalated.

1988 PUREX Plant Closed: Hanford's last reprocessing plant, PUREX, was shut down in 1988. Though a short chemical stabilization run took place between 1989 and 1990, the plant never restarted. Operational and environmental compliance problems were given as the reason.

1989 Tri-Party Agreement: Documentation of Hanford waste sites for potential placement under the cleanup authority of the Comprehensive Environmental Response, Compensation, and Liability Act (CERCLA)[4] began in 1985. In June 1987, the U.S. Environmental Protection Agency initiated actions to place Hanford on the National Priorities List, a list of U.S. waste sites that are potentially the most hazardous in this

(3) As declassification was taking place, a Freedom of Information request was filled in January 1986 for similar information. By February, the previously started declassification process had progressed far enough to release the first batch of documents.

(4) The federal Comprehensive Environmental Response, Compensation, and Liability Act governs the cleanup of hazardous, toxic, and radioactive substances at abandoned or uncontrolled waste sites. A trust fund, known as Superfund, was created to finance investigations and cleanup of some waste sites. See Section 7.1.

country and warrant further investigations. Two months later, the U.S. Department of Energy, U.S. Environmental Protection Agency, and Washington State Department of Ecology began discussions on establishing a cleanup agreement. After much frustration, an agreement among the three parties, the Hanford Federal Facility Agreement and Consent Order or informally the Tri-Party Agreement, was signed in May 1989. Hanford's waste sites were placed on the priorities list in September of the same year. The first public involvement plans were soon developed.

A month-by-month summary of Hanford news events for the first 10 years (1989-1999) following signing of the Tri-Party Agreement was written by the Oregon Office of Energy (1999b).

1990 Tank Watch List: Congress passed a law, known informally as the Wyden Amendment, creating a list, called the "Watch List," of Hanford tanks requiring special safety precautions because of their potential for unwanted chemical reactions or the buildup of flammable gases. Reports circulated about Hanford officials knowing since the late 1970s about the potentially dangerous buildup of hydrogen gas in Tank SY-101 but failing to intervene or inform the public. This fueled concerns about the safety of radioactive waste in Hanford's other 176 underground tanks and the forthrightness of local officials.

6.2 Public Involvement

"The emergence of collaborative planning processes has been an attempt by ordinary citizens, in the midst of myriad demands on their time, to try to get things gone."

—Jim Burchfield, *Across the Great Divide: Explorations in Collaborative Conservation and the American West* (2001, pp. 241-242)

Citizens of the Northwest are leaders in public involvement at Hanford and other nuclear weapons sites.

Hanford was the first site to sign a legally binding cleanup enforcement agreement.

Public, tribal, and local government involvement have significantly influenced Hanford's approach to managing waste and cleaning up contaminants. Both ad hoc and long-term groups were established beginning in the early 1980s. This new approach was supposed to pioneer a change in public interaction by the U.S. Department of Energy from a "decide, announce, and defend" tactic to early and more continuous engagement.

Citizens of the Northwest are leaders in public involvement at Hanford and other nuclear weapons sites. Hanford was the first U.S. Department of Energy site where the public reviewed contamination records and established values for achieving future site uses. It was first to form advisory boards focused specifically on the site; these groups included members from the public, Native American tribes, environmental groups, government agencies, labor, and members from a range of other organizations. The Hanford Advisory Board, formed in 1994, is a leader in helping advisory boards at other U.S. Department of Energy sites establish procedures. Hanford was the first to sign a legally binding cleanup enforcement agreement. It was the first site to establish an onsite U.S. Environmental Protection Agency presence. Hanford was the first to use historical data to reconstruct the dose received decades ago and examine the potential health impacts from these releases. Hanford even permitted public review of operating budgets.

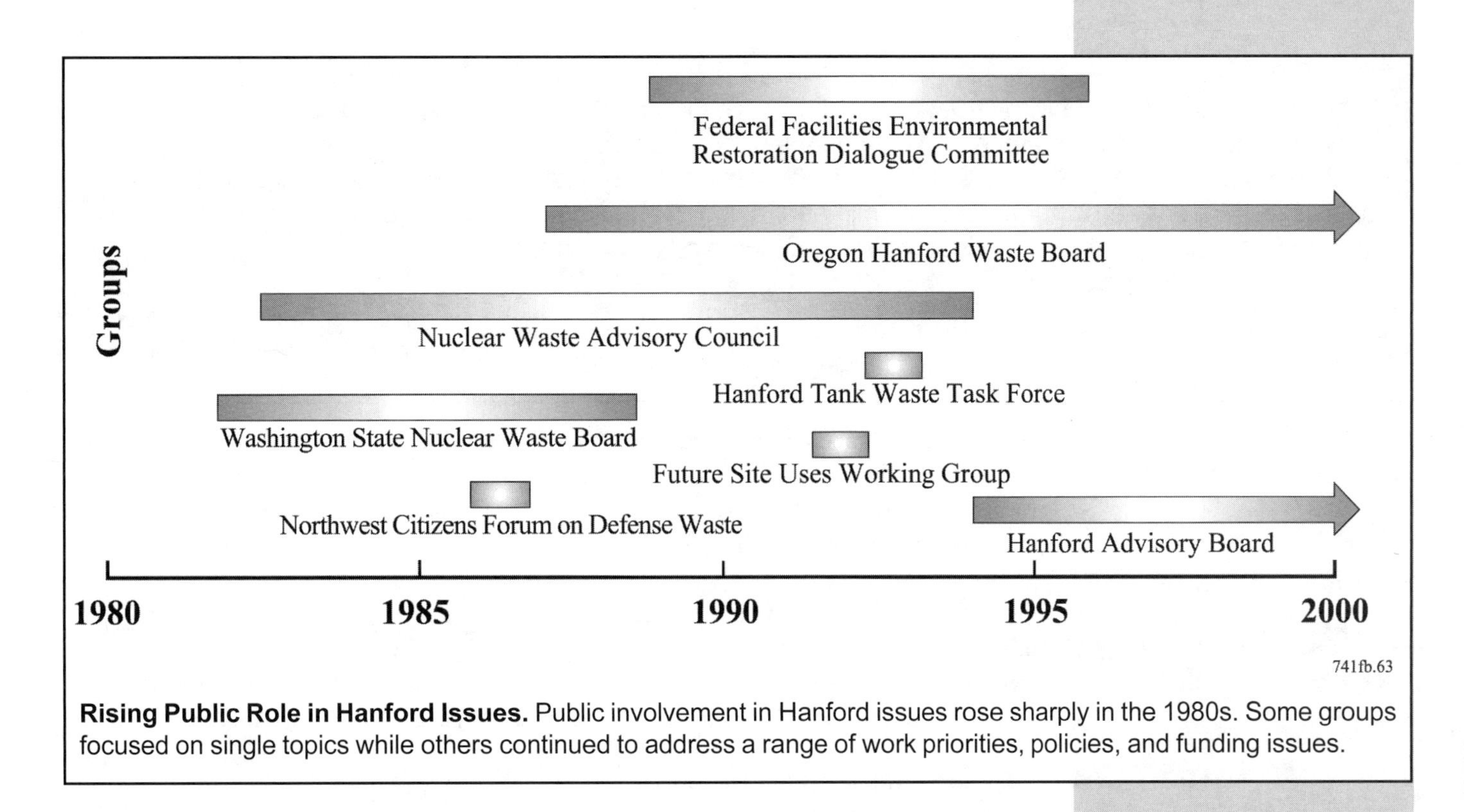

Rising Public Role in Hanford Issues. Public involvement in Hanford issues rose sharply in the 1980s. Some groups focused on single topics while others continued to address a range of work priorities, policies, and funding issues.

Many of these topic-specific and longer standing groups are identified below. All are specific to Hanford or Washington State with the exception of the nationwide Federal Facilities Environmental Restoration Dialogue Committee. A major challenge facing all members of boards, advisory panels, etc., is how to translate their private concerns into a broad social consensus. The two are not equal. All representative roles are delicate, and care must be exercised to represent constituent concerns.

A major challenge facing all members of boards, advisory panels, etc., is how to translate their private concerns into a broad social consensus.

Washington State Nuclear Waste Board and Nuclear Waste Advisory Council: In 1982, Governor John Spellman formed the Washington State Nuclear Waste Board. It was created because there was no single point of contact for the state to deal with a growing number of nuclear issues. Concerns included the potential use of basalt rock beneath Hanford for a geologic repository and the financial troubles facing commercial reactor construction in the Pacific Northwest. Each state agency having nuclear-related jurisdictional responsibilities was made a board member. In 1983, the board was supported by legislative statute and membership expanded to include the academic community plus ex officio legislators. The Nuclear Waste Advisory Council, formed as part of the board, expanded membership to include citizens. From 1983 through 1989, the board and council held public meetings and provided comments to the U.S. Department of Energy on Hanford documents and activities. Recommendations were also given for guiding development of the Tri-Party Agreement.

The board was dissolved after Congress revised the Nuclear Waste Policy Act in 1987 to eliminate Hanford from further study as a repository and with the signing of the Tri-Party Agreement 2 years later. The Nuclear Waste Advisory Council remained for another 5 years. It was disbanded after formation of the Hanford Advisory Board.

> "There is tremendous public distrust of USDOE [U.S. Department of Energy] and deep concerns about the decision-making process. Many people feel the decisions have already been made, the decisions may not have a scientific basis, and that the state and its citizens have little voice in the decisions. Most citizen comments on public health, safety, and environmental issues related to concerns about possible contamination of the Columbia River and the potential for serious impacts to groundwater and agriculture"
>
> —Warren A. Bishop, Chair,
> Washington State Nuclear Waste Board (1986)

Northwest Citizens Forum on Defense Waste: Mike Lawrence—former manager of the Richland office of the U.S. Department of Energy Richland—established this forum in February 1986. Friar Bernard Coughlin, president of Gonzaga University, Spokane, Washington, chaired the forum. Its 28 members, from a range of backgrounds, conducted an independent review of the Hanford waste management program. Special emphasis was on review of a draft environmental impact statement dealing with tank and transuranic waste. Public involvement processes were also examined. The forum held public meetings across the Pacific Northwest. Recommendations were reported to U.S. Department of Energy in August 1986 (Coughlin 1986).

Oregon Hanford Waste Board: Hanford is about 35 miles north of the Washington-Oregon border. In the past, contaminant releases traveled downwind and downstream into Oregon. The citizens of Oregon expressed concerns over potential threats to the environment, public health, and public safety. The Oregon board is their advocate on these issues.

Air- and riverborne contaminants have traveled into Oregon. Oregon's citizens have expressed concerns.

The 20-member board, created in 1987 by the state's legislature, includes citizens and select agency heads, representatives from the legislature, the governor's office, and the Confederated Tribes of the Umatilla Indian Reservation (Oregon Office of Energy 2000a). In addition to an executive committee, the board has three standing committees: waste cleanup and site restoration, transport, and public involvement (Oregon Office of Energy 2000c).

In 1996, Oregon's Governor John Kitzhaber asked the board to aggressively pursue six areas to protect public safety and the Oregon environment as it relates to Hanford (Kitzhaber 1996):

- Challenge Congress to pass legislation to give Oregon a greater say in Hanford cleanup decisions.
- Ensure that actions are taken to protect the Columbia River.
- Ensure that nuclear waste transported through Oregon is moved safely and that the state is prepared for any emergency.
- Monitor any proposals to bring additional waste to Hanford.
- Give Oregonians the opportunity to make their voices heard and give them the information they need to influence cleanup actions.
- Support, whenever possible, the Confederated Tribes of the Umatilla Indian Reservation in exercising their treaty rights without hindrance.

Federal Facilities Environmental Restoration Dialogue Committee: In 1989, a U.S. House of Representatives committee conducted a hearing on the process used by the U.S. Department of Energy to set environmental restoration priorities. This was a concern because cleanup funds were limited though problems seemed unbounded. Besides, the existing approach to allocating funds seemed haphazard. That spring, the governors of 10 states sent the Secretary of Energy a letter calling for federal action to establish a coordinated national cleanup program. Further support was added when 49 attorneys general sent an open letter to the U.S. House of Representatives expressing support for a dialogue on federal facility cleanup issues.

Cleanup funds were limited though problems seemed unbounded.

These events led to establishing the Federal Facilities Environmental Restoration Dialogue Committee, sponsored by the U.S. Environmental Protection Agency. The committee, which first met in 1991, included representatives from tribal governments, citizen groups from across the nation as well as the U.S. Department of Energy, Department of Defense, and U.S. Environmental Protection Agency. The committee identified 14 principles to guide the prioritization of cleanup activities and to ensure that the priorities and concerns of all stakeholders were reflected (Federal Facilities Restoration and Reuse Office 1996). It's not surprising that several of these principles resembled those promoted by other Hanford groups; after all, people involved in Hanford cleanup were part of or consulted by the committee. The principles are paraphrased below:

The federal government is legally and morally bound to clean up its sites.

- The federal government is legally and morally bound to clean up its sites.
- The federal government must continue to complete cleanup at a reasonable and defensible pace.
- Contamination must not unfairly burden underprivileged areas.
- Federal and private sites should be treated the same in terms of cleanup standards.
- Cleanup contracts should reward desired outcomes, not simply efforts or products.
- Funding mechanisms should provide flexibility to ensure efficiency.
- Decision processes must preserve and balance the roles played by many actors.
- Cleanup agreements set priorities and provide a way to balance interdependent roles.
- A balance must be struck between public health and other factors, such as cost, cultural values, legal requirements, and effectiveness of cleanup actions.
- Pollution prevention and control are essential to prevent future problems.
- Reasonably anticipated future land use must be considered in making decisions.
- Cleanup processes should be streamlined to reduce extraneous studies and paper work.
- Cleanup decisions should be made within the broader environmental and cleanup context.
- Involve stakeholders to determine and achieve cleanup goals.

Cleanup processes should be streamlined.

Future Site Uses Working Group: This group was convened in April 1992, by the U.S. Department of Energy, U.S. Environmental Protection Agency, and Washington State Department of Ecology. Recommendations by its 48 members—from federal and state government agencies, agriculture, labor, environmental groups, tribal nations, local cities, and business (Hanford Future Site Uses Working Group 1992)—had a significant impact on how future cleanup activities were undertaken. Their broad concerns included

- Protect the Columbia River.
- Deal realistically and forcefully with groundwater contamination.
- Use Hanford's central plateau (200 Area) for waste management.
- Do no harm during cleanup or when implementing new economic opportunities.
- Clean up areas that will more than likely be used in the future.

- Clean up land to the level necessary to permit future use options.
- Transport waste safely and be prepared for emergencies.
- Capture economic development opportunities.
- Involve the public in future Hanford decisions.

The group was disbanded in December 1992. Success of the Future Site Uses Working Group prompted development of the Hanford Tank Waste Task Force, and their land-use principles served as a foundation for later Hanford Advisory Board actions.

Hanford Tank Waste Task Force: Funded by U.S. Department of Energy, this stakeholder group—with membership in the high twenties—provided values about tank waste cleanup and created general principles for negotiating Hanford cleanup.

In many ways, the task force reinforced and expanded on the principles recommended by the Future Site Uses Working Group. Significant changes to the cleanup approach resulted from the task force's recommendations. One was the termination of the grout facility. The grout facility, in the 200 East Area, was Hanford's first full-scale plant to solidify low-level radioactive waste. After filling one of four grout vaults with phosphate- and sulfate-contaminated waste containing 40 curies of radioactivity from decontamination activities at N Reactor, the facility construction stopped because of concerns about the potential for unacceptable future contaminant releases. Other examples of changes were shifting the near-term emphasis for cleaning up soil and groundwater away from central Hanford (200 Areas) and towards land adjoining the Columbia River and promoting development of an integrated plan for treating waste in both the single- and double-shell tanks. Low-level radioactive waste from the tanks would be incorporated into a retrievable glass form rather than a cement mixture.

U.S. Department of Energy 91031537-23 741fb.85

Grout Vaults. Construction of the first facility in the 200 East Area to solidify low-level radioactive and chemical tank waste into 1.4-million-gallon cement (grout) vaults was halted in 1993 because of citizen concerns about long-term safety and waste retrievability. In 1999, the U.S. Department of Energy decided to retrofit these vaults for the future disposal of less radioactive waste generated from the initial phase of tank waste vitrification.

During its short life (May through September 1993), the task force recommended the following guiding principles (Hanford Tank Waste Task Force 1993):

- Minimize land use for waste management.
- Avoid contaminating uncontaminated land.

- Avoid further harm to cultural resources, natural resources, and the environment, especially critical habitat and groundwater.
- Protect the Columbia River (stop actual and potential future contamination; prevent the migration of contamination beyond the site's borders).
- Don't depend on dilution to achieve safe conditions and/or to avoid legal discharge limits.
- Conserve and reuse resources.
- Recognize the importance of preserving biodiversity on land and in the river.
- Integrate Natural Resource Damage Assessments[5] under the Comprehensive Environmental Response, Compensation, and Liability Act into cleanup milestones.
- Preserve natural resource rights in treaties, and enforce laws protecting natural and cultural resources.
- Include risk assessments for natural and cultural resources in cleanup, waste management, and all other site activities.

Groups with opposing viewpoints can reach consensus on substantive issues.

These principles plus those identified by the Future Site Uses Working Group were later suggested to the signers of the Tri-Party Agreement as guidance for all cleanup activities (HAB 1994, 1996a).

Success of the Future Site Uses Working Group and the Hanford Tank Waste Task Force proved that diverse groups, holding opposing viewpoints, could work together and reach consensus on substantive issues helpful in guiding site cleanup.

Hanford Advisory Board: Following the success of previous public advisory committees, the U.S. Department of Energy, U.S. Environmental Protection Agency, and Washington State Department of Ecology created the Hanford Advisory Board. Members were selected, and the board first convened in January 1994. Though there are some 100 expert panels and advisory groups working at Hanford,

U.S. Department of Energy 02070014-104

Hanford Advisory Board. This board is a broadly representative group of the public, government agencies, and others seeking a regional consensus on a range of policy issues facing Hanford.

(5) The Natural Resource Damage Assessment provides a process for calculating the cost of repairing the natural environment from injuries or compensating the public for such injuries.

there is only one that has such a wide representation that addresses broad public policy issues: the Hanford Advisory Board.

The board is an independent group seeking a regional consensus on Hanford cleanup issues. It makes recommendations to U.S. Department of Energy, the U.S. Environmental Protection Agency, and the Washington State Department of Ecology. The board's 35 primary members, include representatives from local governments, businesses, universities, the Hanford work force, regional citizens, public interest groups, tribal governments, public health organizations, and the states of Oregon and Washington. The board maintains five committees: Budgets and Contracts; Health, Safety, Environmental Protection; Public Involvement and Communication; River and Plateau; and Tank Waste.

Much of the Hanford Advisory Board's advice foretold problems later faced by Hanford officials.

Over the years, the board has issued 137 pieces of "Consensus Advice" (informed recommendations and advice)[6] including recommendations about the Hanford restoration strategies, budget, nuclear material disposition, groundwater cleanup, decision-making, and worker concerns. Much of the board's advice foretold problems such as in tank waste privatization, the shift to a management and integration contract (Project Hanford Management Contract), and uncertainties in Hanford's proposed role to receive, treat, and dispose of low-level and mixed waste (HAB 1995a, 1996b, 1997b). The board provided early recommendations for refocusing work from the 200 Area to more immediate concerns for protecting lands adjoining the Columbia River. The board recommended slant hole drilling beneath tanks that leaked waste so that contaminant movement could be better understood. Their advice has provided U.S. Department of Energy managers with reasons and political support for continuing and starting some projects while scrapping others. One of the many challenges facing board members is the struggle between focusing on the "big ticket" strategic issues versus delving into the details of Hanford's daily activities.

Information about the board, including annual reports, advice, membership, guiding principles, and ongoing activities can be accessed at http://www.hanford.gov/boards/hab/index.htm.

Priorities for the Hanford Advisory Board

The Hanford Advisory Board (2000b) reports that tank waste stabilization and vitrification remain its highest priorities. These priorities are followed by protecting the Columbia River and resolving other urgent risks such as groundwater contamination, 618-10 and 618-11 Burial Grounds (north of the 300 Area), stabilizing plutonium in the Plutonium Finishing Plant, and stabilizing spent fuel in the K Basins.

(6) An average of 17 pieces of "Consensus Advice" is published each year.

Chapter 7

Environmental Regulations and the Cleanup Agreement

"[Cleanup] technology development typically requires several years to produce deployable results, depending on the initial technology status. In contrast, regulations and stakeholder values concerning remediation issues and decisions often change much more rapidly."

—National Research Council,
An End State Methodology for Identifying Technology Needs for Environmental Management, With an Example From the Hanford Site Tanks (1999, p. 6)

Environmental regulations embrace three components. First, congressional legislation establishes the broad intent of Congress. Usually, legislation is initiated by public concerns. Second, another federal agency such as the U.S. Environmental Protection Agency is instructed to analyze potential impacts from the legislation. Examples include the risk to humans, cost, benefits, and technical feasibility. Third, the same agency usually develops and issues rules to implement the legislation. These have the force of law and define whether or not a private or federal organization is in compliance with the law. Compliance requires actions such as maintaining waste in certain chemical or physical conditions, keeping contaminant emissions within prescribed limits, or retaining records of waste inventories. Rules generally describe the types of behavior a regulated party must demonstrate rather than the outcomes to be achieved (Dummer et al. 1998). Failure to comply can have punitive consequences including fines and other penalties.

Regulations generally describe the types of behavior expected rather than outcomes to be achieved.

While some argue the costs for implementing an intense regulatory framework are too high, most agree that the quality of our environment has benefited. The economic and environmental impacts of having no regulations would be higher. Significant progress has been made compared to earlier practices of just dumping raw sewage into rivers or venting unfiltered gases into the atmosphere.

In the late 1960s, the regulatory community grew in response to increasingly unacceptable, and even dangerous, environmental conditions.

At Hanford, contamination problems were exacerbated for years by practices that left waste-handling operations unchallenged. Into the early 1960s, little national-level attention was focused on the dumping of radionuclides and chemicals into the soil, streams, and rivers (Consumer Reports 1960). The exception was strong local public reactions along some East Coast and Gulf of Mexico communities to discharging low-level

> During the first 50 years at Hanford, as much as 450 billion gallons of liquids (mostly uncontaminated water) were released to the soil. How much is this? If this liquid was poured into standard 20,000-gallon railroad tanker cars and each connected, it would form a train 220,000 miles long—a distance stretching from the earth to near the moon.

radioactive waste into the nearby ocean. Federal agencies were exempt from many environmental protection laws until passage of the Federal Facilities Compliance Act in 1992.

Radioactive waste is regulated based on its source. The U.S. Nuclear Regulatory Commission regulates radioactive waste from commercial sources such as hospitals and power reactors. Both hazardous and radioactive wastes are regulated by the U.S. Environmental Protection Agency. Mixed waste is more challenging because the U.S. Department of Energy (DOE) regulates the radioactive portion while the U.S. Environmental Protection Agency regulates the hazardous chemicals. States also have laws governing the treatment, storage, and disposal of hazardous materials.

Contamination problems were exacerbated by practices that left waste-handling operations unchallenged.

In addition, two federal agencies are charged with environmental health standards: the Centers for Disease Control and Prevention and the Agency for Toxic Substances and Disease Registry (ATSDR). These agencies are part of the U.S. Department of Health and Human Services. The Centers for Disease Control and Prevention began in 1946, as part of the U.S. Public Health Service, working with state and local health officials to fight communicable diseases such as malaria and typhus. Today, it is the lead federal agency responsible for the broad mission of protecting and monitoring public health and safety.

On the other hand, the ATSDR is the principal federal public health agency involved with hazardous waste issues. It implements the health-related sections of environmental laws.(1) Congress created ATSDR under provisions of the Comprehensive Environmental Response, Compensation, and Liability Act (CERCLA). The ATSDR is not a regulatory agency though it advises those agencies and makes recommendations when specific actions are needed to protect the public's health.

Today, numerous federal and state requirements and several agreements and orders govern Hanford activities.(2) Many are discussed in the Tri-Party Agreement. A 13-page list of requirements pertaining to soil and groundwater cleanup is included in just one plan (DOE 1999d). Approximately 100 pages of principles and requirements are indexed for conducting an assessment of the environmental impact to the Columbia River from Hanford contaminants (DOE 1998a). While some requirements are not legally binding, their sheer numbers underscore the intense oversight and scrutiny Hanford activities receive. Compliance is a formidable task.

A formidable array of requirements, agreements, and orders govern Hanford's cleanup.

In general, Hanford cleanup is governed through federal natural resource management, pollution control, and cultural resource laws. The two primary federal laws are the Resource Conservation and Recovery Act (RCRA) and the CERCLA. Other federal laws such as the National Environmental Policy Act, Nuclear Waste Policy Act, Clean Air Act, and Clean Water Act also apply.

Examples of state laws that pertain to Hanford when federal law delegates enforcement or implementation authority are the State Environmental Policy Act, Hazardous Waste Management Act, and the Model Toxics Control Act.

(1) Responsibilities of ATSDR include providing public health assessments of waste sites, health profiles of hazardous substances, registries of people exposed to hazardous substances, responses to emergency chemical releases, plus health-related research, education, and training.

(2) A comprehensive list of environmental laws, agreements, and regulations governing Hanford is in the *Final Hanford Comprehensive Land-Use Plan Environmental Impact Statements* (DOE 1999c).

Historic resources and cultural values are subject to several federal laws and Executive Orders. These include the National Historic Preservation Act, Archaeological Resources Protection Act, American Indian Religious Freedom Act, Native American Graves Protection and Repatriation Act, and the Sacred Site Executive Order 13007.

The above laws offer broad guidance, rather than an unchanged prescription, for environmental and human protection. Laws are made and adjusted, as needed. Changes occur by statute or through agency choices. Interpretations take place through court cases, guidance documents, and policy statements. Changes also result from experience, new scientific knowledge, technology improvements, and adjustments in the political climate and public attitudes. Concerns over and improvements in our understanding of risk, cost, and benefit drive many applications. Environmental regulations grow from the emergence of a dominant view of what people believe is important and how they see themselves in relation to the environment.

These laws offer broad guidance, rather than an unchanged prescription, for protection.

The rest of this chapter summarizes some of the key laws, treaties, and the agreement governing Hanford cleanup.

The state of Washington's "involvement in the cleanup of the Hanford Site is based on its legal and regulatory authorities and responsibilities relating to the management of hazardous waste and environmental protection. However, efforts to implement environmental laws and regulations at Hanford are influenced and shaped by a number of important considerations. These include public and tribal concerns and values, environmental conditions, risk considerations, technical issues, fiscal constraints, regulatory overlaps, and regional and national political considerations."

—Washington State Department of Ecology (1993)

7.1 Cleanup of Abandoned Waste Sites

Congress originally passed the Comprehensive Environmental Response, Compensation, and Liability Act in 1980 to establish a national program for the cleanup of abandoned, uncontrolled hazardous waste disposal sites.

Public outcry from such contamination problems as Love Canal, New York; Times Beach, Missouri; and Jacksonville Municipal Landfill, Arkansas, pushed environmental concerns into the forefront of the news and to the top of policy makers' priorities. Nonetheless, the legislative roots of CERCLA lie in previous environmental acts regulating the discharge of pollutants, especially into surface waters, such as the Clean Water Act and the more-than-100-years-old Rivers and Harbors Act.

The CERCLA provides for the identification and cleanup of sites where contaminants were or might have been released into the environment, evaluation of damages to natural resources, and creation of a claims process for parties who have cleaned up sites.

In 1986, CERCLA was expanded by the Superfund Amendments and Reauthorization Act (SARA). This strengthened CERCLA by including changes such as increased public participation; consideration of state cleanup standards; stronger links to other

Regulations grow from a dominant view of how people see themselves in relation to the environment.

The philosophy of SARA is simple: treatment not containment; remedies not transportation.

environmental protection laws; and more stringent cleanup standards with a preference for permanent treatments that reduce the volume, toxicity, or mobility of hazardous substances. The philosophy of SARA is simple: treatment not containment; remedies not transportation. Also, SARA established new programs such as the Emergency Planning and Community Right-to-Know Act, which improved community preparedness for possible contaminant releases and provided the public with better information about hazardous materials located in their community.

Health and environmental risk assessments are woven throughout the CERCLA process, including an initial determination of how hazardous a site is (known as its hazardous ranking), screening of potential cleanup alternatives, selecting a preferred cleanup remedy, and specifying cleanup standards. Defensible standards and credible risk assessments drive CERCLA cleanup.

The U.S. Environmental Protection Agency is required to set cleanup standards based on federal and state standards for ambient water quality, groundwater, and soil. These standards known as "applicable or relevant and appropriate requirements"[(3)] are grouped into three categories (EPA 1988):

- **Chemical-specific** requirements limiting the amount of a chemical remaining at a site. These are health- or risk-based standards. Examples include maximum contaminant levels under the Safe Drinking Water Act and ambient surface water quality standards.
- **Location-specific** requirements restrict activities within certain locations such as floodplains. Examples include requirements under RCRA for determining the location of a waste site or dredging restrictions under the Clean Water Act.
- **Design or performance** requirements for treatment technologies and disposal activities such as landfill restrictions, storage requirements, and air emissions related to the management of hazardous waste.

Where applicable or relevant and appropriate requirements are not available or are not sufficiently protective, the U.S. Environmental Protection Agency generally sets site-specific remediation levels (Walker 2001).

The legislative history of CERCLA indicates that cleanup must balance protection and cost.

The legislative history of CERCLA indicates that cleanup "action at a given site must be that cost-effective response which provides a balance between the need for protection of the public health or the environment...and the availability of funds to respond to the other sites which present or may present a threat to public health or the environment" (HR REP 1980). This resembles wording found in the state of Washington's adoption of their statutory CERCLA law known as the Model Toxics Control Act. It says, "cost shall not be a factor in determining what cleanup level is protective of human health and the environment....[but] cost shall....be considered when selecting an appropriate cleanup action" (Washington State Department of Ecology 2001b).

Cleanup actions under CERCLA "must...utilize, to the maximum extent practicable, permanent solutions and alternative treatment technologies or resource recovery technologies" (HR REP 1986). The act's history provides flexibility for cleanup remedies

(3) "Applicable or relevant and appropriate requirements" are called ARARs.

less than permanent, such as institutional controls (land or water use restrictions) and engineered barriers. Cleanup objectives are not firm goals set at the beginning of a remediation process but are negotiated as a function of available technologies,[4] cleanup standards, land use, risk, money, politics, economic considerations, public preferences, or legal rulings (Hersh et al. 1997).

The state of Washington in its adoption of CERCLA notes that the cleanup levels and evaluation methods for guiding cleanup are designed to be as streamlined as possible "to avoid the confusion and delays associated with the federal Superfund program" (Washington State Department of Ecology 1996). The Model Toxics Control Act also has built-in flexibility to address remediation on a site-specific basis. Under the act, the Washington State Department of Ecology can gain access to properties, enter into settlements, file actions or issue orders to compel cleanups, and impose civil penalties and seek recovery of state-incurred costs.

7.2 Managing Active Waste Sites

The Resource Conservation and Recovery Act was enacted by Congress in 1976 to manage waste at operating and future facilities.

Through RCRA, the U.S. Environmental Protection Agency has the authority to regulate the management of hazardous waste, solid waste, and underground storage tanks such as those buried beneath gasoline stations. This act provides for "cradle-to-grave" tracking of hazardous waste—from generation to disposal. The legislative history of the act indicates that human health and the environment were to be protected regardless of cost (HR REP 1984).

Human health and the environment are protected under RCRA, regardless of cost.

The 1984 Hazardous and Solid Waste Amendments to RCRA required phasing out land disposal of untreated hazardous materials and correcting waste releases that did not comply with the amendments, regardless of when that release occurred. Other mandates include increased enforcement authority for the U.S. Environmental Protection Agency, more stringent hazardous waste management standards, and a more comprehensive underground storage tank program.

The federal RCRA program was delegated to the Washington State Department of Ecology for enforcement in 1986 and was revised to include regulations for mixed chemical and radioactive waste in 1987 (EPA 2000). The state's program is executed through its Hazardous Waste Management Act as promulgated under the Washington State Dangerous Waste Regulations (Washington Administrative Code 1982). The purpose of the Hazardous Waste Management Act is to establish a statewide framework for the planning, regulation, control, and management of hazardous materials to prevent or reduce land, air, and water pollution and conserve the state's natural, economic, and energy resources. Corrective actions might be interim or permanent to reduce or eliminate contaminant releases. State-enacted hazardous waste regulations are consistent with and as (or more) stringent than the federal program.

---

(4) If problems aren't solved using available technologies, then new investments are made to create the required capability.

When DOE applied CERCLA and RCRA to its sites, facilities that were considered operational became RCRA-regulated sites. Other facilities fell under the jurisdiction of CERCLA.(5) Subsequently, as a RCRA facility stops operations, its designation changes from an "operational unit" to a "past-practice site." While the site's designation changes, it remains under RCRA rather than moving to CERCLA authority.

7.3 Treaties with Native American Tribes

"The exclusive right of taking fish in all the streams, where running through or bordering said reservation, is further secured...together with the privilege of hunting, gathering roots and berries, and pasturing their horses and cattle upon open unclaimed land."

—Treaty with the Yakama 1855

The United States interacts with Native American tribes as sovereign nations. The extent to which these interactions are truly government-to-government is open to debate.

For centuries, several tribes of Native Americans roamed eastern Washington, including the lands that would become the Hanford Site. In 1855, the Yakama Indian Nation, the Umatilla Tribe, and Nez Perce Tribe ceded land near the Columbia River in three separate treaties. However, the tribes retain rights to hunt and fish, gather roots and berries, and pasture horses and cattle on open and unclaimed land. Farmers and ranchers later settled on the same land. Small towns grew over the years. In 1943, the Hanford Site was also established on these lands.

The United States recognizes the Native American tribes as sovereign nations and takes a government-to-government approach. Both entities entered into a trust relationship whereby the United States assumes responsibility to preserve the rights and resources of the tribes and to consult with the tribes about decisions that could affect those rights or resources. This relationship is passed along to all federal agencies. As a result, the DOE consults with the Native American tribes about Hanford issues, especially those that could impact the natural resources of the site.

Treaty compliance has been recognized by the federal government as a legal requirement for cleanup. The U.S. Department of Energy has committed to conducting tribal treaty use risk scenarios to establish acceptable cleanup levels. Cleanup levels necessary to achieve treaty compliance have yet to be determined.

7.4 Regulatory Responsibility

Regulatory compliance is a shared responsibility.

Federal and state agencies are responsible for overseeing different parts of the nuclear cleanup efforts around the country. At Hanford, the main players, U.S. Environmental Protection Agency and Washington State Department of Ecology, are responsible for enforcing CERCLA, RCRA, and other regulations (Ecology et al. 1998). In broad terms, the U.S. Environmental Protection Agency regulates under federal statutory requirements; the state regulates under state statutory requirements where Congress and the U.S. Environmental Protection Agency have delegated authority. Table 8 summarizes the areas of regulatory responsibility for various federal and state organizations.

(5) Some other liquid releases are managed under Washington State Waste Discharge permits.

Table 8. Regulatory Responsibilities at Hanford

Activity	Agency
Air emissions	U.S. Environmental Protection Agency, Washington State Department of Ecology
Groundwater cleanup	U.S. Environmental Protection Agency, Washington State Department of Ecology
Inactive facility cleanup	U.S. Environmental Protection Agency, Washington State Department of Ecology
Offsite waste transportation	U.S. Department of Transportation, U.S. Nuclear Regulatory Commission, and states through which waste is shipped
Operating facilities	Washington State Department of Ecology
River emissions	Washington State Department of Ecology
Soil cleanup	U.S. Environmental Protection Agency, Washington State Department of Ecology
Soil emissions	Washington State Department of Ecology
Solid waste cleanup	Washington State Department of Ecology
Tank waste cleanup	Washington State Department of Ecology
Waste treatment and storage	U.S. Environmental Protection Agency, Washington State Department of Ecology

Under the Tri-Party Agreement, the Washington State Department of Ecology oversees cleanup work. The Washington State Department of Health provides review and recommendations to Ecology on cleanup standards. The Health Department is the primary state agency for protection of human health and the environment from ionizing radiation. When promulgated, their regulatory guidance will be an applicable or relevant and appropriate requirement for the development, selection, and implementation of CERCLA cleanup actions (Washington State Department of Health 1997).

7.5 Tri-Party Agreement

"For each facility listed on the National Priorities List....CERCLA requires DOE to enter into an interagency agreement with EPA for the completion of all necessary remedial actions at the facility. These agreements often include the affected states as parties to the agreements. These agreements may be known as Federal Facility Agreements or Tri-Party Agreements."

—General Accounting Office,
Waste Cleanup: Implications of Compliance Agreements on DOE's Cleanup Program (2002b, p. 4)

The Tri-Party Agreement is a foundation upon which to build. In many ways, it pioneered integrating preexisting regulations into a more workable cleanup arrangement for a site of the size and complexity of Hanford. The toughest cleanup tasks and regulatory challenges at Hanford remain in the future.

The toughest cleanup tasks remain in the future.

In 1985, the DOE directed the Pacific Northwest Laboratory[6] to initiate the first remedial investigations of inactive waste sites at Hanford. This consisted of compiling waste site descriptions, disposal histories, and waste inventories (Stenner et al. 1988). The collected information was entered into a computerized data file.

(6) Known today as Pacific Northwest National Laboratory, the Lab is operated by Battelle for the U.S. Department of Energy.

The Tri-Party Agreement is a document struggling to make the government both accountable and efficient...traits not mutually exclusive though sometimes interfering.

Two years later, the U.S. Environmental Protection Agency began efforts to place Hanford on their National Priorities List[(7)], delineating the highest priority sites for further investigation and possible cleanup, using the data supplied by Stenner et al. (1988). The U.S. Environmental Protection Agency, the DOE, and the state of Washington also initiated discussions to establish a Hanford cleanup agreement. Our nation's military defense did not need more plutonium, the last reactor had shut down, and the final plutonium production run was taking place in PUREX. The mission of Hanford was about to change.

Based upon the U.S. Environmental Protection Agency's assignment of potential hazards, Hanford was listed as four sites on the National Priorities List. These were the 100 Area, 200 Area, 300 Area (including solid waste burial sites a few miles north), and the 1100 Area.[(8)] The 1100 Area covered both the southern tip of Hanford used for vehicle maintenance, transportation, and warehousing and the 120-square-mile Fitzner/ Eberhardt Arid Lands Ecology Reserve located around Rattlesnake Mountain. Each site was further divided; smaller divisions were called operable units. These units were groupings of waste sites that shared a common environmental setting, contaminant sources, waste forms, and likely cleanup approach. Originally, 74 operable units were identified along with 4 major groundwater units. Because all operable units can't be simultaneously studied or remediated, they were, and continue to be, prioritized. Hanford is now divided into 42 soil and 10 groundwater operable units (EPA 2001a).

In May 1989, Hanford became the first DOE site to enter into a formal cleanup agreement formally known as the Hanford Federal Facility Agreement and Consent Order—popularly known as the Tri-Party Agreement. The agreement is among Washington State Department of Ecology, U.S. Environmental Protection Agency, and DOE. Though concerns were expressed that existing laws and regulations weren't well designed for tackling Hanford-type cleanup problems, the agreement marked a beginning.

The Tri-Party Agreement signaled a significant departure from previous years of secrecy and self-rule. It sought to build a new language of environmental protection and social accountability rather than governance dominated by plutonium production, using the environment as a dump for contaminants. It's a document struggling to make the government both accountable and efficient—traits that are not mutually exclusive though sometimes interfering. Remediation activities were made into legally enforceable milestones. Institutionally, DOE did not enter into the agreement willingly though there were those who recognized that change was overdue. Only after the state of Washington threatened DOE with a lawsuit did the department sign the agreement.

---

(7) The U.S. Environmental Protection Agency uses their Hazard Ranking System to place abandoned waste disposal sites on the National Priorities List. This ranking system uses a numerical screening process to assess a waste site's potential threat to human health or the environment. After sites are proposed, the agency accepts public comments, responds to comments, and then formally places those waste sites on the list that continue to meet the requirements for a high hazard.

(8) The 1100 Area was remediated and deleted from the National Priorities List in 1996.

Oregon: The Downstream State

The state of Oregon desires a greater say in decisions made about Hanford (Kitzhaber 1996). This extends to becoming a full partner in the Tri-Party Agreement and a Hanford regulator (Oregon Office of Energy 2000b). Oregon lies downstream of Hanford, radioactive wastes are shipped through the state, and several Oregon communities are located within the 50-mile nuclear emergency planning radius of Hanford. The state of Oregon and DOE signed a memorandum in 1997 to keep Oregon apprised of Tri-Party Agreement issues and provide a basis for addressing Oregon concerns (DOE and State of Oregon 1997). In 1999, the Oregon Legislative Assembly passed a joint resolution encouraging Congress to provide Oregon with "legal rights in matters affecting the Hanford Nuclear Reservation that involve handling, storage, and cleanup of radioactive wastes, including party status in the Hanford Tri-Party Agreement or any successor thereto" (Oregon Legislative Assembly 1999). No congressional action has taken place.

The agreement requires DOE to bring Hanford's active waste sites into compliance with provisions of RCRA and the state's Hazardous Waste Management Act. It also establishes a framework and schedule for implementing cleanup actions in accordance with CERCLA. The agreement consists of three parts:

- The **Legal Agreement** describes the roles, responsibilities, and authorities of the three agencies in the cleanup, compliance, and permitting processes. It sets up a dispute resolution process to avoid penalties and litigation and describes how the agreement will be enforced.
- The **Action Plan** describes the methods for implementing the cleanup and waste permitting efforts and includes milestones for initiating and completing specific work for the agencies. The agreement provides the opportunity for milestone changes as new information is gained.
- The **Community Relations Plan** describes how the public will be informed and involved throughout the cleanup process. This includes convening public meetings, providing opportunities for the public to comment on documents, and applying for technical assistance grants available through federal or state governments. The plan also states, "DOE, EPA, and the state of Washington have adopted policies [that] recognize tribal sovereignty and commit to government-to-government relationship with the tribes" (Ecology et al. 1998).

Enforceable schedules and procedural controls are the venue to track cleanup activities.

From a regulatory perspective, the agreement's enforceable schedules and procedural adherence are the legal venue to ensure progress. And from a public perspective, the Tri-Party Agreement is their Maginot line for ensuring health and environmental protection.

Since signing the agreement, many Hanford waste management and safety improvements have taken place. A cleanup agreement was needed, and strong public participation was long overdue. Now comes the long-term balancing of accountability and efficiency in applying environmental laws and work practices to ensure they provide meaningful safeguards.

7.6 A Challenging Future

"The original milestones in the TPA [Tri-Party Agreement] were ambitious—too much so in many cases, and did not sufficiently reflect the complexity and challenges that exist at Hanford."

—Oregon Office of Energy,
Hanford Cleanup: The First Ten Years (1999b)

Thanks to the Tri-Party Agreement and those who have worked to maintain it, Hanford is managed in a more environmentally conscientious manner than before. Those who believe an agreement was unnecessary or too awkward to manage must ask themselves how much progress would have been made without it? Decades of managing waste "as usual" came under a different accountability; a new onsite culture took root. However, most contaminants and nuclear materials remain untouched.[9]

When the Tri-Party Agreement was signed, "expectations were high," and people imagined nearly instant progress (Oregon Office of Energy 1999b). However, milestones were too ambitious and people did not adequately gauge the technical, regulatory, or social complexity faced. This was also reflected in DOE's own concerns that many of the requirements and agreements entered into by the federal government "were established without an understanding of the magnitude of the required cleanup of contaminated sites and the stabilization of nuclear materials…as well as the associated costs….in the process of defining the cleanup program and the approaches to be employed, DOE accepted the existing requirements and agreements in some cases without regard to their appropriateness" (DOE 2002, p. IV-3).

When the cleanup agreement was signed, expectations were high, and people imagined nearly instant progress.

Originally, some senior officials thought cleanup of the nuclear weapons complex would take less than 5 years. Thus, they were not interested in supporting investments beyond the mid-1990s and did not understand the need for a multi-decade cleanup agreement.

Experience taught otherwise. Surprises continue to raise our attention above the background of project baselines. As noted in Section 8.4, since cleanup began several unanticipated events captured Hanford's attention and shifted priorities. These ranged from resolving problems of with the "Watch List" tanks (1990-2001) to the unacceptable storage of spent fuel in the K Basins (1994-present) to the deeper-than-expected movement of leaked contaminants from tanks (1997-present). More surprises, and the need for flexibility, will follow as we deal with the bulk of Hanford contaminants and nuclear materials.

Surprises continue to raise our attention above the background of project baselines.

As at other nuclear weapons sites, Hanford faced poorly known problems, ill-suited work processes, questionable contract strategies, inadequate cleanup technologies, conflicting regulations, and uncertainty about what cleanup meant in the first place. Governance was poorly defined, and cost estimates were a roll of the dice.

(9) This is measured in the amount of chemicals and radionuclides recovered, stabilized, or otherwise treated versus those materials that remain untouched.

"Changes to the compliance agreements for the long-term sites should aim not at relaxing requirements but at recognizing the realities of the situation and the state of current knowledge and capabilities.... But stakeholders and overseers of the cleanup program will not allow significant deviation from current commitments until they are convinced that there are better approaches and that the search for them is motivating the site management to change the compliance status quo." However, "Allowing a compliance agreement based on inadequate knowledge, impossible schedules, or an irrelevant paradigm (cleanup to a 'releasable' state, instead of long-term remediation and stewardship) to drive the application of scarce resources invites failure on a grand scale" (Washington Advisory Group 1999).

The Tri-Party Agreement is hamstrung from an inherent problem that no amount of rewording, or re-baselining, can solve: the primary determinants of cleanup (approaches, final land states, final products, risks, and cleanup levels) remain unresolved and too many people, organizations, and agencies affect cleanup decisions.

Work is heavy on process and light on technical rigor.

Most people support the agreement but wonder how cleanup is built around a continuous stream of renegotiated milestones based upon a questionable understanding of the problems faced and actions taken. They're concerned over the gap separating the power of enforcement from the responsibility for cost, of holding Hanford accountable for past misdeeds while preserving realistic work approaches. Concerns exist about agencies abdicating their responsibility to protect taxpayers by defining the work scope "in ways that cannot be justified on technical or cost grounds" (Blush and Heitman 1995). Regulations and the Tri-Party Agreement make legal and ethical sense but, as implemented, is value commensurate with cost? Where is the line separating environmental safeguards from unfathomably complex work processes and distracting administrative burdens?

The mounds of guidelines directing Hanford work are heavy on process but light on technical rigor (Washington Advisory Group 1999). These include work procedures, regulatory directives, and milestone schedules. Yes, it's essential to hold the government accountable for cleanup commitments. Nonetheless, all organizations and agencies are accountable for streamlining constraints. Michaels (2000) stresses that the real issue is not whether a specific cleanup standard or certain lifetime risk is achieved, for both U.S. Environmental Protection Agency and U.S. Nuclear Regulatory Commission are already regulating to levels far beyond those that can be verified by science. Rather, the real issue is whether the regulatory standards are flexible enough to ensure that they do not bring about consequences not commensurate with benefit.

Cleanup priorities are driven by narrow interpretations of regulations, not necessarily protecting human health and the environment (NRC 1996d).

Dummer and her colleagues (1998) at the Joint Institute for Energy and Environment, Knoxville, Tennessee, argue that sufficient flexibility exists in legislative statutes permitting reform of cleanup practices to deliver more environmental protection earlier and at a lower cost. At Hanford, an example of this flexibility was the Environmental Restoration Refocusing Package, signed as an amendment to the Tri-Party Agreement in 1995, that added emphasis on groundwater and soil cleanup along the Columbia River and created a plan to achieve greater coordination of activities (Community Relations Plan 1999). On a national scale, regulators have been quite flexible in adjusting cleanup milestone dates when DOE could not meet the original deadlines. Ninety-three percent of these milestone changes were approved (GAO 2002b, p. 8).

> "We have argued that a systematic review of the legislative history and environmental legislation leads to the conclusion that Congress intended that protection of the environment and public health and safety be given first priority, but the choices of how specifically to do so be guided by common sense and a recognition of constraints" (Dummer et al. 1998, p. 11).

What is one of these contentious issues where costs and benefits must be carefully weighed? It's the waste in Hanford's 177 tanks. These tanks contain nearly half of all the radioactivity and hazardous chemicals on the Hanford Site. Hanford cleanup means tank cleanup.

An Office of Inspector General report covering the tank program stresses cleanup milestones were established before work do-ability was determined (DOE 2000e). For example, what amount of waste should be or could be recovered? The Tri-Party Agreement specifies that 99% by volume of Hanford's tank waste should be removed. However, "even if 99% retrieval is technically attainable, it may not be desired considering cost, exposure to radiation, and technical practicability…waste volume does not equate to environmental risk." The report states that Washington State, the U.S. Environmental Protection Agency, DOE, and Hanford contractors have acknowledged the long-term tank cleanup milestones are "unrealistic" and not feasible under present safety, operational, and technology constraints. Waste retrieval will likely be lower—perhaps 60 to 80%.

The regulators, DOE, and contractors have acknowledged that the long-term tank cleanup milestones are "unrealistic" (DOE 2000e, p. 11).

Does this mean not trying to retrieve 99% of the tank waste? Certainly not. Rather, a fundamental different question is asked, Do we understand the implications of our milestones? For example, does all tank waste pose the same level of biological risk? If 99% of the waste can't be removed, then what investments in advanced retrieval technology, engineered barriers, or alternative treatment technologies are being pursued—today—and scheduled to ensure that needed capabilities are available? In fact, a U.S. Environmental Protection Agency report (2000) expresses concerns over inadequate funding to develop waste retrieval technology though the tank cleanup program must have that technology in a timely fashion to achieve anything close to 99% removal of tank waste.

Do we understand the implications of our milestones?

In a critique of their nationwide cleanup plans, DOE (2002, p. ES-2) questioned the correctness of their own strategies. "Many wastes are managed according to their origins, not their risk. This approach has resulted in costly waste management and disposition strategies that are not proportional to the risk posed to human health and the environment."[10] Further, DOE stated that many of the necessary "reforms can be achieved within the current statutory framework." If not, changes to carrying out an accelerated, risk-based approach to cleanup will be proposed to Congress. These are contentious issues now debated before judges. However, even some writers who normally take DOE to task on waste management issues have also questioned the adequacy of how waste types are defined (e.g., Hancock and Makhijani 1992; Fioravanti and Makhijani 1997).

(10) The question of waste origins is discussed in Section 8.1.

High-quality milestones are essential to tracking cleanup progress. Concerning soil and groundwater cleanup, the Integration Project Expert Panel (2000a) warns that poorly thought out actions "could cause more harm than good. We are seeing little discussion of these important issues."

"Some of the fundamental inadequacies of the present environmental remediation, management, and disposal of waste" result from the confusing and inadequate classification of waste (Fioravanti and Makhijani 1997, p. 299).

In October 2002, the DOE Office of River Protection announced plans to accelerate the cleanup and closure of 26 to 40 Hanford single-shell tanks. This effort also includes chemically treating 1 million gallons of transuranic-laced tank sludge and removing the final gallons of pumpable liquids from all single-shell tanks. Work is to be accomplished between the years 2004 and 2006 (Stang 2002d).

The proposal took federal and state regulators by surprise. In fact, it came just 2 months after DOE and the Washington State Department of Ecology reaffirmed their cooperation on the tank cleanup program. Technical and regulatory compliance details of this newest proposal are sketchy at best. At this time, no one knows what tank closure means, how the proposal complies with environmental laws, what it will cost, or where the funds will come from. In addition, before the sludge is solidified into some dry form and packaged for shipment to the Waste Isolation Pilot Plant, New Mexico, it must be treated in a new Hanford facility. That's not built. According to state officials, just the environmental studies and permits required to carry out this work will take up to 3 years to complete. Nearly any attempt to accelerate cleanup progress is welcomed.

U.S. Department of Energy 1388

Single-Shell Tank Construction. This 1944 photograph shows one of the first single-shell tank farms under construction at Hanford. Workers are pouring concrete to form a tank dome. These old, leak-prone tanks are the first scheduled for permanent closure.

However, there is concern that this latest announcement is reminiscent of the Department's old tactic of "deciding, announcing, and defending." Cooperation hinges upon forethought, not afterthought. Policies are based upon what agencies do, not what they say.

Historically, government policies are the cause of both problems and solutions.

Andrews (1999, p. x) notes in his discussion of the American environmental movement that policies are commonly manifestations of a continuous stream of debates, refinements, and sometimes radical changes where advocates seek solutions through government intervention. Historically, however, government policies "are often causes of environmental problems as well as solutions to them." The most contentious issues are often broader than regulations, research investments, economics, or restoration actions. Rather, they often center upon governance—upon control.

Chapter 8

Cleanup of Hanford and the Nuclear Weapons Complex

"Most of the weapons complex is not going to be cleaned up in the foreseeable future. Merely stabilizing the wastes is an enormously sophisticated technical enterprise."

—Linda Rothstein,
The Bulletin of the Atomic Scientists (1995, p. 34)

The cleanup of Hanford sets within a national context—cleanup of the nuclear weapons complex. The two are intricately linked by history as well as environmental laws, multi-site dependencies, health standards, citizen values, remediation technologies, national policies, and funding availability.

Hanford contains the mother lode of cleanup problems facing the weapons complex.

Hanford contains one of the largest inventories of radioactive and chemical contamination in the Western Hemisphere. It's the mother lode of cleanup problems facing the nuclear complex. Compared to all other U.S. Department of Energy (DOE) sites, Hanford has

- 30% of 4000 waste storage and release sites (Office of Technology Assessment 1991).
- 40% of 1 billion curies (DOE 1997a).
- 60% of the 91 million gallons of radioactive tank waste.
- 60% of the 4.4 million cubic feet of buried transuranic-contaminated solid waste (DOE 2000f).
- 80% of the 2900 tons of spent fuel (National Spent Nuclear Fuel Program 2002).

Since 1943, 134 sites in the United States were used for uranium mining, nuclear material production, and weapons research and testing (DOE 1998b). At the beginning of 1998, cleanup responsibility for 21 sites, managed under the Formerly Utilized Sites Remedial Action Program, was transferred to the U.S. Army Corps of Engineers (DOE 2000g). Some of these sites were rather small. For example, 3000 cubic yards of uranium- and thorium-bearing solid waste was excavated from a 45-acre site in Albany, Oregon, and shipped to Hanford for disposal (U.S. Army Corps of Engineers 2000).[1] The site in Albany was used for metallurgical operations supporting the Atomic Energy Commission.

The transfer of these sites to the U.S. Army Corps of Engineers left 113 places under the responsibility of the DOE. As of 2002, active cleanup was completed at 74 of these locations (DOE 2002).[2] Most of the settings contained uranium mill tailings,

(1) A volume of 3000 cubic yards fills a cube 14 yards on a side.
(2) Many of these locations will require long-term care because some waste or cleanup products will remain in place or long-term contaminant treatment and monitoring will continue.

Left alone, the situation will not improve.

remediated since 1978 under the Uranium Mill Tailings Remedial Action Project. Cleanup involved stabilizing mill tailings[3] in place by covering them with layers of soil, moving tailings to more remote public lands, or simply restricting land uses. Thirty-nine locations, under DOE jurisdiction, remain to be cleaned (DOE 2000g). These contain an aging infrastructure holding contaminants and nuclear materials increasingly vulnerable to release unless more secure and permanent remedies are carried out. Left alone, the situation will not improve.

The DOE is responsible for 113 sites across the country once engaged in nuclear material or weapons production. Their sizes range from a few acres to hundreds of square miles. They include locations of uranium mining, uranium refining and enrichment, nuclear materials production, weapons manufacturing, research, and weapon testing. Most are no longer active. Combined, these sites are referred to as the DOE nuclear weapons complex.

8.1 Cleanup—What's That?

"I'm often asked why it has taken so long to clean up and stabilize Hanford's radioactive and hazardous wastes. These queries generally come from individuals who have never seen the site, cannot envision the size and complexity of its buildings and their contents, and have little understanding about the short and long-term risks to workers, the public, or the environment."

—Merilyn Reeves, past chair of the Hanford Advisory Board (HAB 1999)

The word "cleanup" is a misnomer. It can mean many things such as contaminant destruction, treatment, containment, relocation, natural attenuation, or simply relying upon institutional controls (fences, gates, and guards) to keep people away.

Contaminants must be dealt with. How this is done is based upon knowledge and capability. Why this is done is based upon societal preferences.

Contaminants must be dealt with and the resulting products managed. How this is done is based upon knowledge (science) and capability (technology) as applied through a regulatory framework. Why this is done and what residual waste forms are acceptable are based upon societal preferences.

How clean is clean? This is not a new issue. It was one of the first questions facing industry and communities starting in the 1960s when Congress began passing a sweeping series of environmental laws. The question is commonly debated in the abstract, but it can't be answered in the abstract. In a practical sense, it's addressed at every waste site,

(3) Uranium mill tailings are sand-like material remaining from the separation of uranium from its ore. More than 99% of the ore taken from a uranium mine becomes mill tailings (DOE 1995a, p. 99).

Three hundred years spans ten half-lives of strontium and cesium. After that time, only 1/1000th of the original amount of these radionuclides remains.[4]

and it must blend local values with national and state health protection standards. Without local values, those living near the site will not support cleanup decisions; without protective standards, those enforcing regulations or providing funding won't support the decisions.

The National Research Council (1995, p. 44) reports: "Although the general goal of environmental restoration is cleanup, there is no universal answer to the question 'How clean is clean enough?'…Standards exist for drinking-water supplies (protection of human health) and surface waters (protection of ecosystems). Few such standards exist for soils, even with respect to hazardous chemicals, and no standards have been designed specifically for cleanup of most radionuclides in soil. The only standards designed for the cleanup of radionuclides are those for land and buildings contaminated by uranium-mill tailings at inactive uranium-processing sites."

"Although the general goal of environmental restoration is cleanup, there is no universal answer to the question 'How clean is clean enough?'" (NRC 1995).

Agreeing that a site should be cleaned and knowing what cleanup means are different. In simple terms, it's like the difference between wanting to take a vacation and actually purchasing tickets and traveling. The first is a general desire that everyone in your family can agree to. The second is based on an understanding of what you want to do, where you want to stay, how you're traveling, and how much you're willing to pay for the experience. You will face many choices in how to spend your time and money. Decisions will be based on values and preferences. Compromise must prevail to meet competing desires of family members. We know problems will arise, such as with rental cars or keeping the kids entertained between stops, yet they won't be experienced until you're on the road. Problems aren't failures; rather, they're the natural outcome of taking action. Your vacation is finished when you have achieved your goals, run out of time, or spent your money.

Problems are the natural outcome of taking action and learning.

One of the key issues in the cleanup debate revolves around categories of land use: residential, recreational, commercial, or industrial. These help anticipate who might be exposed to contaminants and how exposure could take place. Different cleanup standards are applied depending on whether children will play in the dirt or heavy industry and parking lots will occupy the space. Over the long term, it's nearly impossible to predict (or control) land uses. Nonetheless, DOE's Office of the Inspector General cautions about cleaning up any lands to levels "inconsistent with projected land uses" (DOE 1999e).

However, Robert Hersh and his colleagues (1997) reported there was little correlation between land-use preferences selected under the U.S. Environmental Protection Agency's Superfund program with the actual state achieved when remediation efforts ended. Sometimes cleanup was finished when funding was spent, waste hauled elsewhere, or there was general agreement that enough work was accomplished. Decisions, they found, were commonly based on choices of the moment rather than risk reduction or contaminant levels achieved.

---

(4) For a discussion of half-lives and to see the half-lives for certain radionuclides, see Appendix A.

Time	Yesterday	Today	300 Years	1000+ Years
Total Radioactivity	1 Billion Curies	400 Million Curies	1 Million Curies[1]	400,000 Curies
Dominant Radionuclides	Cesium Strontium Short-Lived Radionuclides[2]	Cesium Strontium	Cesium Strontium Long-Lived Radionuclides[3]	Long-Lived Radionuclides[3]
Chemicals and Metals	400,000 to 600,000 Tons			

1 Assumes 99.9% is cesium-137 and strontium-90.
2 Examples include isotopes of argon, krypton, manganese, sodium, and neptunium.
3 Examples include isotopes of plutonium, americium, technetium, and iodine.

741fb.68

Waste Inventory Over Time. Radionuclide decay changes the waste inventory and onsite hazards. Short-lived radionuclides, with half-lives of less than a few years, have decayed away. Longer-lived radionuclides will dominate future risks. Chemicals will persist onsite in concentrated, dilute, or altered forms.

On a national level, many Superfund decisions are based on choices of the moment rather than risk reduction or contaminant levels achieved.

The meaning of cleanup will become clearer as work progresses and expectations are refined. We have only to remember the evolution of Hanford operations over the last 20 years and cleanup approaches taking place since signing of the Tri-Party Agreement in 1989. A quick review of the first "five-year" waste management and restoration plans for Hanford and the DOE weapons complex reveal major differences in work scope and schedules compared to what we have today (DOE 1991a). An upcoming case-in-point took place during late 2001. It involved the 91 million gallons of tank waste.

Burial Grounds Along the River

In September 2000, the DOE, U.S. Environmental Protection Agency, and Washington State Department of Ecology signed an interim agreement to clean up 45 burial grounds along the Columbia River near the 100 Area (Hanford Reach 2000). Exhumed materials would be hauled to an onsite landfill and then the original site would be backfilled with clean soil and planted with native vegetation. In 10 years, the agencies will decide if the area is sufficiently clean to remove it from EPA's list of hazardous sites. During those years, preferred land uses, economics, and other considerations will interplay and influence future decisions.

In November 2001, a memo signed by the DOE Assistant Secretary for Environmental Management announced new priorities for her office (Roberson 2001).(5) Specifically, the memo stated: "HLW [high-level waste] processing is the single largest cost element in the EM [Office of Environmental Management] program today. Eliminate the need to vitrify at least 75% of

(5) As of 2002, the majority of the nation's high-level liquid tank waste (60% or 53 million gallons) is at Hanford. Most of the remaining rests at the Savannah River Site in South Carolina (40% or 38 million gallons). A small amount of unvitrified tank waste remains at West Valley, New York. Also, 140,000 cubic feet of a dehydrated powder of calcined high-level waste lies in six large bins at the Idaho National Engineering and Environmental Laboratory (NRC 1999c). The Idaho site also contains about 900,000 gallons of liquid high-level waste in 11 tanks. All values change from year to year.

Redefining Waste?

In 2002, DOE reported: "Low-activity high-level waste (HLW) is managed as if high-cost retrieval and vitrification were the only option available to protect the public. This waste is less hazardous than some low-level waste (LLW) that is considered acceptable for lower-cost near-surface disposal. This problem arises because HLW is defined based on its source rather than its constituents and their concentrations." Therefore, DOE proposed: "wastes will be classified according to total curie content and the curie content of long-lived isotopes and treated accordingly. Only those wastes with high-curie long-lived isotopes will be vitrified. Alternative processes such as steam reforming, calcination, saltstone, and other grouting techniques should be considered for stabilizing tank waste containing low-activity and TRU [transuranic] wastes" (DOE 2002).

the waste scheduled for vitrification today. Develop at least two (2) proven, cost-effective solutions to every high-level waste stream in the complex." This review may be the first step in a sweeping nationwide overhaul to try refocusing cleanup plans. How it might impact present tank treatment actions at the DOE's Savannah River Site or plans at Hanford or the Idaho National Engineering and Environmental Laboratory is unknown.

Hanford tanks are reported to contain 11 million gallons of high-level radioactive waste and 42 million gallons of less radioactive waste.

How tank waste is to be treated is a contentious topic. Yes, one way to lower waste reprocessing costs is to reduce the amount of waste vitrified. Currently, the process that created it, not the biological hazard, categorizes tank waste. Thus, to significantly reduce the amount of waste vitrified, as proposed in the assistant secretary's memo, some portion would have to undergo less-stringent processing.

The response to this proposal was swift and unequivocal. In February 2002, the Natural Resources Defense Council, Confederated Tribes and Bands of the Yakama Nation, and the Snake River Alliance filed a lawsuit in U.S. District Court against any attempts by DOE to reclassify tank waste so that less-stringent disposal criteria might be applied (NRDC et al. 2002). The states of Washington, Idaho, and Oregon were named as friends of the court in that lawsuit. The Shoshone-Bannock Tribe, who live downstream of the Idaho site, also joined the lawsuit. The plaintiffs want to stop any "open-ended process for exempting high-level waste from the stringent technical and procedural requirements of the NWPA [Nuclear Waste Policy Act]." The lawsuit states that the waste potentially "abandoned in the storage tanks contain equal to or greater concentrations of radioactive elements than the waste removed for disposal in a geologic repository." Concerns focus on the potential risk from radionuclide releases and the subsequent use of "significant land-use restrictions, maintenance, and monitoring in perpetuity" for sites where tank waste might be treated in place.

A lawsuit was filed against any attempt to reclassify high-level radioactive waste so less-stringent disposal criteria might be applied (NRDC et al. 2002).

There are merits to both sides of the argument.

This suit typifies concerns around the traditional definitions of waste that will play out in how cleanup is undertaken. However, this is not just a curie-count issue. It's a biological threat issue. Residual waste could contain a low-curie count made of long-lived transuranic elements (for example, plutonium, neptunium, or americium) or a high-curie load

The attorney general for the state of Washington stated that the tank waste should not be reclassified just "because it is too difficult or expensive to do what is right" (Cary 2002b).

Protection, not percentages, is at the heart of the cleanup game.

Cleanup is a negotiated, conditional state.

of short-lived radionuclides such as strontium and cesium.[6] The first presents a potential long-term biological risk (thousands of years) while the second presents a much shorter risk (hundreds of years). The bulk of the long-lived transuranic-contaminated waste rests in hard-to-remove sludges settled near the bottom of tanks.

Is it premature to skirmish over how much waste, if any, should remain in the tanks? No. The case for biological protection afforded by any waste form must be proven beyond a reasonable doubt—before irrevocable treatment actions are undertaken. Protection, not percentages, are at the heart of the cleanup game. However, today we lack the in-tank knowledge and processing experience to know how all tank waste is best handled.

As noted by Hancock and Makhijani (1992, pp. 22-23), and Fioravanti and Makhijani (1997, pp. 4-5), the present classification of military-generated waste is inadequate for guiding environmental cleanup and disposal decisions where biological risks come into play. Treatment of tank waste is leading this debate.

Yes, the cost of vitrifying all tank waste and sending the glass logs to a geologic repository will be high—so was the cost of producing plutonium. At the same time, the cost and biological threat from this option must be balanced against the expense and risk from alternative treatments.

Regardless how cleanup is achieved, progress will take place one year at a time, one project at a time. Each year's budget and work should leave contaminants in a progressively more stable, contained, and secured state. Even after active cleanup is complete, future landlords will monitor the site for centuries—and will reopen decisions thought settled by previous generations.

8.1.1 Interim and Final States of Cleanup

"In the past, attainment of policy goals and enforcement of environmental regulations were often assessed in terms of bureaucratic endpoints (number of permits issued, reduction in contaminant effluent)...rather than tracking the biological condition of the system being protected."

—Integration Project Expert Panel, *Closeout Report for Panel Meeting Held October 25-27, 2000* (2000b)

Cleanup is a negotiated, conditional state.

Interim waste states address immediate problems and should lower near-term risks until more permanent conditions (known as end points), are reached. To date, most Hanford work has focused on interim actions such as trucking contaminated soil away from the banks of the Columbia River, pumping groundwater, entombing reactors, or preparing spent fuel for improved temporary storage.

(6) Short-lived radionuclides decay quickly by emitting lots of radiation over a brief period of time. Long-lived radionuclides decay much slower as they release their energy over an extended time. This is why you could quickly die from holding a large chunk of strontium-90 (half-life of 29 years) in a gloved hand versus the essentially zero risk from holding a chunk of plutonium-239 (half-life of 24,000 years).

Will Hanford ever reach a final, pre-ordained end state? Most likely not—or only in hindsight after final cleanup goals are redefined to match what was accomplished. Perhaps the term "end state" is best replaced by "exit-point" to denote achieving a condition where people agree that active remediation should stop (NRC 2000b, p. 22).

Will Hanford ever reach a pre-ordained end state? Most likely not.

For protecting the Columbia River and its wildlife, Hanford's Integration Project Expert Panel (2000b) believes that "biological endpoints" and "ecological risk factors" should be incorporated into decisions. The panel recommended that cleanup be defined by what protection is attained, not by what activity or milestones are completed.

Biological end points and ecological risk factors should be incorporated into decisions.

Land end states and waste product end points are moving targets; they are hard to identify because they will be based upon the application of existing knowledge and technical solutions as well as the changing interests from multiple (and often conflicting) authorities.

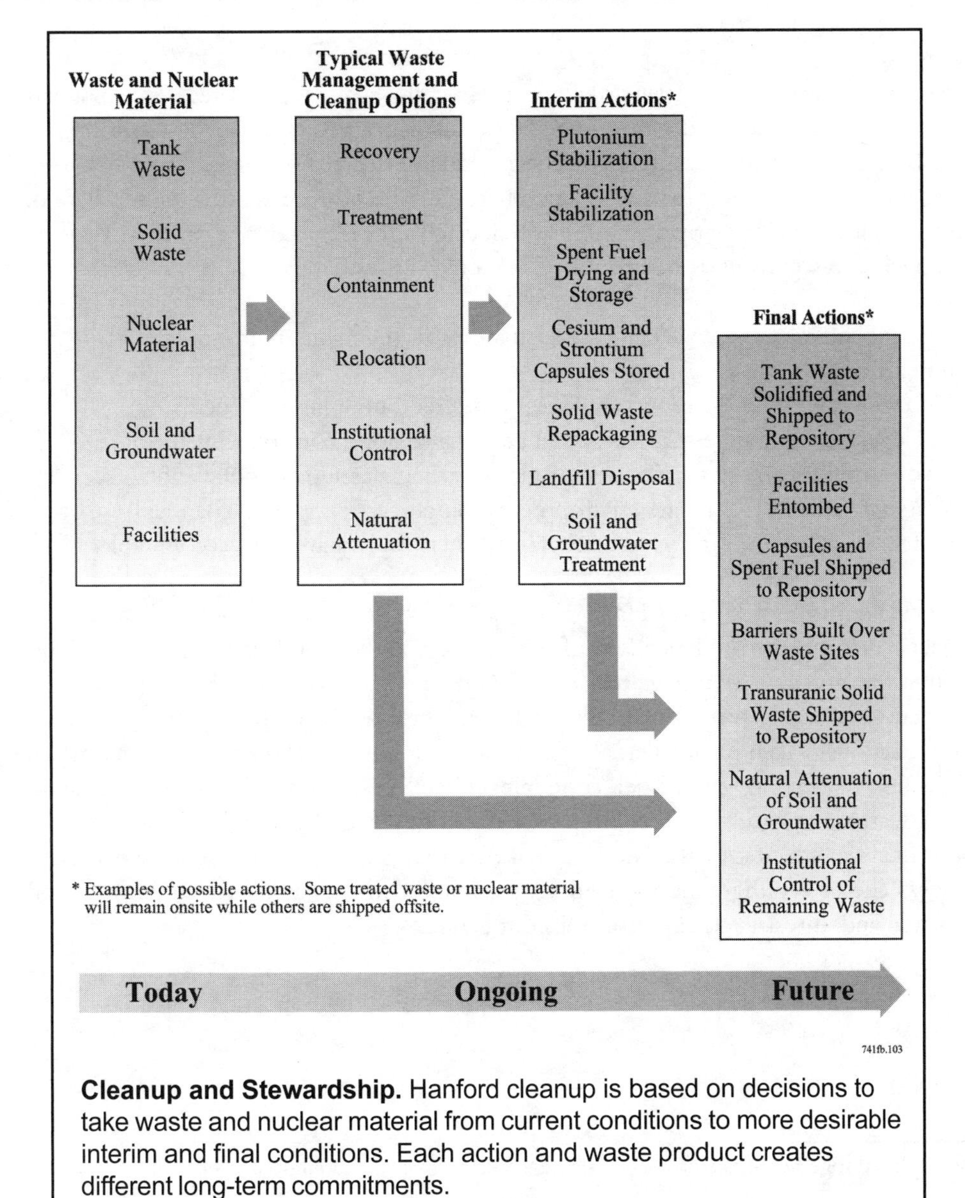

Cleanup and Stewardship. Hanford cleanup is based on decisions to take waste and nuclear material from current conditions to more desirable interim and final conditions. Each action and waste product creates different long-term commitments.

Interim or Final End Point?

Starting the 1960s in a facility near Idaho Falls, Idaho, 140,000 cubic feet of dehydrated, powdery high-level waste, called calcine, was stored in six large bins.[7] This material was classified as high-level waste because it's from the reprocessing of nuclear fuel. The National Research Council (1999c) identified no significant health or environmental hazards in keeping the material inside existing bins, which have a life expectancy of 500 years—should those bins continue to perform as designed. However, DOE is considering retrieving the calcine and immobilizing it into a smaller volume of glass or another solid form and then possibly hauling that material to a geologic repository. Thus, is the calcine in an interim or final end point? The perceived final end point of one generation may in fact turn into an interim end point by another generation.

Land end states and waste product end points are moving targets.

The General Accounting Office (2000c) expressed the challenge of predetermining land end states when it reviewed more than 500 performance goals governing environmental work by the U.S. Environmental Protection Agency. On a national level, only 16% of the agency's performance goals were directly related to end results—that is, the state of the environment, such as river miles restored, discharges reduced, or standards achieved. Why such a low percentage? Several explanations were offered:

- Not enough data are available to quantify improvements in environmental and human health.
- Knowing how to link cleanup activities to desired outcomes is difficult.
- Achieving cleanup milestones is within an organization's control, while being accountable for the state of the environment when cleanup is done is not.
- Years are needed to validate environmental outcomes compared to annual performance reviews for which people and their careers are held accountable.

Achieving a final end state for Hanford is made more difficult because small cleanup projects often proceed independent of each other and independent of cleanup actions taken in the larger surrounding area. The National Research Council (2000b, p. 21) reported, "end states appear at present to be emerging as the de facto result of multiple interim actions" taken on smaller, scattered waste sites. In the 100 Area, for example, the disposition of reactors, spent fuel, contaminated soil, contaminated groundwater, and buried waste are proceeding along separate tracks governed by few final decisions about final remediation goals. The National Research Council (2000b, p. 21) labeled this a "Swiss cheese" configuration—relatively clean and contaminated sites supporting both restricted and unrestricted land uses mixed together.

(7) If placed together, this amount of calcine would form a cube 50 feet on a side.

8.1.2 Cleanup Approaches

"No matter how much money is spent, some hazards will remain at over two-thirds of the [U.S. Department of Energy] sites. The lack of proven technologies to address radioactive contamination, and contaminated soil and groundwater, as well as the fact that many DOE sites will be home to waste storage and disposal facilities, ensures that hazards will remain at these sites for hundreds, if not thousands, of years..."

—Katherine Probst and Michael McGovern, *Long-Term Stewardship and the Nuclear Weapons Complex: The Challenge Ahead* (1998, p. viii)

Cleanup is commonly thought of as completely removing or destroying contaminants, to return a site to its previous, noncontaminated condition. This is sometimes possible when contaminants are concentrated, recovered, or rendered harmless.

Ninety-six percent of U.S. nuclear weapons sites will require long-term care (DOE 2000g).

In reality, returning industrial sites to a pristine condition is rarely possible—let alone a huge industrial site like Hanford. It's a romantic notion not steeped in reality. The U.S. Environmental Protection Agency reports that for many Superfund sites, cleanup takes many years to complete, if ever (EPA 2001b). Therefore partial cleanup and restricted land uses are often the norm.

Cleanup is achieved using technologies to reduce the contaminant toxicity, mobility, and volume. Some technologies are applied at the location where contamination exists, such as in the soil near a crib; these are "in situ technologies." Other approaches require contaminants be moved and treated elsewhere. A hierarchy of technologies and methods exist (Office of Technology Assessment 1989). Contaminant destruction or removal is preferred to containment or long-term institutional control. Naturally, creating fewer contaminants in the first place is best.

Contaminant destruction or removal is preferred to containment or long-term institutional control.

Recovery: Technologies that recover or remove contaminants for safer storage, shipment, or commercial use. Examples include waste site excavation and groundwater pumping and treating.

Treatment: Technologies that use heat, chemical, or biological processes to immobilize, neutralize, destroy, or separate contaminants. Examples include the injection of chemical reagents that can immobilize contaminants or enhance their removal.

Containment: Engineered controls that hold contaminants in place. They include surface and subsurface barriers, ground covers, walls, landfills, and waste repackaging. Containment buys time until more permanent solutions are applied, radionuclides decay, or toxicity is reduced. An engineered barrier[8] now covers a waste disposal crib in the 200 East Area that received strontium- and cesium-bearing liquids. That's containment.

(8) An engineered barrier is a human-made structure, such as a specially tailored earthen mound, used to better isolate or stabilize a waste site.

The use of surface engineered barriers to contain contaminants may increase.

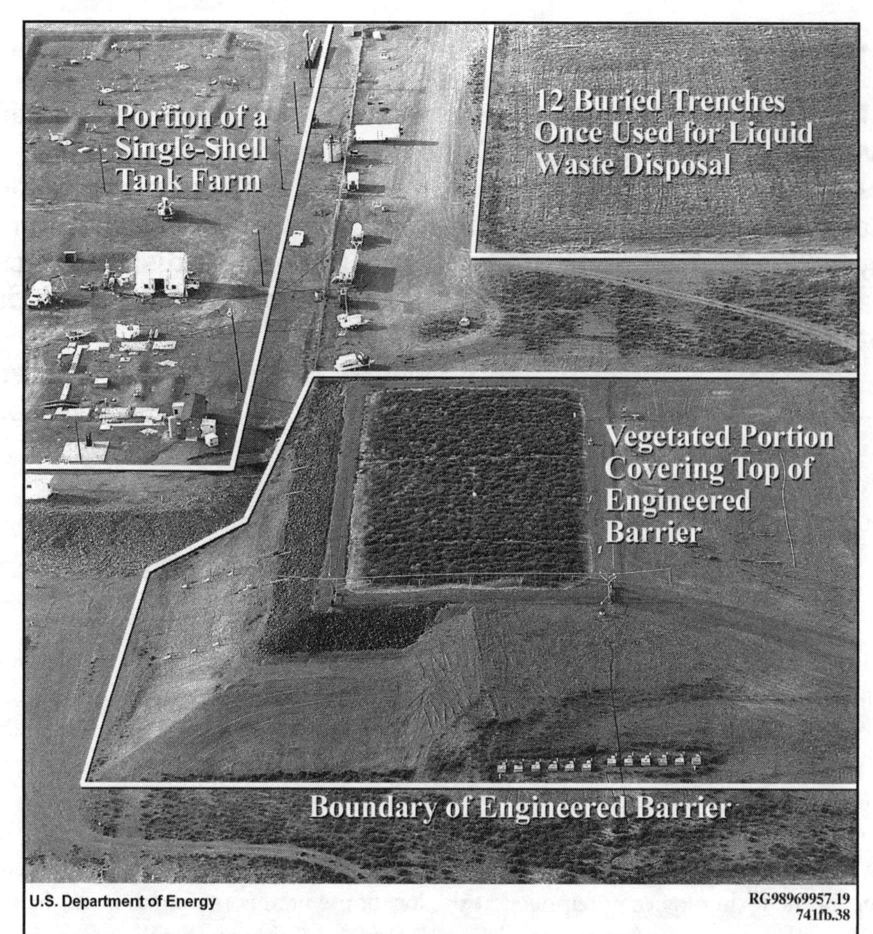

Engineered Barrier. This aerial photograph shows an engineered surface barrier built in 1994 over a liquid waste site in the 200 East Area. It covers 5 acres and is 30 feet high. Such barriers, constructed from layers of natural sediments and human-made materials, control moisture, plant, and animal entry while minimizing erosion. Modern barriers are designed to be nearly maintenance free for hundreds of years.

There is growing recognition that surface and subsurface engineered barriers are needed to minimize further contamination spread in soil and groundwater (EPA 2000). Significant investments have been made in barrier technology since the early 1980s (Rumer and Mitchell 1996). However, implementation is stalled over issues of long-term effectiveness and concerns that barrier use might encourage leaving waste in place rather than exhuming and treating it.

Institutional Control: Legal and administrative restrictions that control resource uses, site access, or activities. They include fences, deed restrictions, onsite security, relocation of people, and alternative water supplies.

Natural Treatment: Natural processes that degrade, dilute, or control the movement of hazardous substances. The dilution of nitrate and the decay of tritium in groundwater are examples of natural treatment.

8.2 Cleanup of the Nuclear Weapons Complex

"EM [Environmental Management] programs are now virtually on auto-pilot, driven by the agreements that govern cleanup standards, techniques, and timetables, and through which EM has almost totally delegated the choice of how much, and in what manner, to spend on cleanup...environmental and other regulations governing risks from toxic material inventories tend to be process-oriented rather than outcome-oriented. That is, while based on the goal of preventing harm...they describe the steps that must be taken to prevent harm, rather than the outcome, measured by risk levels or risk reductions, that must be achieved"

—David Bjornstad et al.,
Implementing Outcome-Oriented Risk Planning (1998, p. 4)

The Office of Environmental Management within DOE is responsible for waste management and cleanup across the nuclear weapons complex. This program began in 1989 with an annual budget of $1.6 billion and grew to a 2002 budget of $6.7 billion.[9] During those years, nearly $75 billion was spent. This makes it the most expensive environmental cleanup project in our nation's history, perhaps in the world. It's twice as large as the yearly public and private expenditures on all nonfederal Superfund cleanup efforts (Probst and Lowe 2000, p. 4).

DOE's cleanup program is the most expensive in our nation's history.

Initial cleanup strategies were based on a process where waste was first characterized, technologies matched against waste types, and cleanup scheduled over 30 years (DOE 1990, 1991a). This seemed like a logical, straightforward cause-and-effect approach to problem solving.

However, cleanup was a desired, but ill-defined, term. Restoration commitments were made with only the haziest notion about waste properties, technical feasibility, cost, and schedule (NRC 1996d).

Some people thought cleanup would take a few years.

In 1990, some people thought cleanup would take just a few years. The enormity of the effort, however, revealed itself as attempts were made to implement the growing number of cleanup agreements. Available regulations and technologies that seemed suitable for small industrial-type waste sites containing a few contaminants were cumbersome and ill suited for large, complex nuclear sites. Concerns over the biological hazards of radiation, not dealt with in the Superfund program, dominated debates.

Cleanup requires commitment from agencies whose membership changes faster than work is accomplished.

Remediation of large DOE sites involved a host of regulators, contractors, and others who were required to reach a quasi-consensus on cleanup approaches. Furthermore, people had to write plans the outcome of which spanned generations. Finally, cleanup required multi-year funding and commitment from agencies whose membership and priorities would change faster than work is accomplished.

(9) In 1989, the program was called the DOE Office of Environmental Restoration and Waste Management.

The initial goal to clean up the nuclear waste in 30 years was unfounded.

At the same time, it was impossible in many cases to separate truly hazardous conditions from those where contamination is measurable but not a serious health risk (NRC 1995). This was a key reason why the Office of Technology Assessment—which once analyzed technical issues for congressional committees—reported that DOE's initial goal to clean up the weapons sites in 30 years was simply "unfounded" (Office of Technology Assessment 1991, p. 6).[10]

In a search for solutions, and in many cases quick fixes, Congress skyrocketed cleanup budgets in the early 1990s. Future funding seemed unconstrained. When needed, dollars appeared. Many thought progress would rely upon off-the-shelf technologies and Superfund-type cleanup approaches. Yet, Superfund cleanup commonly hauled contaminated soil to a landfill, covered contamination with an extra layer of soil, or pumped and treated water for a few years. Such measures are inadequate for dealing with nuclear waste.

By the mid-1990s, there was growing pressure from Congress to demonstrate progress for the billions of dollars spent each year. Though Congress cannot directly manage cleanup projects, if displeased, it can control losses or redirect work by reducing funding (NRC 1999b).

As threats of funding restraints grew, DOE managers worked to more closely align cost, schedule, and work. This led to issuing the Baseline Environmental Management Report in 1995 (DOE 1995b). This was the Department's first systematic attempt to document work scope and life-cycle costs. Projected cleanup costs ranged from $206 billion to $360 billion. A year later, the second "baseline" report lowered cost estimates to between $189 billion and $265 billion (DOE 1996a). Reductions were attributed to proposing less expensive cleanup solutions and sites having gained an improved understanding of their waste problems. These reports were then suspended.

Even lower costs—$40 billion less—were projected 2 years later in DOE's new proposal for cleanup, called the 10-year cleanup plan or *Accelerating Cleanup: Paths to Closure* (DOE 1998b). It touted cleaning up 90% of DOE's waste sites by the year 2006. The majority of these sites were small and had limited contamination. The 10-year plan credited, "a variety of factors for lowering cleanup costs further...including completed cleanup work, reduced overhead and support costs, re-sequenced activities, and improved cross-site integration." The total cost was now estimated at $147 billion. However, this did not include money for managing the "project's technical uncertainties" (DOE 2000g). In other words, there were no backups to baselines or contingencies for solving the inevitable surprises that would arise. Cleanup projects had to march forward uninterrupted and executed perfectly—the first time.

Even senior DOE officials questioned the accuracy of cleanup cost estimates. (Grumbly 1998).

Two years after publishing the 10-year cleanup plan, the estimate for completing cleanup was raised to between $168 billion and $212 billion (DOE 2000g). However, these estimates did not include all future environmental liabilities such as the cost of decontaminating and decommissioning active and surplus facilities (for example, inactive naval

(10) After 23 years of service, the Office of Technology Assessment was disbanded by the 104th Congress in 1995.

reactor facilities and the gaseous diffusion plant at Oak Ridge, Tennessee); disposition of excess plutonium, highly enriched uranium, and depleted uranium; life-cycle costs for a geologic repository to hold high-level radioactive waste; or long-term stewardship for sites cleaned up under the Formerly Utilized Sites Remedial Action Program (NRC 1995; GAO 2000b).

Reminiscent of DOE's onsite investigative Tiger Teams of the late 1980s, a new Assistant Secretary for Environmental Management conducted a top-to-bottom review of DOE's national cleanup program in 2001 (DOE 2002). In releasing the results, Assistant Secretary Roberson (2002a) reported, "The EM program has not lived up to expectations in significant measure because the Department tried to solve the wrong problem and used the wrong set of indicators to measure success….One important example of the problem in this program is our own indicators and milestones." While over 90% of regulatory milestones have been met, little significant progress in cleanup or risk reduction has been achieved. Indicators measure "process, not progress, opinions, not results" (DOE 2002). The Department argued that in spite of well-considered plans, "the program as a whole is not working and will not deliver unless significant reforms are made" (DOE 2002).

U.S. Department of Energy

Workers Handling Rusted Steel Drum. These workers are packaging fragments of rusted steel barrels once containing chemicals. The direct handling of most Hanford waste or nuclear materials is unsafe or otherwise not possible.

Similarly, the General Accounting Office (2002b, p. 7) reported that DOE has completed nearly 80% of its compliance milestones on schedule—but many milestones just satisfy administrative requirements, such as writing a report or submitting a permit, "that may not indicate what, if any, actual cleanup work was performed." Reliance on compliance milestones discourages strategic focus. Their primary value is for tracking activities, rather than gauging cleanup progress.

Reliance on compliance milestones discourages strategic focus.

Budget and Curies

More than 75% of DOE's environmental management budget will go to five sites in five states—Hanford Site in Washington State, Savannah River Site in South Carolina, Rocky Flats Environmental Technology Site in Colorado, Oak Ridge Reservation in Tennessee, and Idaho National Engineering and Environmental Laboratory in Idaho (Probst and Lowe 2000, p. 2).

Combined, Hanford, Savannah River, and Idaho contain 96% of the 1 billion curies of weapons-related radioactivity and 85% of the nearly 7 million cubic feet of transuranic waste in the DOE complex (DOE 1997a, pp. 36, 42, 58).[11]

(11) At these three sites there are 91 million gallons of high-level waste, which would fill a cube 230 feet on a site; 1.4 billion cubic feet of contaminated soil, a cube 1100 feet on a side; and 1.7 trillion gallons of water, a cube 1.2 miles on a side.

While more than 90% of regulatory milestones have been met, little significant progress in cleanup or risk reduction has been achieved (DOE 2002).

All of this was in response to the Secretary of Energy's challenge to accelerate cleanup of the nuclear weapons complex (Abraham 2002). This was part of DOE's reassessment of their nationwide cleanup effort to reduce costs by $100 billion and shorten the 70-year schedule by 30 years.

This was the first plan in years attempting to break through the business-as-usual approach to cleanup. It proposed creating an "expedited cleanup account" of $800 million to $1.1 billion, sliced from DOE's existing $6.7 billion cleanup budget, allocated to sites as sort of an awards package targeted at slashing costs and accelerating schedules.

However, DOE "did not seek input from site regulators or other stakeholders when developing its latest initiative" (GAO 2002b, p. 16). Though DOE worked to find common ground to implement this strategy, regulators from across the nation told the General Accounting Office they would oppose this initiative if it means less funding, delays in completing cleanup, or imposing less-stringent cleanup standards. Besides, the proposed strategy of allocating account funds between sites is risk driven. While that's an admirable goal, all attempts by DOE since the late 1980s to implement a risk-driven decision-making process have failed (see Section 11.7).

How much of this new money might come to Hanford is uncertain, should the cleanup account concept even survive. If none came, the site would face a $262 million cut in cleanup funding (Blumenthal 2002). This is on top of an additional $200 million increase in funding many believe is needed just to keep cleanup on track. One newspaper editorial called the refocusing of funds "inadequate" and "insulting" (*Tri-City Herald* 2002). A senior DOE official was "surprised" by the reaction in Washington State (Roberson 2002b).

Attempts to "alter the funding balance among DOE sites" face stiff opposition (GAO 2002b, p. 4).

Soon after the new funding approach was announced, DOE reported it had reached an agreement with the U.S. Environmental Protection Agency and the Washington State Department of Ecology to work out plans to shorten cleanup schedules to capture more than half ($433 million) of the new account funds (Stang 2002a). Specifics were missing.

How all of this will play out is unknown. Bottom line? On the positive side, DOE has begun trying to refocus its environmental policies on quicker risk reduction efforts and less on administrative process, going so far as to propose renegotiating compliance requirements and signed Records of Decision. Increased attention is being given to national interest instead of just advocated local positions. On the downside is the concern that health and environmental protection standards will be sacrificed on the altar of fast-tracked schedules and cost savings. In such a polarized setting, even potentially good ideas are met with skepticism, if not overt opposition.

"The debate about the nation's environmental policy all but ignores the environmental concerns" (Probst and Lowe 2000, p. 4).

Probst and Lowe (2000, p. 4) from the Center for Risk Assessment write, "the debate about the nation's environmental policy all but ignores the environmental concerns." Why? The contamination and associated risk to workers, the public, and the environment are generally ignored. Three reasons were offered: "a lack of clear policy about the purpose and direction of both DOE in general and the EM [Environmental Management] program in particular....disinterest among senior officials—both in the administration and within Congress—in clarifying EM's goals and assuring taxpayers that EM funds are wisely spent....the continued support for using the EM program as an engine for economic development to keep federal jobs at sites that once created nuclear weapons" (Probst and Lowe 2000, p. 5).

Today, cleanup of the nuclear weapons complex is governed by 70 compliance agreements[12] in effect at 23 separate DOE sites (GAO 2002a, p. 6). These are legally enforceable documents between DOE and it regulators,[13] specifying cleanup actions and milestones. These agreements empower the states to make legal claims on federal responsibility. Tribal nations also have treaty rights to portions of some weapons sites.

"A dearth of data and a confusing internal budgeting process make it almost impossible for anyone outside of DOE to figure out what the [cleanup] money is being used for" (Richanbach et al. 1997).

These agreements reflect site-specific DOE, regulatory, and community priorities. Collectively, they do not provide a means for prioritizing or ranking risks between sites. In lieu of a nationwide ranking, a relatively stable amount of funding is allocated to each site, based upon local priorities. This is one reason why the proposed expedited cleanup account, rewarding sites that cooperate within the context of a national plan, is so threatening.

As part of their annual budget requests, individual DOE sites develop cost estimates for satisfying compliance agreements.[14] This information goes to DOE's Headquarters where it's adjusted to reflect national priorities, competing demands among sites, and congressional discretion (GAO 2002b). Headquarters then reallocates funds to the sites. Each site is responsible to use their funds as best they can to meet compliance agreements and achieve cleanup goals.

As cleanup is completed on smaller sites, plans are to shift funding to the larger and more complex sites (DOE 1998b). The viability of this approach remains untested. Delays in some cleanup activities could prevent funds from shifting—providing these funds are not removed entirely from the DOE budget. For example, delays at Rocky Flats in Colorado could hurt Hanford's chances for receiving additional money (GAO 1999). The General Accounting Office—the investigative arm of Congress—voiced concern over likely multi-billion-dollar cost escalations and overly optimistic planning assumptions for remediating smaller sites such as the uranium enrichment plant at Paducah, Kentucky (GAO 2000d). Other assumptions in DOE's nationwide cleanup plan, such as Hanford receiving waste from other sites, are not covered by existing agreements, adding further cost uncertainty (HAB 1997a). An example of this was highlighted when the environmental group Heart of America Northwest obtained, under a Freedom of Information Act request, information about a draft proposal to ship transuranic and other solid waste from Ohio to Hanford (Stang 2002b). Hanford officials were unaware of this plan.

---

(12) The 70 waste management and cleanup compliance agreements in place to cover DOE sites are of three basic types: (1) twenty-nine are required by the Comprehensive Environmental Response, Compensation, and Liability Act to clean up the potentially most dangerous waste sites or by Resource Conservation and Recovery Act to address the management of mixed radioactive and hazardous waste; (2) six are court-ordered agreements resulting from lawsuits; and (3) thirty-five are other agreements such as state administrative orders (GAO 2002b, p. 2). These compliance agreements contain 7186 enforceable milestones (GAO 2002b, p. 6).

(13) Regulators are federal or state agencies having environmental or health jurisdiction, authority, or responsibility in compliance agreements.

(14) Executive Order 12088 directs federal agencies to request sufficient funds to comply with pollution control standards (GAO 2002b, p. 10). Usually sites submit two budgets. One meets "full requirements" that mark how much money is needed to accomplish work in the most desirable manner. The second is a "target" budget, which is based upon the amount of funds the site normally receives. Targeted funds are the more realistic of the two funding scenarios.

Failing to See the Bigger Picture?

The General Accounting Office reported that local site managers commonly consider the effects of decisions only on their own budgets and not the immediate or future cost of decisions at other sites (GAO 2000b). Such self-interest is natural but could result in higher costs by creating duplicate facilities and capabilities across the nuclear weapons complex. A similar situation occurred with state governments during the 1980s when the federal government encouraged regional "compacts" between states to build disposal sites for holding commercially generated low-level radioactive waste (GAO 1999d). Over most of the last 20 years, only two regional disposal sites operated.

8.3 Building on Conventional Experience?

"Contractors are financially rewarded for pressing forward with a conventional technology acceptable to regulators, whether or not it is effective. In most circumstances, there are strong financial and contractual disincentives to undertaking the serious consideration and developmental engineering of innovative technologies that deviate from the contractual baseline."

—Washington Advisory Group,
Managing Subsurface Contamination: Improving Management of the Department of Energy's Science and Engineering Research on Subsurface Contamination (1999)

A wealth of Superfund experience exists but does not apply.

A wealth of Superfund field experience to permanently clean up waste sites doesn't exist. Even at traditional industrial sites, conventional cleanup methods have met with "limited success" (NRC 1997, p. 32). In many cases, they rely upon temporary remedies such as hauling contaminated dirt to a landfill. This is why nearly 70% of the Superfund sites will require long-term onsite management and monitoring when active cleanup efforts stop (GAO 1995a).

In the cleanup of solvents and metals underlying the nation's military bases, the Department of Defense's Office of Inspector General (DOD 1998)—the independent office that initiates and conducts audits and investigations relating to defense programs—

Conventional Pump-and-Treat Technology

This technology involves drilling wells, pumping contaminated water from underground, and then treating the water. The cleaner water may or may not be returned to the aquifer. The problem with this technology is that it removes only a portion of the contamination, that portion dissolved in the groundwater. Many contaminants are physically trapped or chemically bound underground—or remain untouched in shallow sediments overlying the aquifer, unaffected by pumping groundwater. Pump-and-treat technology is made more effective when combined with other in situ techniques, such as chemical treatment. For more about the experience with one pump-and-treat effort at Hanford, see Section 9.3.1.

reported that "complete restoration of a contaminated aquifer is an unrealistic goal" using existing technology. The use of groundwater pump-and-treat systems, so popular in 1980s, has proven ineffective after about a decade of experience, except for restoring fairly simple waste sites.

At DOE sites, more than 80% of the contaminated groundwater problems use conventional pumping and treating, natural attenuation, and capping and containment (MacDonald 1999). The predominant remedy for contaminated soil is excavation and redisposal. This resembles what's done at Superfund sites.

While conventional cleanup methods have lowered some immediate risk and improved environmental quality at many sites, total cleanup remains an ideal rather than a reality. There are several reasons: contamination is far greater than originally envisioned, technology is inadequate, cost is higher than anticipated, and health and environmental risks are different than expected (Russell 1995).

As noted by the 104th Congress when it began funding a new research program addressing the most perplexing DOE waste problems, "this funding is to be used to stimulate the required basic research, development and demonstration efforts to seek new and innovative cleanup methods to replace current conventional approaches which are often costly and ineffective" (NRC 1996c, p. 10).

0030103-16cn

U.S. Department of Energy

00050028-99 741fb.30

Underground Pipe Removal. An extensive network of underground piping once transferred liquids between facilities and to waste disposal or storage sites. Much of this piping is now rusted and no longer needed. Here, 5-foot-diameter piping is being removed for disposal. These pipes once carried cooling water from the D and DR Reactors for release into the Columbia River.

Yet, cleanup of Superfund sites garners more national public and political attention than federal DOE sites. Why? The answer likely rests in the fact Superfund sites are spread between many communities and congressional districts rather than concentrated in a few, relatively isolated areas of the country.

Cleanup remains an ideal rather than a reality.

The common Superfund news scenario goes like this: industry actively lobbies Congress to change environmental laws as well as points out to the news media unfairness in liability and cleanup standards. Environmental groups then mount an aggressive counter charge against industry's claims. This keeps Superfund problems in the forefront of public attention. With the exception of periodic news highlights, such as tank waste leaks, cleanup at federal sites tends to be treated as a local, unique issue. As a result, few reporters are independently examining the big picture of cleaning up federal sites and even fewer members of Congress care.

The average time to clean up a Superfund waste site has grown from 2.4 years to 10.6 years.

Cost and Schedule for Cleaning Superfund Sites

In the early 1990s, as Congress began questioning the success of DOE cleanup efforts, it was also questioning the U.S. Environmental Protection Agency about the cost and time for cleanup of Superfund sites. The agency spends about $1.4 billion a year on the nation's Superfund program (GAO 1999b).

In 1993, the General Accounting Office estimated there were 130,000 to 450,000 contaminated sites in the United States, and it would cost $650 billion to remediate them (GAO 1993). More than 1200 of the worst of these sites are listed on the National Priorities List. Hanford is on that list.

The General Accounting Office (1997b) reported that the yearly cleanup costs for 47% of Superfund sites were less than $10,000. For 50% of the sites, the yearly cost is between $10,000 and $10 million. For the remaining 3% of the sites, the annual cost is more than $10 million. According to a survey of Fortune 500 companies, legal expenses for the cleanup of industrial sites accounted for 28% to 46% of a company's restoration costs (GAO 1994b).

In 1996, cleanup was averaging 10.6 years a Superfund site compared to 2.4 years in 1986 (GAO 1997a, 1998f, 1999c).[15] Two years earlier, a U.S. Congressional Budget Office (1994) study reported the typical cleanup time was 13 to 15 years.

For federal sites, cleanup times were taking 6.6 years. However, the General Accounting Office (1997a) emphasized that those times would increase because many of the largest federal projects are more complex and remain to be done.

The time and cost for executing permanent solutions are not reflected in the above averages.

In comparison, the annual spending for waste management and cleanup of multiple sites at Hanford is nearly $2 billion each year. Why? First, Hanford contains more than a thousand contaminated sites. Second, the facilities required for and procedural controls applied to radioactive waste are more complex and costly than those faced at more traditional industrial waste sites. Third, cleanup of Superfund sites commonly relies on temporary remedies—measures inadequate or not acceptable for most radioactive waste. And fourth, nearly two-thirds of DOE's site remediation expense is devoted to mortgage costs—maintenance, security, and monitoring type activities.

(15) Here, cleanup identifies the time between when a waste site is placed on the National Priorities List to when cleanup begins. It does not include the years required to operate in-the-field remedies.

U.S. Department of Energy

96070034-1cn
741fb.33

Environmental Restoration Disposal Facility. A truck is unloading slightly contaminated soil into the Environmental Restoration Disposal Facility near the 200 West Area. This facility is a regulatory-permitted landfill. It receives contaminated soil and building debris from various Hanford cleanup activities. The worker in white protective clothing is spraying water for dust suppression. The worker between the vehicles is monitoring radiation levels with a hand-held meter.

People believe cleanup is necessary but disagree about what it means.

8.4 Work Accomplished at Hanford

"Farming looks mighty easy when your plow is a pencil and you're a thousand miles from the corn field."

—Dwight D. Eisenhower, former U.S. President

Cleanup progress is about increments.

As the Cold War was ending, public concerns fueled DOE's efforts to move its former production sites towards a new culture of environmental protection (Chapter 2). Hanford began this shift in the late 1980s. Actions were undertaken to stabilize, treat, or isolate waste and to improve the handling of onsite materials. These efforts reduced immediate threats even as new surprises were uncovered seemingly with each inspection.

There have been many debates about cleanup since the initial cleanup agreement for Hanford was signed in 1989 (Section 7.5). People believe cleanup is necessary, but opinions differ about what it means.

The following are examples of remediation actions accomplished or under way at Hanford:

- Excavating and hauling contaminated soil and building debris from near the Columbia River to a new, safer landfill (Environmental Restoration Disposal Facility) constructed in the 200 Area.

- Pumping and treating groundwater to remove contaminants (such as chromium, nitrate, strontium, technetium, uranium, and carbon tetrachloride).
- Stabilizing plutonium-laden residues inside the Plutonium Finishing Plant. Nearly 20 tons of material are being thermally treated to create a chemically stable powder.
- Removing spent fuel from the water-filled K Basins, drying it in the Cold Vacuum Drying Facility, repackaging it, and storing it in the Canister Storage Facility in the 200 East Area.
- Shipping transuranic-contaminated solid waste to the Waste Isolation Pilot Plant in New Mexico. This began in 2000 and was made possible by building a new solid waste characterization and packaging facility, the Waste Receiving and Processing Facility in the 200 West Area.
- Building and operating two new facilities for treating liquids previously discharged untreated to the ground.
- Resolving numerous safety issues about the waste stored in underground tanks.
- Demolishing old, uncontaminated facilities, deactivating reprocessing plants, and structurally stabilizing old reactors to lower the risk to workers monitoring these facilities and to significantly reduce maintenance costs.
- Constructing the Hanford tank waste treatment plant.[16]

Two reprocessing plants were deactivated early—saving taxpayers $175 million.

The deactivations of the PUREX Plant in 1997, 1 year ahead of schedule, and B Plant in 1998, 4 years early, are examples of where site contractors, agencies, and members of the public worked together to expedite cleanup. These activities saved taxpayers $175 million. In addition, $33 million in maintenance costs were saved each year for each plant (DOE 2000c).

U.S. Department of Energy 99120089-55cn 741fb.37

Liquid Effluent Treatment Facility. Untreated wastewater is no longer released into the ground. Two regulatory-permitted facilities located in the 300 and 200 Areas began operating in the mid-1990s to treat chemical, low-level radioactive, heavy metal, and organic-contaminated liquids. This is the control room in the 200 East Area Liquid Effluent Treatment Facility.

(16) Construction began in 2002.

U.S. Department of Energy 99070190-6cn 741fb36

Reactor Stack Demolition. The demolition of unwanted, old buildings have changed the Hanford landscape. These two 200-foot stacks of the D and DR Reactors were dropped to the ground and buried in 1999. Such activities reduce risks to workers and save money otherwise spent monitoring these facilities.

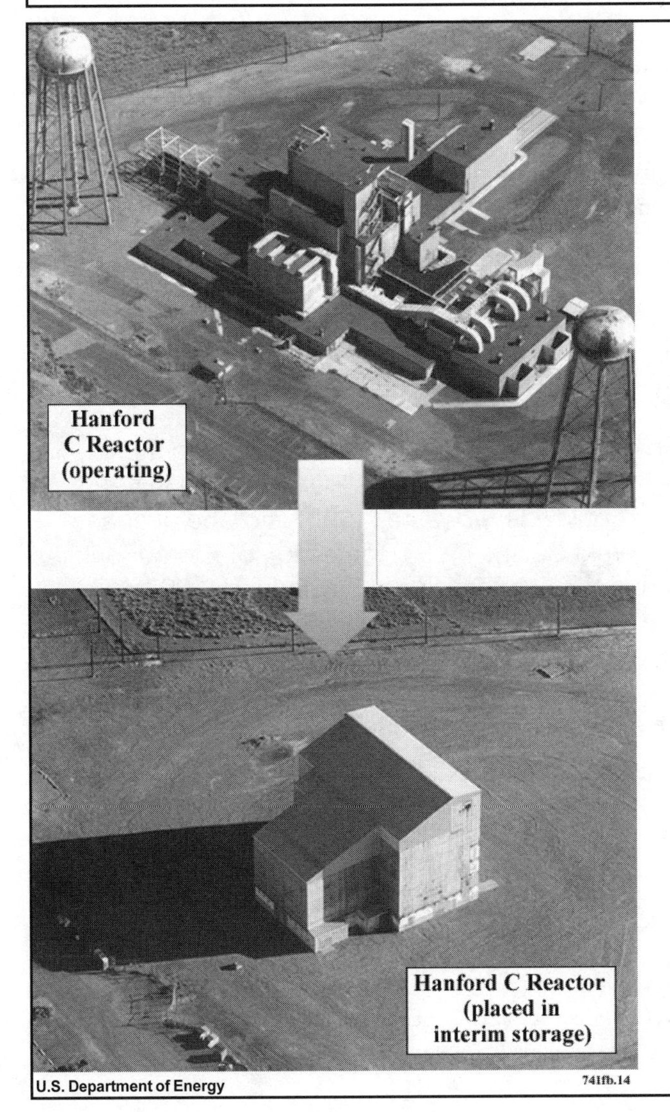

U.S. Department of Energy 741fb.14

C Reactor Interim Storage. In 1998, C Reactor became the first reactor placed into interim (75-year) storage. The reactor was demolished to its inner walls, a weatherproof metal roof was installed, and openings were sealed. All support structures around the reactor were demolished.

Eight of the nine reactors at Hanford will be "cocooned" and placed into interim (75-year) storage.

Another example of progress was placement of C and DR Reactors into a physical condition, called safe interim storage (referred to as "cocooning"), for the next 75 years. The reactors were demolished to their inner walls that housed the operating gallery and core. A weatherproof aluminum-alloy roof was then built, and all openings were sealed. This work reduces a reactors' footprint (the amount of land covered by buildings) by 80% and lowered annual surveillance costs by 90%. With the possible exception of B Reactor, all other reactors will undergo a similar fate.(17)

If Hanford reactors remain where they are, the $200 million cost per reactor to transport them to a burial site in the 200 Area will be reduced to about $20 million per reactor by cocooning them where they are.

However, the bulk of Hanford waste remains where it has been for decades—in tanks, in buildings, and in the ground. Most cleanup remains in the future. This is not to criticize work accomplished but to underscore the continued commitment required.

Cleanup plans seem to come and go. For Hanford, the most enduring document has been the Tri-Party Agreement though 300 milestones have changed since 1989.(18)

Changes to the site's cleanup approach are exemplified by the site's history of trying to initiate tank waste cleanup since the late 1980s.

The most enduring document has been the Tri-Party Agreement.

In 1987, construction of a government-owned and contractor-operated tank waste vitrification plant was to begin in 1990 with operation in 1999 (DOE 1987, 1988b). By 1991, plant startup was listed as "to be determined" by DOE (1991a) though according to Hanford's first five-year cleanup plan, the foundation was to be poured in 1992 with operations starting by 1999 (DOE 1991b). This plant was called the

In recent years, the most dramatic changes in Hanford priorities were barely anticipated.

> Since the mid-1980s, a series of fluctuating priorities, most unanticipated, dramatically shifted Hanford's activities and spending patterns. These included concerns over the earthquake vulnerability of the Plutonium Finishing Plant (1984), shutdown of N Reactor over issues of the Chernobyl reactor explosion (1987), hydrogen releases during the "burping" of tank waste (1989), shutdown of the PUREX Plant in 1990, cancellation of plans to grout low-level waste (1993), storage of spent fuel in the K-East and K-West Basins (1994), detection of leaked tank waste deeper underground than previously acknowledged (1997), the demise of a private financing strategy to build a vitrification plant (2000), and the development of a new proposed plan to accelerate Hanford cleanup (2002). A year before the events took place, they were hardly on the planning screen. This teaches us that change is inevitable.

(17) As the world's first full-scale operating reactor, B Reactor is a National Historic Civil Engineering Landmark and may be preserved for its historic and technological significance.

(18) U.S. Environmental Protection Agency official, testimony before a National Research Council committee, Richland, Washington, April 11, 2000. The number of milestones in the Hanford cleanup agreement changes over time. As of 2002, there were 1080 milestones. Since cleanup began in 1989, 825 milestones were completed and new ones added (GAO 2002b, p. 7).

Hanford Waste Vitrification Plant and was to treat waste from just the 28 double-shell tanks. A $550 million contract was awarded in 1989 for construction. Waste pretreatment—evaporating and separating most of the chemicals from the bulk of radionuclides in tank waste—was to take place in a revamped B Plant beginning in 1993.

U.S. Department of Energy 741fb.86

Tank Farm Workers. Some Hanford duties require a platoon-size crew of workers wearing protective clothing and following step-by-step procedures. This crew is atop what was once the thermally hottest single-shell tank (C-106) at Hanford to sample vapor gases.

By 1994, a new strategy using private contractors and focusing on a limited portion of the total tank waste volume was initiated. Two contracts were awarded in 1996 for planning the new treatment plant. (Plans for the Hanford Waste Vitrification Plant were dropped.) Two years later, a single contractor was chosen to proceed with facility design. Concerns grew about the use of a privatization-type contract to drive down cost in perhaps the most complex, and yet ill-defined, cleanup task ever undertaken (Fioravanti and Makhijani 1997).

In 1998, the tank cleanup project shifted from the DOE Richland Operations Office to the newly formed, congressionally mandated DOE Office of River Protection, also located in Richland. In early 2000, the single private contractor chosen in 1998 was fired because of high cost projections. Another team took over. This ended the 6-year effort to privatize tank waste vitrification and brought back a government-owned and contractor-operated management approach. Plant startup is to begin in 2007 with full-scale operation by 2011.

Assuming that full-scale operation begins on schedule, this is two decades after signing the Tri-Party Agreement. An additional 20 to 30 years are needed to remove, treat, and solidify waste from all 177 tanks, should the present cleanup strategy be pursued. Years of onsite interim waste storage, closure of tanks,[19] waste transport to an offsite geologic repository (should one open), and monitoring will follow.

The DOE maintains offices "in the field" as well as its headquarters in Washington, D.C. The DOE Richland Operations Office and the DOE Office of River Protection are the local offices for Hanford.

(19) Tank closure involves removing as much waste as technically and economically feasible and then backfilling the tank with bulk material to give it structural and chemical stability. All closure actions must ensure the tank and any residual material meet long-term health and safety standards.

Comparing the Schedule of Vitrification Plants

How does Hanford's proposed schedule for their tank waste treatment plant compare to other DOE sites? Planning, permitting, building, and testing of the Defense Waste Processing Facility, vitrifying tank waste at Savannah River Site, South Carolina, took 18 years. Just facility construction and testing lasted 13 years—from 1983 to 1996.[20] Sixteen years passed as planning and testing were completed at the West Valley Demonstration Project, New York.[21]

One of the two most recent plans for accelerating cleanup work along the Columbia River was proposed in the year 2000 (DOE 2000a). It centered on restoring Hanford land near the river, preparing and using the central plateau for long-term waste management, continuing to address urgent concerns such as the removal of spent fuel from the K Basins, and putting Hanford's assets to work for the future. Protecting and restoring the river corridor was built upon past values expressed by members of the public and others who worked on the Future Site Uses Working Group, Hanford Tank Waste Task Force, and the Hanford Advisory Board. For more information on these groups, see Section 6.2.

This plan sets out to complete key pieces of cleanup along the river by 2012. It required Congress to appropriate additional cleanup funding or DOE shifting funds from other projects. When discussed before the public, concerns were raised about cleanup standards, delays in addressing more contaminated soil sites, and whether the federal government was tackling easy projects rather than the more difficult and higher risk problems (Stang 2000h). The Hanford Advisory Board expressed worries about available funding not enabling sufficient groundwater treatment and remediation to release some river corridor land for unrestricted uses (HAB 2000b). The U.S. Environmental Protection Agency reminded all agencies "to be accurate in discussing expectations for the accomplishments that are realistic for the plan" (EPA 2001b, p. 2). Similarly, the Washington State Department of Ecology (2001a) stressed that increased attention to and funding for groundwater cleanup and technology development are needed if the plan is to be accomplished.

A second plan proposed to accelerate cleanup at Hanford from being completed in 2070 to 2035 or "as soon as 2025" was drafted in 2002 (DOE 2002d). This is to be a risk-reducing cleanup strategy aimed at

- Restoring Hanford lands around the Columbia River by remediating 869 waste sites and buildings over the next 10 years.
- Revamping the tank cleanup program through several actions including increasing the capacity of the tank waste treatment plant, accelerating tank waste closures, and demonstrating alternative tank waste treatment and immobilization solutions.

---

(20) The Defense Waste Processing Facility produces 2 metric tons of glass each day. The planned peak capability of the Hanford plant is 6 metric tons of glass each day.

(21) In 1980, the West Valley Demonstration Project Act was signed directing the DOE to solidify 550,000 gallons of radioactive waste held in one tank. This waste remained from the processing of about 640 tons of spent fuel. The first canister of nonradioactive glass was poured in 1984, after 12 years of testing and construction. Vitrification of radioactive waste began in 1996. It's nearing completion.

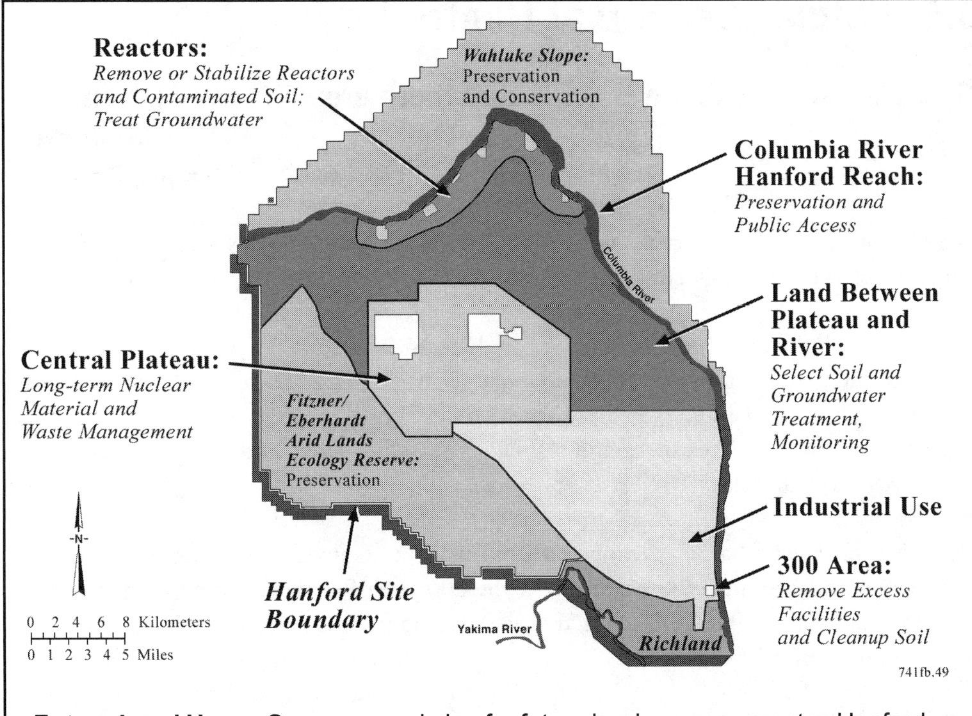

Future Land Uses. One proposed plan for future land uses segregates Hanford into areas for conservation, preservation, industrial use, and waste management. Uses will evolve as cleanup experience is gained and preferences change.

- Accelerating programs to remove spent fuel and sludge from the K Basins, stabilize and store plutonium now in the Plutonium Finishing Plant, and move all cesium and strontium capsules from wet storage to onsite dry storage.
- Accelerating the treatment and disposal of mixed chemical and low-level radioactive waste (including offsite shipment of transuranic waste to the Waste Isolation Pilot Plant in New Mexico) and taking care of all the 1200 waste sites now in and around the 200 Areas.
- Dealing with more than 900 unwanted "excess" facilities including the five reprocessing canyons.

These initiatives would be accompanied by significant changes in how business is done onsite in the areas of "contracting, project management, budgeting, requirements reductions, and infrastructure management" (DOE 2002d). All of this is in response to the Secretary of Energy's challenge to speed up cleanup of DOE sites (Abraham 2002).

Ground zero for cleanup debates? Budgets and milestones. Attempts to accelerate cleanup and lower costs are applauded—as long as biological health and human safety are protected.

(22) Spent fuel is uranium fuel that was irradiated inside a nuclear reactor, removed from the reactor, and is now stored or ready for storage. Hanford's spent fuel is located inside two concrete basins at the KE and KW Reactors (see Section 5.4).

The U.S. Environmental Protection Agency and the Washington State Department of Ecology are increasingly concerned about possible delays that could increase the risk for more waste leaks from aging tanks (EPA 2000).

Spent Fuel Storage Project: Shorter Schedule, Easier Problem

The design through testing of the spent fuel[22] handling, drying, and packaging facilities for Hanford's K Basins spent fuel project took 6 to 7 years before operations began in 2000. This was a challenging task but one far simpler and much less demanding than a treatment plant for tank waste.

8.5 Future Cleanup at Hanford

"The trouble with the future…is that there are so many of them."
—Robert Park,
Voodoo Science (2000, p. 87)

Cleanup of the nuclear weapons complex began in the late 1980s. No schedule has remained intact. Work will continue to adjust to discoveries and changing expectations.

Senator Patty Murray of Washington State called any proposal that threatens Hanford cleanup funding and milestones "dead on arrival" (Blumenthal 2002).

There is no blueprint or proven know-how that can be pulled from a shelf, dusted off, and implemented. Neither will problems just go away. "The Hanford Advisory Board has persistently emphasized that there are no quick fix solutions to the vast amount of legacy defense wastes. Treatment and stabilization will be costly and there are technical challenges and many risks" (HAB 1998).

Near-term work emphasizes cleanup of land near the Columbia River; control of contaminant releases into the river; use of the 200 Area for management of nuclear materials and waste; indefinite restricted use of groundwater; and restoration of portions of Hanford for economic, research, cultural, and recreational uses (DOE 1996b, 1998i, 1999a, 1999c, 2001).[23] Some waste forms and contaminants will remain at Hanford. How much is unknown. Land use will be determined through the Comprehensive Environmental Response, Compensation, and Liability Act (CERCLA) Record of Decision process (Washington State Department of Health 1997). This use will influence how clean segments of Hanford real estate might become.

"In the absence of well-defined end states, the Hanford cleanup program appears to operate on the philosophy that [it] is better to take a step in approximately the right direction than to know exactly where it is going" (NRC 2001b, p. 144).

The 100 Area contains 400 waste sites and 9 reactors with ancillary buildings. An initial land-use plan was developed (DOE 1999c). In the late 1990s, DOE and the regulators were revisiting cleanup standards to determine if any should be adjusted from residential standards proposed in the initial Records of Decision.[24] Later, DOE reported that the 100 Area would be remediated to meet less stringent industrial standards (DOE 2000g).

Plans call for land surrounding the 200 Area to be designated as industrial-exclusive and devoted to waste management (DOE 2002d). Cleanup standards would be the least stringent in that area. Approximately 700 waste sites, including all waste tanks, most onsite solid waste, and five reprocessing plants are located there. Waste is from onsite and offsite sources (see Chapter 5). Remediation is expected through a combination of demolition, excavation, contaminant treatment, offsite shipment, engineered barriers, and access control. Any waste storage facilities would be closed in a clean state under provisions set forth in the Resource Conservation and Recovery Act (RCRA).

---

(23) Cooperating government agencies and consulting tribal governments developed six land-use alternatives (DOE 1999c). With the exception of a no-action alternative, each alternative was the preferred alternative of one of the participants. The DOE preferred alternative emphasized (1) consolidating waste management operations within the 20-square-mile central plateau (where the 200 Area is located), (2) allowing industrial development in the eastern and southern portions of Hanford, (3) increasing recreational use and access to the Columbia River corridor, and (4) consolidating wildlife reserve areas north and west of Hanford.

(24) Fairly clean residential standards are default restoration values, used by the U.S. Environmental Protection Agency, where no publicly reviewed land-use plan exists.

The 300 Area and surrounding region will be remediated to meet industrial cleanup standards (DOE 2000g). It contains about 190 buildings and support structures plus some 50 waste sites including 2 former waste ponds, 7 trenches, 2 burn pits, and 9 burial grounds (Hanford 2000). There are more buildings per acre in the 300 Area than anywhere else on Hanford. Contamination rests under many buildings. Facilities not turned over to a private or government contractor will be demolished unless needed for an ongoing or future mission.

The Columbia River corridor will have a variety of uses. These include preservation to protect cultural and ecological resources and recreation. According to one DOE land-use plan (DOE 1999c), the southern portion of Hanford will be converted to industrial and research purposes supporting future DOE missions or independent commercial development.

Experience demonstrates that most schedules are uncertain and normally based upon optimistic expectations.

Hanford's long-term cleanup approach includes the activities identified below. Few dates are listed. Experience demonstrates that most schedules are uncertain and are normally based upon optimistic expectations including technology availability and performance. No one really knows how long it will take to complete many of the active phases of cleanup. For this reason, when dates are noted, they are offered to give only the most general context.

Cesium and Strontium Capsules: The long-term fate of the 1936 cesium- and strontium-filled capsules stored at the Waste Encapsulation and Storage Facility in the 200 East Area is not determined. Options include sale and transfer to an offsite commercial user, repackaging and shipment to a geologic repository, or vitrification followed by interim storage at Hanford and later shipment to a geologic repository. The DOE (2002d) suggests that capsules will be placed in long-term dry storage at Hanford. They can later be repackaged (or vitrified) and shipped to a geologic repository.

Facilities: Reactors, reprocessing plants, and other facilities will be deactivated and cleaned sufficiently to enable their removal, destruction, entombment, or use for solid waste disposal. Interim stabilization will allow additional time for radionuclide decay. The graphite cores of all reactors (except perhaps B Reactor) are scheduled for interim stabilization for the next 75 years (DOE 1998c). At that time, plans call for their removal and burial in the central part of Hanford. Whether or not this takes place will be left to another generation. A canyon disposition initiative is examining how the lower levels of the reprocessing plants might be used to dispose of some types of solid waste, assuming the decision is to decommission those facilities in place (HAB 1998). The upper outer walls would be collapsed and the remaining debris covered with an engineered barrier or cap (NRC 2001b; DOE 2002d). Excess facilities will be torn down.

Groundwater: Groundwater use beneath large portions of Hanford will remain restricted—for as long as we can imagine. Final cleanup levels and containment approaches have not been determined. Treatment or hydraulic control methods will be used to protect the Columbia River. This protection takes the form of reducing the amount of contamination that enters the groundwater through the soil, and controlling

U.S. Department of Energy 00120016-29

Multiple-Canister Overpacks for Spent Fuel. The Cold Vacuum Drying facility, built near the K Reactors, packages and dries uranium fuel corroded from years of underwater storage. The fuel is then shipped in a 13-foot-long, multi-canister overpack (shown) to the Canister Storage Facility in the 200 East Area where it's lowered into 40-foot-long tubes for storage.

the migration of some plumes.[25] Other treatments will continue on a case-by-case basis. Monitoring will last for years because contaminants will remain in the aquifer and the overlying sediment after active treatment is finished. Contaminants remaining in the soil are potential sources of new contamination entering the groundwater.

Irradiated Uranium Fuel: Irradiated fuel rods now located near the Columbia River in two water-filled basins at the 100 Area KE and KW Reactors are being removed, dried, repackaged, and stored in the Canister Storage Facility, in the 200 East Area.[26] This nuclear material will remain at Hanford until space is available in a geologic repository, should the decision be made to move the material there.

Soil: Some contaminated soil will be treated to levels supporting alternative land uses. These levels have not yet been agreed upon. Because of the large volume of contaminated soil, much of it will likely remain untouched (NRC 2001b). Studies are under way to determine the potential long-term risk from movement of radionuclides and chemicals through the soil (DOE 1999d). The results will help plan future cleanup work.

Containment will likely play a key role in soil contamination management. Various surface caps and engineered barriers may be built atop some waste sites, and in situ treatment undertaken at others. The use of barriers to contain contaminants is growing in acceptance and importance (Rumer and Mitchell 1996; NRC 2001b). Contaminated soil in some areas, such as along the Columbia River, will be treated or removed to the 200 Area.[27] Land restoration, especially in the 200 Area, will not allow unrestricted access to most locations.

Solid Waste: Retrievable solid waste contaminated with large amounts of transuranic elements will be packaged and shipped offsite to the Waste Isolation Pilot Plant, New Mexico. Transuranic solid waste shipments from Hanford began in 2000 and will continue until 2035. Approximately 3% (700,000 cubic feet) of the 26 million cubic feet of solid waste now at Hanford will be shipped to the Waste Isolation Pilot Plant. Some waste buried after 1970 may be retrieved, characterized, repackaged, and buried in the 200 West Area.

The disposition of solid waste buried before 1970 is uncertain. These materials contain transuranic elements, hazardous chemicals, and a mixture of poorly characterized debris buried in collapsed or corroded containers. Some will

(25) Groundwater plumes are those areas in an aquifer where groundwater is contaminated with radionuclides or chemicals above natural background levels or drinking water standards (see Section 5.6.3). They originate at a source of contamination and spread out in the direction of groundwater movement.

(26) The sludge, mostly from K-East Basin, will be packaged inside canisters and stored at T Plant. Eventually, it may be shipped to the underground Waste Isolation Pilot Plant in New Mexico (Stang 2001d).

(27) Soil restoration activities are scheduled for completion in the 100 and 300 Areas before the 200 Area.

Why is 1970 Important for Buried Solid Waste?

Before 1970, transuranic-contaminated solid waste was not considered separately from low-level radioactive waste. For decades, both transuranic and non-transuranic waste were mixed together and buried in shallow trenches. In 1970, the Atomic Energy Commission (a predecessor to DOE) determined that transuranic elements were associated with higher biological risks and that transuranic-containing waste should be stored in facilities with higher levels of confinement. So, after 1970, transuranic waste was stored in burial grounds in such a way to make later retrieval easier.

remain in the ground while the rest is characterized, packaged, and stored onsite or shipped to New Mexico. Restoration of solid waste burial sites is scheduled for completion over the next three decades. All burial sites will be monitored for a long time.

Tank Waste: What will happen to all the tank waste? For years, the plan was to retrieve, treat, and convert it into a glass form ("logs"). The potential benefits and liabilities of alternative treatment methods and solidification approaches for some of the lowest risk tank waste will likely be re-examined. Most tank chemicals, containing low levels of non-transuranic radioactivity, will be destroyed or immobilized, perhaps in a solid form and disposed onsite. The most dangerous portions will be immobilized as glass. Initially, glass logs will be stored at Hanford. Eventually, they will be shipped to a geologic repository should one become available (see Section 8.5.2). All tank waste immobilization is scheduled for completion by 2028 (Ecology et al. 1998; DOE 2001). Closure of the tanks (for example, filling them with some type of stable material, removing connecting pipes, covering with surface caps or barriers) would follow waste removal. If any appreciable amount of the waste is treated and remains in a tank, it would be monitored. There are no plans to excavate the tanks.

U.S. Department of Energy 00060063-49cn 741fb.96

Packaging Solid Waste. Steel barrels containing transuranic-contaminated solid waste are lowered inside a 10-foot-tall stainless steel cylinder called a TRUPACT. As many as 14 barrels are placed inside each cylinder. The solid waste is prepared inside the Waste Receiving and Processing Facility before shipment to the Waste Isolation Pilot Plant in New Mexico.

Currently, neither the storage capacity nor agreements are in place to dispose of all the waste forms that might leave Hanford. The consolidation of facilities across the nuclear weapons complex, to avoid capability duplication, lower costs, and ensure better security, is a growing concern (DOE 2002). Hanford will be competing with the other sites for the same disposal facilities, transport carriers, and shipping containers. The transportation of radioactive materials through communities is unpopular. In some cases, the risk of road accidents is higher than the risk of leaving contaminants onsite. If transport across state boundaries is not likely or is delayed, extended onsite storage is necessary. Whatever the outcome, decisions about waste relocation and treatment across the nuclear weapons complex will play a significant role in Hanford's future.

8.5.1 Cleanup Cost at Hanford

"No amount of money can return all the land and water under DOE facilities to their original condition, though this may prove feasible in some cases. Once long-lived radioactive wastes are created, nothing practical can be done to make them 'go away'."

—Stephen Schwartz,
Atomic Audit: The Costs and Consequences of US Nuclear Weapons Since 1940 (1998, p. 374)

Most estimates for cleaning up Hanford range between $80 billion and $100 billion.

Planning had better be indisputably good for the site will inherit the best and worst of its character.

Mortgage costs consume most money earmarked for cleanup.

The cost to cleanup Hanford is difficult, if not impossible, to calculate. Most estimates range between $80 billion and $100 billion though costs and schedules are "highly uncertain" (Roberson 2002a).[28] One should remember that more than $20 billion has already been spent—and most cleanup remains in the future.

"In many ways, attempting to estimate the magnitude of these costs with precision is similar to someone 20 years ago trying to estimate the dramatic changes in costs of computers today" (DOE 2000g). The same is true for other large nuclear weapons sites. In many instances, long-term waste stewardship and operational assumptions are not factored into publicized cost estimates.

Most expenses depend on how and in what way waste is handled. When is the best time to save money? It's during upfront planning. Once buildings are under construction or waste treatment initiated, momentum is created to follow that course of action—regardless. Therefore, planning had better be indisputably good, for the site will inherit the best and worst of its character.

Cleanup is only one part of the environmental legacy of the nation's nuclear weapons production sites. A much larger part is called "mortgage," referring to fixed commitments such as maintenance of infrastructure, security, and other support services that must be paid regardless if and how cleanup is undertaken. Estimates about the amount of mortgage vary. In 1997, DOE believed that as much as 50% of DOE's nationwide cleanup budget goes to mortgage costs (Alm 1997). Schwartz (1998, p. 374) reported that 75% of environmental management funds goes to mortgage expenses. At the DOE Rocky Flats site in Colorado, two-thirds of the site's annual budget of nearly $700 million is spent to maintain the site in a relatively safe and secure state, with the rest going to cleanup (GAO 1999a). In 2002, DOE reported, "only about one-third of the EM [the DOE Office of Environmental Management] program budget today is going toward actual cleanup and risk reduction" (DOE 2002). This is close to the Schwartz estimate. While the numbers vary, the fact is that mortgage costs consume a big chunk of available money. Even if cleanup activities were delayed, significant ongoing expenses remain. There is no free or low-cost walk-away scenario.

One challenge facing taxpayers is understanding how funds are spent. What is being purchased? Many believe DOE's yearly environmental restoration budget of nearly $7 billion is targeted for cleanup. After subtracting mortgage costs, a much-reduced portion remains for in-the-field applications and environmental compliance. Increasingly,

(28) Hanford cleanup costs cast doubt on the entire nuclear weapons complex being remediated anywhere close to the lower half of the present $150 billion to $350 billion estimate.

there is little discretionary money, and thus little flexibility, to provide the knowledge and capability to solve new problems and to improve treatment processes. In 2000, DOE (2000g) reported that approximately 85 to 90% of the cleanup budget is spent satisfying mortgage and regulatory compliance costs.

There is a widening gap between yearly budgets and the funding required to satisfy legal obligations. The hundreds of millions of dollars needed each year for the Hanford vitrification plant forced this issue to the forefront (Stang 1999b). Productivity adjustments, new contractual agreements, and work rescheduling simply delayed its arrival.

Hanford is not alone in this quandary. News reporter Stang (1999a) writes: "The rest of DOE's sites can expect to face the same massive cleanup budget shortages as Hanford does. Those sites likely will be as anxious as Hanford to scrounge up extra dollars."

U.S. Department of Energy 7813231-26cn

Concrete Shell Surrounding Tanks. Steel-reinforced concrete, shown in this 1978 photograph, is applied during the final construction phase for six 1.1-million-gallon double-shell tanks built in the AW Tank Farm, 200 East Area. The tanks were later covered with soil. During Hanford cleanup, tanks could have all waste removed for a "clean" closure or some portion left inside. Each type of closure has unique risk, cost, technology, and long-term care-taking requirements.

After examining the Hanford program to understand groundwater and contaminant movement around tank farms, one expert panel wrote about their concerns that not enough attention (and money) was devoted to characterizing the subsurface including contaminant movement underground (Integration Project Expert Panel 2000a). "We understand that [field] characterization is costly, but the cost of mistakes in remediation decisions are potentially far greater."

"We understand that characterization is costly, but the cost of mistakes in remediation decisions are potentially far greater" (Integration Project Expert Panel 2000a).

When cleanup began in 1989, Hanford federal and contractor employment was 12,700. This peaked at 18,700 in 1994. It dropped to just over 10,000 by the year 2002 (DOE 1997c; communication, Michael J. Scott, Pacific Northwest National Laboratory, June 2002). A smaller workforce focused more dollars on cleanup.

As of 2002, more than $20 billion was spent at Hanford to pay for cleanup and other costs. This averages to about $1.6 billion each year. The budget for 2002 was $1.8 billion. The Hanford budget is 20% of DOE's nationwide cost for environmental management.

Sometimes, estimated cleanup costs are moving targets. Three examples are given.

Cleanup costs a moving target for tanks: Life-cycle costs for cleaning up just the site's tank waste vary from $30.5 billion to $52 billion (DOE 1998c; Lawrence 1999; DOE 2000e). Why such a range? It's because of large uncertainties in the tank waste cleanup baseline (approved engineering approach), volume of waste to be turned into glass, composition of various waste types that impacts waste treatment, costs used to calculate expenses, construction expense, waste form production rates, etc. (Perez et al. 2001).

Examples of Expenses

U.S. Department of Energy 00030089-4cn 741fb.107

Slant Hole Drilling. Drilling holes at an angle (slant) enable investigators to collect contaminated sediment from beneath a tank. The samples are sealed and shipped to a laboratory for analysis.

Tanks: The cost of collecting a grab sample of waste from one location in one tank is about $125,000.(29) The cost for gathering a single vertical core sample through several feet of waste is about $800,000. While these costs are approximations (each tank or sample has unique collection requirements), they are based on experience. Performing chemical analyses on these samples is extra—$100,000 to several million dollars a tank.

During the year 2002, approximately $2.9 million a day was spent dealing with Hanford tank waste (tank waste maintenance, characterization, monitoring and preparation for the treatment plant to solidify).(30)

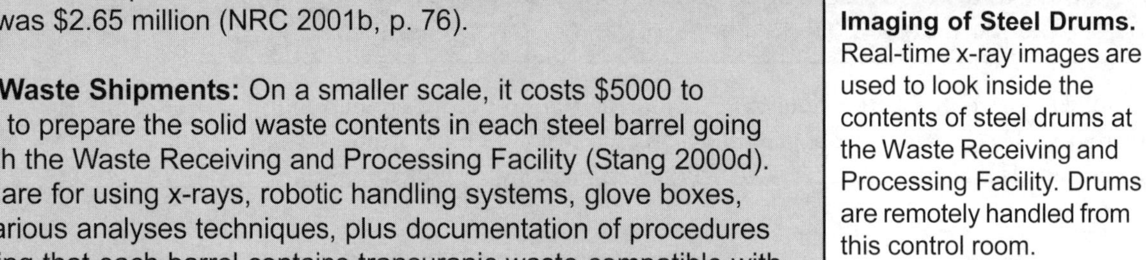

U.S. Department of Energy 99010017-48cn 741fb.48

Imaging of Steel Drums. Real-time x-ray images are used to look inside the contents of steel drums at the Waste Receiving and Processing Facility. Drums are remotely handled from this control room.

K Basins: Operations supporting the storage, handling, cleaning, packaging, and drying of spent fuel from the K Basins cost about $4 million a week.(31)

Well Drilling and Sampling: Depending on borehole depth, design, contaminants encountered, and procedures followed, drilling and sampling of contaminated sediment can range from $1 million to more than $2 million a borehole. The cost to drill, retrieve, and chemically analyze soil samples from one slant hole drilled beneath the SX Tank Farm was $2.65 million (NRC 2001b, p. 76).

Solid Waste Shipments: On a smaller scale, it costs $5000 to $7000 to prepare the solid waste contents in each steel barrel going through the Waste Receiving and Processing Facility (Stang 2000d). Costs are for using x-rays, robotic handling systems, glove boxes, and various analyses techniques, plus documentation of procedures certifying that each barrel contains transuranic waste compatible with the waste receipt requirements of Waste Isolation Pilot Plant in New Mexico. More than 100,000 barrels, packaged in stainless steel TRUPACT containers, will be packaged and hauled to New Mexico.

---

(29) Karyn Wiemers, Pacific Northwest National Laboratory, communication, March 15, 2000.

(30) Based upon a fiscal year 2002 budget of $1.062 billion—$690 million for the waste treatment plant and $372 million for the tank farms (Stang 2001e).

(31) Conversation with spent fuel project staff during an August 2001 tour of K Basins by the author.

During the first months of 2000, the expense for completing the first phase of tank cleanup carried out under a privatization contract (contractor-owned plant)—ranged between $6.9 billion and $15.2 billion (Stang 1999d, 2000f).[32] This phase would have treated just 10% of the tank waste volume and 25% of its radioactivity. These costs covered initial financing, building, and operating the waste treatment plant. They do not include the cost of alternative treatment options. Dealing with the remaining 90% of the tank waste was scheduled to take place by the year 2028—at a yet-unspecified expense.

New tanks will be needed if waste solidification is delayed or if more tanks leak.

New double-shell tanks are likely needed to store semi-solids and sludge removed from single-shell tanks in preparation for vitrification (Stang 2000b). Additional tanks will cost $60 million to $100 million each (Boston 2001). By the year 2006, the existing 2 million gallons of available double-shell tank space will be filled with single-shell tank waste being prepared for solidification. No surplus tank space exists.

Small amounts of toxic chemicals, such as polychlorinated biphenyls (PCBs), lacing any tank waste could trigger additional waste treatment requirements, covered under the Toxic Substances Control Act. Encountering any exotic chemicals in the tanks or unexpected chemical reactions, as experienced at Savannah River, could contribute to higher-than-projected tank waste costs[33] (Blumenthal 2000).

Cleanup costs for K Basins rose $1 billion over first estimate: According to the General Accounting Office (1999e), cleanup of irradiated nuclear fuel rods and radioactive sludge from the K Basins is to be completed by 2007 at a cost of $1.7 billion. This is about 6 years and $1 billion over the original plans of 1995, which were strongly promoted but on closer scrutiny found to be overly optimistic.

Tritium changes focus and costs regarding some burial grounds: Cleanup of the 9-acre 618-11 solid waste burial ground, located north of the 300 Area, is estimated to cost $328 million. The U.S. Environmental Protection Agency wants to ensure that this site and the nearby 6-acre 618-10 burial ground are remediated no later than 2018 (EPA 2001b)[34] A few years ago, these burial grounds were hardly footnotes in cleanup discussions, let alone factored into budgeting. Then, high levels of tritium were found in the local groundwater. Now, these burial grounds command high attention. Are other costly surprises awaiting discovery?

---

(32) Costs did not include waste retrieval and chemical preparation for the vitrification plant.

(33) An in-tank chemical process was used at Savannah River to precipitate cesium, strontium, and the actinide elements from tank waste. However, the addition of the chemical sodium tetraphenylborate generated unwanted amounts of toxic and flammable benzene gas (GAO 1999). Savannah River suspended the process in 1998 and began searching for options. This failure could delay the cleanup of high-level waste at Savannah River by 8 years and cost an additional $2.3 billion to $3.5 billion—though life-cycle cost impacts remain unknown. Each year the program slips its schedule has the potential to add $400 million to the tank cleanup project merely from the expense of maintaining tanks in their current condition (NRC 2000c).

(34) Why was cleanup set for possibly 30 years after signing the Tri-Party Agreement? Three reasons are commonly given. First, "no definitive solutions have been established" to safely excavate, treat, package, and ship this waste (Klein 2000). Second, the Hanford budget is stretched to the limit. Third, there are higher cleanup priorities. The Hanford Advisory Board endorses quicker action because radioactive tritium, which is very mobile in groundwater, is leaking from the 618-11 burial ground. The board states that this "is a warning that other contaminants may be migrating from the burial ground" (HAB 2000a). These two burial grounds are described in Section 5.5.

8.5.2 Geologic Repository for High-Level Radioactive Waste

"There is worldwide scientific consensus that deep geologic disposal...is the best option for disposing of high-level radioactive waste."

—National Research Council,
Rethinking High-Level Radioactive Waste Disposal (1990b, p. vii)

A geologic repository is critical to the long-term cleanup plans of Hanford.

Efforts to identify a suitable location for a high-level waste geologic repository show how national strategies and estimated costs and schedules change. Whether or not a repository opens has major implications to the long-term storage of waste at Hanford.

The search for a geologic repository site in the United States began nearly 50 years ago. In 1955, the Atomic Energy Commission asked the National Research Council to examine the issue. Salt deposits were generally favored though significant research was required to assess the feasibility and safety of geologic disposal (NRC 1957, 1970). The commission sponsored investigations of large salt deposits underlying the Gulf Coast, southwest, and northeast states.

By 1970, an abandoned salt mine at Lyons, Kansas, was selected as a potential repository. Bodansky (1996) reports this site was chosen in response to pressures from Idaho's governor and senators to find a location for waste recently transferred to Idaho. Two years later, in the face of unfavorable site conditions and local opposition, the Lyons site was judged unacceptable. The Energy Research and Development Administration[35] then proposed development of a retrievable surface storage facility at Hanford. That proposal was also dropped amid concerns that it would detract from efforts to find a more permanent solution to waste storage.

A geologic repository is made of a series of tunnels mined deep underground where waste packages containing dangerous amounts of radionuclides are permanently placed. A repository can eventually be backfilled, sealed, and monitored to ensure contaminants remain contained or are seeping out at levels not posing unacceptable biological risks.

Deep underground burial is the ultimate out-of-sight, out-of-mind, and end-of-the-line waste disposal action. The properly selected rocks can provide long-term isolation including shielding from radiation, chemical sorption for radionuclides, stability, and slow-moving groundwater that should keep contaminants from venturing far from their sources.

In 1977, the National Waste Terminal Storage Program began. It focused on lands underlain by salt plus other rock types beneath federally owned property. This initiated site screening at Hanford and the Nevada Test Site. An interagency review group, initiated by President Carter, recommended the nation proceed with geologic disposal and a wider

(35) The Atomic Energy Commission was dissolved in 1974, and two new organizations were created. The Energy Research and Development Administration was responsible for developing nuclear energy and operating the nuclear weapons complex. The U.S. Nuclear Regulatory Commission assumed the oversight role for regulating nuclear energy and nuclear materials. In 1978, the U.S. Environmental Protection Agency took on the role of developing radiation protection standards for nuclear waste (EPA 1985).

consideration of geologic formations. It also attempted to develop a consensus between federal agencies on the qualification requirements of a repository (Jacob 1990, p. 80). Preliminary screening of several rock types started in 17 states.

U.S. Department of Energy

Aerial View of Yucca Mountain. The proposed geologic repository site for storage of DOE high-level radioactive waste and commercial reactor spent fuel is beneath Yucca Mountain, 90 miles northwest of Las Vegas, Nevada. Waste packages would be stored 1000 feet below ground level in an ancient volcanic rock called tuff.

By 1982, Congress imposed a legislative solution to the 15-year agony of trying to identify a suitable geologic site by passing the Nuclear Waste Policy Act. The act created the Office of Civilian Radioactive Waste Management within DOE to manage the repository program. Geologic disposal was again established as the nation's long-term strategy for dealing with highly radioactive commercially and defense-generated waste and nuclear material. By 1983, DOE had selected nine sites in six states as potentially acceptable. By 1986, these sites were narrowed to three—basalt beneath Hanford, salt in Texas, and a volcanic rock (tuff) in Nevada. In addition, Congress included provisions for siting an aboveground monitored retrievable storage facility at Oak Ridge, Tennessee. A geologic repository was expected to be operating by the end of the twentieth century.

The Nuclear Waste Policy Act was overhauled in 1987. The three sites under study were narrowed to one—Yucca Mountain, Nevada. The act established an office in DOE to develop the repository and required the program's civilian costs to be covered by a fee (known as the "Nuclear Waste Fund") applied to nuclear-generated electricity. Also, the act contained a timeline for the creation of a storage facility, outlined a selection process, and specified selection criteria. The search for a second repository site was terminated as well as studies of the monitored retrieval storage facility. The office of a Nuclear Waste Negotiator was established to seek a state or native tribe willing to host a temporary aboveground waste repository—commonly called a monitored retrieval storage facility.[36] This is an important issue because several nuclear power utilities are having problems storing their growing inventory of spent fuel onsite. The negotiator's authority expired in 1995, and the office was closed.

Commercial nuclear fuel is now stored at power plant sites. Defense waste is at DOE sites—most of it in underground tanks awaiting treatment.

The federal government missed its commitment to begin receiving spent fuel from the commercial nuclear power industry by January 31, 1998, as mandated by the Nuclear Waste Policy Act. Also missed was having the proposed repository licensed and operating by 1998. No reliable date is now available.

(36) In 1992, a grant was awarded to the Skull Valley Band of the Goshutes to investigate the possible siting of a monitored retrieval storage facility on the Skull Valley Reservation, Utah—35 miles south of Salt Lake. The state of Utah is adamantly opposed to the idea.

Reliance on the unique geohydrologic properties of the Yucca Mountain site to isolate waste has taken a back seat to building multiple engineered barriers around the waste (Ewing 1999). In the meantime, the estimated cost for constructing and operating a repository has risen from $1 billion in the late 1970s to $60 billion.[37] As of 2001, $7 billion has been spent on characterization to assess the site's viability and safety (*Nature* 2001). In the year 2002, the Secretary of Energy reported the Yucca Mountain site was suitable for moving ahead with licensing by the U.S. Nuclear Regulatory Commission. President Bush also approved the site, and Congress voted to override Nevada's objections to having the repository in their backyard. The next battlegrounds are in the licensing arena and whether or not the public feels safer with glassified nuclear waste and packaged spent fuel being concentrated at one location or remaining scattered across the nation.

Reliance on the unique geohydrologic properties of the Yucca Mountain site to isolate waste has taken a back seat to building engineered barriers around the waste (Ewing 1999).

After more than 40 years, much is known about the potential performance of the candidate geologic rock in Nevada and engineered barrier systems that would surround the emplaced waste. However, the "situation is now widely recognized to be more complex than originally expected" (NRC 2001a, p. 13).

The geologic disposal program in the United States is said to be "making a last stand at Yucca Mountain" (Carter and Pigford 1999). In 1970, the repository was expected to be operating in just 10 years (Secretary of Energy Advisory Board 1993). When the Nuclear Waste Policy Act was enacted in 1982, the repository was to be open in 1998. Today, operation remains at least 10 to 15 years away (GAO 2001b). Technical, financial, legal, and licensing hurdles appear to keep the repository at arm's length. According to the same General Accounting Office report, the nation does not have "a reliable estimate of when, and at what cost, such a repository can be opened."

"Safety is in part a social judgement, not just a technical one" (NRC 1990b, p. 3).

One option to initially emplacing waste underground is developing surface facilities for storing materials until a repository is constructed. Similarly, the current single-use cycle of uranium (mine it, use it, discharge it as waste) maximizes waste generation and minimizes energy generation.[38] This "once-through" fuel cycle places the highest waste-receipt burden on a repository. As noted by James Lake, former president of the American Nuclear Society, most other countries with large commercial nuclear power programs recycle irradiated fuel by reprocessing it into new fuel (Lake et al. 2002). Uranium and other select isotopes are recovered for use or storage. "This effort doubles the amount of energy recovered from the fuel and removes most of the long-lived radioactive elements from the waste that must be permanently stored...The removal of cesium, strontium and the actinides from the waste shipped to a geologic repository could increase its [the repository's] capacity by a factor of 50." Whether or not the cost and risk tradeoffs of these options, or others, are acceptable will be determined by the public and politicians.

(37) The General Accounting Office (2001b) reports that the total cost of the repository program over its 100 plus year lifetime is $58 billion (in 2000 dollars).

(38) An "open" nuclear fuel cycle uses uranium just once inside a reactor. Afterwards, it is discharged and treated as waste. A "closed" fuel cycle reuses spent fuel by recycling it—dissolving the used fuel, chemically extracting usable uranium and other isotopes, and making new fuel from the recovered material. The down side of the "closed" cycle is that the recycled fuel is more expensive the second time around and there are concerns about the potential diversion of plutonium for nuclear weapons.

If the Yucca Mountain project fails, years may pass before Congress tackles the problem again. No nation has succeeded in building a permanent repository for spent fuel or highly radioactive waste (NRC 2001a). For Hanford, this could mean glassified tank waste and other highly radioactive materials would remain onsite for a long time. Sites with cleanup commitments extending for more than 20 years "should establish long-term on-site interim or monitored retrievable storage plans, pending permanent disposal" of their wastes according to a recent DOE report (DOE 2002).

If the Yucca Mountain project fails, years may pass before Congress tackles the problem again.

Confidence-Building Measures

In a review of the nation's high-level waste repository program, the National Research Council (2001a, p. 29) stressed "much of the scientific and technical work has included too few confidence-building measures such as peer review, formalized quality assurance, transparent documentation, large-scale demonstration, and—of great importance—development of processes to assure open discussion among all involved parties."

High-level waste repository efforts "generally have not developed scientifically sound and objectively balanced considerations of alternatives to geologic disposal. Perceptions by the public and some members of the technical community of bias or predisposition toward the geological option probably have offset the credibility of much scientific effort" (NRC 2001a, p. 29).

Perceptions of bias or predisposition have offset scientific credibility (NRC 2001a).

8.6 Site Stewardship

"The greatest immediate need is to recognize the importance of factoring the future into present decision making. This will require developing an effective decision-making process that is strong enough to overcome the inertia of inaction and produce real results; that is enduring enough to survive the short attention span of the political system in which it must function; and is democratic enough to earn the confidence and engage the participation of the people who are affected and those on whom responsibility for intergenerational obligations ultimately rests."

—National Academy of Public Administration, *Deciding for the Future: Balancing Risks, Costs, and Benefits Fairly Across Generations* (1997, p. 3)

Imagine oversight spanning hundreds to tens of thousands of years. That's the challenge facing the stewardship of Hanford land and many other nuclear waste sites. Currently, no processes exist ensuring institutions have the required controls and incentives in place (Dummer et al. 1998), and existing regulations do not prescribe a process for long-term stewardship. Besides, you can't bind a future Congress, let alone future generations, to present commitments.

You can't bind future generations to present commitments.

This places a tremendous burden on our generation to ensure we understand the consequences of our actions because a fair amount of residual contamination, making up some future inherited risk, will remain at Hanford. Fortunately, future risks are shaped, not foreordained.

Future risks are shaped, not foreordained.

Long-term stewardship refers to all activities protecting people and the environment from unacceptable hazards posed by residual materials remaining onsite after active cleanup is complete. It includes physical controls, barrier upkeep, institutional constraints, land-use restraints, information management, and monitoring (GAO 1995a; NRC 2000b).

The benchmark of long-term stewardship is to maximize future choices and minimize future burdens.

Outside of protecting basic human health and environmental resources, it's presumptuous for our generation to anticipate the desires of future generations. That is why the benchmark of long-term stewardship hinges on passing along knowledge, technical capabilities, and best solutions—to maximize future choices and minimize future burdens.

Stewardship is not an independent post-remediation action, decoupled from decades of cleanup decisions. Rather, what we do today is intrinsically linked to the liabilities others will inherit. This is why there is a tremendous co-dependence among contamination reduction, contaminant isolation, and stewardship measures in executing a defensible approach to the long-term waste management.

Werner (1999, presentation at Energy Communities Alliance Fall Conference) estimated that two-thirds of DOE's nuclear weapons sites would require long-term stewardship. This is a similar percentage to that experienced in the Superfund program over the last 20 years (GAO 1995a). The DOE suggests 96% of all nuclear weapons sites will require some form of long-term care (DOE 2000g). Whatever the percentage, humanity is committed to most nuclear weapons sites.

Until permanent solutions are developed, irreversible actions should be avoided (NRC 1995, p. 40).

The cost of extended care is projected to be $10 billion through the year 2070. In fiscal year 2000, DOE spent $64 million on long-term stewardship costs at 58 cleaned or partially cleaned sites (DOE 2002). This annual cost is expected to rise to $100 million by 2050. However, these cost estimates "will likely change over time as sites move further into the cleanup process and are able to define the nature and scope of stewardship requirements more clearly" (DOE 2000g). Actual expenses for long-term care are unknown.

"Most land containing fission product radionuclides with long half-lives can be considered unusable for agricultural purposes for centuries. Although most of these radionuclides probably could be separated from the land, reduction of the contamination to a level which would permitted [sic] unrestricted use undoubtedly would cost more than the value associated with normally expected uses...Land containing transuranic materials, particularly plutonium, can be considered unusable for any purpose for hundreds of thousands of years."

—*Accelerating Closure: Paths to Closure, Hanford Site* (DOE 1998c)

The requirements of stewardship depend on what remains behind. Some contaminants will be treated and then transported offsite. Some percentage will remain onsite, especially that released underground.

When contaminants are left behind, those responsible for the site bear the liability should unacceptable levels of contamination escape (NRC 1997, p. 20). While most cleanup activities are focusing on cleanup remedies, less attention has been paid to ensuring continued institutional control over time (DOE 1999b; NRC 2000b). One of the biggest challenges today is preserving decision processes that "maintain choice and that are open, transparent, and collaborative with independent scientists, critics, and members of the public" (NRC 2001a, p. 30). Perhaps the greatest long-term challenges are societal. Will future societies maintain waste storage facilities to ensure containment integrity and knowledge communication to the next generation?

Our generation holds in trust the interest of those yet born. Our ability to alter nature and impact the environment requires we face issues of intergenerational equity—the challenge of balancing risk, cost, benefit, and care-taking responsibilities across generations.

Intergenerational equity requires a fair allocation of resources and responsibilities grounded on maximizing the choices available to future generations while minimizing the burden imposed by our choices. It's critical to consider how today's decisions will affect tomorrow's stewardship needs though future generations will exercise their own preferences.

When is Site Cleanup Appropriate?

"Waste site remediation is appropriately left to future generations if risks are low, if it is impractical with currently available technology, or if it would impose unacceptable costs on society were it to be undertaken today. Remediation is inappropriately left to future generations if the risks are such that what is a tractable remediation problem today becomes much less so in the future as a result of events or changes in conditions that could reasonably have been foreseen" (NRC 2000b, p. 4). The council stressed improving the function of management systems and scientific knowledge to reduce cost and risk to future societies. It also cautions about a regulatory framework that "appears to encourage a constrained and piecemeal approach" to the needs of stewardship.

How does one generation equitably account for the interests of future generations? There are no clear answers though the following principles are a starting point (National Academy of Public Administration 1997, pp. 9-12).

Trustee Principle: Every generation has obligations as trustee to protect the interests of future generations.

Sustainability Principle: No generation should deprive future generations of the opportunity for a quality of life comparable to its own.

Chain-of-Obligation Principle: Each generation's primary obligation is to provide for the needs of the living and succeeding generations. Near-term concrete hazards have priority over long-term hypothetical hazards.

Precautionary Principle: Actions that pose a realistic threat of irreversible harm or catastrophic consequences should not be pursued unless there is some compelling counter need to benefit either current or future generations.

The desirability of one cleanup approach over another depends on how we discount the future.

The desirability of one cleanup approach over another depends on how we discount the future. One might argue that Hanford cleanup, at this point in history, is too expensive, too risky, or too unknown to set out on a course of action (NRC 1996a). Waste

Love Canal, New York: A Recent Failure in Site Stewardship

The story of Love Canal near Niagara Falls, New York, left a deep impression on the American public. It prompted Congress to pass the Comprehensive Environmental Response, Compensation, and Liability Act (CERCLA), now known as Superfund, in 1980. Love Canal also underscores the potential fallibility of institutional controls—even over the short term.

The lesson learned is that "the working assumption of DOE planners must be that many contamination isolation barriers and stewardship measures at sites where wastes are left in place will eventually fail, and that much of our current knowledge of the long-term behavior of wastes in environmental media may eventually be proven wrong. Planning and implementation at these sites must proceed in ways that are cognizant of this potential fallibility and uncertainty" (NRC 2000b, p. 5).

Love Canal was a 60- to 100-foot-wide and 3000-foot-long canal built in the late 1800s to connect two portions of the Niagara River. The project was never completed. The canal was sold at public auction, and after serving as a swimming hole and ice rink, it was used as a municipal and chemical dump. From 1942 to the early 1950s, Hooker Plastics and Chemical Corporation dumped 20,000 to 25,000 tons of chemical waste (for example, dioxin, metals, solvents, and pesticides) into it. The U.S. Army and city of Niagara Falls also released hazardous and chemical wastes into it.

Image used with permission of the University Archives, State University of New York at Buffalo

Love Canal Aerial Photo. This aerial photograph of the Love Canal area, Niagara Falls, New York, was taken in 1978. During the 1950s, homes were built adjacent to the canal and an elementary school's playground was constructed on top of it. Prompted by public concerns, President Carter declared a federal emergency for the area in 1978.

In 1953, Hooker Plastics and Chemical Corporation sold the site to the local school board for $1. Before transfer, the corporation received assurances from the board that no construction would take place where wastes were dumped. With full, open, and written knowledge of the site's history, the property was transferred with a deed to hold the company harmless. The deed was not heeded. Soon after, the land changed owners, and homes were built. By 1955, an elementary school was constructed on top of the dump. By the late 1950s, residents were complaining about a host of illnesses, odors, and dark sludge oozing from the ground. Chemical burns were found on children. By 1978, 25 years after industrial waste dumping had stopped, contamination had migrated into the basements of homes and had seeped into the local schoolyard. Local citizens began their own investigation into the site's history and potential health impacts. Initially, local officials balked at recognizing citizen concerns. The site was declared a public emergency by the state of New York, and national attention was focused on the problem of abandoned waste sites.

Public-generated political pressure forced a government lawsuit, home closures, and a school closure. Nine hundred families were relocated. Drainage ditches were built, a clay cap was installed, and portions of the canal were cleaned. Waste that had migrated beneath neighborhoods remains there.

One stumbling block to agencies acting more quickly was the lack of independently reviewed scientific studies on the health effects to children and pregnant women living near the site 24 hours a day compared to risk studies for more traditional and limited adult workers exposed to toxins over just a 40-hour work week. Such oversight would have caught study weaknesses and had them corrected before results were released to the public.

—Adapted after Russell et al. (1992), Hersh et al. (1997), Gibbs (1998), National Research Council (2000b)

should be carefully managed as it decays, degrades, is repackaged, or disperses into the environment. After all, radioactivity decay is now decreasing the curie load onsite at a rate of about 10 million curies each year; in three centuries, 99% of all radioactivity now at Hanford will no longer exist.

Our generation holds in trust the interest of those yet born.

Such a philosophy would minimize today's expenditures and lower risk to workers. But is this wise? Today's savings could be offset by much greater long-term cost and risk. A delay in cleanup would abdicate environmental responsibility, potentially placing another generation at greater risk. Besides, over time, more contaminants will escape from tanks, burial grounds, and other locations making them increasingly difficult to deal with. Minimizing our dealing with contamination in the near term typically increases longer-term hazards. Thus, a balance of both long- and short-term tradeoffs weighs into wise decision-making.

"It appears that each level of government is under the impression that ensuring compliance with institutional controls is someone else's responsibility" (Probst 1999).

Who is responsible for stewardship? "According to surveys by state and local government associations, it appears that each level of government is under the impression that it's someone else's responsibility" (Probst 1999). Perhaps the bottom-line question is whether the financial resources needed for stewardship will be available once the spotlight of cleanup is dimmed (Bauer and Probst 2000). Business-driven market forces alone won't provide protection and equity.

Settlement of Stewardship Lawsuit

In their Baseline Environmental Management Reports of the mid-1990s, DOE acknowledged that residual materials will remain at most of their sites (DOE 1995b, 1996a). In fact, cleanup would leave some sites closer to "brownfields" (partially clean) than "greenfields" (totally clean). Therefore, long-term care taking is needed. The report *From Cleanup to Stewardship* summarized these stewardship responsibilities (DOE 1999b).

In 1997, the Natural Resources Defense Council acting on behalf of 38 non-government groups filed suit (*Natural Resources Defense Council v. Richardson*, Civ. No. 97-936 (SS)) against DOE and several senior officials alleging the agency had violated a 1990 consent order[(39)] by failing to prepare a programmatic environmental impact statement covering the national environmental restoration and waste management program. The defendants alleged they were not in violation of the order. A good faith settlement[(40)] was reached in 1998. That settlement contained several features including:

- Creation of a central database, available to the public through the Internet, providing details about contaminants, nuclear materials, facilities, and waste at DOE sites (http://cid.em.doe.gov/).
- Sponsorship by DOE of at least two national stakeholder forums to address database issues.
- Preparation by DOE of a study of its long-term stewardship activities.
- Establishment of a $6.25 million citizen monitoring and technical assessment fund.
- Implementation of a dispute resolution process.

In addition, DOE prepared a final *Long-Term Stewardship Study* to comply with the terms of settlement (DOE 2001a). Today, what is the programmatic health of the federal government's stewardship mission?

During the first 13 years of Hanford cleanup, five Secretaries of Energy (Watkins, O'Leary, Peña, Richardson, and Abraham) and two acting Secretaries of Energy (Stuntz and Curtis) were responsible for DOE. During this same time, five officials (Duffy, Grumbly, Alm, Huntoon, and Roberson) managed the DOE's Office of Environmental Management and one person (Owendoff) was an acting manager. Each official is in office for about 2 years. Each administration brings unique priorities. Seemingly, no administration lasts long enough to gain deep traction. This demonstrates why cleanup and stewardship must be grounded in more enduring knowledge, compliance agreements, social values, and organizational cultures that might transcend political appointments and temporary emphasis.

(39) Stipulation and Order issued October 22, 1990, in *Natural Resources Defense Council v. Watkins*, Civ. No. 89-1835 (SS).

(40) Accessed at http://www.nrdc.org/nuclear/9812doe.asp#I.

8.7 Preserving Site Memory[41]

"Today, information often is not linked to key decisionmaking, nor is the process of using data and information to make decisions well understood."

—U.S. Department of Energy,
Strategic Plan for Hanford Site Information Management (1994b, p. 5)

One aspect of site stewardship receiving little attention deals with preserving data, records, and interpretations. Will people years from now struggle to reconstruct today's cleanup decisions as our generation struggled to rebuild yesterday's waste history? Are important data, and information inferred from that data, lost because they are not of immediate value?

One aspect of site stewardship receiving little attention deals with preserving data, records, and memories.

When the National Cancer Institute (1997) sough to reconstruct critical bomb testing and fallout data from just 50 years ago to estimate the radioactive iodine doses to the American people, much of information was "unavailable or of questionable reliability" (NRC 1999f, p. 5). Reconstructing Hanford's history of contaminant releases to the Columbia River and the atmosphere took nearly 6 years of concerted record reconstruction and analysis (see Section 4.3). Health analyses took another 8 years.

Today's knowledge is tomorrow's historical record.

Today's knowledge is tomorrow's historical record. Access to data and the rapid searching (or "mining") of information sources is critical to public openness and review of Hanford activities (Hanford Openness Workshops 1998). The public needs clear, concise, and readable information while in-depth reviewers require hard-core facts and technical analyses.

An integrated approach to information management delivers the right data and information, in a usable form and at an acceptable cost, to people who need it, where they need it, and when they need it. This necessitates investments in data preservation and in computational tools permitting data and record management.

The *Long-Term Stewardship Study* by DOE stresses that "stewardship needs public awareness and institutional openness to facilitate continued protection of human health and the environment states" (DOE 2001a, p. 95). This is built upon a system of information preservation and access.

Environmental assessments, litigation, and special data requests increasingly require integrated data, from multiple cleanup missions. Hanford has many critical but disparate databases. An onsite study (Porter 2000) found 14 separate databases tracking just hazardous chemicals. The National Research Council (2001b) reported on more than a dozen databases maintained by contractors plus a large set of documents that contained chemical and radionuclide inventory information. Although these databases achieve their original purposes, they are inconsistent in format and content, hindering comparisons. Blush and Heitman (1995) spoke of a "fragmented, incoherent, incomplete, and seemingly always out-of-date compilation of analyses" available to analyze site risks.

(41) Written with input from Paula Cowley, computer scientist, Pacific Northwest National Laboratory, Richland, Washington.

A strategic plan for Hanford information management was developed (DOE 1994b). Washington State Department of Ecology, DOE, U.S. Environmental Protection Agency, U.S. Army Corps of Engineers, and contractor personnel jointly wrote the document. The plan was built upon a vision to create a system of data access, comparison, and analysis across all Hanford missions. The plan was issued and met a Tri-Party Agreement milestone. It was never implemented.

An incredible amount of institutional memory (popularly known as tribal knowledge) exists about Hanford. Some is stored onsite in vaults or offsite in warehouses. Other memory exists in electronic databases, spreadsheets, and documents in offices. There is also the personal memory of those who worked and lived here. Unless preserved, this memory is lost as staff and contractors change. And in the years since cleanup began, the ebb and flow of onsite personnel has accelerated beyond anything the site experienced in its first 50 years.

Pacific Northwest National Laboratory

Information Management. Information of known quality (pedigree) must be available in a usable form when people need it and where people need it. This is a challenge facing projects spanning centuries. Years from now, could future generations decipher what our generation knew about Hanford?

Institutional memory is still being created. Are contractor turnover and some business practices creating a working environment where preserving this memory is not a rewarded activity? Perhaps from a pure business perspective, it's not cost effective to maintain data infrequently used or not of immediate benefit. However, some data are essential to the preserving site knowledge.

One example of how institutional memory is preserved is the Tank Waste Information Network System[43] (TWINS). This computer system provides access to tank waste data, documents, drawings, photos, videos, maps, and references. It's organized as a user-friendly library for browsing. This system is routinely updated, and engineers access it to obtain the official and most accurate information about tank waste.

The release of formerly secret reports continues; it began in 1986.[42] These records are invaluable for reconstructing past waste releases and focusing cleanup activities. A moratorium against destroying old records related to Site operations was instituted in 1991 (Highland 1990). This moratorium remains in place.

There has been progress in protecting institutional memory, but the sheer volume is overwhelming, its value varies, and the quality of preservation differs between contractors and agencies.

A strategic plan for Hanford information management was developed but never implemented.

Also consider that much of the digital file formats and magnetic tape technology of the last few decades are no longer compatible with today's computers or are breaking down from age. Even the original priceless magnetic tapes from the 1976 *Viking* spacecraft landings on Mars can't be read. Digital photos are already yellowing. When was the last time you tried to open an electronic file recorded on some old software package? Easy? Imagine reconstructing it in another 20 or 100 years. Institutions from government agencies to courts are struggling to avoid a digital dark age where critical records are lost.

In the few years it took to research and write this book, several of the original Web-based sources of information grew obsolete or simply vanished. In fact, the most enduring source of information used the oldest of technologies—paper. Thank goodness for inquisitive staff tucking away faded files atop dusty shelves. The DOE Hanford Declassified Document Retrieval System (http://www2.hanford.gov/declass/) continues to mature into an excellent source of historical information. It and complimentary files must be kept current. Will future cleanup contractors, let alone future generations, be capable of deciphering what our generation did, what was left behind, and why?

Data lost or not readily available takes time and resources to reconstruct recollect.

The need for effective data management reaches beyond Hanford. It's a DOE-wide problem dating back decades. In 1978, the General Accounting Office reported on the "lack of effective management processes to plan and control the Department's information resources" including tracking nuclear health, safety, and environmental restoration activities (GAO 1978). Years later, the General Accounting Office (1992) reiterated that information management problems remained because "staff have not been assigned clear responsibility or sufficient authority," and it lacked "top management attention." The existing Central Integrated Database[44] for the nuclear weapons complex is a "starting point for developing a more comprehensive referencing system for long-term stewardship data" (DOE 2001a, p. 101).

(42) These records are accessed through the Hanford Declassified Document Retrieval System at http://www2.hanford.gov/declass/.
(43) The Tank Waste Information Network System is accessible at http://twins.pnl.gov:8001/.
(44) See http://cid.em.doe.gov/

Chapter 9

Exploring Choices and Decisions

"How can one conceive of a one-party system in a country [France] that has over two hundred varieties of cheese?"

—Charles De Gaulle,
former President of France

This chapter explores the issues of choices and decisions: how cleanup actions must be taken in spite of less than perfect knowledge, how trust and credibility are gained, what creates healthy decisions, and how a problem-solving language is built so that people can find common ground to work together.

In the late 1980s, a nearly bewildering array of decisions faced those trying to achieve the vague goal of cleaning up a big chunk of the nation's radioactive waste remaining at Hanford. It seemed like every organization brought a unique perspective to the cleanup table. These ranged from "fencing the site" to removing every atom of contaminant. Reality hid between these extremes.

When plutonium production ended, along with it went the clearly defined goals of site operations. The fundamentals of chemical and nuclear engineering were now steeped in social preferences, regulations, and uncertain cleanup goals. Oversight grew, and public input became commonplace. As discussed in Chapter 2, the onsite culture had changed dramatically.

Why is making a decision in this new world so tough? Placing to the side the fact many people fear nearly anything "nuclear" as well as the host of conflicting agendas organizations bring to the negotiation table, a deeper issue is how much confidence in outcomes do we need to make a decision? How clearly can we discern the future? Lacking perfect discernment, "prudence must always substitute to some extent for proof" (Andrews 1999). To err on the side of prudence is important because the adequacy of many cleanup decisions may not be known for generations—long after organizations and companies are held accountable. Nonetheless, experience demonstrates that decisions built on technically sound information, and made in cooperation with potentially affected people, are the most durable.

Cleanup decisions should be made in a deliberate manner where logic is traceable and lines of accountability are clear. This has been a challenge across the weapons complex. As noted by the National Research Council (1999b, p. 9), "too many people in DOE [U.S. Department of Energy] act as if they were project managers for the same project, and too many organizations and individuals outside of official project organizations and lines of accountability can affect project performance."

When plutonium production ended, the fundamentals of chemical and nuclear engineering became steeped in social preferences, regulations, and uncertain cleanup goals.

"The potential impact of a technically flawed strategy is significant" (Grumbly 1996).

Cleanup decisions should be made in a deliberate manner where logic is traceable and lines of accountability are clear.

"Too many people in DOE act as if they were project managers for the same project, and too many organizations and individuals outside of official project organizations and lines of accountability can affect project performance" (NRC 1999b, p. 9).

One of the greatest stumbling blocks to problem solving is miscommunication. Kaplan (1997) proposed two reasons (called theorems) for why people can't understand each other:

"Theorem 1: 50% of the problems in the world result from people using the same words with different meanings.

Theorem 2: The other 50% comes from people using different words with the same meaning."

Strom and Watson (2002) added a creative third theorem:
"Theorem 3: There are a significant number of problems in the world caused by people using words with no idea what they mean."

At Hanford, there is a consensus about big-ticket items like protecting the Columbia River or the health of future generations. However, the devil is in the details. And these details are often clouded by what we don't know.

9.1 Unknowns Are Constant Companions

"Never try to walk across a river just because it has an average depth of four feet."

—Martin Friedman, writer

Couple unknowns about the long-term behavior of natural systems and engineered materials with public distrust and one begins to understand why people are conservative in their approach to making cleanup decisions.

Unknowns outline the limits of our knowledge.

Our knowledge of the world remains less than perfect. We always face missing facts, subjective interpretations, non-linear extrapolations, and incomplete theories. What's important is not to get mired down and agonize over our lack of perfection but to use our best knowledge to estimate the consequence of action. By understanding and bounding possible impacts, we can dodge, or at least rule out, severe unwanted consequences. Perhaps this is why Kenneth G. Johnson wrote that education is the process of moving from "cocksure ignorance to thoughtful uncertainty" (Johnson date unknown). The admission of ignorance keeps us from remaining trapped, recycling inside the confines of overly confident thinking rather than unmasking reality. Accounting for critical unknowns should be "foremost" in our minds (HAB 1999).

By understanding and bounding possible impacts, we can dodge, or at least rule out, unwanted consequences.

The primary question centers on whether learning more about a particular situation (that is, reducing unknowns) will lessen our chance of making an incorrect decision. Knowledge should be adequate to assess the range of "potential future behaviors with sufficient confidence to allow the appropriate societal decisions to be made" (NRC 2001a, p. 86). In so doing, we'll recognize that "neither a 100 percent level of safety nor 100 percent confidence" in the reliability of performance projections is possible (NRC 2001a, p. 47). This is as true for waste cleanup options as it is for any technical or societal endeavor. Making decisions is a step-by-step process with confidence built upon knowledge and experience. There is no other way to assign credible boundaries to our expectations.

Ideally, in a dollar-conscious world, the cost of reducing unknowns, such as an improved characterization of tank waste or sediment properties, should not exceed the cost of living with an unacceptable consequence from not having that additional knowledge. If the consequence is too high, then more investment is required.

Making decisions is a step-by-step process with confidence built upon experience.

Some level of unknown is imbedded in every regulation, research finding, cleanup approach, legal brief, or public opinion. Attempts to quantify unknowns are commonly associated with technical issues such as measurement or sampling errors. However, though not often acknowledged, "the limitations of 'hardware' systems and supporting scientific understanding are amplified by the inherent fallibility of the human and organizational systems" upon which we rely for making decisions (NRC 2000b, p. 4). A case in point was the measurements of tank waste levels inside Tank T-106 back in 1973. The data were collected, but no one paid any attention to what the plotted numbers revealed—dropping liquid levels. The result? The largest tank leak in the history of the nuclear weapons complex (see Section 5.6.5). The technology was adequate. However, the human element failed.

The long-term behavior of natural and engineered systems is not easily simulated and is open to interpretation. This introduces uncertainty. For example, at the proposed geologic repository site for high-level waste inside Yucca Mountain, Nevada, regulations and stakeholders are demanding to know that contaminant releases will remain within prescribed limits for 10,000 years or more. However, the National Research Council (1990b, pp. 2-3) said, "The government's HLW [high-level waste] program and its regulation may be a 'scientific trap' for DOE and the U.S. public alike, encouraging the public to expect absolute certainty about the safety of the repository for 10,000 years and encouraging DOE program managers to pretend that they can provide it…This 'perfect knowledge' approach is unrealistic…it runs the risk of encountering 'show-stopping' problems and delays that could lead to a further deterioration of public and scientific trust."

"The limitations of 'hardware' systems and supporting scientific understanding are amplified by the inherent fallibility" of humans and organizational systems (NRC 2000b, p. 4).

Years later, the National Research Council continued to voice similar concerns: "The essential problem with using models to predict the behavior of environmental systems is that the scale of interest for predictions is rarely, if ever, the scale for which information is available to construct and validate the model" (NRC 2001b, p. 171).

Over the short term, testing and monitoring data are used to validate[1] descriptive and predictive models of complex systems as well as recalibrate those same models as more is learned. Long-term monitoring provides the information backbone to more convincingly demonstrate system performance—and the potential risk of system failure. Such knowledge is essential to gain the evidence needed for public and regulatory acceptance of cleanup actions or creation of certain waste products.

Decision-makers must recommend actions despite limited knowledge. Yet, what is adequate knowledge when unknowns might mask subtleties and assumptions dominate comparisons? It is the duty of those making decisions to struggle with potentially conflicting recommendations "without being captured by...specific legions" promoted by advocacy groups (Burchfield 2001, p. 238).

(1) A validated model is one that accurately describes the characteristics and behavior of the system under study. In other words, it gives reasonably accurate answers.

Environmental protection is about achieving an adequate understanding of problems, based upon a "serviceable truth" (Burchfield 2001, p. 238).

When a person lacks an adequate understanding of the facts, they are more likely to make a poor choice (Resnik 1997).

Researchers reduce complex systems to a handful of variables that can be observed, analyzed, and extended to other behaviors. Small errors in knowledge about key characteristics can lead to large errors in predictions.

More information is always preferable, but the absence of complete information is not an excuse for delaying important actions. After exhaustive consideration, one must commit rather than remain caught in cycles of second-guessing. Continued "reconsideration of most cleanup decisions often results only in delays to important cleanup work" (Martin 2001).

Here are examples of how people and institutions sometimes deal with unknowns. Some approaches are healthy; others are not.

Acknowledge Unknowns: *Search for what is not understood.* In one extreme, this leads to studying a problem to death. In the other extreme, unknowns are acknowledged but never addressed. The wisest approach ensures actions are undertaken to resolve unknowns critical to making decisions.

Avoid Unknowns: *Keep your fingers crossed and hope for the best.* This is a refusal to examine problems from more than one perspective—usually your own. New findings raise uncomfortable questions and cast doubt on existing beliefs. The less we know, the better we are. This is like shooting an arrow and then drawing in the bull's-eye. In such an environment, it's nearly impossible to initiate investigations because additional knowledge might derail plans. This approach means investigations are pushed forward only in response to a crisis rather than a natural outgrowth of problem solving.

Exaggerate Unknowns: *Toss the baby out with the bath water.* Unknowns are used to discredit even what is understood. Program momentum is hard to maintain because of constant justifications and having to defend oneself. Second-guessing runs rampant. On the outside, learning may be encouraged, but new findings are used to undercut rather than strengthen decisions.

Dismiss Unknowns: *I would never have seen it if I hadn't believed it.* While healthy skepticism is important, this treatment of unknowns diminishes concerns without unbiased examination. Predetermined beliefs drive actions.

In the late 1940s when reviewing the safety of operations at the Clinton site (now Oak Ridge) in Tennessee, a safety and industrial health board sponsored by the Atomic Energy Agency reported, "A faith method of operation, indigenous to Tennessee, was detected in many hazard problems. When questioned about certain hazard procedures, the answer was 'I believe that this is safe.' This answer was given in cases which other observers currently consider hazardous, and in other cases where experimentation was possible but never attempted. We deduce that this is an intellectual version of the less hazardous rattlesnake-handling cult also indigenous to the region. It has already spread to other locations [in the weapons complex], especially in disposal problems, and is a very dangerous philosophy" (Parker in Williams 1948). This is an example of keeping your fingers crossed and hoping for the best.

Unknowns can be addressed and managed. An example is the soil permeability beneath Hanford. Permeability is one factor that determines how quickly groundwater and contaminants move. While many measurements of permeability exist, uncertainty remains in how it varies from place to place (DOE 1999d). Field testing and modeling reduces uncertainty. Another example is uncertainty about the long-term effect of radiation-caused damage inside glass or other waste forms (Weber et al. 1999). Such damage can lower the ability of waste products to contain contaminants. Uncertainty is reduced by research.

However, sometimes, there are things we don't even know we don't know. These are surprises, potential showstoppers. They can't be quantified, for we may not know they exist. When encountered, the experience is humbling, highlighting the limits of our knowledge. By disturbing the status quo, surprises create powerful opportunities for re-energizing our knowledge and motivations. One example at Hanford was the unexpectedly complex rock structures randomly encountered in exploratory wells drilled deep into basalt for geologic repository studies (Gephart et al. 1983). Others could be any unanticipated behavior of tank waste treated before vitrification or the existence of small mobile particles, called colloids, facilitating contaminant movement from waste sources. (Kersting et al. (1999) reported such colloid-aided movement of plutonium below the Nevada Test Site.)

By disturbing the status quo, surprises create powerful opportunities for re-energizing our knowledge and motivations.

Beware of proclamations to "completely characterize" something, as if all unknowns can be resolved. This is especially true for the waste in Hanford tanks. Researchers can sample only a small portion of the whole and then extrapolate to a larger volume, over time, or to different circumstances.

What characterization means depends on what decision is faced. For example, if all tank waste is processed into glass, then the waste must be characterized to the extent its properties are adequately understood to safely process glass. If some waste is to remain within tanks, then characterization is focused on ensuring safe in-place storage. Likewise, each option for dealing with underground contamination (for example, monitoring, in situ treatment, natural attenuation, or pumping and treating) raises different characterization issues. What characterization means centers around purpose—the use of knowledge.

What characterization means centers around purpose—the use of knowledge. Large complex systems, such as Hanford tank waste, can't be "completely" characterized.

Sometimes, waste properties change. Hanford's storage tanks act as slowly evolving chemical reactors where waste characteristics vary between tanks, within a tank, as well as over time. Some waste properties have changed over the years. For example, 150 tons of potentially explosive ferrocyanide[(2)]—used to precipitate the cesium from the tank liquids, allowing the liquid to be disposed of directly into the soil—have degraded chemically into less reactive compounds.

Are we adequately addressing critical unknowns at Hanford important to future decisions? Here is one example.

For years, the sorptive ability of sediments was expected to hold most leaked waste high above the water table (GAO 1989, 1998a). Therefore, the U.S. Department of Energy maintained there was little threat of further tank leakage, and the potential environmental impacts from past leaks were low (DOE 1987). This was questioned because evidence did not support the belief (GAO 1989; NRC 1996a).

(2) Ferrocyanide has the potential to ignite or explode when heated above 450°F. In 1996, DOE announced that tank waste characterization and analyses showed that the ferrocyanide had decomposed and was no longer a threat (DOE 1996).

Traditionally, the modeling of contaminants leaked from old single-shell tanks did not factor in how their unique chemistry could alter the natural chemistry of the underlying sediments—though the presence of large amounts of waste was known for years to appreciably reduce cesium sorption on soil particles (McHenry 1954). Radiation specialist Herbert Parker (1954a) reported, "If complexing substances are present, however, unusual results may be expected depending on the composition of the [chemical] solution and the concentration of the various components." Leaked tank waste is very caustic (high pH) and contains fluids with high ionic strength, high density, and formerly high temperature. This waste created a distinctive anomalous chemical environment where some radionuclides, normally of low mobility, could move more easily.

On closer examination of groundwater chemistry, contaminant distributions beneath single-shell tank farms, and geophysical data collected in wells, DOE acknowledged that cesium-137, technetium-99, and cobalt-60 had migrated deeper than was previously expected (DOE 1999d). The groundwater also contains other tank-originated metals such as chromium and sodium.

Reviewers had suspected this for years (GAO 1998a). The announcement drew national attention. It caused program reorganizations, created new projects and expert panels, and rejuvenated the site's focus on contaminant behavior in the subsurface. Was this finding a surprise? To some, yes. Did it disturb the status quo? Certainly. Could it have been avoided? Yes, by acknowledging unknowns and ensuring official agency positions were backed by evidence in hand—not just the lack of evidence. Environmental writer Josh Silverman (2000) cautions about organizations building a "culture of complacency" that does not acknowledge the unexpected. Eventually, the physical, chemical, and biological realities of the world reveal flaws in even our most ardent beliefs.

"If the decision is made to continue the trend away from a minimum credible level of [subsurface] characterization, it should be made with the full realization that it is not supported by a scientific understanding of the issues" (Integration Project Expert Panel 2000a).

And yet, while agencies recognize the need to better understand the behavior of contaminants leaked from tanks, is enough attention being paid to this problem (EPA 2000)? As of the year 2002, only three boreholes have been drilled to facilitate closer study of contaminant movement beneath Hanford's 177 waste tanks. By comparison, some sixty wells are available for groundwater sampling near the tanks, and 750 shallow wells are used for geophysical surveys.(3)

According to the Integration Project Expert Panel—a panel of experts that provide DOE with recommendations on reducing Hanford Site vadose zone and groundwater contamination—current plans for studying waste behavior underground will not produce enough information to support critical cleanup decisions in a timely manner (Integration Project Expert Panel 2000a). The panel continued, "This trend away from adequate characterization comes at the same time that stunning discoveries are being made including high concentrations of ^{99}Tc [technetium-99] found in a groundwater well, high concentrations of tritium and gross beta found in another, and ^{137}Cs [cesium-137] formerly thought to be essentially immobile in soil, found much deeper in the vadose zone than predicted. If the decision is made to continue the trend away from a minimum credible level of characterization, it should be made with the full realization that it is not supported by a scientific understanding of the issues." The Integration Project Expert Panel was disbanded the following year—in 2001.

(3) Geophysical surveys are done by lowering instrument probes down wells to indirectly estimate the properties of the nearby soil, moisture, and contaminants.

This concern was reinforced by the National Research Council (2001b, p. 75): "the planned rate of characterization is not sufficient to establish, even approximately, the current distribution, speciation, or potential for transport in the subsurface or important subsurface properties." There is little information on the distribution of contaminants beneath Hanford tanks, almost no data for the deep vadose zone, and skepticism about "unrealistic" experimental times scheduled to measure the properties and processes controlling contaminant movement—5 to 10 years are sometimes needed rather than the 1 to 2 years allotted. This information is essential to making credible cleanup and site stewardship decisions.

"Early on [Nobel Laureate Kenneth] Arrow became convinced that most people overestimate the amount of information that is available to them...His experience as an Air Force weather forecaster during the Second World War added the news that the natural world was also unpredictable...One incident that occurred while Arrow was forecasting the weather illustrates both uncertainty and the human unwillingness to accept it. Some offices had been assigned the task of forecasting the weather a month ahead, but Arrow and his statisticians found that their long-range forecasts were no better than numbers pulled out of a hat. The forecasters agreed and asked their superiors to be relieved of the duty. The reply was 'The Commanding General is well aware that the forecasts are no good. However, he needs them for planning purposes'" (Bernstein 1996, p. 203).

The implications of uncertainty must be understood. As more investigations are carried out, we'll increasingly push the limits of our knowledge. This is especially true in the complex chemical mixtures found in tanks and the natural variability of sediments deposited over millions of years. These are not finely engineered systems whose designs, components, and performance under a variety of stresses are well documented.

No science, law, regulation, or milestone can ever prove a system will perform as predicted. At best, knowledge reveals the boundaries around likely performance.

Then, in the face of such uncertainties, what guarantees do people have towards achieving satisfactory remediation progress? An attainable goal is reached by doing and learning (step-by-step progress) to make sure that the likelihood of "serious unforeseen events" is minimal (NRC 1990b, p. 3). Assurance rests on the credible application of scientifically sound principles with uncertainties assigned to the results uncovered. In so doing, it's critical to investigate controlling characteristics and processes rather than scampering after merely interesting ones.

An open and informed discussion of Hanford's cleanup problems—what is known, unknown, and assumed—is in the best interest of everyone. Unknowns temporarily side stepped will eventually return to haunt presumed assurances.

"The Commanding General is well aware that the forecasts are no good. However, he needs them for planning purposes" (Bernstein 1996).

Unknowns temporarily side stepped will eventually return to haunt presumed assurances.

Public Distrust. Distrust of Hanford runs high. The sign's other side read, "We believed and were betrayed." In 1999, the Fred Hutchinson Cancer Research Center in Seattle and the Centers for Disease Control and Prevention presented their preliminary findings to the public about potential health impacts from radioactive iodine releases blown downwind of Hanford. (Photo from January 29, 1999, *Tri-City Herald*, Kennewick, Washington).

9.2 Trust and Credibility

"Public fears and opposition to nuclear-waste disposal plans can be seen as a 'crisis in confidence', a profound breakdown of trust in the scientific, governmental, and industrial managers of nuclear technologies."

—Paul Slovic, *Risk Analysis* (1993)

Trust is the prerequisite to change and credibility.

The lack of public trust in Hanford officials is a major impediment to reaching consensus not only on the type, degree, and schedule for remediation, but also on the very process to reach decisions (NRC 1994b). No amount of scientific analyses or engineering comparisons can allay public concerns unless the system, its information, and its messengers are trusted.

Trust is fragile, gained slowly, and damaged easily. One or two negative events influence public opinion more powerfully than a truckload of positive gains. Regretfully, relationships are sometimes so poisoned that trust may never be regained in the near term but must wait for another generation.

Public confidence in Hanford was shaken in 1986 when the federal government began declassifying documents revealing that large amounts of radiation was released during early years of site operations. Even some of Hanford's most stalwart supporters were surprised by these revelations. In spite of the need for a certain amount of operation secrecy, people felt they had the right to be informed about contaminants possibly causing personal or environmental harm. After all, even during the earliest years, officials knew that Hanford's waste management practices were inadequate and growing more worrisome. Today, many believe withholding such information was ethically wrong (Oughton 1996).

No amount of analyses can allay public concerns unless the decision-making system, its information, and its messengers are trusted.

After the initial release of declassified documents, the next major step in openness took place with the signing of the Tri-Party Agreement in 1989 (see Section 7.5). It committed the government to cleaning up Hanford and abiding by federal and state environmental laws. Afterwards, efforts began to understand public concerns and regain their trust.

In 1991, Secretary of Energy James Watkins created a task force to look into the issue of earning public trust (Secretary of Energy Advisory Board 1993). This task force, known as the Secretary of Energy Advisory Board, found widespread distrust that elaborate agency reorganizations would not solve. After all, organizations are simply surface artifacts of more deep-seated values, beliefs, and assumptions that drive how institutions think and behave (Schein 1999). Ingrained values must change before behavior will change.

Beginning in 1993, Secretary of Energy Hazel O'Leary embraced the philosophy of public openness with a series of initiatives centered on publishing formerly secret information (for example, the U.S. plutonium stockpile records and evidence about past human-radiation experiments), re-examining the rules governing information classification, and establishing an Openness Advisory Panel under the Secretary of Energy Advisory Board. O'Leary also held numerous press conferences and public, town hall meetings across the nation.

The Secretary of Energy Advisory Board (1993) said that DOE leaders must give everyone opportunities for involvement and demonstrate fairness. Recommendations included

- Providing early and continuous public involvement with officials who give well-informed, candid responses to questions.
- Committing to agreements unless modified through an open process.
- Pursuing technical options and strategies whose consequences are communicated and understood.
- Rewarding honest self-assessments and questioning that permits contractors to get ahead of problems.
- Developing tough oversight processes including public review.

The most important part of building trust? It's public participation.

O'Leary and past Assistant Secretary for Environmental Management Thomas Grumbly stated that public participation, as equal players, is the most important component in making decisions and in building trust (Hanford Openness Workshops 1997).

The Oregon Office of Energy facilitated creating the Hanford Openness Workshops in 1997. This was a collaborative effort among the Consortium for Risk Evaluation with Stakeholder Participation (CRESP), DOE, the Washington Department of Ecology, the Oregon Office of Energy, regional tribal representatives, and citizens. The workshops were configured to promote communication and access to Hanford information that "enables citizens to contribute to decisions meaningfully and powerfully" (Hanford Openness Workshops Fact Sheet 1999b). Workshop summaries, meeting minutes, fact sheets, and mission-related information are available online at http://www.hanford.gov/boards/openness/index.htm. The last workshop meeting was held in early 2000.

In the year 2000, Secretary of Energy Bill Richardson asked the workshop's advisory panel to review and assess DOE's relationships with the communities surrounding its facilities (Openness Advisory Panel 2000). As before, findings recognized the continuing "legacy of public distrust" and offered suggestions for improving communication.

Distrust is fueled by our society's tendency to manage risk in an adversarial manner.

Regardless of Hanford history, public distrust of government and industry has grown during the last few decades. Several factors contributed including the increased number of right-to-know laws related to hazardous materials, more knowledge about exposure to hazards, and greater coverage of environmental and health issues by the news media. Distrust is also fueled by our society's tendency to manage risk in an adversarial manner, pitting expert against expert, institution against institution, and belief against belief.

Institutions have integrity but are no longer viewed as infallible.

Once initiated, distrust reinforces itself and is difficult to stop. Regretfully, it inhibits the kinds of personal experiences necessary for healing and colors interpretation of subsequent events (Slovic 2001).

This is frustrating to many officials, especially those involved in the profession of information sharing and risk communication. While these specialists strive to lessen communication blunders, there is little evidence that risk communication has significantly reduced the gap between risk estimations and risk perceptions (Slovic 1993). If the public trusts someone, or the institution they represent, they are more likely to listen and be open to what they say. If people don't trust the messenger, no communication style or training will help (Fessendon-Raden et al. 1987).

People listen to those who they believe are credible. Research indicates that four criteria are used to evaluate a person's credibility: perceived caring and empathy, competence and expertise, honesty and openness, and dedication and commitment (Covello 1992, 1993). Of the four, caring and empathy are most important, accounting for 50% of the trust and credibility earned. Each of the remaining factors account for 15 to 20%. In the medical profession, a patient's judgment about how caring a doctor is takes a short time to form—sometimes just 30 seconds (Covello 1993). Once made, these judgments are resistant to change. Now imagine standing in front of an audience and talking about nuclear waste! What's the half-life of that person's perceived credibility?

In the medical profession, a patient's judgment about how caring a doctor is takes a short time to form—sometimes just 30 seconds (Covello 1993).

Many institutions are looked on as unwilling to acknowledge problems, share information, allow meaningful public participation, or fulfill their responsibility to protect the environment (Covello 1992). Yet, at the same time, people view government and industrial officials as among the most knowledgeable sources for information. Agencies are knowledgeable but not trusted. Institutions have integrity but are no longer viewed as infallible.

A common myth is that so-called activist groups are responsible for stirring up unwanted concerns and public distrust (Chess et al. 1988). However, many activists focus public anger rather than create it. Distrust was preconditioned. They tend to flag problems to focus public awareness and work towards solutions. Besides, if concerns raised by small groups are powerful enough to derail multi-billion dollar plans, then perhaps the plans were weak in the first place.

If concerns raised by small groups of activists are powerful enough to derail multi-billion dollar plans, then perhaps the plans were weak in the first place.

How can the working environment between government agencies and the public be detoxified? People trust and support what they understand, are involved in, and benefit from. Power sharing, meaningful participation, candid discussions, genuine care, and commitment follow-through are essential. Questioning and litigation are tools of change. Just as in research, informed skepticism strengthens.

Building public trust is a process. It's initiated, not decreed. Experience allows trust to grow. An airplane passenger trusts that the plane is well built and the pilots are well trained and will take actions in the best interest of passengers. On-time schedules, lack of lost luggage, and the absence of accidents build trust. As Hanford cleanup progress is made and as milestones are achieved, trust should gradually return.

9.3 What Makes a Healthy Decision?

"Nothing chills nonsense like exposure to air."

—Woodrow Wilson,
former U.S. President

We use available knowledge, experience, and preferences to make our best decisions. Several options are normally available. This is especially true during the early stages of a project when uncertainty is greatest. When deciding among options, decisions can be fair (healthy) or unfair (unhealthy). The following items are characteristics of a healthy decision (Presidential/Congressional Commission on Risk Assessment and Risk Management 1997):

- Solves problems and achieves goals within context of potential human health and environmental impacts.
- Emerges from an open decision process including the views of those potentially affected.
- Uses scientifically sound, technically defensible, and independently peer-reviewed evidence.
- Results from examining a range of alternatives.
- Can be implemented and is biased towards taking action.
- Is self-correcting—decisions are revisited as new information becomes available without hindering progress.
- Reduces or eliminates risk in a way that
  - Relates benefits to costs in a reasonable way.
  - Gives priority to preventing rather than just controlling risk.
  - Is sensitive to political, social, legal, and cultural considerations.
  - Includes incentives for innovation, critical thinking, and research.
  - Accounts for multiple waste sources, multiple contaminants, and multiple risk situations.
  - Protects current and future generations.

What makes a healthy decision sometimes hinges upon what we think the problems are.

However, what makes a healthy decision sometimes hinges upon what we think the problems are. In a 2-year assessment of the resources required to complete cleanup of contaminated Superfund sites, Russell et al. (1991) found that the identification of problems varied among stakeholders, agencies, and sites. For instance:

Residents living near a waste site were concerned about health effects from contamination and releases from proposed remedies, disruption of personal lives, availability of information, and their inclusion in the remediation process. People who owned land worried about property values. Some residents emphasized problems while others downplayed them, depending upon what was at stake—for them personally.

Local officials and business owners wanted to know that remediation was under way, their ability to use or sell land was preserved, and adverse economic effects from community association with a waste site were mitigated.

Elected representatives reflected public concerns—being assured that cleanup progress was under way.

Environmental groups focused on cleanup standards, suitability of proposed remedies, and long-term health effects and ecological damage.

Contractors wanted a clear chain of command and guidance on cleanup standards.

State and U.S. Environmental Protection Agency officials were concerned about meeting cleanup standards and milestones, keeping the community informed as well as themselves being apprised or involved in cleanup activities.

Natural resource trustees[(4)] focused on preventing additional ecological harm and reconstructing fish and wildlife habitats.

Everyone approaches problems differently. The next section discusses two past remediation actions and asks the reader whether or not the decisions made were wise.

9.3.1 Healthy Decisions or Not? You Decide

We act on the best information and judgment available.

Constraints are limitations to what can be done or understood. Common constraints are funding, schedules, regulations, work approaches, scientific knowledge, and technical capability. All constraints are flexible except knowledge and capability. Funds can be stretched, schedules adjusted, preferences changed, and regulations modified. However, problems can't be solved unless the right knowledge and capabilities exist. Like gravity, reality won't be fooled.

Problems can't be solved unless the right knowledge and capabilities exist.

Issues associated with the pumping and treating of radioactive strontium (strontium-90) near N Reactor and the early shutdown of the PUREX Plant are good examples of the interplay in solving problems while dealing with limitations in our knowledge. Were healthy decisions made? It depends upon your perspective.

Strontium Removal from Groundwater Near N Reactor: The following example illustrates attempts to reduce, or at least control, strontium-contaminated groundwater in an area near N Reactor called N-Springs.

Starting in the 1990s, studies of vegetation along the Columbia River near N Reactor raised concerns over strontium-contaminated groundwater entering the riverbed.

Starting in the 1990s, studies of vegetation along the Columbia River raised concerns about strontium-contaminated groundwater entering the riverbed and the fall chinook salmon nests (Peterson and Poston 2000).[(5)] While the problem of strontium in the Hanford environment is not new, sampling showed groundwater near N-Springs had the highest strontium concentrations of any riverbank at Hanford—up to 2400 times above drinking water standards (Peterson and Poston 2000).

This strontium came from water discharged from N Reactor's cooling system that became contaminated when the metal coating enclosing irradiated fuel slugs failed. The contaminated water was pumped into nearby cribs and trenches. These disposal systems received contaminated water from 1963 to 1993. About 89 curies (0.02 ounce) of

(4) Agencies or groups of people acting as an advisory board for resource protection.
(5) The upwelling of contaminated groundwater into the riverbed, use of bio-indicators, and offshore environmental sampling are under way along the Hanford Reach of the Columbia River to identify any life forms under stress.

strontium is in the aquifer: 88.0 curies are chemically bound to the soil and 1 curie is dissolved in the groundwater. Another 1800 curies (0.5 ounce) lie in the soil above the water table (Serne and LeGore 1996).

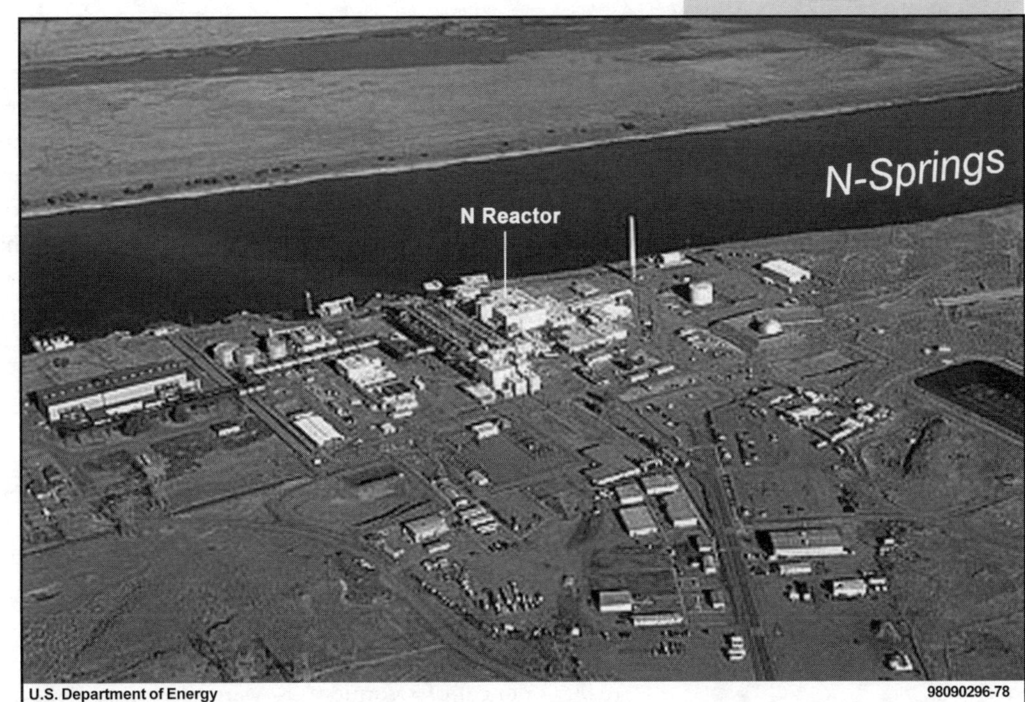

N Reactor and N-Springs. Strontium-contaminated groundwater from N Springs seeps into the Columbia River just downstream from N Reactor. This radioisotope entered the reactor's cooling water when the metal coating enclosing irradiated fuel slugs failed. That water was pumped into cribs and trenches. They were built along the area shown on the right edge of this photograph near the river. Today, the springs extend over a smaller area than in the past.

To reduce the amount of strontium from seeping into the river, in 1991 an engineering firm compared six groundwater containment options. Recommended options included building a slurry or cryogenic wall between the disposal sites and the river. Alternatives not recommended included chemical injection, surface sealing and capping, and a hydraulic wall formed from groundwater withdrawal and injection. Pumping and treating the groundwater was not favored because of projected high costs, low effectiveness, and likely construction and operation problems.

Three years later in 1994, a board of geohydrologic consultants and civil engineers wrote a second review of 12 treatment options (Advanced Sciences Inc. 1994). They evaluated the cost-effectiveness and technical defensibility of remedial actions being considered. Board members supported the goal of reducing strontium entering the river and echoed concerns in the 1991 study—pumping and treating strontium "would result in insignificant total mass removal due to the natural immobility of ^{90}Sr [strontium-90]" (Advanced Sciences Inc. 1994). The board stressed that the "effectiveness of pump-and-treat alternative was incorrectly assessed" by agencies involved in deciding whether DOE should apply this technology. Issues of high costs for the 10 years of planned remediation plus the lack of clear goals driving strontium removal were mentioned. Advanced Sciences Inc. preferred a vertical barrier using a slurry wall. That option had the least amount of technical and cost uncertainty while at the same time slowing down groundwater and strontium movement so that more radioactive decay could take place before it entered the Columbia River.

Pumping and treating the strontium-laced groundwater was not recommended because of projected high costs, low effectiveness, and likely construction and operation problems.

In September 1994, the Washington State Department of Ecology and the U.S. Environmental Protection Agency issued an "action memorandum" directing DOE to immediately begin actions to (1) reduce strontium-90 concentrations in the groundwater near N-Springs, (2) evaluate commercially available treatment options for strontium-90, and (3) provide data necessary to set demonstrable strontium-90 cleanup standards

It will take 75 to 90 years to remove 90% of the strontium-90 in the aquifer by groundwater pumping. Natural radioactive decay will remove 90% of the strontium in 95 years without any groundwater extraction or expense (Serne and LeGore 1996).

(Butler and Smith 1994). This memorandum noted there was a "majority [of public] interest in proceeding with pump and treat" based upon the letters written and comments received. Though no formal risk assessment had been performed regarding the strontium-90, there was concern that the average strontium-90 concentration in groundwater was far above the drinking water standard. The agencies concluded that DOE should combine a groundwater pump-and-treat system with a vertical barrier made of metal driven into the ground. The barrier would have a minimum length of 3000 feet and a depth of about 50 feet. It would form a low permeability wall cutting off (or at least slowing down) strontium-laden water. The cost for groundwater pumping and treating was estimated at between $5.9 million and $22.4 million. The DOE was told to have the barrier wall installed by June 1995 and an operating pump-and-treat system by September 1995.

What was done and learned?

- Installation of the vertical barrier failed. The underlying sediment was too hard to drive the metal piling through. This approach was abandoned in December 1994 (Washington State Department of Ecology 1995).
- After 6 years of pumping and treating 160 million gallons of water, 1.1 curies (0.0003 ounce) of strontium-90 were recovered (Hartman and McMahon 2002). During this same 6 years, about 12 curies (0.003 ounce) of strontium-90 naturally decayed away—10 times more than the strontium recovered after 6 years of pumping.
- It will take 75 to 90 years to remove 90% of the strontium-90 in the aquifer at the existing or increased pumping rate. On the other hand, natural radioactive decay will remove 90% of the strontium in 95 years without any groundwater extraction or expense (Serne and LeGore 1996).
- As of the end of the year 2000, the construction, maintenance, and operation cost of the pump-and-treat system was $18 million.[6] The annual operating cost is $800,000 (Integration Project Expert Panel 2000b).
- Each year, about 8.8 curies (0.002 ounce) of strontium flows down the Columbia River from northern Washington and into the Hanford Reach (Patton and Poston 2000). Hanford groundwater containing strontium seeps into the river and contributes an additional 0.3 curie (0.00008 ounce) of strontium-90 each year. If one assumes this 0.3 curie comes from the N Reactor area, then 97% of the strontium-90 in the Columbia River is from other sources (primarily radioactive fallout) and are unaffected by any Hanford cleanup efforts.
- Salmon do not spawn near N Reactor because riverbed conditions are not suitable. However, small bottom-dweller sculpins do. The radiation dose to sculpin near N Reactor is 0.008% of the guidance established by DOE Order 5400 (Peterson and Poston 2000, p. 5.4).
- The major ecological risk at the site appears associated with the ingestion of strontium-90 by plant-eating ducks (Peterson and Poston 2000).

Recently, the cost and effectiveness of strontium removal are being questioned. "Such an operation may have to run for about 300 years" to reduce the radioactivity present by a factor of 1000 (NRC 2000b). A panel of groundwater experts said, "alternative

(6) Communication from Garrett Day, Bechtel Hanford, Inc., February 20, 2001.

remediation strategies for the ^{90}Sr plume should be evaluated" (Integration Project Expert Panel 2000b). The U.S. Environmental Protection Agency (2001a, p. 100-20) wrote that groundwater pumping and treating "does not appear to be an effective method for reducing Strontium-90 concentrations in the aquifer relative to natural decay." They recommend finding alternatives while continuing groundwater withdrawal.

Groundwater pumping and treating "does not appear to be an effective method for reducing Strontium-90 concentrations in the aquifer relative to natural decay" (EPA 2001a).

Across Hanford, five pump-and-treat operations, costing $4 million to $4.5 million a year to run, have slowed the movement of some chromium, carbon tetrachloride, strontium, uranium, and technetium plumes though the "sources and bulk of these plumes remain untouched" (Stang 2000a). Concerns are being voiced about the wisdom of using conventional treatment technologies.

Worries over the effectiveness of groundwater pumping and treating are not confined to nuclear weapon sites. After evaluating the escalating costs and general ineffectiveness of 75 major groundwater pump-and-treat systems[7] located on military bases, the Department of Defense (1998) reported that "much of the regulatory guidance for groundwater cleanup was written before 1989, when the limitations of pump-and-treat systems were not fully appreciated...[we] recommend that a systematic approach be developed in cooperation with environmental regulators, the scientific community, and the public to determine more effective alternative methods for future groundwater clean up."

With each nuclear disarmament agreement, the United States grew less reliant on a large stockpile of nuclear weapons. The PUREX Plant had served its purpose.

PUREX Plant Shutdown: With each nuclear disarmament agreement and the end of the Cold War, the United States grew less reliant on a large (let alone growing) stockpile of nuclear weapons. Plutonium, chemically teased out of fuel rods that had been irradiated in a reactor's core, was no longer needed. The PUREX Plant had served its purpose. It had removed plutonium from nearly 70% (73,100 tons) of Hanford's spent fuel.

When the PUREX Plant was not operating from 1972 and 1983, spent fuel from N Reactor was stored in two water-filled concrete basins near the KE and KW Reactors. The last fuel was placed there in 1989. Eventually, 200,000 fuel elements weighing 2300 tons and containing 55 million curies of radioactivity (decayed as of 2001) ended up in these basins.

The PUREX Plant was operated in 1989 and into early 1990 to chemically stabilize 100 tons of irradiated fuel remaining from when the plant was shut down by Hanford officials in December 1988 because of management and safety concerns (DOE 1989). In early 1990, three environmental groups notified DOE of their intention to sue if the PUREX Plant was restarted without preparing a supplemental environment impact statement (GAO 1990). It became increasingly difficult, if not impossible, for the PUREX Plant to satisfy the waste management provisions of the Resource Conservation and Recovery Act (Saleska and Makhijani 1990). The plant was built to standards unacceptable in the regulated world of the late twentieth century.

(7) Some of these groundwater pump-and-treat systems have run since 1981. Fifty percent of the 75 systems operated for contaminant containment, and 50% functioned for containment and treatment.

The PUREX Plant was built to standards unacceptable in the regulated world of the late twentieth century.

To process the K Basin and other onsite spent fuel, DOE's Richland office wanted to restart the PUREX Plant the next year in 1991 (GAO 1990) and then permanently shut the plant down around the year 2000. At that time, the plant's annual operation cost was $90 million to $100 million.

The General Accounting Office (1990) cited several reasons why the plant should not restart: (1) dwindling need for more weapons-grade plutonium, (2) lack of an update to a 1983 environmental impact statement for changing the PUREX mission from plutonium production to waste management,[8] (3) loss of trained staff, and (4) uncorrected safety deficiencies. The restart plan was not "supported by any detailed technical, engineering, or cost analyses that fully demonstrate that PUREX is the best option." Secretary of Energy Watkins announced in late 1990 that the PUREX Plant would be placed on standby and, if restarted, an environmental impact statement would be prepared.

In late 1992, DOE issued a final closure order. Public concerns for shutdown centered on the same topics identified by the General Accounting Office but also included the lack of facilities to vitrify the extra high-level waste created from reprocessing the remaining spent fuel at Hanford.

U.S. Department of Energy 8405512-5cn

PUREX Plant and AP Tank Farm. The PUREX Plant was shut down in 1990. This left unprocessed spent fuel stored in basins next to the Columbia River for another 15+ years as new fuel handling, drying, and storage facilities were built and operated. This 1984 photograph shows the PUREX Plant behind eight double-shell tanks being constructed in the AP Tank Farm.

(8) Plans were to process spent fuel stored in the K Basins, the commercial Shippingport Reactor in Pennsylvania, and the experimental Fast Flux Test Facility at Hanford.

Arguments were made comparing the costs of operating the PUREX Plant for another year or two to process the remaining spent fuel versus building new facilities to stabilize and temporarily house that fuel. Assuming an average processing rate of 10 tons of fuel a day, the 2300 tons of K Basin fuel would be reprocessed in about 1 year. That year of PUREX operation might have cost $100 million to $200 million. Storage and disposal of the resulting waste would cost extra. Processing might have generated 600,000 gallons of additional tank waste.[9] This is about 1% of the 53 million gallons of waste now in Hanford's tanks.

By 1994, in a post-PUREX era, the K Basin spent fuel project began and various cleanup and storage alternatives were examined (Fulton 1994). Forums were held, and values were heard from stakeholders, tribal nations, and government agencies. An environmental impact statement, supporting a preferred cleanup option of removing, treating, and storing of dried spent fuel, was issued (DOE 1996c).

The deactivation of the PUREX Plant was completed in 1997 for about $145 million (Bailey and Gerber 1997). The facility is in a safe, low-maintenance mode until future decisions are made. This shutdown reduced the annual maintenance and monitoring cost for PUREX from $35 million to $1 million.

The question of PUREX Plant shutdown revolved around timing.

Today, the projected construction, treatment, and near-term operation cost for dealing with the K Basin spent fuel is $1.6 billion. The project is scheduled for completion in 2006 with temporary onsite storage until the year 2030. While Hanford tank waste will be solidified, the spent fuel from K Basins will be stored onsite—for perhaps 30 years—pushing long-term mortgage costs and potential risks to another generation. Whether it must be converted into still another product before shipment to a geologic repository is unknown. Each storage and treatment step adds more cost.

The question of PUREX Plant shutdown revolves around timing. Early shutdown resulted in 80% of the DOE's nationwide inventory of spent fuel remaining another 15 to 20 years in aging concrete basins along the banks of the Columbia River—and it costing at least $1.6 billion to provide a temporary solution. Was the wait worth the risk? After all, this spent fuel remained "one of Hanford's most dangerous risks to people and the environment" (HAB 1997a).

Just the dollars...dealing with the remaining spent fuel from Hanford

Chemically reprocessing spent fuel in the PUREX Plant: $100 to $200 million

Packaging, drying, and storing K Basin spent fuel: $1.6 billion

In both cases, future storage and disposal costs of the resulting waste, waste forms, and nuclear materials are extra and largely unknown.

(9) Number calculated from a 500 gallon a ton PUREX high-level waste generation rate given by Agnew (1997). This was typical for the late 1950s through 1972. Total waste volume was then reduced 50% by evaporation—a percentage based upon historical records summarized by Agnew.

9.4 Building a Problem-Solving Language

"Our knowledge does not become a communicated idea if it must push through a briar patch of sticky words."

—Alton Blakeslee, *Worlds Apart: How the Distance between Science and Journalism Threatens America's Future* (Hartz and Chappell 1997, p. 110)

Richard Andrews (1999), professor of environmental policy at the University of North Carolina at Chapel Hill, writes that creating a common environmental policy and vision "involves negotiating conflicts among mutually exclusive preferences for the use of indivisible resources."

Even differences in how people express themselves create conflict as interpretations are juggled and a common language sought.

Decisions do not require, and are rarely based on, consensus. The best one can hope is to reduce conflicts.

Decisions do not require, and are rarely based on, consensus. "In the real world, an insistence upon consensus is a recipe for inaction—or actions at the level of the lowest common denominator" (Alvarez 2000). In discussing American environmental policy, Richard Andrews writes, "consensus seems intuitively the most desirable form of political decision-making. In practice, however it rewards the most extreme form of minority tyranny by allowing holdouts to demand extra individual benefits....as the price of their approval" (Andrews 1999, p. 7). The best one can hope is to reduce conflicts. Although good decision processes do not always lessen conflict, bad practices predictably increase conflict (NRC 1996b).

Another challenge facing decision-makers is deciding what, how much, and the quality of information needed for taking action (see Section 9.1). As Albert Einstein once said, "not everything that can be counted counts, and not everything that counts can be counted." Sometimes decision-making is held hostage to the hope that the weight of

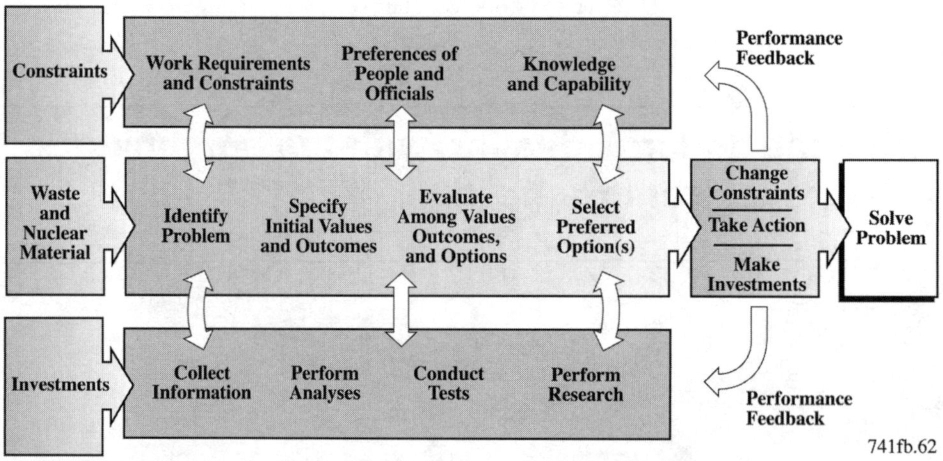

Constraints
Work Requirements and Constraints
Preferences of People and Officials
Knowledge and Capability
Performance Feedback
Waste and Nuclear Material
Identify Problem
Specify Initial Values and Outcomes
Evaluate Among Values Outcomes, and Options
Select Preferred Option(s)
Change Constraints
Take Action
Make Investments
Solve Problem
Investments
Collect Information
Perform Analyses
Conduct Tests
Perform Research
Performance Feedback
741fb.62

Making Decisions. Decision-making takes many forms depending on the problems faced, work constraints, and investments required to achieve goals. It's an iterative, deliberative process grounded in performance feedback and learning from experience.

more information will provide indisputable evidence. That rarely happens. Making decisions involves choices and the possibility of being wrong. This frightens some organizations and leads to a protracted, conservative approach to making decisions.

No single approach or model to problem solving guarantees success. Typical actions include (modified after Keeney 1992 and NRC 1996b):

- Identifying the problem.
- Specifying initial objectives, values, and goals.
- Evaluating solution alternatives.
- Iterating between objectives, values, goals, and alternatives.
- Selecting preferred alternative(s) and beginning work.
- Solving the problem.

While these actions seem obvious, completing them is not easy. Sometimes, just agreeing what the problem is, much less when a problem is solved, is difficult. These actions assume we have

- Engaged critical partners.
- Overcome funding and technical constraints to problem solving.
- Tracked progress.
- Involved independent oversight.
- Forged clear lines of authority and accountability.

Each action is described.

Identify the Problem: The first challenge in building a problem-solving language is to define the right problem. Problems are usually general statements identifying concerns. Radioactive waste that leaked from tanks, contaminants found in the soil, nuclear materials stored inside aging buildings, or contaminated groundwater seeping into the Columbia River are examples of problems—if they present unacceptable threats to health or values. Problem identification must be rooted in reality—that is, can something be done to improve the situation? No reason to spend time on an unsolvable problem or one for which benefit is not quasi-commensurate with effort.

Most problems also exist inside a larger context. Cleaning up one waste site inside a host of others may not make sense. Hanford's oldest solid waste burial grounds hold mixtures of transuranics, combustibles, and other potentially dangerous materials. While it's desirable to remediate these sites, the risk of exhuming materials might be too high unless new worker safety and waste-handling technologies are developed. Similarly, the removal of non-liquid waste from tanks requires introducing fluids to create a slurry that can be pumped. However, adding fluids could lead to more unwanted tank leakage.

Specify Initial Objectives, Values, and Goals: Objectives help us answer the "why" for cleanup, values help us decide what is important, and goals tell us whether our objectives are achieved. Agreeing on these keeps work from drifting in a decision vacuum.

Sometimes decision-making is held hostage to the hope that the weight of more information will provide indisputable evidence. This rarely happens.

Making decisions involves choices and the possibility of being wrong.

Objectives: An objective is a fundamental statement of what one desires to achieve. It establishes "a sound basis" and a "stable reference point" for making decisions (Keeney 1992). Objectives give context to decisions and answer the question, "why is money being spent?" They determine the capabilities required, investments needed, and help identify who the key decision-makers are.

In an everyday example, one objective in purchasing a new car might be safety. Others include style, personal comfort, utility, or gas mileage. Objectives are a matter of choice.

Values help us decide what is important. They are the principles used for making judgment.

One reported objective for environmental management at Hanford "is to safely cleanup and manage the site's legacy waste" (DOE 1999a). The primary objectives from DOE's plan to accelerate cleanup and reduce the size of Hanford are "restoring the Columbia River corridor, transforming the Central Plateau, and preparing for the future" (DOE 2000a).

An objective such as minimizing contaminant releases could emphasize the installation of barriers separating the public or the Columbia River from contaminants. An objective of maintaining public risk at acceptable levels might involve leaving contamination where it is and monitoring it or engaging in a treatment effort. A range of actions could achieve the same objective.

One must be cautious not to confuse cleanup activities with cleanup objectives. Is data collection an objective? No. Information is input to problem solving. Is regulatory compliance an objective or a means to an end? There is a difference.

Values: Values are the principles used for making judgments. They encompass our feelings about potential consequences. Values may originate from ethical standards, cultural backgrounds, theological beliefs, or attitudes towards the acceptance of risk.

Public values tend to be personal, while those expressed by agencies commonly stress economic outcomes or avoiding risk. We instinctively use values to help make decisions, though identifying them requires close attention. Values might include protecting children or workers, surface waters, farmland, aquatic life, business opportunities, or agency reputations.

Is regulatory compliance an objective or a means to an end? There is a difference.

Goals: Goals should be identified early in the decision-making process and be open to change as experience is gained. They allow us to track progress and to know when work is complete.

The knowledge and technology needed to attain goals are quite diverse. One can easily estimate a building's completion by comparing what's been constructed to blueprints. It's also simple to collect a groundwater sample and compare its analyses to a standard. On the other hand, how does one gain consensus about whether the amount of a contaminant estimated to move through the ground thousands of years from now is acceptable?

Goals allow us to track progress and to know when work is complete.

Sometimes Hanford restoration goals are given in cubic yards of contaminated soil moved or gallons of groundwater pumped (DOE 1999a). However, such goals are meaningless without reporting the amount of *actual* contamination recovered compared to that remaining behind. When contaminant amounts are impractical to measure, other goals such as reduction in radiation dose or chemical exposure might be used. Regardless, the most useful goals are traceable and measurable.

A challenge facing decision-makers stems from choosing between or balancing different goals. Examples include

- **Risk goals**—reducing or eliminating risk from exposure to contaminants. Risk goals range from minimizing immediate risk to lessening risk to future generations.
- **Environmental protection goals**—preserving plant and animal species, reducing loss of sensitive ecosystems, or minimizing contaminant flow into the Columbia River.
- **Contaminant recovery goals**—removing all or some percentage of contamination from a waste source.
- **Mortgage reduction goals**—reducing the cost of maintaining old facilities, monitoring waste, or managing nuclear materials.
- **Economic goals**—preserving jobs, achieving cleanup at the lowest possible cost, or developing Hanford for future uses.
- **Regulatory goals**—following regulatory requirements and meeting milestones.
- **Land-use goals**—diversifying land use or preserving certain areas from development.

Goals can be updated. A good example is found in Hanford's approach to reactor dismantling. The original environmental impact statement and Record of Decision called for dismantling eight of the nine Hanford reactors (DOE 1992c). At that time, the 12,000-ton reactor cores were to be loaded onto massive crawler-type vehicles and moved as far as 15 miles for burial. This expensive undertaking was re-examined. Subsequent engineering studies led to the concept of encasing the reactor cores and storing them in place for 50 to 75 years.

By 2002, C and DR Reactors were successfully placed in interim storage. Work is under way on most other reactors. "Enclosing the structures and letting the radioactivity decay away dramatically reduces the money required for building maintenance, avoids disturbing uncontaminated Hanford Site land, and protects workers and the environment from removal of a still-radioactive core" (DOE 1998g). These actions satisfy several restoration goals.

Evaluate Solution Alternatives: This is the diagnostic stage of decision-making where information is gathered, analyzed, and compared to potential cleanup alternatives. What makes one approach, or combination of approaches, preferable? The public, engineering community, researchers, and others must help decision-makers identify workable and preferred alternatives.

This is also where personal or organizational agenda setting ("stacking the deck") can take deep root, especially when comparing costs. "Code" words such as quick, simple, popular, or inexpensive tend to describe pre-favored alternatives while complicated, lengthy, controversial, or costly are associated with less-favored approaches. Comparisons should be technically defensible and honestly done.

Total system life-cycle costs should be compared; that is, the cumulative cost of construction, operation, product disposal, maintenance, and monitoring using one alternative versus another.

The examination of alternatives is carried out within the context of legal, social, cultural, and political expectations. Alternatives not scientifically sound or that violate ethical principles should be rejected.

Code words of bias such as "quick," "simple," "popular," or "inexpensive" tend to describe pre-favored alternatives while "complicated," "lengthy," "controversial," or "costly" are associated with less-favored approaches.

Total cleanup life-cycle costs of alternatives should be compared.

Comparing alternatives may lead to redefining cleanup objectives and goals. This is part of learning when first opinions may not necessarily be the most informed opinions.

Iterate Between Objectives, Values, Goals, and Alternatives: During problem solving, it's important to periodically examine where we are and whether we wish to remain there. Superfund cleanup experience demonstrates that much is learned about a waste site as experience is gained. This is part of a self-correcting process to ensure the best alternatives are preserved and associated benefits received. It is especially important to secure agreement on what the end point of a cleanup action is versus working under an uneasy truce that collapses once problem closure is sought.

Looking at Alternatives for Tank Waste

Closer examination of potential cleanup alternatives for Hanford's tank waste has been proposed (NRC 1996a, 1999a). This does not suggest any delay in preparing for long-overdue tank waste solidification. Rather, complementary approaches are a sign of a healthy decision process when unknowns run high. And because the first gallon of tank waste has not been turned into glass, there are a lot of unknowns.

About 99% of the long-lived radionuclides such as plutonium-239, technetium-99, or americium-241 that pose the greatest long-term risk are found in 40 of Hanford's 177 tanks (DOE 1998e). Assuming this is correct, then the potential environmental risk in all tanks is not equal. Perhaps this is why, in 2001, DOE's Headquarters began floating ideas on how to significantly reduce the cost and time for cleanup of the nuclear weapons complex, particularly in tank waste, as it is the single largest expense in the environmental management program (Roberson 2001; Stang and Cary 2001). Alternatives to vitrifying all tank waste (perhaps reducing by 75% the volume of tank waste vitrified) were raised.

"Informed, defensible decisions on how to retrieve tank wastes and close the tank farms cannot be made until the risks to the vadose zone, groundwater, Columbia River, and, ultimately, human and ecological health and safety are understood" (HAB 1997c).

This is not new news. In 1993, the Hanford Tank Waste Task Force cautioned about relying on a single cleanup approach: "Keep a folio of technological options and make strategic investments over time to support a limited number of promising options…when a better option becomes known through an open and credible systems design and R&D [research and development] process, be willing to adopt it" (Hanford Tank Waste Task Force 1993).

After reviewing the draft environmental impact statement for Hanford tank cleanup, the National Research Council (1996a, pp. 2-3) wrote: "Not enough is known at this time to choose a final, long-term strategy for management of all of the Hanford Site tank wastes…Significant uncertainties exist in the areas of technology, cost, performance, regulatory environment, future land use, and health and environmental risks…additional alternatives for management of the tank wastes need to be explored in parallel [to waste vitrification], using a phased decision strategy…Among additional options that should be analyzed are (1) in-tank stabilization methods that are intermediate between insitu vitrification and filling of the tanks with gravel, (2) subsurface barriers that could contain leakage from tanks, and (3) selective partial removal of wastes from tanks with subsequent stabilization of residues, using the same range of treatment technologies as in the alternatives involving complete removal of wastes."

Three years later, the National Research Council (1999a, pp. 5, 7) reflected upon their previous concern: "At present, many public stakeholders at Hanford apparently want DOE to follow the current compliance-driven Hanford baseline approach [for tank cleanup], and they view investment of significant resources in technology development for alternative scenarios as a diversion from that effort…The committee recommends a limited parallel pursuit of alternatives (including end states) to the current strategy…The uncertainties and costs are great and the chances for unacceptable results are too high not to pursue viable alternatives."

The Hanford Advisory Board raised another cautionary note: "informed, defensible decisions on how to retrieve tank wastes and close the tank farms cannot be made until the risks to the vadose zone, groundwater, Columbia River, and, ultimately, human and ecological health and safety are understood" (HAB 1997c). As noted in Chapter 10, this is not off-the-shelf knowledge.

After examining alternatives for processing tank waste at the Savannah River Site, South Carolina, and following 4 years of operating the site's vitrification plant (Defense Waste Processing Facility), the National Research Council (2000c, p. 84) wrote, "instead of blending tank wastes to produce a feed that might allow all tank contents to be treated by a single process…would it be advantageous to tailor processing based on chemical and radionuclide contents of individual tanks?…Although this tailored approach might require additional regulatory approval and perhaps some facility modifications (e.g., the construction of additional waste transfer lines), it might allow the tank wastes to be processed on a faster schedule, thereby reducing costs and freeing up tank storage space."

Hanford faces similar issues.

Backup plans and alternative solutions should be kept at hand to provide an in-depth defense.

Select Preferred Alternative(s) and Begin Work: Once an alternative is chosen, work begins. Progress is constantly reviewed. Rarely are choices obvious or tradeoffs easy, especially about controversial issues viewed with passionately held beliefs. Selecting a single alternative may be contentious.

First opinions may not necessarily be the most informed opinions.

Normally, a one-size-fits-all approach to problem solving is risky and invites failure. This is why financial advisors diversify their investments to achieve the greatest return while lessening risk. Cleanup is similar. Because perfect knowledge does not exist, and many cleanup problems are unique, backup plans and alternative solutions should be kept at hand to provide an in-depth defense. The absence of contingencies is a recipe for disappointment and frustration.

Experience is part of a self-correcting process to ensure the best alternatives are preserved.

Solve the Problem: This is the moment of truth. If there was any illusion lingering over objectives, values, progress, or goals, they'll manifest themselves when someone asks others to "sign on the bottom line" and stop work. Problem closure must be demonstrable to all key parties. If conflicts were swept under the rug, this is when they'll return in full force.

At first, some problems seem impossible to solve. As former Atomic Energy Commission Chair David Lilienthal said, "In the course of observing *what* men have done, I have had a chance to observe and reflect upon *how* they get things done. I have found it is by what I call the technique of the manageable job, in short the art of achieving the possible, and thereby of bringing the impossible progressively closer to our grasp…the manageable job…is not that of seeking some single, final solution. It is rather the far more arduous and earthy one of seizing the many opportunities to act" (Lilienthal 1963, pp. 7-8).

If there were any illusions, they'll manifest themselves when someone asks others to "sign on the bottom line" and stop work.

If after review, it is the judgment of the lead regulatory agency that additional remedial action is appropriate, then that agency may require more work (Washington State Department of Ecology et al. 1998). Therefore, though agreed-on actions are undertaken, a continuing re-evaluation process is built into regulations that can call into question past decisions and cleanup actions.

Precautionary Principle

In the 1990s, much was written about the precautionary principle—the need to take action in spite of unknowns. There are times when society cannot wait for all questions to be answered. The environment must be protected, courts must resolve disputes, regulations must be enforced—all based upon best available evidence (Park 2000, p. 45). We choose to take risk, to select a path forward, hoping our course of action is correct but knowing that it might result in a costly or harmful mistake.

The precautionary principle promotes solving environmental problems recognizing that "(1) science will not always provide the information needed in a timely manner, (2) the cost of preferred actions may be prohibitive, and (3) biological resources will be lost as one waits for more knowledge to make the decisions" (Risk Excellence Notes 2001). Taking actions when outcomes are unclear requires follow-up research to determine the action's effectiveness and provide lessons learned. However, the "more precautionary we are, the more likely we will be wrong" and have missed taking the most appropriate action if more knowledge had been available (Goldstein 2001).

There are times when society cannot wait for all questions to be answered. The environment must be protected, courts must resolve disputes, regulations must be enforced—all based upon best available evidence (Park 2000, p. 45).

Chapter 10

Science: Partner in Choices

"Perhaps the biggest environmental problem to be solved...is our collective inability to apply good science to policy making."
—Michael Carlowicz, *EOS Transactions* (1996)

Science is an indispensable partner in making informed choices. Those versed in the sciences have several responsibilities. First, they are providers of factual information; second, they explain what this information means; and third, they are informed citizens working with the public and policy-makers.

Science is more than just the producer of knowledge. It's a way of thinking. Science's foundation of testing, fact-finding, and critical thinking is a powerful ally for making decisions. Great science extends understanding beyond the obvious or accepted patterns of thought. If it can't be tested or if it does not make the world more predicable, then it's not science.

The value of science is in the knowledge it provides and the critical thinking it encourages.

However, science does not assign value, does not evaluate the moral desirability of conditions, and does not tell us what we should do. Those are personal decisions where choices may not match the weight of scientific evidence. The application of science and technology is deeply influenced by social, cultural, and religious dispositions (Bosch 1991). Science rests within the context of interpretation. Experts can more effectively serve society when they boldly step inside the world of public communication and politics.

Science can explain why an engineered system, such as a waste form, performs inside or outside of expectations but not whether those expectations are desirable. Science can describe the potential health impacts from exposure to toxins; however, it does not choose whether or not a given impact is acceptable.

The technically trained person has both a professional and an informed citizen role to play. Professionally, they should seek the underlying facts and reasonable extrapolations of those facts. As an informed citizen, experts should be engaged in important social issues taking care to distinguish between facts and personal opinions. Jim Burchfield, author and professor, drives these distinctions home when discussing the separate roles of researchers and policy makers: "Researchers should not be accountable for the consequences of actions, since they should not recommend actions in the first place....In a painful, repeated, and embarrassing display of timidity, elected officials and their hired administrators have done everything in their power to avoid making choices. Instead of doing their jobs, they pass the buck to science...From the point of view of administrators, it's been a sweet deal. Gone is the accountability derived from making choice" (Burchfield 2001, p. 241).

Elected officials and their hired administrators have done everything in their power to avoid making choices (Burchfield 2001).

Penny Colton, Pacific Northwest National Laboratory

Talking with People. There is no substitute for visiting Hanford, talking with people, and understanding cleanup from a range of voices. This is the author speaking to visitors atop Gable Mountain in the middle of the Hanford Site.

The integrity of science is anchored in the willingness of its practitioners to test their ideas against the knowledge of others. However, not everyone is happy in a world where available facts are laid on the table, where uncertainty is admitted, and agendas are uncovered. It's a world where claimants must produce evidence rather than remaining submerged in assumptions and personal judgments. The result is a much refined and workable understanding of the problems faced. Some findings are proven wrong. Others are supported. The best ideas move forward.

The integrity of science is anchored in the willingness of its practitioners to test their ideas against the knowledge of others.

Physicist Robert Park (2000) coined the terms fringe science, junk science, and voodoo science when referring to a host of ailments involving the foolish or even fraudulent misuse of science. For example, the highest quality science is "argued in the halls of research institutions, presented at scientific meetings, published in scholarly journals. Voodoo science, by contrast, is usually pitched directly to the [news] media, circumventing the normal process of scientific review and debate" (Park 2000, p. 26). Thus, the media become judge and jury. Without media visibility, voodoo science means nothing and its promoters lose influence and public attention. Adherents of voodoo science commonly isolate themselves from informed skeptics.

The highest quality science is "argued in the halls of research institutions, presented at scientific meetings, published in scholarly journals...Voodoo science, by contrast, is usually pitched directly to the [news] media" (Park 2000, p. 26).

Regretfully then, who does the public believe especially when a disproportionate share of the so-called science they see in the news media is flawed? Journalist Jim Hartz and scientist Rick Chappell (1997, p. viii) echo this concern: "Adequate coverage of science stories is rare, found in only a handful of news outlets." They call for journalists and scientists to work together, taking a new look at science and science communication. This is the only way the public can become "better equipped to understand and participate in the growing debates" that face society, that require looking critically at issues rather than accepting opinions at face value.

The scientific method provides an invaluable guidepost for rigorous deliberation and in keeping policy-making from floating in a sea of currents, shifting with each actor and the accepted norms of the moment (Park 2000). Critical thinking helps transcend the flaws of individuals and organizations.

Critical thinking helps transcend the flaws of individuals and organizations.

Nonadvocacy peer reviews are a vital part of maintaining a credible decision-making process. The National Research Council says peer reviews should be conducted by people with technical expertise in part or all of the subject matter in question "who are independent of and external to the program of work being reviewed" (NRC 1999e, p. 8). Colson et al. (1997, p. 1.6), who proposed a risk-based decision process for Hanford tank waste characterization, caution about the same organization requesting a project review also accepting or rejecting the very recommendations that could affect their business. "This is a clear conflict of interest, making it difficult to implement changes, especially fundamental changes, in how an organization conducts business." The General Accounting Office (1998a) stresses not only reliance on independent reviews but also greater diligence in responding to and implementing expert panels' recommendations.

"The new [U.S. Department of Energy] culture…will emphasize an open door philosophy and demand professional excellence in both government and contractor performance, and it will be a culture wherein constructive criticism from any source, external as well as internal, is encouraged and rewarded" (DOE 1990, p. 56).

For years, experts were seen as infallible holders of immutable truths (Lilienthal 1963, p. 63). People were impressed by those whose words and logic were not widely understood. Obviously, they must know what they're talking about, whether in matters of their specific discipline or the broader arena of public affairs. To avoid introducing nonscientific elements into the equation of Hanford operations, public concerns and discourse were kept at a comfortable distance, outside the curtain of informed consent. Waste handling, contaminant releases, and risk evaluations were left to expert judgment (see Chapter 2). This began to change in the United States, and at Hanford, in the mid-twentieth century. One of the public's first rallying points was radioactive fallout from the atmospheric testing of nuclear weapons.

Atmospheric testing of nuclear weapons caused a sharp increase in radioactivity raining down on the world.

In the late 1950s, congressional debates raged about the potential health risk from fallout. Atmospheric testing of nuclear weapons was near its peak.[1] This caused a sharp rise in radioactivity raining down on the world, especially in the Northern Hemisphere. Hearings were held and news bulletins issued about high levels of radiation, especially strontium, found in milk, bread, and garden vegetables. Public concern grew. People were advised not to drink rainwater or children to eat freshly fallen snow. This controversy led to the tightening of radiation exposure levels and the Limited Test Ban Treaty of 1963. An excerpt from one testimony, written by Barry Commoner,[2] former professor of botany at Washington University in St. Louis, addressed the role of scientists in this public policy issue (Commoner 1959, pp. 2575-2576). If one reads this quote in a fuller context (thinking of environmental cleanup rather than fallout and of a range of experts rather than just scientists), it uncannily parallels many of the science, decision making, and public engagement challenges surrounding today's cleanup work.

(1) Plutonium production and waste generation at Hanford were also peaking.
(2) Barry Commoner was one of the founders of the Scientists' Institute for Public Information that promoted greater public awareness of technological hazards.

"What is the source of public confusion on the fallout problem? In the past few years, and especially during the last presidential campaign, the public has become aware of a political cleavage on the wisdom of continued testing of nuclear weapons. Political controversy is a natural, expected, and welcome part of public affairs in this country. What appears to trouble the public is not that political opponents have disagreed on the nuclear test issues but that the opinions of scientists have been marshaled on both sides of the debate. This appears to violate science's traditional devotion to objectively ascertainable truth.

"This division is in part due to the factual uncertainties that have already been discussed, and the public concern may reflect an awareness of these uncertainties. However, this difficulty results as well from some confusion concerning the scientists' two roles in these matters. As a student and interpreter of nature, the scientist can explain to the public what consequences may result from a given policy that affects nature. As an informed citizen, the scientist has the right and the obligation shared by all citizens to form and express an ethical judgment on the wisdom of enduring that policy. Estimation of the probable damage to health that might result from the continuation of nuclear weapons tests is a scientific question. But there is, I believe, no scientific way to balance the possibility that a thousand people will die from leukemia against the political advantages of development of more efficient retaliatory weapons. This requires a moral judgment in which the scientist cannot claim a special competence which exceeds that of any other informed citizen…

"…the public must be given enough information about the need for testing and the hazards of fallout to permit every citizen to decide for himself whether nuclear tests should go on or be stopped. It is the natural task of the scientists and their professional organizations to bring the necessary facts and the means for understanding them to the public…

"What we need now is to marshal the full assemblage of facts about fallout, their meaning and uncertainties, and report them to the widest possible audience. This is not an easy task. It is much simpler to publicize conclusions alone, and have them accepted not because their factual origin is fully understood but because they carry the authority associated with science.

"It seems to me that we dare not take the easy way out. Unless the public has sufficient information to provide a reasonable basis for independent judgment, them [sic] oral burden for the future effects of nuclear testing will rest on some smaller group. And no such group alone has the wisdom to make the correct choice or the strength to sustain it. Unless the public is made aware of the gaps and uncertainties in our present knowledge about fallout, we cannot expect to support the expensive research needed to minimize them. Without public understanding and support, no Government policy can long endure."

"Without public understanding and support, no Government policy can long endure" (Commonor 1959, p. 2576).

Scientifically based information can help accelerate the decision-making process (NRC 2001b). Similarly, openly conducted reviews by qualified experts are the hallmark of high-quality research. Overall, the system works amazingly well because "good work eventually rises to the top" while the rest "remains manageable" (Park 2000, p. 15).

10.1 Need for Innovative Knowledge and Capability

"Even though the department [Department of Energy] spends between $5.6 billion and $7.2 billion per year on cleanup, Congress has seen fit to give it a pitiful sum for research on more effective technologies...officials have relied on traditional methods such as excavation of contaminated soil or pumping out contaminated water. But they often don't work...If the contaminated sites are ever to be cleaned up, it must be made easier to bring into play new technologies that show a reasonable chance of success."

—*Seattle Post-Intelligencer* (June 27, 1999)

No new science and technology are needed to clean up the nuclear weapons complex unless we care about costs, schedules, safety, and understanding the hazards left to future generations.

Now this might sound radical, but no new science and technology are needed to clean up the nuclear weapons complex. That's true *unless* we care about reducing cost, accelerating schedules, ensuring safety, and understanding the hazards left to future generations.

During the years of plutonium production, our nuclear arsenal grew in response to immediate, military threats. This fueled research investments such as larger reactors, more efficient reprocessing plants, and weapons of megaton yields. It also pioneered world-class research into the biotoxicity of radiation and the behavior of contaminants in the environment (see Sections 3.1 and 4.1). Today, the site-specific, internally imposed mandates of environmental protection drive investments. Many threats faced by this and future generations are not overt but are steeped in probabilities and assumptions.

Technology is the capability created from knowledge.

Innovative science and technology are needed to clean up the nuclear weapons complex (Western Governors' Association et al. 2000). This concern is echoed by the League of Women Voters (1993, p. 110), "the challenge of cleaning up at Hanford lies not only in the vast amount of waste involved, but also in the lack of proven, economical technologies available to address many of Hanford's problems...contaminants will have to be contained and monitored until technologies are developed to remediate these sites." The National Research Council (2001b, p. 2) reports, "the knowledge and technology needed to address the most difficult [subsurface cleanup] problems...do not yet exist." The Congressional Budget Office (Glass and Snyder 1995) reported that using nonconventional methods for site investigations and cleanup at military defense sites could cut costs by 50% or more.

Yet, environmental statues and federal regulations do not encourage the development of innovative technologies (NRC 1997). Rigid milestones and statutes often lack flexibility and lag behind needs.

Regretfully, research is commonly the last item added to cleanup budgets and the first item cut.

Regretfully, research is commonly the last item added to cleanup budgets and the first item cut. It's often viewed as a luxury with questionable or only long-term benefits rather than a necessity; a bandage applied when problems bruise plans. Why? Reasons include budget shortfalls, absence of contract incentives, skittishness about implementing new technologies, and perhaps a lack of cleanup milestones elevating research visibility. Sometimes, science requires long-term investments with benefits gained during another administration's watch or another contract extension. Who is the watchdog for these investments and is accountable for their use (NRC 2001b, p. 4)?

Neither a zero-risk-of-failure ("wait until all needed data are in hand") option nor a high-risk-of-failure ("go quickly") option should prevail in selecting technologies (NRC 2000c, p. 86). The best approach is to couple cleanup and research in a hand-in-glove partnership to realistically schedule work while minimizing life-cycle costs and risks. Research spans both the short and long term. This includes solving immediate problems as well as examining the range of conditions likely encountered to ensure work-around solutions are available before problems halt work.

Here are four examples of innovative science and technology investments that are now paying rich dividends at Hanford or other sites:

Research has both immediate and long-term payoffs.

- Today's knowledge about glassifying high-level waste, on a large scale, was launched by research at Hanford starting back in the late 1960s (McElroy et al. 1976, 1982). These far-sighted investments enabled the Savannah River and West Valley sites and countries such as England, France, and the former Soviet Union to later deploy vitrification technology.
- The idea of chemically manipulating an aquifer's oxidation-reduction[3] potential to create a reduction zone that can treat redox-sensitive contaminants was proposed in the early 1990s (Fruchter et al. 1992). The concept involved injecting a nontoxic chemical (sodium dithionite) into wells penetrating a groundwater plume contaminated with toxic and mobile hexavalent chromium. By 1995, a proof-of-principle field experiment was deployed in the 100-D Area of Hanford. The treatment immobilized the hexavalent chromium to nonmobile trivalent chromium. In 1998, this treatment site was transferred to a Hanford contractor for use. Today, it is one of Hanford's most successful in situ groundwater cleanup projects.
- From 1996 to 1999, the effect of radiation on tank waste solutions was studied to deliver fundamental chemical data that helped resolve Hanford tank waste safety issues such as the generation of explosive gases (Orlando et al. 1999). It's now available for studying tank waste treatment options.
- Experimental research into the sorption chemistry of radioactive cesium-137 leaked from Hanford tanks began in 1997. Laboratory studies, coupled with sediment sampling collected from beneath tanks, are now providing improved models and descriptions of the field-scale behavior of cesium used to support regulatory decisions involving the movement of cesium in the subsurface (Zachara et al. 2002).

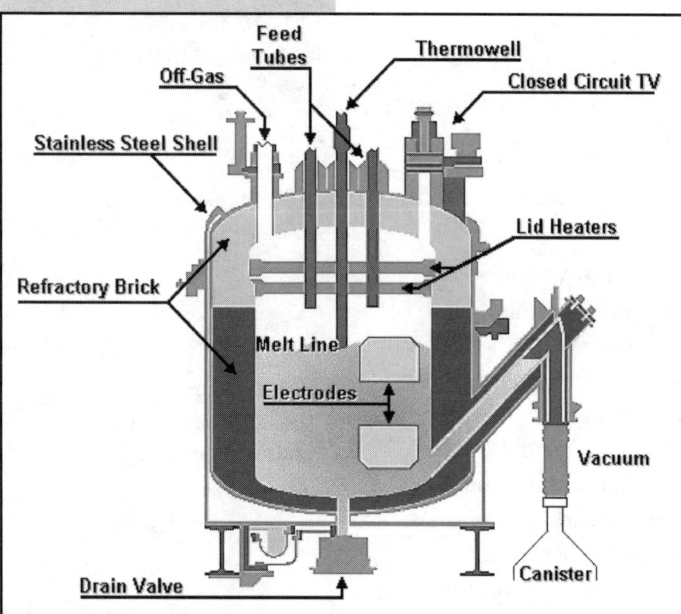

Vitrification Technology. In 1996, the vitrification of tank waste began at the Savannah River Site in South Carolina. This illustration shows their melter design where a mixture of waste and glass-forming chemicals are electrically heated into a thick liquid. The melted material is then poured into a stainless steel waste canister and stored. Research leading to this technology was conducted at Hanford in the 1960s.

(3) Oxidation-reduction (or redox) is a chemical reaction in which there is either a transfer or sharing of electrons to form bonds between atoms. The atom that loses an electron(s) is said to be oxidized. The atom that gains an electron(s) is said to be reduced.

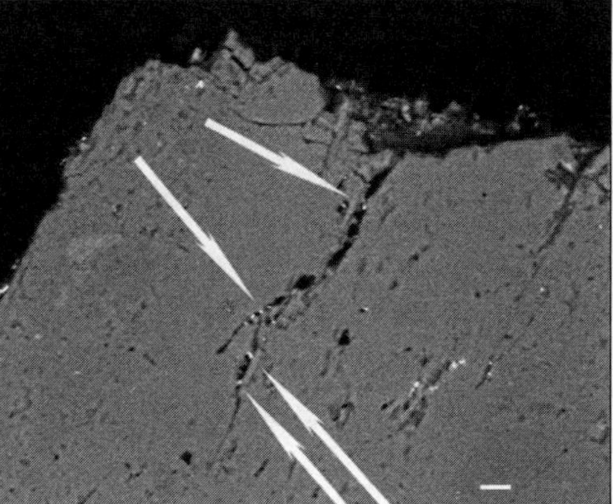

James McKinley and John Zachara, Pacific Northwest National Laboratory

Uranium Precipitates Beneath Hanford Tank. These are scanning electron microscope images of two types of minerals (mica on the left and quartz on the right) found to contain a form of uranium (uranium +6) sampled 136 feet beneath Tank BX-108. The uranium exists as precipitates in minute mineral cavities and fractures. These precipitates are shown as small white dots at the end of each arrow. Their formation slows uranium migration underground. Such information aids our understanding of which contaminants might move in the environment to pose future health risks. The white bars on the lower right of each image represent 10 microns in length (one-tenth the width of a human hair).

On a site-by-site basis one finds other technology deployment accomplishments. Examples from Hanford are reported in *Hanford Technology Deployment Accomplishments* (DOE 2002e).

However, in reviewing the U.S. Department of Energy's nationwide program for environmental science and technology development, the National Research Council (1999e, p. 112) commented that most efforts are "not directed at techniques outside the current [baseline]." Furthermore, there is a "significant disincentive to change [existing cleanup approaches]...and to work on different approaches" even though, where justified, replacement technologies could significantly reduce costs, schedules, and risk.

On a national level from 1989 to 2000, the U.S. Department of Energy made $3.4 billion in science and technology investments to reduce the estimated $200-billion to $250-billion cost of cleanup. What was gained? The U.S. House of Representatives writes about a "substantial graveyard" of funded technologies that failed to be deployed even though a number of them are viable and available (U.S. House of Representatives 2000). Of the 918 technologies funded since the late 1980s, 31 of them (3% of total) were used in the field more than 3 times, and only 4 technologies were used 10 times. Of the technologies applied, more than half were used only once. Interviews with contractors revealed few planned uses of the U.S. Department of Energy-funded technology.

The underground treatment of chromium is one of Hanford's most successful in situ groundwater cleanup projects.

Funding research is like investing in your retirement, that is, the best choices for achieving financial goals are based on being well informed, thinking critically, and managing investment risk. While a few people might strike it rich using a single high-risk investment scheme, the proven approach involves continued investments into a balanced and diverse portfolio.

Knowing how and if certain contaminants can move through the subsurface beneath tanks can have multi-billion-dollar impacts. Studies of contaminant distribution and retention beneath single-shell tanks, such as documented for cesium-137 by McKinley et al. (2001), are one focal point for scientific input into decisions. They help nail down the answer to the question of what will happen to leaked waste. Can waste be hydraulically removed (sluiced) from aging tanks without releasing unacceptable amounts of contaminants into the soil or chemicals that might remobilize contaminants from previous leaks? How much waste might remain inside tanks and still protect the environment? Are engineered barriers needed?

"DOE's historical problems in deploying innovative remediation technologies...have parallels in other sectors" such as the U.S. Environmental Protection Agency's Superfund program (NRC 1999d, p. 203).

Existing technologies are used in spite of the potential benefit of new technology.

Why such a dismal record? Reasons include:

- Technology planning occurs in a complex, ever-changing, and politicized environment fueled by multiple customers.
- Portions of the development expenditures are subject to the variations of congressional budgets.
- Site managers and contractors are under no obligation to use these technologies.
- Technologies were often developed without considering the needs at individual sites.
- It is difficult for private companies to penetrate the institutional barriers protecting the status quo of the nuclear weapons complex to provide improved technologies (NRC 1999e; U.S. House of Representatives 2000).

Separate from these studies, the National Research Council (1997, 1999d) addresses basic obstacles faced by U.S. Department of Energy, as well as private industry, to develop and deploy new technologies:

Lack of customer demand: The primary obstacle is the lack of customer (end user)[4] demand. Contractors are not required nor rewarded to search for cost-effective solutions. Though contractors can receive limited funding for conducting technology demonstrations, incentives are often lacking to implement cleanup using innovative technologies.

Public and regulatory resistance: "Site regulators and vocal members of the public have also limited the application of innovative remediation technologies at DOE sites....The regulatory requirements for selection of cleanup remedies under CERCLA and the Resource Conservation and Recovery Act (RCRA) also have been faulted for limited use of innovative technologies" (NRC 1999d, p. 206). Existing technologies are used in spite of the potential benefit of new technology. There is a natural hesitancy to avoid the risk of being the first to try something new. Who is liable should it fail—especially in a compliance-driven environment? Conventional methods are more comfortable even if they're inefficient. Thus, the regulatory community favors traditional technologies. When selecting technologies, criteria require pre-existing cost and performance data. This restricts the use of innovative technologies in favor of conventional approaches where performance is documented. Besides, changing a cleanup remedy is a cumbersome process—potentially attracting a lot of unwanted attention.

(4) An "end user" is the customer for an innovative technology—a company or organization committed to using the new capability.

Funding through the "valleys of death": Little investment is available to shepherd innovative ideas through the distinct phases of research, development, field testing, and deployment. Each step is a "valley of death" where funding is rarely available to bridge the interfaces. Many technology development companies fail because moving innovative capabilities across these gaps is herculean (NRC 1997).

Little investment is available to shepherd ideas from research to deployment.

In recent years, Congress has slashed the U.S. Department of Energy's budget for scientific studies and technology developments for cleanup across the nuclear weapons complex. In 1995, its budget was $410 million (NRC 1999d, p. 20). By 1998, this dropped to $274 million. A senior U.S. Department of Energy official expressed confidence this was an appropriate level of funding and that efficiency improvements would make up the difference (Owendoff 1998). The budget has now dropped below $100 million.

"It is time to create a 'learning culture' which empowers those with a stake in successful cleanup at Hanford, including workers, to constantly evolve new, applicable, and efficient management policies and technologies that lead to even more environmentally sound cleanup. It is time to 'get on with it'....Getting on with it means that we make use of available technology and resources now, and that we do so without precluding future application of emergent technology. We must do well all that we know how to do, and we must persist in seeking answers for the questions that remain...that which is unknown must be acknowledged so that our research and development energies might be clearly focused and wisely applied" (Hanford Future Site Uses Working Group 1992).

The National Research Council (2000a, p. 116) reports that the U.S. Department of Energy's own research program supporting cleanup "does not appear to be a high priority, judging by EM's [Office of Environmental Management] budget requests to Congress."

Some reviewers of Hanford projects recommend a sustained long-term research and development program be instituted with expenditures keyed to a percent of life-cycle cost savings based on an assessment of problems faced and the potential benefit of delivering new capabilities (Integration Project Expert Panel 2000a).

The U.S. Department of Energy's own research program supporting cleanup "does not appear to be a high priority" (NRC 2000a, p. 116)

In 2002, the U.S. Department of Energy Office of Environmental Management announced plans to refocus science and technology investments "to directly address the specific, near-term applied technology needs for cleanup and closure. Longer-term or more basic research and technology activities...should be transferred to other DOE [U.S. Department of Energy] programs" (DOE 2002). The workability of such an approach is unproven.

Compliance with existing cleanup agreements and contracts using existing capabilities has become a de facto operating principle. This problem is not new. It has plagued cleanup programs for years (Office of Technology Assessment 1989, 1991; EPA 1991).

Chapter 11
A Matter of Risk

"Deciding what level of risk ought to be acceptable is not a technical question but a value question."

—Vincent Covello,
Journal of Occupational Medicine (1993, p. 19)

Risk, like beauty, is in the eye of the beholder.

Risk is the chance of injury, harm, or loss. Knowing risk is like strolling under a streetlight: it illuminates our path but doesn't dictate where we walk. We can use or ignore the light it offers. Zero risk is a myth, and while believing it feels good, it's a fantasy.

Risk is like a streetlight: it illuminates our path but doesn't dictate where we walk.

An understanding of risk helps us sequence cleanup activities and allocate funds (Fields 1998). Its use is practical (CRESP 1999, p.4). Risk is combined with other considerations—such as regulatory requirements, social preferences, ecological impacts, costs, treaty rights, and the future value of affected resources—to guide decision-making.

What is risk, how is it perceived, how is it faced in our everyday lives, what is the difference between risk assessment and risk management, and how well has risk been handled as a decision-aiding tool across the nuclear weapons complex? These topics are discussed.

11.1 What is Risk?

"Life is a weird mixture of bottled water, whole-wheat bread, and complex dietary supplements broken up by bouts of coffee drinking, dessert consumption and car travel."

—Mark Kingwell,
The New York Times Magazine (1999)

Risk is estimated by knowing the likelihood and consequence of something unwanted happening.

Nearly anything we do, eat, or breathe is potentially harmful.

Sometimes we confuse risk with the mere existence of something dangerous. They are not the same. Under certain conditions, nearly anything we do, eat, or breathe is potentially harmful.

For example, most common chemicals, from salt to vitamins, are hazardous if delivered in large enough doses. Every day we come into contact with them without suffering harm. A lake is potentially dangerous because someone might drown. However, the person must first enter the lake. Contact, not just existence, can place someone in harm's way. If contact is prevented, risk is avoided. If a person knows how to swim, wears a lifejacket, or wades only in shallow water, then risk is managed.

Contact, not just existence, places someone in harm's way. If contact is prevented, risk is avoided.

Likewise, contaminants pose a threat if they are inherently hazardous and come into contact with humans in harmful concentrations. Destroying contaminants or chemically binding them inside solid waste forms is one risk-reduction solution. Institutional control of contaminants (for example, inside buildings) is adequate for the short term but questionable over the long haul. One can also reduce risk by controlling or cutting off a contaminant's pathway to a receptor. Risk is lowered by reducing the amount of contaminants available for release—like removing unwanted combustibles to reduce the severity of a fire. Firefighters have a host of options to extinguish flames or lower risk of fire. So do waste managers (see Section 8.1.2).

> Risk is also applied in ways other than depicting potential harm to health. What is the probability of losing money from investing in future commodity prices (financial risk), a project failing to achieve goals (programmatic risk), a technology not working (technical risk), or a legally mandated milestone not achieving its goal (regulatory risk)?

An evaluation of risk requires analysis and deliberation. It's estimated from knowing the probability (likelihood) that harm will take place and the consequences of that harm. Informally, a third input is sometimes factored into risk—outrage. These factors are multiplied together to estimate risk.

> **Risk** = **probability** (likelihood) *x* **consequence** (harm) *x* **outrage** (fear)

Probability: Probability gives the numerical odds of an event happening. It expresses scientific meaning in the form of numbers. When a coin is flipped, the probability it will land on heads is one out of two, or 50%. We are never certain what will happen next; we only know the odds and accept uncertainty as part of the game. The better the odds are known, the better you'll play the game.

Consequences: Built into negative consequences are estimates of exposure and hazard. A contaminant must reach a person to expose them plus it must be inherently hazardous at the concentration delivered.

Choosing a basis for assessing consequence may sound simple but is surprisingly complex because of personal values. While death is a commonly used consequence, others include injury or impacts to property values, aesthetics, recreational opportunity, job availability, or cultural heritage.

Moral outrage is the silent partner in risk.

Outrage: Moral outrage is a strong force. This silent partner in defining risk can outweigh either probability or consequence. Outrage includes those human intangibles causing fear, anger, or frustration. Contamination of the Columbia River is a prime example. While probability and consequence can be assessed based on expert judgment and scientific information, outrage emerges from deeply held feelings. A lack of trust and perceived political injustice fuels outrage.

11.2 Risk: Part of Everyday Life

"Risk-perception research documents that people view medical technologies based on use of radiation and chemicals (i.e., X-rays and prescription drugs) as high in benefit, low in risk, and clearly acceptable. However, they view industrial technologies involving radiation and chemicals (i.e., nuclear power, pesticides, industrial chemicals) as high in risk, low in benefit, and unacceptable. Although X-rays and medicines pose significant risks, our relatively high degree of trust in the physicians who manage these devices makes them acceptable."

—Paul Slovic, *Risk Analysis* (1993, p. 676)

We live in a "sea of common risks" with many risks, some of them high, considered acceptable, and beneficial (Washington State Department of Ecology 1991). Often risks of low probability are feared most. It's hard for people to understand, let alone distinguish, between risks, especially risks of small probability.

Often, it is the risks with the least likelihood of happening that cause the greatest fears.

We unconsciously make risk-based decisions from the time we get out of bed (possibility of slipping or having a heart attack), brush our teeth (exposure to fluoride), eat breakfast (consume fat, cholesterol, and preservatives), drive or bike to work (possibility of an accident), drink tap water (ingest chloroform, a by-product of chlorination), sip wine (contains ethyl alcohol) with dinner, or walk into the family room (contains radioactive radon gas) to watch television (exposure to radiation from TV set).

Risk-controlling strategies range from fundamental protections guaranteed under the Bill of Rights to speed bumps.

Health risks are woven around all activities whether we recognize them or not. Most risks are small and not worth fretting about. We instinctively take precautions and rely on experience and judgment. Only when we undertake a new activity, such as rafting down Class 4 white waters or seating our family inside a helicopter to fly over the Grand Canyon, might we consider this new risk rising above the baseline of our routine lives. There is no right or wrong answer. Just preference based upon perceived benefit.

Every day, we encounter risk-controlling strategies imposed by others seeking some greater good while maintaining personal freedom and ethical standards of conduct. These range from fundamental protections guaranteed under the Bill of Rights to speed bumps. Just look at the half-dozen rearview mirrors on a mail truck. Does your child wear a helmet when riding a bike? Why do we stop at a red light? Why install door locks? Such risk-control strategies provide a means of protecting us and providing common ground for people of diverse habits and interests to cooperate. How risk is managed is a social contract between the public and those who regulate. Ignoring certain risks can impact our lives and our health.

How risk is managed is a social contract between the public and those who regulate.

11.3 The Changing Face of Risk

"During a 20 year period during which our society has grown healthier and safer on average and spent billions of dollars and immense effort to become so, the American public has become more—rather than less—concerned about risk. We have come to perceive ourselves as increasingly vulnerable to life's hazards."

—Paul Slovic, *Risk Analysis* (1993, p. 675)

The risk landscape has changed radically during the last few decades (Ross 1999). Previous generations never faced decisions about whether to have surgery, undergo diagnostic imaging, or therapeutic treatment based on the statistical probability of success. We go to the health clinic for a flu shot and sign a notice about the possibility of having an allergic reaction. The chance is small, but it exists. However, the opportunity to miss the next flu outbreak by receiving the inoculation is large. For a moment, we weigh these thoughts and then sign—a risk-based decision.

Risk was once thought of as the playground of the gods (Bernstein 1997).

Risk was once thought of as the playground of the gods (Bernstein 1997). Today, we live in a cause-and-effect world where investments and insurance policies are undergirded by actuarial tables, percentages, and probabilities. With sufficient information, we can understand potential positive as well as unwanted impacts from natural resource use, waste disposal, and the production of biological agents and nuclear materials.

When are potential risks serious enough to alter our behavior or rethink a decision?

New technologies such as genetic engineering are raising complex ethical dilemmas reaching beyond our everyday experience. When are potential risks serious enough to alter our behavior or rethink a decision? These are hard choices, especially when risk is not obvious and answers lie hidden inside assumptions and approximations, inside "what ifs."

How people react to unexpected events reveals their tolerance for risk. Complex machines, such as the space shuttle or large-scale industrial processes, are inherently risky regardless of the precautions taken. Risk assessments are used to assess whether an engineered system will perform as expected. Potential failures and consequences are studied, and then systems are rebuilt or procedural controls are set in place to better manage risk.

Outrage influences why some people believe waste sites are more threatening than other environmental problems such as naturally occurring radioactive radon in their homes (Covello et al. 1989). This is not to say waste sites are not dangerous—some are lethal. This also does not suggest that feelings aren't important. They are. Rather, it underscores the mismatch that exists between perceptions and impacts, between the cost effectiveness of some public health programs compared to lives saved, and between the decrease of life expectancy resulting from personal habits versus the impact of natural diseases (Covello et al. 1988). The cornerstone issue appears to be whether people voluntarily accept a certain risk and whether they receive a benefit from it.

Fright Factors

Understanding what may trigger concern is critical to communicating information about potential harm. Research has established some rules of thumb (Bennett 1997; Fischhoff et al. 1981). Risks are generally less acceptable and more worrisome if perceived to be

- Involuntary, such as exposure to air pollution, rather than voluntary, such as smoking.
- Inequitably distributed (some people benefit while others suffer).
- Inescapable.
- Arising from an unfamiliar source.
- Resulting from human rather than natural sources.
- Causing hidden or irreversible damage, such as illness years after exposure.
- Posing danger to children or to a future generation.
- Threatening a form of death, illness, or injury arousing dread.
- Damaging identifiable rather than anonymous victims.
- Poorly understood by science.
- Subject to contradictory statements from responsible sources.
- Discussed by an untrusted person.

People are much less fearful if they voluntarily accept a certain risk or receive a benefit from it.

The public's apathetic response to radon gas appears to result from the fact it is of natural origin, exposure occurs in the comfortable natural setting of one's home (with no one to blame), and it can never be totally eliminated (Slovic 1996).[1] On the other hand, opposition to the proposed burial of low-level radioactive waste near a community comes from the fact this hazard is imported, an industry is to blame, exposure is involuntary, and just stopping the project can eliminate any impacts.

The President of the International Society for Risk Analysis, Roger E. Kasperson, and his colleagues (1988) use the phrase "social attenuation" to describe the public's relatively low level of interest in risks posed by such well-documented and significant hazards as smoking, drinking, and driving without seatbelts. For example, cigarette smoking is responsible for approximately 500,000 premature deaths each year in this country (Napier 1998). Can you image the uproar arising if any industry would recommend introducing a product that would kill half a million people each year? If a person would die after smoking just a single pack of cigarettes, how many people would

"When inoculations, antibiotics, and other recent medical inventions effectively knocked out infectious diseases from the top spots in the mortality tables, a new order of risks took their place: heart disease, cancer, cirrhosis of the liver, and stroke. As opposed to infections that come from nature, and often kill quickly, these new killers are chronic and derive largely not from nature but from habits, lifestyles, and behaviors of modern life...roughly 85% of known diseases causing premature death can be attributed to lifestyle choices" (Ross 1999, p. 61).

(1) Indoor radioactive radon gas is the second leading cause of lung cancer after cigarette smoking (NRC 1998a, p. 19). Epidemiological evidence examined by the sixth Committee on Biological Effects of Ionizing Radiations suggests 3000 to 33,000 people die from radon-induced lung cancer each year in the United States (NRC 1998a, pp. 18-19).

The concept of risk is tricky when it comes down to personal choice, tax revenues, and jobs.

smoke? Likely fewer because cause and effect would be undeniable and immediate—quickly demonstrated on the evening news—and not hidden behind years of silent disease growth. Yet, the concept of risk is tricky when it comes down to personal choice, tax revenues, and jobs.

11.4 An Acceptable Level of Risk?

"Establishing an acceptable level of protection requires consideration of human values and does not lend itself easily to mechanical calculations. The acceptable level of protection is the essential policy question for site cleanups, and is therefore part of risk management..."

—Washington State Department of Ecology,
Responsiveness Summary on the Amendments to the Model Toxics Control Act Cleanup Regulation Chapter 173-340 WAC (1991)

In a stigmatized setting, can real threats be separated from shared imagery?

No matter how hard proponents try to separate the peaceful use of nuclear technology from the frightening images of destruction, mutation, and invisible peril, the connection is firmly rooted in the public's consciousness (Slovic et al. 1991). Perceptions about nuclear waste are even more negative than nuclear power (Slovic 1996). Today, people distrust Hanford and most of the news coming from officials in the nuclear weapons complex. They want to hear that dangers are gone, not just temporarily reduced or sidestepped. In such a stigmatized setting, how can real threats be separated from perceived risks or just some shared imagery? What are the real dangers faced at Hanford and other nuclear weapons sites? Regretfully, such risk-based comparisons do not exist.

"New beliefs are selected by the brain to be consistent with beliefs already held, but they are generated without any particular regard for what is true or what is not" (Park 2000, p. 35).

Perhaps these questions are unanswerable, for in risk communication perception shares equal or greater weight with reality. "New beliefs are selected by the brain to be consistent with beliefs already held, but they are generated without any particular regard for what is true or what is not" (Park 2000, p. 35). Besides, people are very risk averse when dealing with an uncertain or highly emotional issue. This is why comparisons of risk from nuclear waste to driving a car, riding a bicycle, or flying in an airplane make statistical sense but are rejected by the public.

The challenge faced is that numbers, such as tons of chemicals released or curies of radiation received, do not speak for themselves—they give a one-dimensional view of risk. A more dynamic view comes when preferences, trust, and equity are introduced. Stressing only the technical aspects of risk is too ambiguous to serve as the defining yardstick for making policies (Kasperson et al. 1988). In such a world, risk estimates do not define solutions but rather they should nudge those making decisions towards more informed choices.

Numbers give a one-dimensional view of risk.

Yet, is the distinction between real and perceived dangers important? Yes, for two reasons. First, impressions dictate cleanup work and cost. Second, humans may be persuaded by perceived threats, but true biological health and environmental protection benefits are gained by avoiding real threats.

Public View of Risk

"A few years ago, sociology professor William Freudenburg gave 300 of his students at the University of Wisconsin at Madison a survey that asked whether they would support the establishment of a hazardous waste facility in their hometowns. The students responded overwhelmingly in the negative. The survey then informed them that the chances of dying as a result of living near such a facility were about one in a million. Only a few students changed their minds. Then they were told that the risk of death was less than that from smoking cigarettes. This information made them oppose the facility even more.

"The final response made Freudenburg curious. 'Why?' he asked them. 'Because we've heard that before,' said one cynical student, to the delight of others in the room, who burst into laughter.

"Not only does this small example illustrate that the public perceives risk differently than the experts who compute the probability figures; it also indicates that trust is an important influencing factor in the public's perception of risk.

"The significance of trust may seen obvious, but the prevailing theory within the scientific community used to be that lack of technical and scientific knowledge was the main factor preventing the public from seeing eye to eye with risk experts. 'By and large, the scientific community used to think—and still does—that if only the public had the same scientific background, they would agree on risk,' says Freudenburg. But a number of studies conducted over the past two decades have shown the lack of scientific understanding is only a small part of the problem, and probably not the most essential part. Rather, the key difference between the public view of risk and the expert view is that the public incorporates subjective values or issues—such as trust, equity, and whether a risk is new or familiar—into its perception of various hazards."

— Leslie Lamarre, *Electric Power Research Institute Journal* (1992)

People are willing to pay considerably more to ensure against objectively less-probable but psychologically more-probable events (Johnson et al. 1993). In cleanup, as in all other human activities, real and perceived risks are intertwined. Real risk can be reduced or controlled by applying knowledge and capability. Building trust and keeping meaningful public participation in making decisions reduces perceived risk.

People are willing to pay considerably more to ensure against objectively less-probable but psychologically more-probable events (Johnson et al. 1993).

Factoring risk into decisions is a matter of choice or chance. When risk is understood, its acceptance is a matter of informed choice. When we are misinformed or ignorant about risk, then risk is a matter of chance. Failing to factor risk into decisions is like trying to build a house while leaving essential tools back in the garage because they are too unfamiliar or uncomfortable to handle.

Factoring risk into decisions is a matter of choice or chance.

In one form or another, nearly every waste management and cleanup action is based on some type of risk—real or perceived.

In Washington State, the Model Toxics Control Act cleanup regulation (chapter 173-340 of the Washington Administrative Code) details how risk assessments are carried out and factored into setting cleanup levels, demonstrating cleanup compliance, and even defining natural background concentrations for potentially hazardous substances (Washington State Department of Ecology 2001b). The act states, "for individual carcinogens…cleanup levels are based upon the upper bound of the estimated excess lifetime cancer risk of one in one million" (WAC 173-340-700). For multiple contaminants or multiple pathways of exposure, the act notes, the total excess lifetime cancer risk for a site should not exceed "one in one hundred thousand" (WAC 173-340-700). For individual noncarcinogenic substances, cleanup levels are set by the act at concentrations anticipated to result in no acute or chronic toxic effects on human health.

The U.S. Environmental Protection Agency is implementing Superfund reforms to facilitate waste site cleanup and streamline program accountability. One reform is for "the use of risk-based priority setting for determining Federal Facility clean-up milestones" (Fields 1998). This reform is based upon recommendations developed by the Federal Facilities Environmental Restoration Dialogue Committee (see Section 6.2). The reform promotes a more understandable and consistent approach to setting cleanup priorities, implements the inclusion of risk in making decisions, and provides direction for the use of risk capabilities developed by other agencies.

"Regulatory actions are often taken when the probability of risk of cancer is within the range of 1 in 1,000,000 to 1 in 10,000" (EPA 2002).

The U.S. Environmental Protection Agency has generally set a risk of 1 in 1 million that an individual will develop cancer in a lifetime as a goal for waste site cleanup and has considered a risk of greater than 1 in 10,000 to be potentially excessive (GAO 2000f; Walker 2001).[2] Risk levels for noncarcinogens are established so that no adverse effects are experienced.

Site-specific cleanup levels at Superfund sites—for all exposure pathways and all contaminated media—are generally expressed at the risk levels noted above. This is why knowing risk is essential to determining compliance with environmental protection laws. Yet, as the U.S. Department of Energy acknowledges, their own cleanup strategy "is not risk-based" (DOE 2002, p. II-3). Often, wastes are managed according to their origin rather than their risk and interpretations of work orders; regulations and cleanup agreements have resulted in diverting resources from high-risk to low-risk activities.

Herein lies a dilemma facing Hanford and other nuclear waste sites—can we predict risk with confidence? Are the factors underlying risk understood well enough to characterize the extent and severity of threats of one cleanup recommendation versus another?

This is where opinions collide. One camp focuses on uncertainties and discounts the usefulness of using risk. In the extreme, if cleanup milestones don't call for an assessment of risk, then risk isn't believed useful. The other camp stresses the need for risk-influenced decision-making to avoid simply rolling the dice and hoping for the best. Despite uncertainties, risk offers insight into how systems might behave and what level of cleanup is needed. However, when unknowns run high, the only true statement is that risk lies somewhere between zero and a worst-case estimate. Under such conditions, risk comparisons, rather than absolute numbers, provide the most usable insight.

(2) The following comparisons depict one in a million (after Kamrin et al. 1995): 1 inch in 16 miles, 1 minute in 2 years, 1 cent in $10,000, or 1 car in bumper-to-bumper traffic stretched between Ohio and California.

How are estimates of risk used today? Many regulations are framed around reducing risk to acceptable levels. However, are risks overstated? This would help preserve cash flow or attract the news media's attention. Are they understated? This would comfort a concerned public but mask real threats. If risks are not well understood, cleanup drifts in a vacuum of uncertain health-protection priorities.

Activity Performed Each Year	Risk of Death (number of chances in one million)
Smoking 1 pack of cigarettes each day	3600
Drinking 1 can of beer or 4 ounces of wine each day	100
Firearm and sporting accidents	10
Pleasure boating accidents	6
Drinking chlorinated tap water	3
Riding or driving in a car (300 miles)	2

highest risk ↕ lowest risk

Likelihood of Death From Common Activities. Many risks are taken as a matter of choice. Risks are sometimes accepted because of the personal benefit they provide. These values are generally accepted approximations for the risk encountered from various lifestyle activities (after Dirkes et al. 1999).

For years, environmental protection at Hanford was hampered by its low priority compared to preserving plutonium production. David Lilienthal, former chairman of the Atomic Energy Agency, stated that waste management practices were a "crude method of dealing with these poisons" (Lilienthal 1963, p. 135). Potential risks went unspoken or were minimized through reporting just officially sanctioned information (see Chapter 2). Hanford historian Michele Gerber (1997) concluded that Hanford officials took advantage of secrecy without exercising responsibility to inform the public about contaminant releases. This was done with the understanding that more effective safety and pollution control measures would come along to "ameliorate these risks and lessen the hazardous conditions formerly created" (Wolman and Gorman in Williams 1948, p. 67). Some actions were taken to reduce risk, though most were driven to preserve production goals and maximize the use of the Hanford environs for contaminant disposal.

If risks are not well understood, cleanup drifts in a vacuum of uncertain health-protection priorities.

Herein lies a dilemma facing Hanford and other nuclear waste sites: can we predict risk with confidence?

At Hanford, risks need to be examined in two ways. First, at the waste site level; second, at the broader Hanford-wide level. The need for an integrated, continuously updated assessment—risk profile—is not new. A study of the total "effluent problem" at Hanford was recommended 50 years ago (Parker 1951). We must be cautious of only managing individual cleanup projects at the expense of managing risk from an integrated set of cleanup actions.

Risk operates along a timeline. An *acute* risk takes place when cause and effect are immediate or short term. A serious car accident is an acute risk. A *chronic* risk happens when harm is separated from cause. Examples include the years between drinking alcohol to excess and developing cirrhosis of the liver, or the emergence of malignant melanoma from overexposure to sunlight. Cause and effect are easily witnessed in accidents; chronic ailments are trickier to track.

We must be cautious of managing risk from individual cleanup projects at the expense of managing risk from an integrated set of cleanup actions.

The Hanford Advisory Board (1996c) also reports there is a "need for better sitewide coordination of and consistency between the risk and impact assessment approaches used for all the projects and programs at the Hanford Site." There is concern that individual cleanup projects are making decisions without fully understanding the cumulative effect across the Hanford Site. The result is a patchwork of actions with little end-result coordination.

11.5 Risk Assessment

Mastering the complexities of risk involves building a strong science foundation set within a social context underlain by trust.

"To be decision-relevant, risk characterizations must be accurate, balanced, and informative. This requires getting the science right and getting the right science."

—National Research Council, *Understanding Risk: Informing Decisions in a Democratic Society* (1996b, p. 7)

Estimating the potential for danger is called risk assessment. Acting upon that information is risk management.

Risk assessment is an organized analysis process. Assessments are built on linkages between contamination sources, environmental pathways, exposure, and human or environmental response to carcinogenic exposure. Models analyze these interactions and estimate risks. Those models can range from simple back-of-the-envelope calculations to sophisticated computer-generated estimates—the choice depends upon the rigor needed and unknowns faced. Risk assessments help us conserve resources by establishing priorities and comparing alternatives to discriminate between the best risk-reduction options.

All risk analyses have limitations because our understanding of the world is imperfect.

In 1993, the National Research Council held a workshop to review the management of risk in the U.S. Department of Energy's environmental restoration program (NRC 1994b). Members of the public, government agencies, and industry participated. It was concluded that risk assessments of possible future outcomes in the U.S. Department of Energy complex

- Are highly desirable parts of making decisions especially as they demonstrate how information is gathered, uncertainty is dealt with, impacts are explored, and outcomes are communicated.
- Are useful when information is limited, as long as purposes and limitations are defined.
- Should link analyses and experience.
- Are part of a consensus-building process.
- Must be carried out by credible organizations having expertise and acceptance.

Assessments can be retrospective or prospective. Retrospective assessments examine the potential impact from past contaminant exposure. Radioactive iodine (iodine-131) releases from Hanford or worker exposure to hazardous materials are examples. Similarly, risk assessment can be prospective—evaluating possible impacts from future actions.

Estimates of risk must be used with caution. Risk numbers are sometimes invoked to advance predetermined agendas and sway public opinion. The selective use of data and assumptions can minimize risk or make situations sound more fearful. Therefore, the public must be cautious in what they accept as truth and who is making risk-based claims. David Ropeik, Director for Risk Communication for the Harvard Center for Risk Analysis, wrote that perhaps it's time to create truly independent groups of experts, outside of the government, to produce the "credible, reliable science to help develop policymaking that looks beyond our fears," and beyond agenda setting (Ropeik 2000). While independence is laudable, policy making must always rest within the boundaries of informed public debate.

All risk analyses have limitations because our understanding of the world is imperfect; sometimes even unpredictable randomness dominates at a molecular or larger scale. Uncertainties often give rise to heated controversies in the scientific community that, in turn, could lead to public confusion if not carefully handled (Upton 2001). This is why risk communication is so important yet fraught with danger. No matter how sound the underlying science, the values, assumptions, and tradeoffs imbedded in risk assessments remain as subjective inputs open to questioning.

The scientific defensibility of risk assessments must be carefully reviewed. This includes examining key assumptions and data uncertainty. Errors reveal themselves in surprise findings or a fundamental lack of knowledge about physical, chemical, or biological processes (see Section 9.1). In some instances, the use of formal risk assessments may be inappropriate or even impossible when data uncertainty is too large. In those instances, less rigorous, perhaps relative assessments might be suitable. Besides, most risk assessments are good to no better than a factor of 10 (order of magnitude).

Risk communication is fraught with danger.

Assessments are conducted to estimate risk below the range of observable events people witness in their everyday lives or scientists can study in the laboratory. Normally, health risks are low compared to more easily countable effects such as car accidents or occupational injuries. Estimates of small risks are speculative and can't be verified (Presidential/Congressional Commission on Risk Assessment and Risk Management 1997). Expressing a small risk in numbers, especially a single number, is misleading. It falsely conveys accuracy. This is why a range of uncertainty is assigned to events having a low chance of occurring.

It's a challenge to measure, let alone estimate, small changes in disease incidence from low-level exposure to many chemicals or radionuclides. All the while, people who were exposed to past contaminant releases at Hanford not only want to know if those releases increased health problems, but they particularly want to know if their health problems resulted from these releases (Section 4.3). In most instances, no one will ever be certain.

A new type of study being factored into environmental issues is called ecological risk assessment. This evaluates the likelihood for adverse effects of contaminants on animals or plants or physical damage to sensitive habitats. Effects may range from best judgments to quantitative descriptions. As in human health, ecosystem risk studies are hampered by poorly understood contaminant exposure-to-effect studies from more than one stressor (NRC 2001b). Nonetheless, the U.S. Environmental Protection Agency notes it is better to convey conclusions and associated uncertainties qualitatively than to ignore them because they are not easily estimated (EPA 1998).

A new type of study being factored into environmental issues is called ecological risk assessment.

The current practice in ecological assessment relies on simple exposure-effects-extrapolation models targeted at individual organisms. Further development of population- and ecosystem-level models is needed. "There are significant gaps, some substantial, in our understanding of the environmental fate, bioaccumulation, and ecotoxicity of chemicals in terrestrial and aquatic ecosystems. The science behind ecological modeling continues to evolve at a rapid pace" (Wenning 2001). Factoring ecological insults into decisions is particularly important to the tribal nations because of their close cultural connection with the land. The U.S. Environmental Protection Agency guidelines for conducting ecological assessments are found at http://www.epa.gov/ncea/. While there is wide use

Often, people try to put a monetary value on the Hanford Site using the price of similar real estate. That method does not take into consideration the cultural value of the land.

The Hanford Site is remarkably rich in cultural resources, defined as any prehistoric or historic district, site, building, structure, or object considered important to a culture, subculture, or community for scientific, traditional, religious, or other reasons. In addition to the adjoining Columbia River, considered by many to be the crown jewel of Hanford's cultural resources, Hanford contains numerous, well-preserved archaeological sites representing both prehistoric and historic time periods. Hanford also retains many resources associated with the Manhattan Project and Cold War eras, both of which were significant in the history of the United States and the world. Finally, Hanford is still thought of as a homeland by many Native American people; as a result, various traditional land-use areas, tribal rights, and spiritual places are highly valued. Depending on which topic and which audience you are talking about, Hanford resources are important for their spiritual, cultural, scientific, historical, humanistic, or interpretive values.

—Darby Stapp, archeologist
Co-author of *Tribal Cultural Resource Management: The Full Circle of Stewardship*, Altamira Press

Does a cultural-based susceptibility to certain risks exist?

of the agency's framework, as documented in *Framework for Ecological Risk Assessment* (EPA 1992), different state and federal agencies tailor risk methodologies to match their unique regulatory responsibilities.

The assessment of risk also raises the issue of whether it should be calculated uniquely for different populations of people. Does a cultural-based susceptibility to certain risks exist? (See cultural legacies by Stuart Harris in Appendix B.) Information about Hanford cultural and historic resources, resource protection, and public involvement programs are accessed at http://www.hanford.gov/doe/culres/index.htm. It's ironic that as the practice of using risk assessments has steadily increased to meet public demands for a safer and healthier environment, many people have become more, rather than less, concerned about risk (Kunreuther and Slovic 1999, p. 274). This is particularly true for involuntary exposure to contaminants, which the public readily associates with danger and death.

The application of risk for assessing changes in institutional and social systems is a less-mature science compared to understanding human harm from carcinogens.

11.6 Risk Management and Communication

"The essence of risk management lies in maximizing the areas where we have some control over the outcomes while minimizing the areas where we have absolutely no control over the outcomes."

—Peter Bernstein, *Against the Gods: The Remarkable Story of Risk* (1996, p. 197)

If outcomes were obvious and all people held the same values, then making decisions would be simpler. If public trust was high, actions recommended to manage risk would be more accepted. Hanford exists in a different world.

Risk management uses risk assessment results, along with supportive information such as cost and values, to guide decisions. It allows alternatives to be compared, examined, and negotiated for achieving a mutually agreed-on outcome. Risk management attempts to introduce risk reasoning into decision-making.

Except for the most egregious health problems, there are few opportunities where risk reduction can be measured and compared to risk management goals. Cancer or other diseases may not occur for decades following exposure to toxins. Even then, a clear relationship between exposure and harm may not exist because other environmental factors or human habits mask it.

Risk information must be translated into a form people can use and understand. This is a challenge because "oversimplifying the science or skewing the results through selectivity can lead to the inappropriate use of scientific information in risk management decisions, but providing full information, if it does not address key concerns of the intended audience, can undermine the audience's trust in the risk analysis" (NRC 1996b, p. x).

Good risk management requires keen communication skills and finesse to meet the information needs of audiences (Lundgren and McMakin 1998). Officials must be sensitive to how risk is perceived and communicated. Risk communication is about how risk-related information is exchanged between parties (Covello 1993). Communicators must listen to public concerns; speak clearly and compassionately; tailor discussions to audience needs; and be honest, frank, and open—and truthful (Covello and Allen 1988). Similarly, the public needs to become more fluent in the language of risk. Otherwise, they'll be unable to understand risk input to decision-making and cede opportunities to ask the right questions and participate meaningfully in cleanup discussions.

One challenge facing all communicators is the meaning people assign to words. What one says echoes within a larger context. Words have dual meanings: *actual* meanings based on dictionary-type definitions and *inferred* meanings built from experience. We react most emotionally and personally to our own interpretations.

The development and adoption of quantifiable approaches to risk management, including setting work priorities and judging the value of information collected, are challenges facing Hanford. Some completed studies were published such as those by Hesser et al. (1995), Colson et al. (1997), and the National Research Council (1999a).

For example, in Colson et al. (1997), management decisions about collecting additional waste samples from Hanford tanks were based on the expected economic value of that new information compared with the cost (and risk) of acting on the waste in the absence of new information. Colson asked some fundamental questions: what is the basis for making decisions and how much data are enough to take action?

"Mountains of data are collected in the management of hazardous, toxic and radioactive wastes. Unfortunately, the majority of these data have little or no value except to maintain a prosperous three-ring binder industry in our country. With shrinking resources for waste disposal and cleanup, we can no longer afford to collect data for the sake of collecting data" (Gilbert 1997).

Risk management attempts to introduce risk reasoning into decision-making.

Good risk management requires keen communication skills and finesse.

The public needs to become more fluent in the language of risk.

11.7 Attempts at Risk-Based Decisions

"DOE [U.S. Department of Energy] has found none of the various approaches it has explored for prioritizing its environmental management activities to be entirely satisfactory....each approach has been abandoned before it could develop adequately, owing largely to lack of confidence in the approach by DOE and site personnel, and/or lack of support for it by other stakeholders."

—Consortium for Risk Evaluation with Stakeholder Participation, *Peer Review of the U.S. Department of Energy's Use of Risk in Its Prioritization Process* (1999, p. 3)

The schedules and sequencing of compliance activities reflect concerns that may or may not be risk based.

Having said all of this about risk, we should remember that risk is only one factor influencing waste management decisions and milestones written into compliance agreements.

One organization has helped the U.S. Department of Energy's Office of Environmental Management examine risk in cleanup: The Consortium for Risk Evaluation and Stakeholder Participation (CRESP). In the past, the Center for Risk Excellence also assisted but that center was closed in 2002.(3)

In 1995, CRESP was created to provide a broader understanding of risk issues affecting cleanup of nuclear weapons facilities. It is a university-based national organization headed by the Environmental and Occupational Health Sciences Institute (EOHSI) in New Jersey and the University of Washington in Seattle. The University of Medicine and Dentistry of New Jersey, Rutgers University, and the State University of New Jersey lead EOHSI. Because of the proximity of the University of Washington to Hanford, its School of Public Health and Community Medicine is responsible for distributing CRESP-generated information about Hanford. The consortium is not a decision maker; rather, it's an independent institution to develop data and methodology to make risk a part of the decision-making process. More information about CRESP can be found at http://www.cresp.org/home.html.

Several times the U.S. Department of Energy has tried to develop and implement specific risk-based methodologies to discern between competing nationwide cleanup priorities. None have succeeded. Examples include (after DOE 1989, pp. 26-27; GAO 1995b, pp. 40-41; CRESP 1999; GAO 2002a, pp. 19-20):

1986—A 3-year effort began to identify and prioritize environmental problems. A certain pollution assessment model was used to show how contaminants were released, moved through the environment, and potentially placed humans at risk. The U.S. Department of Energy was criticized for not involving the public in model development as well as inadequate documentation of findings. The project was terminated in about 1989.

---

(3) The Center for Risk Excellence was established in 1997. Located at the U.S. Department of Energy's Chicago Operations Office, the center provided risk expertise and resource coordination. Its goal was to develop and implement policy, practices, and other risk-related support. The center also examined risk issues associated with Hanford's study of groundwater and vadose zone contamination.

1989—A risk-based prioritization methodology, the Environmental Restoration Priority System, used a formal analytical decision-aiding model to assist in budget allocations across the nuclear weapon sites. Inputs included health and safety risks and social, technical, economic, and policy issues. Many input values were elicited from U.S. Department of Energy managers. In 1991, reviews of the model concluded it had major limitations even if the input had been perfect. The approach was not appropriate for guiding budget allocations, lacked understandability, and was suspect by stakeholders. Work was suspended in 1992.

No risk method was pursued long enough to develop into a useful tool for making decisions.

1995—Based upon public and worker safety, risk data sheets were developed to provide numerical scores for rating environmental management activities. This was an evolutionary leap in the U.S. Department of Energy developing a risk-based approach to prioritize activities. However, the effort suffered from limited data, poor definitions, inconsistent scoring, questionable links between scores and activities, single- versus multiple-year assessments, and inadequate involvement of stakeholders. The effort was abandoned in 1997.

1997—Project baseline summaries, using risk classifications of urgent, high, medium, low, or not applicable, were developed over the lifetime of environmental management projects to demonstrate the U.S. Department of Energy was addressing the most urgent risks first. This effort failed because the department did not have a clear basis for understanding or classifying risks, inconsistencies in implementation, and lack of acceptance from the field offices. Rather than comparing risks between field offices, this tool was more useful in comparing the value of projects at a single site. Project summaries were not generated after 1998.

1999—Risk profiles were tested at 10 field offices. This was to be an improvement over previous efforts because it involved more collaboration with site personnel, provided documentation of public health risks, and generated risk results at a site level. These profiles failed to track contaminants through the environment and did not adequately address public, environmental, and worker exposures or health risks. The effort was not supported after 2001.

The average life expectancy for each approach to risk-based decision-making was 2 years.

In 1999, the U.S. Department of Energy sponsored the CRESP to undertake a study as to why past efforts to establish a risk-based decision-making process failed (CRESP 1999). Problems were plentiful. Many of the reasons are captured in the previous bulleted items. Each risk methodology was found to have "embodied promising features," but none were pursued long enough to develop into a useful and acceptable tool. Three years later, the General Accounting Office (2002a, p. 20) also emphasized that U.S. Department of Energy had failed to integrate any of its risk-based approaches into a usable decision-making process. If critical cleanup decisions are to be risk-based (or at least risk-influenced), then a stable, technically sound, and accepted risk framework is required.

Chapter 12

Building a Path Forward

"While everyone can appreciate that a complex, highly sophisticated engineering is required to safely store nuclear materials for thousands of years, few have appreciated the political requirements necessary to design and implement such a solution."

—Gerald Jacob, *Site Unseen: The Politics of Siting a Nuclear Waste Repository* (1990, p. 164)

While the knowledge and capability to clean up nuclear waste are complex, so are the political processes and administrative procedures.

The decisions we make in our everyday lives center on values and benefits—from the cars we buy to how we spend our time. We purchase products for many reasons—price, quality, manufacturer, convenience, or simply out of habit. Your decisions are not the same as your neighbor's. Just glance into someone else's shopping cart the next time you're in the grocery store; it's amazing how we differ. Now, imagine purchasing a product costing a hundred billion dollars, taking decades to achieve, and centuries to care for. That's Hanford.

"The politics of nuclear waste involves hundreds of issues, actors, agencies, institutions, and jurisdictions" (Jacob 1990, p.164).

Many well-intentioned people hold different opinions about how to demonstrate cleanup progress, what the barriers to progress are, and even whether or not progress has been achieved. In fact, one person's barrier might be another person's goal. It's a conflicting world.

Hanford cleanup is approaching a transition point. More than a decade of experience has sharpened thinking and sobered reality. Gone are the days of rhetorical solutions and wishful fixes. People are tracking in-the-field performance. In 2002, the U.S. Department of Energy wrote, "It is clear that with the current path, the cost of the program will continue to increase, with the real possibility that the ultimate cleanup and closure goal will never be met" (DOE 2002). Those words were unheard of until recently. Will cleanup remain on its present course, self correct, or be forced to take a dramatically different route?

Hanford needs forceful, national-level advocacy and local dedication to accelerate cleanup progress, embrace innovative cleanup solutions, and reduce biological risks. To regain this level of attention, fundamental changes are needed in the program's contracts, work processes, accountability—and perhaps environmental laws, or at least our application of those laws. It's time to challenge the status quo. Otherwise, if history is any guide, congressional intervention will result during the flurry of a crisis (Probst and Lowe 2000, p. 31).

It's time to challenge the status quo.

12.1 Benefit Sought

"The best way to achieve an environmental policy goal, such as... preserving ecosystems or landscapes, might often be to improve coordination of conflicting policies across multiple sectors, rather than merely to add a single new law involving one agency."

—Richard Andrews,
Managing the Environment, Managing Ourselves: A History of American Environmental Policy (1999, p. 5)

If a single benefit was chosen, it would couple reducing biological threats and improving the quality of decisions.

For years, a variety of benefits were targeted for each cleanup dollar. These spanned economic diversification of communities long dependent on a defense-based nuclear industry to actually cleaning up contaminants.

One can argue about the legitimacy of one benefit over another, though most people's expectation is that cleanup funds should be spent on in-the-field work.

If a single overriding benefit was chosen, one consistent with the spirit of environmental laws and possessing that gut feeling of rightness, it would couple reducing biological threats and improving the quality of decisions. Living systems are the final concern in virtually all environmental legislation and regulation (Integration Project Expert Panel 2000b).

If this is an acceptable premise, then the prioritization of tasks, resource allocations, and our measurement of progress should be integrated into a framework that reduces biological risks and decision uncertainty.

Although the goals of many individual remediation actions are quasi-defined, the collective, integrated end states for Hanford cleanup are lacking, including how the site will be used in the future and where the boundaries are drawn for estimating regulatory compliance (Integration Project Expert Panel 2000a). Is this important? Yes, because the sum of individual cleanup actions for tanks, groundwater, solid waste, etc., may not result in a desired land end state for the entire Hanford Site.

Sometimes, tensions arise because different unspoken benefits are sought.

In the context of the Hanford cleanup agreement, a report for the U.S. Senate (Blush and Heitman 1995) raised a similar concern: "There is a real danger that commitments associated with the TPA [Tri-Party Agreement] will be assigned a higher priority than all other laws, regulations, policies, and administrative requirements, not because good risk-related reasons exist for doing so but because TPA commitments are accorded a status by DOE [Department of Energy] and the contractors that is out of proportion to their importance to the cleanup."

If cleanup is not risk based, a different framework is required—and that basis for making decisions and allocating money should be clearly spoken.

12.2 Constraints to Progress

"EM [DOE's Office of Environmental Management] can accurately report that over 90% of our regulatory milestones have been met and that our contractors routinely receive over 90% of the available fee. Yet, these 'successes' take place without significant progress in cleanup or risk reduction."

—Jessie Roberson, *Memorandum for the Secretary* (2002a)

Today, we must remain cautious about those who believe only they can speak about Hanford.

Whatever the benefit sought, Hanford contaminants and nuclear material must be well managed for the sake of this and future generations.

Is Hanford executing a cleanup strategy honed on final land end states and waste products? Or on activities that define work processes? One must not confuse the two.

Nearly 40 years passed before the tightly controlled decision process and information release culture of Hanford was opened to broad external scrutiny. During those years, the waste management status quo was rarely challenged (see Chapter 2). In retrospect, this seems bewildering but historically that was how government agencies and industry functioned. Public trust remained high. Today, we also must remain cautious about any culture of status quo protection, guarded by those who believe only they can speak.[1] Cultures self-protect rather than self-correct. "The problem is not so much the people but an outdated and rigid system that rewards those who support the status quo" (Alvarez 2000).

Cultures self-protect rather than self-correct.

Hanford is buffeted from many sides: Congress criticizes operations while imposing new spending limits, local stakeholders demand more remediation while preserving local economies, the news media sensationalizes problems while bypassing accomplishments, and agencies decry rising costs but multiply work requirements. This leads to headlines such as "Hanford cleanup results underwhelm panel [Defense Nuclear Facilities Safety Board]" (Cary 2000b). Workers are amazed at how many people and organizations have formal or informal veto power over Hanford activities.

Experience shows it takes too much to do something and too little to stop it.

Progress seems elusive. There are no quick fixes to magically align expectations, budgets, schedules, regulations, and capabilities. The National Research Council (1999b) reports that problems are pervasive and culturally engrained in the institutions and organizations influencing cleanup.

A number of reviewers have described barriers to progress. A few examples include those published by the General Accounting Office (1994a, 1996, 1998b, 1998c, 1998d, 1999, 2001c), National Research Council (1995, 1996a, 1996c, 1996d, 1998, 1999b, 2000b), Secretary of Energy Advisory Board (1995), U.S. Congress (Blush and Heitman 1995), U.S. House of Representatives (2000), Alvarez (2000), Washington Advisory Group (1999), National Conference of State Legislatures (1996), Fallon et al. (1991), and U.S. Department of Energy (1993a, 2002, 2002d).

(1) In the extreme, this is a new variation on information and communication control that dominated the old Hanford plutonium culture. See Chapter 2.

Agencies are not held accountable for implementing change.

Some findings are controversial. Others are widely accepted. Most studies repeatedly identify the same root problems though the problem context and reviewers change. However, agencies are not held accountable for implementing change. Reports are published, there is a momentary news flash, and then attention fades. The next reviewer stumbles across the same issue. After reading these reports, one wonders, "How many more times and ways can the same problems be identified without someone being accountable for changing the status quo?"

Typical barriers raised include

- Poor contracting practices.
- Changing cleanup visions and approaches.
- Separate, unlinked processes supporting the cleanup triage—work, budget, and schedule.
- Quality of cleanup standards and uncertainty over cleanup levels.
- Lack of high-quality project baselines and contingencies.
- Incentives rewarding work process versus risk-reduction progress.
- Lack of incentives to test and implement improved solutions.
- Questionable cost and benefit comparisons.
- Mismatch between funds required and funds available.
- Lack of rigor in making decisions.
- Practices driven by narrow interpretations of work processes, regulations, and out-of-date agreements.

After years of study, the underlying institutional, regulatory, and business incentive systems remain largely unchanged and perhaps too cumbersome to efficiently solve cleanup problems. The U.S. Department of Energy (2002) writes, "process rather than cleanup results has become the basis for performance metrics, contracts, cleanup approaches, and agreements." Another DOE report laments the need to "break the mold and find new ways to get the job done" (DOE 2002d, p. 8).

The National Research Council (1997) attributes many problems to insufficient incentives, such as the power of economic self-interest, to prompt cleanup. The Hanford Advisory Board echoes this sentiment: "Contract incentives, both positive and negative, are what drive progress" (HAB 1998).

Process rather than results has become the basis for performance (DOE 2002).

The willingness to change and to streamline work requirements and regulatory applications, while preserving environmental protection, is perhaps the simplest litmus test gauging Hanford's ability to rejuvenate high-impact work.

Regardless of opinions, if dramatic actions are not taken, the cleanup program may limp along until Congress calls a halt and takes draconian actions likely not in the best interest of local communities or the Northwest. It's time to raise controversial questions and suggest creative solutions.

> "A major failing of national policy in creating a healthy market for environmental remediation technologies is the lack of sufficient mechanisms linking the prompt cleanup of contaminated sites with the financial self interest of the organization responsible for the contamination...Rather than being driven by environmental regulations alone, organizations responsible for contaminated sites need to be motivated to pursue remediation for financial reasons. Making this transition to a market-oriented system for remediation will require that environmental regulators allow organizations with contaminated sites the freedom to choose how they will accomplish the required remediation end-points; it will require organizations responsible for contaminated sites to honestly evaluate and disclose the full costs of site remediation."
>
> —National Research Council, *Innovations in Ground Water and Soil Cleanup: From Concept to Commercialization* (1997, pp. 3, 75)

12.3 Bridge-Building Philosophies

"Men build too many walls and not enough bridges."

—Sir Isaac Newton

While reviewing studies about barriers to progress, the words simplification, basic principles, and eliminating constraints kept surfacing.

In spite of barriers, some improved waste management practices and cleanup are under way. We do not lack ideas and must not shy from difficult tasks. After all, "People in all walks of life are expert in the art of saying something can't be done" (Lilienthal 1980, p. 18).

In spite of barriers, some improved waste management practices and cleanup are under way.

Today, Hanford is safer and is being managed in a more environmentally protective manner than when cleanup began. And some barriers are being addressed. This includes development of a program to better understand the movement of groundwater and contamination beneath Hanford (DOE 1999d). In 1996, the U.S. Department of Energy initiated the Environmental Management Science Program, a research program "to stimulate the required basic research, [technology] development and demonstration efforts to seek new and innovative cleanup approaches to replace current conventional approaches which are often costly and ineffective" (NRC 1996c, p. 10).[2] In 1998, Congress created the Hanford Office of River Protection to focus more attention on the management and cleanup of Hanford's tank waste.

The National Conference of State Legislatures (1996) points to improvements in administrative and regulatory processes governing cleanup.

The Washington Advisory Group (1999) recognized that regulatory and stakeholder deterrents to using new technologies are also changing. Greater flexibility is being given to demonstrate innovative technologies having the potential to outperform baseline technologies.

The bridge-building philosophies noted below are offered as philosophies to stimulate discussions. Each of us would likely create our own list. None of these philosophies are offered with the intent of anyone avoiding responsibility for site cleanup—but rather

(2) In 2002, the Environmental Management Science Program was moved from the U.S. Department of Energy's Office of Environmental Management to the Office of Science.

with the aim of improving our approach to cleanup. Ultimately, the U.S. Department of Energy, U.S. Environmental Protection Agency, and the Washington State Department of Ecology are collectively responsible for overcoming constraints.

Accelerate Cleanup Experience. There is much merit to getting on with cleanup, demonstrating progress, lowering risk, learning, and building trust.

Accelerating cleanup requires focusing on progress rather than process; outcomes rather than inputs. And this progress should target big-ticket actions designed to significantly reduce risk to workers, the public, and the environment. The construction and testing of pilot- to full-scale operations should be embraced. There is no substitute for getting waste out of tanks or the ground and grappling with it. This quickly focuses attention on the adequacy of plans, suitability of technologies, achievability of schedules, and reasonableness of goals. Debates about treatment plans or efficiencies take a back seat to practical experience. Besides, as expressed by Merilyn Reeves (2000), past chair of the Hanford Advisory Board: "Delays are costly. Pay now or pay more later."

Debates over plans take a back seat to practical experience, to outcomes rather than inputs.

Understand What Needs to be Known. Complete understanding of the properties and processes controlling contaminant behavior and potential impacts is impossible to gain. Rather, we only need a workable knowledge to support decisions, enough information to avoid unacceptable outcomes. What are these controlling characteristics and the contaminants of concern? Are our programs focused on them? Acquire this knowledge as we move forward one step at a time.

There is no shortcut to building credibility.

Simplify Governance and Work Practices. A bias for action hinges on simplified governance and work practices.

Significant cost and schedule savings come from a critical examination of the requirements dictating work. This includes streamlining labor rules, managing Hanford under a simplified regulatory setup, and re-examining how much conservative redundancy is enough. Overlapping requirements and multiple layers of oversight contribute to higher costs and protracted schedules. The earlier changes are incorporated; the more likely they will be seen as informed adjustments rather than simply circumvention of existing work processes.

All procedures and requirements should be linked to practical outcomes while preserving safety and protection. Unlinked practices are wasteful. This requires a culture of accountability—from government agencies through advisory groups. The need exists to establish who is in charge, hold them accountable, and let them do their job—with strong oversight but minimal interference.

Simplifying governance and work processes does not mean gutting environmental laws or essential safety procedures.

To what extent is cleanup driven by safety philosophies inherited from the nuclear power industry or work practices engrained during plutonium production days? According to the U.S. Department of Energy, 85 to 90% of its environmental management budget is directed towards ensuring that its operations comply with a large number of legally enforceable cleanup and compliance agreements (GAO 1999f). These agreements and the U.S. Department of Energy's own rules form a complex web of requirements upon which contractors impose new, sometimes more stringent policies. What is required to simply do the job well?

The U.S. Environmental Protection Agency attributes their own nationwide administrative reform efforts to making cleanup efforts faster and fairer (GAO 1999g).

Know Health Effects. Many environmental standards are decades old, even though modern biology has undergone a revolution in understanding the molecular basis for life and disease. Scientifically defensible cleanup levels are needed. This issue crosses the entire nuclear weapons complex and Superfund cleanup program. Are current standards too lenient? Too stringent? Just right? Without this knowledge, those making decisions and the public have limited assurance that the right cleanup work is under way.

What are the realities of biological risks posed by contaminants?

Protection should extend to all—children, adults, workers, native cultures, the environment, as well as the generations to follow.

Establish Credible Plans and Alternatives. Credible project plans, supported by viable alternatives (options) and directed at clearly defined outcomes, are the only means to achieve timely progress. They support preparing defensible budgets and establishing credible indicators of performance.

A pragmatic set of integrated baselines should be anchored in the environmental sciences, balancing biological impacts with social benefits and cost. Alternative approaches provide defense in depth—capability backup to solve inevitable surprises as they arise.[3] We should not confuse having a collection of individual waste site cleanup plans linked with critical paths with having an integrated, sitewide cleanup strategy.

Match Contracts with Outcomes and Incentives. Meaningful work performance is linked to achieving realistic outcomes that demonstrate hazard reduction. What gets measured gets rewarded. Therefore, it's important to ensure the right performance metrics are tracked. We need a business mindset focused on completing work within clearly defined metrics of progress—such as contaminants recovered, destroyed, or stabilized compared to the amount remaining behind or risk reduced compared to that first faced. Capability development and demonstration, in preparation for reducing hazards, are legitimate short-term outcomes.

Activities, such as gallons of groundwater pumped or tons of soil hauled, demonstrate work performed not progress towards risk reduction goals.

Compare Cost and Benefit. People need to know what the most important problems are and the merit of one solution over another. In so doing, not all benefits are measured in dollars saved, though it's the most frequently used commodity. Protection of certain cultural resources and landforms, such as the Columbia River, are priceless.

Nonetheless, when guided by social preferences and risks, cost-benefit analyses provide invaluable input to decisions. They help answer, "What is our currency of progress?" and "What price are we willing to pay?" Today, it is uncertain how cost and benefits are compared in Hanford cleanup.

(3) Groundwater hydraulic controls, in situ chemical manipulation, natural attenuation, and land-use restrictions are examples of alternative groundwater management actions. For tank cleanup, waste vitrification, in-place chemical treatment and stabilization, and engineered barriers can achieve various risk-reduction goals. Technologies are tailored to risks faced and people's preferences.

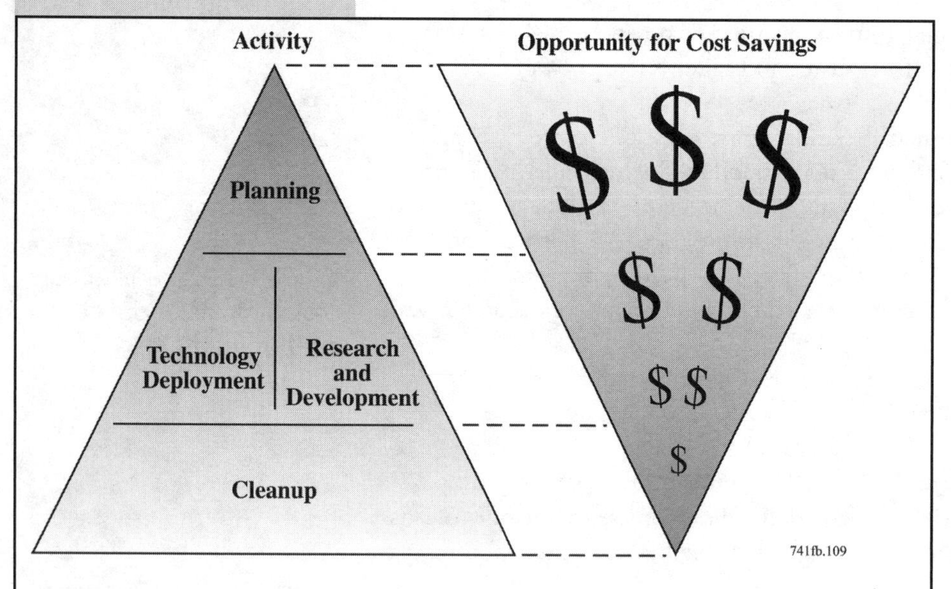

Opportunity for Cost Savings. The basis for most future expenses is set at the planning stage—when decisions are made about cleanup approaches. These choices set future expenses into motion.

Support Strategic Investments. Given the long lead times to achieve many cleanup tasks, the opportunity exists to undertake well-focused investments to find answers to problems. These investments come in two forms. First is the upfront planning, and second is work execution. The greatest dollar savings result from high-quality planning. This is where initial options are weighed and courses of action selected. In business, the most cost-effective option is normally chosen to produce a product that meets customer needs. Existing technology and know-how creates the product. Adjustments and improvements (research and development) improve the product over time and correct production problems—or are used to manufacture an entirely different product. However, the decision to produce a given product in the first place sets into motion most future expenses.

New technologies must offer more than simple substitutes to existing baselines. They should demonstrate significantly better performance, lower costs, lower risks, or enable a difficult task to proceed (NRC 1999e).

Analyses should compare the cost of proceeding with existing capability versus reaching an incorrect decision if new investments are not made.

Cleanup contracts should be written to encourage learning, testing innovative solutions, and ensuring critical knowledge is available before it's needed.

Integrating research and technology demonstration milestones into the Hanford cleanup agreement might be considered. This could create more incentives for improving project baselines and advocating a funding base for investments.

Increase Quality of Information Shared with the Public. To be informed consumers of Hanford news, the public and their representatives must understand the issues faced and the consequences of options. Their attention should focus on big-ticket problems such as land end states, cleanup actions, waste forms, or waste transportation rather than the myriad of day-by-day details needed to operate the site. Information must be tailored to a public audience while preserving integrity.

Particularly important is the public receiving information without spin or bias. The goal is public education, not proselytizing or promoting a single point of view. Hanford's Integration Project Expert Panel (2000b) speaks about the failure to discuss the consequences of waste management and cleanup actions "in terms that are widely understood, technically defensible, and comprehensive."

12.4 Conclusions

"In the middle of difficulty lies opportunity."

—Albert Einstein

Today, Hanford is driven towards securing the safer management and cleanup of contaminants. How this will take place is evolving. How much it will cost and when it will be complete remains uncertain. What is known is that cleanup is a journey where the best solutions will be framed around informed debate, ethical standards, social consensus, and high-quality knowledge.

At some point, expectations, physical reality, and technical abilities must coincide.

At some point, people's expectations, physical reality, and technical abilities must coincide. We all play a role in helping Hanford reach that point as quickly as possible.

Other Sources of Information About Hanford

The U.S. Department of Energy, U.S. Environmental Protection Agency, and Washington State Department of Ecology are working together to clean up Hanford contaminants. To get on the mailing list for public meetings and to receive information about Hanford cleanup, call the Hanford Tri-Party Agreement toll-free hotline (1-800-321-2008), access the Tri-Party Agreement electronic homepage at http://www.hanford.gov/tpa/tpahome.htm, or write to

Hanford Mailing List:
Informational Mailings
P.O. Box 1970 B3-35
Richland, WA 99352

or

Hanford Update
Department of Ecology
P.O. Box 47600
Olympia, WA 98504

To learn more about Hanford, access the U.S. Department of Energy, Richland Operations Office home page at http://www.hanford.gov/. Other sources of online information include the homepages for the U.S. Environmental Protection Agency—Region 10 (http://www.epa.gov/r10earth/index.htm), Washington State Department of Ecology (http://www.ecy.wa.gov/), Washington State Department of Health (http://www.doh.wa.gov/), and the Oregon Office of Energy (http://www.energy.state.or.us/).

A non-federal government view of Hanford is found at a private Web site maintained by Hanford Watch (http://www.hanfordwatch.org/).

Agencies, organizations, and Indian Nations to call, write, or visit about Hanford issues include the following:

Federal and State Organizations

Oregon Office of Energy
Nuclear Safety Division
Ken Niles
(509) 378-4906
or in Oregon 1-800-221-8035
625 Marion Street NE
Salem, OR 97310

U.S. Department of Energy
Yvonne Sherman
(509) 376-6216
P.O. Box 550
Richland, WA 99352

U.S. Environmental Protection Agency
Dennis Faulk
(509) 376-8631
712 Swift Blvd., Suite 5
Richland, WA 99352

Washington State Department of Ecology
Mike Wilson
(509) 735-7581
1315 West 4th Ave.
Kennewick, WA 99336

Washington State Department of Health
Debra McBaugh
(360) 236-3251
1112 SE Quince St.
P.O. Box 47890
Olympia, WA 98504-7890

Environmental and Professional Organizations

American Nuclear Society
(509) 735-3581
Floyd Ivey
Liebler, Ivey & Connor
1141 North Edison
Kennewick, WA 99336

Columbia Riverkeeper
Cyndy DeBruler
(509) 493-2808
P.O. Box 912
Bingen, WA 98605

Heart of America Northwest
Gerald Pollet
(206) 382-1014
1305 4th Ave., Suite 208
Seattle, WA 98101

Oregon Hanford Waste Board
c/o Oregon Office of Energy
Susan Safford
625 Marion Street NE, Suite 1
Salem, OR 97301-3742
(503) 373-7400 or in Oregon
1-800-221-8035

Oregon Physicians for Social
Responsibility
Jennifer Aldrich
921 SW Morrison, Suite 500
Portland, Oregon 97205
(503) 274-2720

Washington Physicians for Social
Responsibility
Ruth Yarrow
(206) 547-2630
4554 12th Ave., NE
Seattle, WA 98105

Indian Nations

The Confederated Tribes and Bands of
the Yakama Indian Nation
Russell Jim
(509) 865-5121
P.O. Box 151
Toppenish, WA 98948

Confederated Tribes of the Umatilla
Indian Reservation
Bill Burke
(541) 966-2414
P.O. Box 638
Pendleton, OR 97801

Nez Perce Tribe
Patrick Sobotta
(208) 843-7375
P.O. Box 365
Lapwai, ID 83540

Wanapum Tribe
Virginia Wyena
(509) 754-3541
P.O. Box 878
Ephrata, WA 98823

Resources

For more information on Hanford history and cleanup, please consult the following public reading rooms:

Branford-Price Millar Library
(503) 725-3690
Portland State University
SW Harrison and Park
P.O. Box 1151
Portland, OR 97207

DOE Public Reading Room
(509) 372-7443
2770 University Drive
CIC, Room 101L
P.O. Box 999, MS H2-53
Richland, WA 99352

Foley Hanford Health Information
Archives Library
(509) 323-3847
Gonzaga University
E. 502 Boone
Spokane, WA 99258-0095

Suzzallo Library
(206) 543-1937
Government Publications
Box 352900
University of Washington
Seattle, WA 98195-2900

References

40 CFR 61. U.S. Environmental Protection Agency, "National Emission Standards for Hazardous Air Pollutants." U.S. Code of Federal Regulations.

40 CFR 141. U.S. Environmental Protection Agency, "National Primary Drinking Water Regulations; Radionuclides." U.S. Code of Federal Regulations.

Abraham, S. February 10, 2002. "New Cleanup Plan Emphasizes Accountability." *Tri-City Herald*, p. F3, Kennewick, Washington.

Adley, F.E., and W.K. Crane. April 21, 1950a. *Meeting of the Columbia River Advisory Group: March 6-7, 1950.* HW-17595, General Electric, Hanford Works, Richland, Washington.

Adley, F.E., and W.K. Crane. April 21, 1950b. *Meeting of the Columbia River Advisory Group: November 21-23, 1949.* HW-15861, General Electric, Hanford Works, Richland, Washington.

Advanced Sciences Inc. February 22, 1994. *Final Report: Independent Technical Review of N Springs Expedited Response Action Proposal Hanford Site.* Westinghouse Hanford Company, Richland, Washington.

Agency for Toxic Substances and Disease Registry (ATSDR). 2002. *ToxFAQs™ for Carbon Tetrachloride.* U.S. Department of Human and Health Services, Washington, D.C. Available URL: http://www.atsdr.cdc.gov/tfacts30.html#health.

Agnew, S.F. 1997. *Hanford Tank Chemical and Radionuclide Inventories: HDW Model Rev. 4.* HNF-3233, Rev. 0, Fluor Daniel Hanford, Inc., Richland, Washington.

Agnew, S.F., and R.A. Corbin. August 1998. *Analysis of SX Farm Leak Histories—Historical Leak Model (HLM).* LA-UR-96-3537, Los Alamos National Laboratory, Los Alamos, New Mexico.

Alm, A.L. March 12, 1997. U.S. Congress, U.S. House Committee on Appropriations, Energy, and Water Development Appropriations for 1998. *Hearings.* 105th Cong., 1st session. U.S. Government Printing Office, Washington, D.C.

Alvarez, R. 2000. "Energy in Decay." *The Bulletin of the Atomic Scientists* 56(3):24-35.

Anderson, J.D. 1990. *A History of the 200 Area Tank Farms.* WHC-MR-0132, Westinghouse Hanford Company, Richland, Washington.

Andrews, R.N.L. 1999. *Managing the Environment, Managing Ourselves: A History of American Environmental Policy.* Yale University Press, New Haven, Connecticut.

Antonio, E.J., and K. Rhoads. 2000. "Potential Radiological Doses From 1999 Hanford Operations." *Hanford Site Environmental Report for Calendar Year 1999*, eds. T.M. Poston, R.W. Hanf, and R.L. Dirkes. PNNL-13230, Pacific Northwest National Laboratory, Richland, Washington.

Antonio, E.J., K. Rhoads, L.H. Staven, and W.M. Glines. 2001. "Potential Radiological Doses From 2000 Hanford Operations." *Hanford Site Environmental Report for Calendar Year 2000.* PNNL-13487, Pacific Northwest National Laboratory, Richland, Washington.

Apley, W. March 3, 2001. *PNNL Science Helping Create New Future for Hanford.* Science and Technology Progress 2001 Edition, *Tri-City Herald*, p. D10, Kennewick, Washington.

Argonne National Laboratory (Argonne). 2001. *Human Health Fact Sheet: Strontium.* Argonne National Laboratory, Argonne, Illinois. Available URL: http://riskcenter.doe.gov/docs/cre/factsheets/Strontium.pdf. (Accessed: May 29, 2002.)

Associated Press. January 30, 2000. "Radiation Exposure Admitted." *Tri-City Herald,* Kennewick, Washington.

Atomic Energy Commission. 1973. *Report on the Investigation of the 106 T Tank Leak at the Hanford Reservation, Richland, Washington.* TID-26431, Richland, Washington.

ATSDR. See Agency for Toxic Substances and Disease Registry.

Bailey, R.W., and M.S. Gerber. October 1997. *PUREX/UO_3 Facilities Deactivation Lessons Learned History.* HNF-SP-1147, Rev. 2, Babcock & Wilcox Hanford Company, Richland, Washington.

Bauer, C., and K. Probst. December 2000. *Long-Term Stewardship of Contaminated Sites: Trust Funds as Mechanisms for Financing and Oversight.* Discussion Paper 00-54, Resources for the Future, Washington, D.C.

Bechtel Hanford. 1998. *Hanford Site Atlas.* BHI-01119, Rev. 1, Bechtel Hanford, Inc., Richland, Washington.

Bechtel Hanford. 2001. *Groundwater/Vadose Zone Integration Project System Assessment Capability.* Available URL: http://www.bhi-erc.com/projects/vadose/sac.htm. Last updated February 2001.

Becker, C.D. 1990. *Aquatic Bioenvironmental Studies: The Hanford Experience 1944-84.* Elsevier, New York.

Bennett, P. 1997. *Communicating About Risks to Public Health: Pointers to Good Practice.* Economics and Operational Research Division, Department of Health, Skipton House, London.

Bernstein, P.L. 1996. *Against the Gods: The Remarkable Story of Risk.* John Wiley and Sons, Inc., New York.

Berry, E.C., and J.F. Cline. January 20, 1950. *First Annual Report of the Botany Field Station.* HW-16056, Hanford Works, Richland, Washington.

Bessell, W.W. Jr. August 13, 1942. *General Order No. 33.* Issued by the War Department, Office of Chief of Engineers, Washington. Available in U.S. Department of Energy Public Reading Room, Richland, Washington.

Bishop, W.A. July 15, 1986. "Testimony of WA Bishop, Chair, Washington State Nuclear Waste Board." U.S. Department of Energy, Public Hearings on Defense Waste Environmental Impact Statement. The testimony is contained in: U.S. Department of Energy (DOE). 1987. *Final Environmental Impact Statement: Disposal of Hanford Defense High-Level, Transuranic and Tank Waste, Hanford Site, Richland, Washington.* DOE/EIS-0113, Vol. 5, p. 505. U.S. Department of Energy, Washington, D.C.

Bjornstad, D.J., D.W. Jones, M. Russell, and C.L. Dummer. 1998. *Implementing Outcome-Oriented Risk Planning: An Overview.* JIEE 98-2, Joint Institute for Energy and Environment, Knoxville, Tennessee.

Blumenthal, L. March 25, 2000. "Hazard Adds to Hanford Concerns." *Tri-City Herald*, Kennewick, Washington.

Blumenthal, L. February 10, 2002. "Hanford Budget Cut Big Blow for State Lawmakers." *Tri-City Herald*, p. F1, Kennewick, Washington.

Blush, S.M., and T.H. Heitman. 1995. *Train Wreck Along the River of Money: An Evaluation of the Hanford Cleanup.* U.S. Senate Committee on Energy and Natural Resources, Washington, D.C.

Bodansky, D. 1996. *Nuclear Energy: Principles, Practices, and Prospects.* American Institute of Physics Press, Woodbury, New York.

Bond, S.L., and T.J. Rodovsky. 1998. *Final Hazard Classification and Auditable Safety Analysis for the DR Interim Safe Storage Project.* BHI-01150, Rev. 0, Bechtel Hanford, Inc., Richland, Washington.

Bosch, D.J. 1991. *Transforming Mission: Paradigm Shift in Theology of Missions.* Orbis Books, Maryknoll, New York.

Boston, H. March 5, 2001. "Boston: ORP Ready to 'Move Forward.'" *The Hanford Reach*, U.S. Department of Energy, Richland, Washington.

Bradley, D.J. 1997. *Behind the Nuclear Curtain: Radioactive Waste Management in the Former Soviet Union.* D.R. Payson, ed., Battelle Press, Columbus, Ohio.

Brevick, C.H. 1994. *Historical Tank Content Estimate for the Northeast Quadrant of the Hanford 200 East Areas.* WHC-SD-WM-ER-349, Westinghouse Hanford Company, Richland, Washington.

Brevick, C.H. 1995a. *Historical Tank Content Estimate for the Southwest Quadrant of the Hanford 200 West Area.* WHC-SD-WM-ER-352, Westinghouse Hanford Company, Richland, Washington.

Brevick, C.H. 1995b. *Historical Tank Content Estimate for the Northwest Quadrant of the Hanford 200 West Area.* WHC-SD-WM-ER-351, Westinghouse Hanford Company, Richland, Washington.

Brevick, C.H. 1995c. *Historical Tank Content Estimate for the Southeast Quadrant of the Hanford 200 East Areas.* WHC-SD-WM-ER-350, ICF Kaiser Hanford Company, Richland, Washington.

Briggs, W. January 30, 1988. "Lawmakers React With Doubt, Anger." *Tri-City Herald*, Kennewick, Washington.

Brown, R.E., and H.G. Ruppert. 1950. *The Underground Disposal of Liquid Wastes at the Hanford Works Washington, Interim Report Covering the Period up to January 1, 1950.* HW-17088, General Electric Company, Hanford Works, Richland, Washington.

Brown, R.E., D.W. Pearce, J.H. Horton Jr, and C.M. Patterson. 1958. "Experience in the Disposal of Radioactive Wastes to the Ground at Two Production Sites." In: *Proceedings of the Second United Nations International Conference on the Peaceful Uses of Atomic Energy*, Vol. 18. pp. 95-100. September 1-13, 1958, Geneva, Switzerland. United Nations, Geneva.

Brown, D.J., R.E. Brown, and W.A. Haney. August 21, 1962. *Appraising Hanford Waste Disposal by Integration of Field Techniques.* HW-SA-2707, Hanford Atomic Products Operation, General Electric Company, Richland, Washington.

Brownell, L.E. August 29, 1958. *Instability of Steel Bottoms in Waste Storage Tanks.* HW-57274, General Electric, Hanford Atomic Products Operation, Richland, Washington.

Bunker, B., J. Virden, B. Kuhn, and R. Quinn. 1995. "Nuclear Materials, Radioactive Tank Wastes." *Encyclopedia of Energy Technology and the Environment*, Vol. 3 (G-P), pp. 2023-2032. John Wiley & Sons, Inc.

Burchfield, J. 2001. "Finding Science's Voice in the Forest." *Across the Great Divide: Explorations in Collaborative Conservation and the American West*, eds. P. Brick, D. Snow, and S. Van De Wetering, Island Press, Washington, D.C.

Burris, K. September 2, 1999. *Hanford Evaporator Campaign a Million-Gallon Success.* Press release, FDH-9909-024.WMP, Fluor Daniel Hanford, Inc., Richland, Washington.

Butler, D., and R.F. Smith. September 23, 1994. *Action Memorandum: N Springs Expedited Response Action Cleanup Plan, U.S. Department of Energy Hanford Site, Richland, WA.* Letter to R. Izatt. U.S. Department of Energy, Richland Operations, Richland, Washington.

Campbell, L.R. 1999. *B Plant Hazards Assessment.* HNF-SD-PRP-HA-008, Rev. 3, Fluor Daniel Hanford, Inc., Richland, Washington.

Cantril, S.T., and H.M. Parker. August 24, 1945. *The Status of Health and Protection at Hanford Engineer Works.* HW-7-2136, Richland, Washington.

Cardis, E., E.S. Gilbert, L. Carpenter, G. Howe, I. Kato, B.K. Armstrong, V. Beral, G. Cowper, A. Douglas, J.J. Fix, S.A. Fry, J. Kaldor, C. Lave, L. Salmon, P.G. Smith, G.L. Voelz, and L.D. Wiggs. 1995. "Effects of Low Doses and Low Dose Rates of External Ionizing Radiation: Cancer Mortality Among Nuclear Industry Workers in Three Countries." *Radiation Research* 142(2):117-132.

Carlowicz, M.J. April 30, 1996. "It's Time to Assess Environmental Policy." *EOS Transactions* 77(18):173-174.

Carson, R. 1994. *Silent Spring.* Houghton Mifflin Company, Boston.

Carter, L.J., and T.H. Pigford. January/February 1999. "The World's Growing Inventory of Civil Spent Fuel." *Arms Control Today*, 29(1):8-14.

Cary, A. April 11, 1999a. "Study Reveals More Hanford-Related Ills." *Tri-City Herald,* Kennewick, Washington.

Cary, A. July 6, 1999b. "Scientists Scrutinize Low-Level Radiation." *Tri-City Herald,* Kennewick, Washington.

Cary, A. June 9, 2000a. "Study Finds Fewer Deaths Among Women Workers at Nuclear Sites." *Tri-City Herald,* Kennewick, Washington.

Cary, A. January 25, 2000b. "Hanford Cleanup Results Underwhelm Panel." *Tri-City Herald,* Kennewick, Washington.

Cary, A. June 19, 2002a. "Court Revives Downwinder Claims." *Tri-City Herald,* Kennewick, Washington.

Cary, A. July 17, 2002b. "State Tries to Join Suit Against DOE." *Tri-City Herald,* Kennewick, Washington.

Centers for Disease Control and Prevention (CDC) and Fred Hutchinson Cancer Research Center. June 21, 2002. *A Guide to the Hanford Thyroid Disease Study—Final Report.* Atlanta, Georgia.

Cheng, E. Summer 1990. "Comment: Lawmaker as Lawbreaker: Assessing Civil Penalties Against Federal Facilities under RCRA." 57 *University of Chicago Law Review* 845, University of Chicago, Chicago, Illinois.

Chess, C., B.J. Hance, and P.M. Sandman. 1988. *Improving Dialogue With Communities: A Short Guide to Government Risk Communication.* New Jersey Department of Environmental Protection, Trenton, New Jersey.

Clarke, R.H. February 2001. "History of ALARA." *Risk Excellence Notes* 3(1):7. U.S. Department of Energy, Argonne, Illinois. Available URL: http://riskcenter.doe.gov.

Clean Water Act. 1977. Public Law 95-217 as amended, 33 USC 1251 et seq.

Colson, S.D., R.E. Gephart, V.L. Hunter, J. Janata, and L.G. Morgan. 1997. *A Risk-Based Focused Decision-Management Approach for Justifying Characterization of Hanford Tank Waste.* PNNL-11231, Pacific Northwest National Laboratory, Richland, Washington.

Commoner, B. 1959. "The Fallout Problem." Reprinted from *Science* 127(3305): 1023-1026, May 2, 1958. U.S. Congress, Joint Committee on Atomic Energy, Special Subcommittee on Radiation. *Hearings.* Vol. 3, pp. 2573-2577. 86th Cong., 1st Session on Fallout from Nuclear Weapons Tests. U.S. Government Printing Office, Washington, D.C.

Community Relations Plan. 1999. *Section 2: Description of the Hanford Site and the Activities Carried Out on the Site.* Available URL: http://www.hanford.gov/crp/section2.htm. Last updated February 19, 1999.

Conklin, A.W. 1986. *Summary of Preliminary Review of Hanford Historical Documents—1943 to 1957.* Office of Radiation Protection, Department of Social and Health Services, State of Washington, Olympia, Washington.

Conklin, A.W., R.R. Mooney, and J.L. Erickson. 1990. "Environmental Monitoring at Hanford by the State of Washington." In *Environmental Monitoring, Restoration, and Assessment: What Have We Learned?* Proceedings for the Twenty-Eighth Hanford Symposium on Health and the Environment, ed. R.H. Gray, pp. 177-181. Pacific Northwest Laboratory, Richland, Washington.

Consortium for Risk Evaluation with Stakeholder Participation (CRESP). December 15, 1999. *Peer Review of the U.S. Department of Energy's Use of Risk in its Prioritization Process.* New Brunswick, New Jersey.

Consumer Reports. 1959. "The Milk We Drink." Reprinted from *Consumer Reports.* U.S. Congress, Joint Committee on Atomic Energy, Special Subcommittee on Radiation. *Hearings.* Vol. 2, pp. 1940-1953. 86th Cong., 1st Session on Fallout from Nuclear Weapons Tests.

Consumer Reports. February 1960. "The Huge and Ever-Increasing Problem of Radioactive Wastes." *Consumer Reports* 25(2):66-67.

Cooper, C.M. December 31, 1943. *Proposed Program for Study of Stack Dispersion at Hanford.* Letter to R.S. Stone. DUH-47, U.S. Department of Energy Public Reading Room, Richland, Washington.

Cooper, A.T., and R.K. Woodruff. 1993. *Investigation of Exposure Rates and Radionuclide and Trace Metal Distributions Along the Hanford Reach of the Columbia River.* PNL-8789, Pacific Northwest Laboratory, Richland, Washington.

Coopey, R.W. 1948. *Preliminary Report on the Accumulation of Radioactivity as Shown by a Limnological Study of the Columbia River in the Vicinity of Hanford Works.* U.S. Atomic Energy Commission Report, HW-11662, Hanford Atomic Products Operation, Richland, Washington.

Corley, J.P. July 18, 1960. *Reactor Effluent Outfall Structures: Status and Potential Problems.* HW-65989, General Electric, Hanford Atomic Products Operation, Richland, Washington.

Corley, J.P. August 1973a. *Environmental Surveillance at Hanford for CY-1970 Data.* BNWL-1669 ADD, Pacific Northwest Laboratory, Richland, Washington.

Corley, J.P. September 1973b. *Environmental Surveillance at Hanford for CY-1970.* BNWL-1669, Pacific Northwest Laboratory, Richland, Washington.

Coughlin, B.J. August 5, 1986. *Northwest Citizens Forum on Defense Waste Report on the U.S. Department of Energy Draft Environmental Impact Statement on Defense Waste.* Letter to M Lawrence, Manager. U.S. Department of Energy, Richland, Washington.

Covello, V.T. 1992. "Trust and Credibility in Risk Communication." *Environment Digest* 6(1):1-3.

Covello, V.T. 1993. "Risk Communication and Occupational Medicine." Editorial. *Journal of Occupational Medicine* 35(1):18-19.

Covello, V.T., and F.W. Allen. 1988. *Seven Cardinal Rules of Risk Communication.* U.S. Environmental Protection Agency, Office of Public Liaison, Washington, D.C.

Covello, V.T., D.B. McCallum, and M.T. Pavlova, eds. 1989. *Effective Risk Communication: The Role and Responsibility of Government and Non-Government Organizations.* Contemporary Issues in Risk Analysis, Vol. 4. Plenum Press, New York.

Covello, V.T., P.M. Sandman, and P. Slovic. 1988. *Risk Communication, Risk Statistics and Risk Comparisons: A Manual for Plant Managers.* Chemical Manufacturers Association, Washington, D.C.

CRESP. See Consortium for Risk Evaluation with Stakeholder Participation.

D'Antonio, M. 1993. *Atomic Harvest: Hanford and the Lethal Toll of America's Nuclear Arsenal.* Crown Publishers, Inc., New York.

Danielson, R.A. December 1999. *Oversight at the 310 Treated Effluent Disposal Facility (TEDF) 1995-1999.* WDOH/320-022, Washington State Department of Health, Olympia, Washington.

Danielson, R.A., and R.E. Jaquish. 1996. *100-D Island Radiological Survey.* WDOH/ERS-96-1101, Washington State Department of Health, Olympia, Washington.

Davis, G.A. Spring 1986. "Mixed Messages on Mixed Waste: Continued Debate Over the Regulation of Mixtures of Radioactive Waste and Hazardous Chemical Waste." 53 *Tennessee Law Review* 585, Tennessee Law Review Associates, Inc., University of Tennessee, Knoxville, Tennessee.

Davis, J.J., R.W. Perkins, R.F. Palmer, W.C. Hanson, and J.F. Cline. 1958. "Radioactive Materials in Aquatic and Terrestrial Organisms Exposed to Reactor Effluent Water." In: *Proceedings of the Second United Nations International Conference on the Peaceful Uses of Atomic Energy*, Vol. 18. pp. 423-428. September 1-13, 1958, Geneva, Switzerland.

DeNeal, D.L. 1965. *Historical Events, Reactors and Fuels Fabrication.* RL-REA-2247, U.S. Atomic Energy Commission, Richland, Washington.

Devesa, S.S., D.G. Grauman, W.J. Blot, G. Pennello, R.N. Hoover, and J.F. Fraumeni Jr. 1999. *Atlas of Cancer Mortality in the United States 1950-94.* Publication No. 99-4564, U.S. Government Printing Office, Washington, D.C.

Dickson, P. 2001. *Sputnik: The Shock of the Century.* Walker & Company, New York.

Dirkes, R.L., and R.W. Hanf. eds. 1995. *Hanford Site Environmental Report for Calendar Year 1994.* PNL-10574, Pacific Northwest Laboratory, Richland, Washington.

Dirkes, R.L., R.W. Hanf, and T.M. Poston. eds. 1999. *Hanford Site Environmental Report for Calendar Year 1998.* PNNL-12088, Pacific Northwest National Laboratory, Richland, Washington.

DOD. See U.S. Department of Defense.

Dodd, E.N. 1999. *Plutonium Uranium Extraction (PUREX) End State Basis for Interim Operation (BIO) for Surveillance and Maintenance.* HNF-SD-CP-ISB-004, Rev. 0, B&W Hanford Company, Richland, Washington.

DOE. See U.S. Department of Energy.

Dresel, P.E., D.B. Barnett, F.N. Hodges, D.G. Horton, V.G. Johnson, R.B. Mercer, R.M. Smith, L.C. Swanson, and B.A. Williams. 2002. "Section 2.8. 200 West Area." *Hanford Site Groundwater Monitoring for Fiscal Year 2001*, eds. M.J. Hartman, L.F. Morasch, and W.D. Webber, PNNL-13788, Pacific Northwest National Laboratory, Richland, Washington, pp. 2.135-2.218.

Dresel, P.E., D.B. Barnett, F.N. Hodges, V.G. Johnson, R.B. Mercer, L.C. Swanson, and B.A. Williams. 2001. "Section 2.8, 200 West Area." *Hanford Site Groundwater Monitoring for Fiscal Year 2000*, eds. M.J. Hartman, L.F. Morasch, and W.D. Webber. PNNL-13404, Pacific Northwest National Laboratory, Richland, Washington.

Dresel, P.E., B.A. Williams, J.C. Evans, R.M. Smith, C.J. Thompson, and L.C. Hulstrom. May 2000. *Evaluation of Elevated Tritium Levels in Groundwater Downgradient From the 618-11 Burial Ground: Phase 1 Investigations.* PNNL-13228, Pacific Northwest National Laboratory, Richland, Washington.

Drew, T.B. June 26, 1943. *Admixture of Pile Effluent With River.* Letter to H Worthington. DUH-10548, U.S. Department of Energy Public Reading Room, Richland, Washington.

DuPont (E.I. DuPont de Nemours & Co). 1945. *Memoranda for the File, Book 12.* DuPont, Hanford Engineer Works, Richland, Washington.

DuPont (E.I. DuPont de Nemours & Co). 1946. *Operations of Hanford Engineer Works, History of the Project, Book 11.* HAN-73214, DuPont, Hanford Engineer Works, Richland, Washington.

Dummer, C.L., D.J. Bjornstad, and D.W. Jones. 1998. *The Regulatory Environment Guiding DOE's Cleanup: Opportunities for Flexibility.* JIEE 98-4, Joint Institute for Energy and Environment, Knoxville, Tennessee.

Ecology. See Washington State Department of Ecology.

E.I. DuPont de Nemours & Co. See DuPont.

Eldridge, E.F., M.E. Curtiss, R.R. Harris, and E.C. Jensen. April 19, 1950. *Proposed Stream Survey at Columbia River.* Letter to F.C. Schlemmer, attached to April 14, 1950, Meeting Minutes of the Columbia River Advisory Group issued under Kornberg (1950) in HW-17732.

Eliason, J.R. 1966. *Earth Sciences Waste Disposal Investigations July-December, 1965.* BNWL-CC-574, Pacific Northwest Laboratory, Richland, Washington.

EPA. See U.S. Environmental Protection Agency.

ERDA. See U.S. Energy Research and Development Administration.

Essig T.H., G.W.R. Endres, J.K. Soldat, and J.F. Honstead, 1973. "Concentrations of ^{65}Zn in Marine Foodstuffs and Pacific Coastal Residents." In *Radioactive Contamination of the Marine Environment*, Proceedings of Symposium on the Interaction of Radioactive Contaminants with the Constituents of the Marine Environment, IAEA/SM-158/43. pp. 651-688. International Atomic Energy Agency, Seattle, Washington. July 10-14, 1972. International Atomic Energy Agency, Vienna, Austria.

Ewing R.C. October 15, 1999. "Less Geology in the Geological Disposal of Nuclear Waste." *Science* 286(5439):415-416.

Fallon W.E., J.M. Gephart, R.E. Gephart, R.D. Quinn, and L.A. Stevenson. May 1991. *Regulatory and Institutional Issues Impeding Cleanup at U.S. Department of Energy Sites: Perspectives Gained From an Office of Environmental Restoration Workshop*. PNL-7692, Pacific Northwest Laboratory, Richland, Washington.

Farris W.T., B.A. Napier, J.C. Simpson, S.F. Snyder, and D.B. Shipler. 1994. *Columbia River Pathway Dosimetry Report, 1944-1992*. PNWD-2227 HEDR, Pacific Northwest Laboratory, Richland, Washington.

Feder T. December 2000. "Japan Arrests Six in Nuclear Accident that Killed Two." *Physics Today* 53(12):61-62.

Federal Advisory Committee Act (FACA). 1972. Public Law 92-463, 86 Statute 770.

Federal Facilities Compliance Act (FFCA). 1992. Public Law 102-386, 106 Statute 1505.

Federal Facilities Restoration and Reuse Office. April 1996. *Final Report of the Federal Facilities Environmental Restoration Dialogue Committee: Consensus Principles and Recommendations for Improving Federal Facilities Cleanup*. Federal Facilities Restoration and Reuse Office, Washington, D.C.

Federal Radiation Council. 1961. *Background Material for the Development of Radiation Protection Standards*. Report No. 2, Federal Radiation Council, Washington, D.C.

Fermi R., and E. Samra. 1995. *Picturing the Bomb*. Harry N. Abrams, Inc., New York.

Fessendon-Raden J., J.M. Fitchen, and J.S. Heath. 1987. "Providing Risk Information in Communities: Factors Influencing What is Heard and Accepted." Chapter 12 in *Science Technology and Human Values*, pp. 94-101. SAGE Publications, Inc., Thousand Oaks, California.

Fields T. August 21, 1998. *Interim Final Policy on the Use of Risk-Based Methodologies in Setting Priorities for Cleanup Actions at Federal Facilities*. Letter to S.A. Herman. U.S. Environmental Protection Agency, Washington, D.C.

Fioravanti, M. and A. Makhijani. October 1997. *Containing the Cold War Mess: Restructuring the Environmental Management of the U.S. Nuclear Weapons Complex*, Institute for Energy and Environmental Research, Takoma Park, Maryland. Available online: http://www.ieer.org/reports/cleanup/index.html.

Fischhoff B., S. Lichtenstein, S.L. Derby, P. Slovic, and D. Keeney. 1981. *Acceptable Risk*. Cambridge University Press, Cambridge, Massachusetts.

Fleming, D. 1972. "Roots of the New Conservation Movement." In *Perspectives in American History*, Vol VI, eds. D. Fleming and B. Bailyn, pp. 7-91, Charles Warren Center for Studies in American History, Harvard University, Cambridge, Massachusetts.

Foster, R.F. 1959. "Radioactive Waste Management Operation at the Hanford Works, Part 2, Release of Reactor Cooling Water to the Columbia River." U.S. Congress, Joint Committee on Atomic Energy, Special Subcommittee on Radiation. *Hearings*. Vol. 1, part 2, pp. 242-269. 86th Cong., 1st Session on Industrial Radioactive Waste Disposal. U.S. Government Printing Office, Washington, D.C.

Foster, R.F. 1972. "The History of Hanford and Its Contribution of Radionuclides to the Columbia River." *The Columbia River Estuary and Adjacent Waters, Bioenvironmental Studies*, eds. A.T. Pruter and D.L. Alverson, pp. 3-18, University of Washington Press, Seattle, Washington.

Foster, R.F. 1990. "The Embryogenesis of Dose Assessment at Hanford." In *Environmental Monitoring, Restoration, and Assessment: What Have We Learned?* Proceedings for the Twenty-Eighth Hanford Symposium on Health and the Environment, ed. R.H. Gray, pp. 169-175. Pacific Northwest Laboratory, Richland, Washington.

Foster, R.F., and J.J. Davis. 1956. "The Accumulation of Radioactive Substances in Aquatic Form." In *Proceedings of the International Conference on the Peaceful Uses of Atomic Energy*, Vol. 13, pp. 364-367. August 8-20, 1955. United Nations, New York.

Foster, R.F., J.J. Davis, and P.A. Olson. October 11, 1954. *Studies on the Effect of the Hanford Reactors on Aquatic Life in the Columbia River*. HW-33366, General Electric, Hanford Atomic Products Operation, Richland, Washington.

Fred Hutchinson Cancer Research Center. 1999. *Hanford Thyroid Disease Study: Results Booklet (Draft)*. Fred Hutchinson Cancer Research Center, Seattle, Washington.

Freer, B.J., and C.A. Conway. 2002. "Chapter 2: Section 4: Chemical Separations." *Hanford Site Historic District: History of the Plutonium Production Facilities 1943-1990*. Battelle Press, Columbus, Ohio.

Fruchter, J.S., C.R. Cole, M.D. Williams, V.R. Vermeul, J.E. Amonette, J.E. Szecsody, J.D. Istok, and M.D. Humphrey. Spring 2000. "Creation of a Subsurface Permeable Treatment Zone for Aqueous Chromate Contamination Using In Situ Redox Manipulation." *Groundwater Monitoring and Remediation* 20(2):66-77.

Fruchter, J.S., J.M. Zachara, J.K. Fredrickson, C.R. Cole, J.E. Amonette, T.O. Stevens, D.J. Holford, L.E. Eary, G.D. Black, and V.R. Vermeul. 1992. *Manipulation of Natural Subsurface Processes: Field Research and Validation. Pacific Northwest Laboratory Annual Report for 1991 to the DOE Office of Energy Research, Part 2: Environmental Sciences*. PNL-8000, Pacific Northwest Laboratory, Richland, Washington. pp. 88-106.

Fulton, J.C. October 1994. *Hanford Spent Nuclear Fuel Project Recommended Path Forward*. WHC-EP-0830, Vol. 2, Westinghouse Hanford Company, Richland, Washington.

Gamarekian, E. May 7, 1959. "AEC Reveals New York Bread Exceeded Strontium 90 Limit." *Washington Post and Times Herald.* In U.S. Congress, Joint Committee on Atomic Energy, Special Subcommittee on Radiation. *Hearings.* Vol. 2, pp. 1559-1560. 86th Cong., 1st Session on Fallout from Nuclear Weapons Tests. U.S. Government Printing Office, Washington, D.C.

Gamertsfelder, C.C. March 11, 1947. *Re: Effects on Surrounding Area Caused by the Operations of the Hanford Engineer Works.* HW-7-5934, General Electric, Hanford Atomic Products Operation, Richland, Washington.

GE. See General Electric Company.

Gee, G.W., M.J. Fayer, M.L. Rockhold, and M.D. Campbell. 1992. "Variations in Recharge at the Hanford Site." *Northwest Science* 66(4):237-250.

Geiger, H.J., and D.G. Kimball. May 1992. "Health Effects of Nuclear Weapons Production: DOE Research." *Facing Reality: The Future of the U.S. Nuclear Weapons Complex.* Nuclear Safety Campaign, Seattle, Washington.

General Accounting Office (GAO). May 1978. *Department of Energy's Consolidation of Information Processing Activities Needs More Attention.* GAO/EMD-78-60, Washington, D.C.

General Accounting Office (GAO). 1989. *Nuclear Waste: DOE's Management of Single-Shell Tanks at Hanford, Washington.* GAO/RCED-89-157, General Accounting Office, Washington, D.C.

General Accounting Office (GAO). June 1990. *DOE Has Not Demonstrated That Restarting PUREX is a Sound Decision.* GAO/RCED-90-207, General Accounting Office, Washington, D.C.

General Accounting Office (GAO). September 1992. *Department of Energy: Better Information Resources Management Needed to Accomplish Missions.* GAO/IMTEC-92-53, Washington, D.C.

General Accounting Office (GAO). 1993. *Superfund: Cleanups Nearing Completion Indicate Future Challenges.* GAO/RCED-93-188, General Accounting Office, Washington, D.C.

General Accounting Office. 1994a. *Nuclear Cleanup: Completion of Standards and Effectiveness of Land Use Planning Are Uncertain.* GAO/RCED-94-144, General Accounting Office, Washington, D.C.

General Accounting Office. 1994b. *Superfund: Legal Expenses for Cleanup-Related Activities of Major U.S. Corporation.* GAO/RCED-95-46, General Accounting Office, Washington, D.C.

General Accounting Office (GAO). 1995a. *Superfund: Operation and Maintenance Activities Will Require Billions of Dollars.* GAO/RCED-95-259, General Accounting Office, Washington, D.C.

General Accounting Office (GAO). March 1995b. *Department of Energy: National Priorities Needed for Meeting Environmental Agreements.* GAO/RCED-95-1, General Accounting Office, Washington, D.C.

General Accounting Office (GAO). 1996. *Department of Energy: Contract Reform is Progressing but Full Implementation Will Take Years.* GAO/RCED-97-18, General Accounting Office, Washington, D.C.

General Accounting Office (GAO). 1997a. *Superfund: Times to Complete the Assessment and Cleanup of Hazardous Waste Sites.* GAO/RCED-97-20, General Accounting Office, Washington, D.C.

General Accounting Office (GAO). 1997b. *Superfund: Trends in Spending for Site Cleanups.* GAO/RCED-97-211, General Accounting Office, Washington, D.C.

General Accounting Office (GAO). 1998a. *Nuclear Waste: Understanding of Waste Migration at Hanford is Inadequate for Key Decisions.* GAO/RCED-98-80, General Accounting Office, Washington, D.C.

General Accounting Office (GAO). 1998b. *Department of Energy: Lessons Learned Incorporated Into Performance-Based Incentives Contracts.* GAO/RCED-98-223, General Accounting Office, Washington, D.C.

General Accounting Office (GAO). 1998c. *Department of Energy: DOE Lacks an Effective Strategy for Addressing Recommendations From Past Laboratory Advisory Groups.* GAO/T-RCED-98-274, General Accounting Office, Washington, D.C.

General Accounting Office (GAO). 1998d. *Nuclear Waste: Department of Energy's Hanford Tank Waste Project—Schedule, Cost, and Management Issues.* GAO/RCED-99-13, General Accounting Office, Washington, D.C.

General Accounting Office (GAO). 1998e. *Nuclear Waste: Management Problems at the Department of Energy's Hanford Spent Fuel Storage Project.* GAO/T-RCED-98-119, General Accounting Office, Washington, D.C.

General Accounting Office (GAO). 1998f. *Superfund: Times to Complete Site Listing and Cleanup.* GAO/T-RCED-98-74, General Accounting Office, Washington, D.C.

General Accounting Office (GAO). 1998g. *Superfund: Analysis of Contractor Cleanup Spending.* GAO/RCED-98-221, General Accounting Office, Washington, D.C.

General Accounting Office (GAO). 1999. *Major Management Challenges and Program Risks.* GAO/OCG-99-6, General Accounting Office, Washington, D.C.

General Accounting Office (GAO). 1999a. *Department of Energy: Accelerated Closure of Rocky Flats: Status and Obstacles.* GAO/RCED-99-100, General Accounting Office, Washington, D.C.

General Accounting Office (GAO). 1999b. *Superfund: EPA Can Improve Its Monitoring of Superfund Expenditures.* GAO/RCED-99-139, General Accounting Office, Washington, D.C.

General Accounting Office (GAO). 1999c. *Superfund: Half the Sites Have All Cleanup Remedies in Place or Completed.* GAO/RCED-99-245, General Accounting Office, Washington, D.C.

General Accounting Office (GAO). 1999d. *Low-Level Radioactive Waste: States Are Not Developing Disposal Facilities.* GAO/RCED-99-238, General Accounting Office, Washington, D.C.

General Accounting Office (GAO). 1999e. *Nuclear Waste: DOE's Hanford Nuclear Spent Fuel Storage Project—Cost, Schedule, and Management Issues: Report to the Chairman, Committee on Commerce, House of Representatives.* GAO/RCED-99-267, General Accounting Office, Washington, D.C.

General Accounting Office (GAO). 1999f. *Nuclear Waste: DOE's Accelerated Cleanup Strategy Has Benefits but Faces Uncertainties.* GAO/RCED-99-129, General Accounting Office, Washington, D.C.

General Accounting Office (GAO). 1999g. *Superfund: Progress and Challenges.* GAO/RCED-99-202, General Accounting Office, Washington, D.C.

General Accounting Office (GAO). 2000a. *Toxic Chemicals: Long-Term Coordinated Strategy Needed to Measure Exposures to Humans.* GAO/HEHS-00-80, General Accounting Office, Washington, D.C.

General Accounting Office (GAO). 2000b. *Low-Level Radioactive Waste: Department of Energy has Opportunities to Reduce Disposal Costs.* GAO/RCED-00-64, General Accounting Office, Washington, D.C.

General Accounting Office (GAO). 2000c. *Managing for Results: EPA Faces Challenges in Developing Results-Oriented Performance Goals and Measures.* GAO/RCED-00-77, General Accounting Office, Washington, D.C.

General Accounting Office (GAO). 2000d. *Nuclear Waste Cleanup: DOE's Paducah Plan Faces Uncertainties and Excludes Costly Cleanup Activities.* GAO/RCED-00-225, General Accounting Office, Washington, D.C.

General Accounting Office (GAO). 2000e. *Department of Energy: Uncertainties and Management Problems Have Hindered Cleanup at Two Nuclear Waste Sites.* GAO/T-RCED-00-248, General Accounting Office, Washington, D.C.

General Accounting Office (GAO). 2000f. *Radiation Standards: Scientific Basis Inconclusive, and EPA and NRC Disagreement Continues.* GAO/RCED-00-152, General Accounting Office, Washington, D.C.

General Accounting Office (GAO). 2001a. *Radiation Exposure Compensation: Analysis of Justice's Program Administrations.* GAO/01-1043, General Accounting Office, Washington, D.C.

General Accounting Office (GAO). 2001b. *Nuclear Waste: Technical, Schedule, and Cost Uncertainties of the Yucca Mountain Repository Project.* GAO/OCG-02-191, General Accounting Office, Washington, D.C.

General Accounting Office (GAO). 2001c. *Department of Energy: Fundamental Reassessment Needed to Address Major Mission, Structure, and Accountability Problems.* GAO/02-51, General Accounting Office, Washington, D.C.

General Accounting Office (GAO). May 2002a. *Waste Cleanup: Status and Implications of DOE's Compliance Agreements.* GAO-02-567, General Accounting Office, Washington, D.C.

General Accounting Office (GAO). July 19, 2002b. *Waste Cleanup: Implications of Compliance Agreements on DOE's Cleanup Program.* GAO-02-852T, General Accounting Office, Washington, D.C.

General Electric Company (GE). April 15, 1952. *Biology Research-Annual Report 1951,* HW-25021, Hanford Works, Richland, Washington.

General Electric Company (GE). 1960. *Manual of Radiation Protection Standards.* HW-25457, Rev. 2, General Electric Company, Hanford Atomic Products Operation, Richland, Washington.

Gephart, R.E. July 1998. *The Science of Cleaning Up Military Waste in the Western World.* Presented at the Gordon Research Conference, Rhode Island. PNNL-SA-30231, Pacific Northwest National Laboratory, Richland, Washington.

Gephart, R.E., and R.E. Lundgren. 1998. *Hanford Tank Cleanup: A Guide to Understanding the Technical Issues.* Battelle Press, Columbus, Ohio.

Gephart, R.E., S.M. Price, R.L. Jackson, and C.W. Myers. November 1983. "Geohydrologic Factors and Current Concepts Relevant to Characterization of a Potential Nuclear Waste Repository Site in Columbia River Basalt, Hanford Site, Washington." In *Scientific Basis for Nuclear Waste Management Vol. VII, Materials Research Society Symposia Proceedings*, Vol. 26, pp. 85-103. Materials Research Society, Boston, Massachusetts.

Gerber, M.S. 1993. *Summary of the 100-B/C Reactor Operations and Resultant Wastes, Hanford Site.* WHC-SD-EN-RPT-004, Westinghouse Hanford Company, Richland, Washington. Available online as *History of 100-B/C Reactor Operations, Hanford Site.* Available URL: http://www.b-reactor.org/hist-toc.htm.

Gerber, M.S. September 1993a. *Brief History of the PUREX and U03 Facilities.* WHC-MR-0437, Westinghouse Hanford Company, Richland, Washington.

Gerber, M.S. 1997. *On the Home Front: The Cold War Legacy of the Hanford Nuclear Site.* University of Nebraska Press, Lincoln, Nebraska.

Gerber, M.S. 1999. "Hanford Gas Releases Subject of Thyroid Study." *Hanford Reach*, U.S. Department of Energy, Richland Operations, Richland, Washington.

Gerber, M.S. 2000. *PUREX Plant: The Hanford Site's Historic Workhorse.* WHC-MR-0435, Westinghouse Hanford Company, Richland, Washington. Available URL: http://www.hanford.gov/history/0435/0435-8th.htm. Last updated December 3, 1999.

Gerber, M.S. 2002. "Chapter 2: Section 5: Plutonium Finishing." *Hanford Site Historic District: History of the Plutonium Production Facilities 1943-1990.* Battelle Press, Columbus, Ohio.

Gerber, M.S, D.W. Harvey, and J.G. Longenecker. 1997. "5.0 The Manhattan Project and Cold War Eras, Plutonium Production at the Hanford Site, December 1942-1990." *National Register of Historic Places Multiple Property Documentation Form Historic, Archaeological and Traditional Cultural Properties of the Hanford Site.* DOE/RL-97-02, Rev 0. Available URL: http://www.hanford.gov/doe/culres/mpd/sec5.htm. Last updated January 24, 1997.

Gibbs, L.M. 1998. "Learning From Love Canal: A 20th Anniversary Retrospective." *Orion Afield* 2(2).

Gilbert, E.S. 1989. "Issues in Analyzing the Effects of Occupational Exposure to Low Levels of Radiation." *Statistics in Medicine* 8:173-187.

Gilbert, E.S. and S. Marks. 1980. "An Updated Analysis of Mortality of Workers in a Nuclear Facility." *Radiation Research* 83(3):740-741.

Gilbert, E.S., D.L. Cragle, and L.D. Wiggs. 1993a. "Updated Analyses of Combined Mortality Data for Workers at the Hanford Site, Oak Ridge National Laboratory, and Rocky Flats Weapons Plant." *Radiation Research* 136(3):408-421.

Gilbert, E.S., E. Omohundro, J.A. Buchanan, and N.A. Holter. June 1993b. "Mortality of Workers at the Hanford Site: 1945-1986." *Health Physics* 64(6):577-590.

Gilbert, R.B. 1997. "Think Before Testing." *Practice Periodical for Hazardous, Toxic & Radioactive Waste Management* 1(3):90-91.

Gillespie, B.M. 2001. "Section 4.1 Air Surveillance." *Hanford Site Environmental Report for Calendar Year 2000*, eds. T.M. Poston, R.W. Hanf, and R.L. Dirkes, PNNL-13487, Pacific Northwest National Laboratory, Richland, Washington.

Glass, W., and S. Snyder. 1995. *Cleaning Up Defense Installations: Issues and Options.* U.S. Congressional Budget Office, Washington, D.C.

Goldbery, S. 1998. "General Groves and the Atomic West: The Making and Meaning of Hanford." *The Atomic West*, B. Hevly and J.M. Findlay eds., pp. 39-90. Center for the Study of the Pacific Northwest in association with the University of Washington Press, Seattle, Washington.

Goldstein, B.D. February 2001. "Philosophies of Risk: The Precautionary Principle." *Risk Excellence Notes* 3(1):6. U.S. Department of Energy, Argonne, Illinois. Available URL: http://riskcenter.doe.gov.

Gorman, A.E. February 3, 1950. *Columbia River Advisory Board Conference.* Letter to LR Rafstad. DDTS-559, Richland, Washington.

Grady, J.A. February 15, 1943. *Site "W" Description.* Letter to R Williams. DUH-10038, General Electric Company, Hanford Atomic Products Operation, Richland, Washington.

Gray, P. (editor). May 1992. "Overview." *Facing Reality: The Future of the U.S. Nuclear Weapons Complex.* Nuclear Safety Campaign, Seattle, Washington.

Gray, R.H. (ed). 1990. *Environmental Monitoring, Restoration, and Assessment: What Have We Learned?* Proceedings for the Twenty-Eighth Hanford Symposium on Health and the Environment, pp. 342. Pacific Northwest Laboratory, Richland, Washington.

Grossman, D. 1994. "Hanford and Its Early Radioactive Atmospheric Releases." *Pacific Northwest Quarterly* 85(1):6-14, University of Washington, Seattle, Washington.

Groves, L.R. 1962. *Now It Can Be Told: The Story of the Manhattan Project.* Harper and Brothers, New York.

Grumbly, T.P. April 15, 1996. Letter to W.J. Madia, Director of the Pacific Northwest National Laboratory, Richland, Washington.

Grumbly, T.P. 1998. "Talking Points for Forbes Magazine and PASHA Publications Speech: The Future of the U.S. and International Environmental Industry." *Federal Facilities Environmental Journal* 8(4):7-10.

H.R. REP. 1980. No. 96-1016, 96th Congress, 2nd Session. Reprinted in 1980 U.S.C.C.A.N. 6132.

H.R. REP. 1984. No. 98-198, 98th Congress, 2nd Session. Reprinted in 1984 U.S.C.C.A.N. 5576, 5622.

H.R. REP. 1986. No. 99-253, 99th Congress, 2nd Session. Reprinted in 1986 U.S.C.C.A.N. 2835, 2839.

H.R. REP 4205. 2000. *Enactment of Provisions of H.R. 5408, the Floyd D. Spence National Defense Authorization Act for Fiscal Year 2001.* Conference Report to accompany H.R. 4205, Sec. 3602.

H&R Technical Associates, and N.R. Kerr. 2000. *U Plant Facility Safety Analysis Report.* BHI-01157, Rev. 2, Bechtel Hanford, Inc., Richland, Washington.

Hall, R.B. 1963. *Environmental Effects of a Fuel Element Failure.* HW-79073, General Electric, Hanford Atomic Products Operation, Richland, Washington.

Hancock, D. and A. Makhijani. May 1992. "Radioactive Waste Storage and Disposal." *Facing Reality: The Future of the U.S. Nuclear Weapons Complex.* Nuclear Safety Campaign, Seattle, Washington.

Hanford. 2000. *300 Area History Hanford Site.* Available URL: http://www.hanford.gov/history/300area/300-3rd.htm. Last updated December 3, 1999.

Hanford Advisory Board (HAB). June 2-3, 1994. *Advice to the Tri-Parties.* Consensus advice #01, issued by the Hanford Advisory Board. Available URL: http://www.hanford.gov/boards/hab/index.htm.

Hanford Advisory Board (HAB). April 7, 1995a. *HAB Consensus Advice #18, Privatization.* Issued by the Hanford Advisory Board. Available URL: http://www.hanford.gov/boards/hab/index.htm.

Hanford Advisory Board (HAB). October 5-6, 1995b. *HAB Advice on TWRS Privatization.* Consensus advice #32, issued by the Hanford Advisory Board. Available URL: http://www.hanford.gov/boards/hab/index.htm.

Hanford Advisory Board (HAB). 1996a. *Hanford Advisory Board Progress Report Fiscal Year 1996: A Summary of Stakeholder Accomplishments and Expectations.* Available URL: http://www.hanford.gov/boards/hab/index.htm.

Hanford Advisory Board (HAB). November 8, 1996b. *Project Hanford Management Contract (PHMC).* Consensus advice #55, issued by the Hanford Advisory Board. Available URL: http://www.hanford.gov/boards/hab/index.htm.

Hanford Advisory Board (HAB). December 5, 1996c. *Columbia River Comprehensive Impact Assessment.* Consensus advice #61, issued by the Hanford Advisory Board. Available URL: http://www.hanford.gov/hab/advice/advice61.htm.

Hanford Advisory Board (HAB). 1997a. *Hanford Advisory Board Progress Report Fiscal Year 1997: A Summary of Stakeholder Accomplishments and Expectations.* Available URL: http://www.hanford.gov/boards/hab/index.htm.

Hanford Advisory Board (HAB). April 4, 1997b. *Ten Year Plan.* Consensus advice #68, issued by the Hanford Advisory Board. Available URL: http://www.hanford.gov/boards/hab/index.htm.

Hanford Advisory Board (HAB). December 5, 1997c. *TWRS Vadose Zone Characterization.* Consensus advice #83, issued by the Hanford Advisory Board. Available URL: http://www.hanford.gov/boards/hab/advice/advice83.htm.

Hanford Advisory Board (HAB). 1998. *Hanford Advisory Board Progress Report Fiscal Year 1998: A Summary of Stakeholder Accomplishments and Expectations.* Available URL: http://www.hanford.gov/boards/hab/progress/hab1998.pdf.

Hanford Advisory Board (HAB). 1999. *Hanford Advisory Board Progress Report Fiscal Year 1999: Managing the Focus on Cleanup in a Time of Transition.* Available URL: http://www.hanford.gov/boards/hab/progress/hab1999.pdf.

Hanford Advisory Board (HAB). April 7, 2000a. *618-10 and 618-11 Burial Grounds.* Consensus advice #106, issued by the Hanford Advisory Board. Available URL: http://www.hanford.gov/boards/hab/advice/advice106.htm.

Hanford Advisory Board (HAB). December 8, 2000b. *Hanford 2012: Accelerating Cleanup and Shrinking the Site (Hanford 2012).* Consensus advice #113, issued by the Hanford Advisory Board. Available URL: http://www.hanford.gov/boards/hab/advice/advice113.htm.

Hanford Environmental Dose Reconstruction Project (HEDR). 1994. *Summary: Radiation Dose Estimates From Hanford Radioactive Material Releases to the Air and the Columbia River.* Technical Steering Panel of the Hanford Environmental Dose Reconstruction Project, Washington State Department of Ecology, Olympia, Washington.

Hanford Future Site Uses Working Group. 1992. *The Future for Hanford: Uses and Cleanup: Summary of the Final Report of the Hanford Future Site Uses Working Group.* Hanford Future Site Uses Working Group, Richland, Washington.

Hanford Natural Resource Trustee Council. 2001. *The Hanford Natural Resource Trustee Council.* Available URL: http://www.hanford.gov/boards/nrtc/index.html. Last updated July 24, 2001.

Hanford Openness Workshops. October 8, 1997. *Hanford Openness Workshop 1 Summary October 8, 1997.* Available URL: http://www.hanford.gov/boards/openness/workshops.htm. Last updated August 18, 1998.

Hanford Openness Workshops. 1998. *Hanford Openness Workshop October 1997- May 1998 Final Report—Executive Summary.* Available URL: http://www.hanford.gov/boards/openness/trac-0818/execsum.htm. Last updated February 19, 1999.

Hanford Openness Workshops Fact Sheet. 1999a. *Document Review.* HOW-980815-2. Available URL: http://www.hanford.gov/boards/openness/factsheets.htm.

Hanford Openness Workshops Fact Sheet. 1999b. *Background.* HOW-980815-1. Available URL: http://www.hanford.gov/boards/openness/factsheets.htm.

Hanford Reach. October 30, 2000. "DOE Reaches Decision on Burial Grounds." *Hanford Reach,* U.S. Department of Energy, Richland, Washington.

Hanford Tank Waste Task Force. 1993. *Final Report, Hanford Tank Waste Task Force.* Hanford Tank Waste Task Force, Richland, Washington.

Hanford Works. March 27, 1947. *The Trend of Contamination Observed in the Air, Columbia River, and Vegetation, at the Hanford Engineer Works for 1946.* HW-3-5402, Richland, Washington.

Hanford Works. January 21, 1948. *Hanford Works Monthly Report December, 1947.* HW-08438-DEL. General Electric Company, Hanford Works, Richland, Washington.

Hanford Works. April 22, 1965. *Chemical Processing Department Monthly Report for March, 1965.* General Electric Company, Hanford Atomic Works Products Operation, Richland, Washington.

Hanlon, B.M. February 1999a. *Waste Tank Summary Report for Month Ending December 31, 1998.* HNF-EP-0182 Rev. 129, Fluor Daniel Hanford, Inc., Richland, Washington.

Hanlon, B.M. May 1999b. *Waste Tank Summary Report for Month Ending February 28, 1999.* HNF-EP-0182 Rev. 131, Fluor Daniel Hanford, Inc., Richland, Washington.

Hanlon, B.M. January 31, 2001. *Waste Tank Summary Report for Month Ending January 31, 2001.* HNF-EP-0182 Rev. 154, CH2MHILL Hanford Group, Inc., Richland, Washington.

Hanlon, B.M. September 2002. *Waste Tank Summary Report for Month Ending July 31, 2002.* HNF-EP-0182 Rev. 172, CH2MHILL Hanford Group, Inc., Richland, Washington.

Hanson, W.C., and R.L. Browning. 1952. "Biological Monitoring." *Biology Research-Annual Report 1951,* HW-25021, Hanford Works, Richland, Washington, pp. 202-203.

Hanson, W.L., and H.A. Kornberg. 1956. "Radioactivity in Terrestrial Animals Near an Atomic Energy Site." In *Proceedings of the First International Conference on Peaceful Uses of Atomic Energy* (Geneva). Vol. 13, pp. 385-388, August 8-20, 1955, Geneva, Switzerland. United Nations, New York.

Hartman, M.J., and P.E. Dresel, eds. 1998. *Hanford Site Groundwater Monitoring for Fiscal Year 1997.* PNNL-11793, Pacific Northwest National Laboratory, Richland, Washington.

Hartman, M.J., and J.P. McDonald. March 2002. "Section 2.1. Overview of Hanford Site Groundwater." *Hanford Site Groundwater Monitoring for Fiscal Year 2001*, eds. M.J. Hartman, L.F. Morasch, and W.D. Webber, PNNL-13788, Pacific Northwest National Laboratory, Richland, Washington, pp. 2.3-2.19

Hartman, M.J. and W.J. McMahon. March 2002. "Section 2.4. 100 N Area." *Hanford Site Groundwater Monitoring for Fiscal Year 2001*, eds. M.J. Hartman, L.F. Morasch, and W.D. Webber, PNNL-13788, Pacific Northwest National Laboratory, Richland, Washington, pp. 2.72-2.73.

Hartman, M.J., L.F. Morasch, and W.D. Webber, eds. 2000. *Hanford Site Groundwater Monitoring for Fiscal Year 1999.* PNNL-13116, Pacific Northwest National Laboratory, Richland, Washington.

Hartman, M.J., L.F. Morasch, and W.D. Webber, eds. 2001. *Hanford Site Groundwater Monitoring for Fiscal Year 2000.* PNNL-13404, Pacific Northwest National Laboratory, Richland, Washington.

Hartman, M.J., L.F. Morasch, and W.D. Webber, eds. March 2002. *Hanford Site Groundwater Monitoring for Fiscal Year 2001.* PNNL-13788, Pacific Northwest National Laboratory, Richland, Washington.

Hartz, J., and R. Chappell. 1997. *Worlds Apart: How the Distance Between Science and Journalism Threatens America's Future.* The First Amendment Center, Nashville, Tennessee.

Haushild, W.L., H.H. Stevens, J.L. Nelson, and G.R. Dempster, Jr. 1973. *Radionuclides in Transport in the Columbia River From Pasco to Vancouver, Washington.* United States Geological Survey Professional Paper 433-N, U.S. Government Printing Office, Washington, D.C.

Healy, J.W. May 1, 1946. *Vegetation Contamination for First Quarter of 1946.* Memo to file, HW 3-3495, General Electric Company, Hanford Atomic Products Operation, Richland, Washington.

Healy, J.W. 1948. *Long-Lived Fission Activities in the Stack Gases and Vegetation at the Hanford Works.* HW-10758, General Electric Company, Hanford Atomic Products Operation, Richland, Washington.

Healy, J.W. 1953. *Release of Radioactive Wastes to Ground.* HW-28121, General Electric Company, Hanford Atomic Products Operation, Richland, Washington.

Healy, J.W. October 12, 1956. *Reactor Effluent Monitoring.* HW-45725, General Electric Company, Hanford Atomic Products Operation, Richland, Washington.

Healy, J.W., B.V. Andersen, H.V. Clukey, and J.K. Soldat. 1958. "Radiation Exposure to People in the Environs of a Major Production Atomic Energy Plant." In *Proceedings of the Second United Nations International Conference on the Peaceful Uses of Atomic Energy*, Vol. 18. pp. 309-318. September 1-13, 1958, Geneva, Switzerland. United Nations, Geneva.

HEDR. See Hanford Environmental Dose Reconstruction Project.

Heeb, C.M. 1994. *Radionuclide Releases to the Atmosphere From Hanford Operations, 1944-1972.* PNWD-2222 HEDR, Pacific Northwest Laboratory, Richland, Washington.

Heeb, C.M., and D.J. Bates. 1994. *Radionuclide Releases to the Columbia River From Hanford Operations, 1944-1971.* PNWD-2223 HEDR, Pacific Northwest Laboratory, Richland, Washington.

Herde, K.E. March 1, 1946. *I^{131} Accumulation in the Thyroid of Sheep Grazing Near H.E.W.* Letter to J.W. Healy. H.W. 3-3455, E. I. DuPont de Nemours & Company, Hanford Atomic Products Operation, Richland, Washington.

Herde, K.E. May 14, 1947. *Radioactivity in Various Species of Fish From the Columbia and Yakima Rivers.* HW-3-5501, General Electric Company, Hanford Atomic Products Operation, Richland, Washington.

Herde, K.E. October 19, 1948. *Radioactivity of Pile Effluent and Its Biological Significance.* HW-11509, Richland, Washington.

Herde, K.E., R.L. Browning, and W.C. Hanson. August 21, 1951. *Activity Densities in Waterfowl of the Hanford Reservation and Environs.* HW-18645, General Electric Company, Nucleonics Division, Richland, Washington.

Hersh, R., K. Probst, K. Wernstedt, and J. Mazurek. 1997. *Linking Land Use and Superfund Cleanups: Uncharted Territory.* Center for Risk Management, Resources for the Future, Washington, D.C.

Hesser, W.A., P.A. Baynes, P.M. Daling, T.F. Demmitt, R.D. Jensen, L.E. Johnson, L.D. Muhlstein, S.M. O'Toole, A.L. Pajunen, M.B. Triplett, J.L. Waite, and T.M. Wintczak. 1995. *Development of a Risk-Based Approach to Hanford Site Cleanup.* PNL-10651, Pacific Northwest Laboratory, Richland, Washington.

Hevly, B.W., and J.M. Findlay. 1998. *The Atomic West.* Center for the Study of the Pacific Northwest, University of Washington Press, Seattle, Washington.

Highland, N.M. December 12, 1990. *Moratorium on Destruction of Records.* Letter from U.S. Department of Energy-Richland Operations to Hanford contractors.

Hodges, F.N. January 1998. *Results of Phase I Groundwater Quality Assessment for Single-Shell Tank Waste Management Areas T and TX-TY at the Hanford Site.* PNNL-11809, Pacific Northwest National Laboratory, Richland, Washington.

Holifield, C. January 1958. "Congressional Hearings on Radiation Fallout." *Bulletin of Atomic Scientists* 14(1):52-54.

Honstead, J.F. July 21, 1954. *Columbia River Survey 1951, 1952, 1953.* HW-32506, General Electric Company, Hanford Atomic Products Operation, Richland, Washington.

Honstead, J.F. October 11, 1955. *Disposal of Reactor Effluent Through an Inland Lake System.* HW-39465, General Electric Company, Hanford Atomic Products Operation, Richland, Washington.

Hornstead, J.F. 1967. *A Program for Evaluating Environmental Radiation Dose to Children.* BNWL-SA-1288, Battelle Northwest Laboratory, Richland, Washington.

Honstead, J.F., R.F. Foster, and W.H. Bierschenk. 1960. "Movement of Radioactivity Effluents in Natural Waters at Hanford." In *Disposal of Radioactive Wastes II. Proceedings of the Scientific Conference on the Disposal of Radioactive Wastes*, Vol. 2, pp. 385-399. November 11-21, 1960, Monaco. International Atomic Energy Agency, Vienna, Austria.

Hughes, G. 1994. *Special Report: Radon in Washington.* Washington State Department of Health, Olympia, Washington.

Hughes, M. March 3, 2001. "Bechtel's Hanford Cleanup Advances Steadily, Safely." *Tri-City Herald*, p. D4, Kennewick, Washington.

Institute for Energy and Environmental Research (IEER) and International Physicians for the Prevention of Nuclear War (IPPNR). 1992. *Plutonium: Deadly Gold of the Nuclear Age, The Health and Environmental Problems of Plutonium Production and Disposal.* International Physicians Press, Cambridge, Massachusetts.

Integration Project Expert Panel. April 7, 2000a. *Closeout Report for Panel Meeting Held January 26-28, 2000.* Prepared for Bechtel Hanford, Inc. and U.S. Department of Energy, Richland, Washington.

Integration Project Expert Panel. December 29, 2000b. *Closeout Report for Panel Meeting Held October 25-27, 2000.* Prepared for Bechtel Hanford, Inc. and U.S. Department of Energy, Richland, Washington.

Irish, E.R. July 31, 1959. *Handling Failed Fuel Elements at Hanford.* HW-61357, General Electric Company, Hanford Atomic Products Operation, Richland, Washington.

Irish, E.R. August 2, 1961. *A Comparison of Ground Waste Disposal Status at Hanford: 1959-1961.* HW-SA-2236, General Electric Company, Hanford Laboratories Operation, Richland, Washington.

Isochem Inc. April 20, 1967. *Chemical Processing Division Monthly Report for March, 1967.* ISO-709-DEL, Isochem, Inc., Richland, Washington.

Jacob, G. 1990. *Site Unseen: The Politics of Siting a Nuclear Waste Repository.* University of Pittsburgh Press, Pittsburgh, Pennsylvania.

Jeppson, D.W. 1973. "Tritium Effluents During Zirconium Alloy-Clad Fuel Processing." *Tritium Control Technology*, pp. 412-416, T.B. Rhinehammer, and P.H. Lamberger, eds. WASH 1269, U.S. Atomic Energy Agency, Washington, D.C.

Jerman, P.C., W.N. Koop, and F.E. Owen. May 27, 1965. *Release of Radioactivity to the Columbia River From Irradiated Fuel Element Ruptures*. RL-REA-2160, General Electric, Hanford Atomic Products Operation, Richland, Washington.

Johnson, K.G. date unknown. "Quotes for Teachers." Complied by Richard D. Zakia, Rochester Institute of Technology. Available URL: http://www.rit.edu/~andpph/text-quotations.html.

Johnson, W.E. December 27, 1962. *Control of Information on Columbia River.* HW-76022, General Electric, Hanford Atomic Products Operation, Richland, Washington.

Johnson, E.J., J. Hershey, J. Meszaros, and H. Kunreuther. 1993. "Framing, Probability Distortions, and Insurance Decisions." *Journal of Risk and Uncertainty* 7(1):35-51.

Kamrin, M.A., D.J. Katz, and M.L. Walter. 1995. *Reporting on Risk: A Journalists Handbook on Environmental Risk Assessment.* 2nd ed. Foundation for American Communications and National Sea Grant College Program, Ann Arbor, Michigan.

Kaplan, S. 1997. "The Words of Risk Analysis." *Risk Analysis* 17(4):407-417.

Kasperson, R.E., O. Renn, P. Slovic, H.S. Brown, J. Emel., R. Goble, J.X. Kasperson, and S. Ratick. 1988. "The Social Amplification of Risk: A Conceptual Framework." *Risk Analysis* 8:177-187.

Keeney, R.L. 1992. *Value Focused Thinking: A Path to Creative Decision Making*. Harvard University Press, London.

Kerr, N.R. 1998. *Surplus Reactor Auditable Safety Analysis.* BHI-01172, Rev. 1, draft, Bechtel Hanford, Inc., Richland, Washington.

Kerr, N.R., H&R Technical Associates, and M.H. Chew & Associates. 2000. *REDOX Facility Safety and Analysis Report.* BHI-01142, Rev. 2, Bechtel Hanford, Inc., Richland, Washington.

Kersting, A.B., D.W. Efurd, D.L. Finnegan, D.J. Rokop, D.K. Smith, and J.L. Thompson. 1999. "Migration of Plutonium in Ground Water at the Nevada Test Site." *Nature* 397(6714):56-59.

Kincaid, C.T., M.P. Bergeron, C.R. Cole, M.D. Freshley, N.L. Hassig, V.G. Johnson, D.I. Kaplan, R.J. Serne, S.P. Streile, D.L. Strenge, P.D. Thorne, L.W. Vail, G.A. Whyatt, and S.K. Wurstner. 1998. *Composite Analysis for Low-Level Waste Disposal in the 200 Area Plateau of the Hanford Site.* PNNL-11800, Pacific Northwest National Laboratory, Richland, Washington.

Kincaid, C.T. et al. July 2001. *Appendix A: Inventory Data for Initial Assessment Performed With the System Assessment Capability (Rev.0).* Updated appendix published for *System Assessment Capability (Revision 0: Assessment Description, Requirements, Software Design, and Test Plan,* May 2000, BHI-01365 Draft A, Bechtel Hanford, Inc., Richland, Washington.

Kingwell, M. August 8, 1999. "The Way We Live Now: A Shock to the System." *The New York Times Magazine.* Section 6, p. 15.

Kitzhaber, J. February 28, 1996. Letter to the Oregon Hanford Waste Board, Salem about its role to protect public safety and the Oregon environment. Salem, Oregon.

Klein, K.A. June 22, 2000. *Response to the Hanford Advisory Board (HAB) Advice on the 618-10 and 618-11 Burial Grounds.* Letter to M.B. Reeves. Available URL: http://www.hanford.gov/boards/hab/response/106.htm. Last updated August 1, 2000.

Klepper, E.L., K.A. Gano, and L.L. Cadwell. 1985. *Rooting Depth and Distribution of Deep-Rooted Plants in the 200 Area Control Zone of the Hanford Site.* PNL-5247, Pacific Northwest Laboratory, Richland, Washington.

Kornberg, H.A. April 27, 1950. *Special Meeting of the Columbia River Advisory Group April 14, 1950-Richland, Washington.* HW-17732, General Electric Company, Richland, Washington.

Kornberg, H.A. 1958. "Radiation Biology as a Supporting Function for Atomic Energy Installations." In: *Proceedings of the Second United Nations International Conference on the Peaceful Uses of Atomic Energy*, Vol. 18. pp. 329-335. September 1-13, 1958, Geneva, Switzerland. United Nations, Geneva.

Koster, G.L. 1998. *Auditable Safety Analysis and Final Hazard Classification for the 105-N Reactor Zone and 109-N Steam Generator Zone Building.* BHI-01179, Rev. 0, Bechtel Hanford, Inc., Richland, Washington.

Kunreuther, H., and P. Slovic. 1999. "Coping With Stigma: Challenges and Opportunities." *Risk: Health, Safety, & Environment* 10:269-280.

Lake, J.A., R.G. Bennett, and J.F. Kotek. January 2002. "Next-Generation Nuclear Power." *Scientific American* 286(1):72-81.

Lamarre, L. 1992. "What Are You Afraid Of?" *Electric Power Research Institute (EPRI) Journal* 17(7):22-27.

Lawrence, M. December 12, 1999. "Disposal Can Wait No More." *Tri-City Herald,* Kennewick, Washington.

League of Women Voters. 1993. *The Nuclear Waste Primer.* The League of Women Voters Education Fund, Washington, D.C.

Lee, M. November 29, 2000. "Seattle Wants It Both Ways With Dams." *Tri-City Herald,* Kennewick, Washington.

Lide, D.R., ed. 1995. *CRC Handbook of Chemistry and Physics: A Ready Reference Book of Chemical and Physical Data.* 76th ed. CRC Press, Boca Raton.

Lilienthal, D.E. 1963. *Change, Hope, and the Bomb.* Princeton University Press, Princeton, New Jersey.

Lilienthal, D.E. 1980. *Atomic Energy: A New Start.* Harper & Row, Publishers, New York.

Lind, P. January 2, 2002. *Poisoned Waters: Pesticide Contamination of Waters and Solutions to Protect Pacific Salmon.* The Washington Toxics Coalition, Seattle, Washington. Available URL: http://www.watoxics.org/pdffiles/PoisonedWaters.pdf.

Lindberg, J.W., and P.E. Dresel. March 2002. "Section 2.12. 300 Area." *Hanford Site Groundwater Monitoring for Fiscal Year 2001*, eds. M.J. Hartman, L.F. Morasch, and W.D. Webber. PNNL-13788, Pacific Northwest National Laboratory, Richland, Washington, pp. 2.305-2.330.

Lindberg, J.W., P.E. Dresel, D.B. Barnett, J.P. McDonald, R.B. Mercer, S.M. Narbutovskih, R.E. Peterson, M.D. Sweeney, and B.A. Williams. 2001. "Section 2.9. 200 East Area." *Hanford Site Groundwater Monitoring for Fiscal Year 2000,* eds. M.J. Hartman, L.F. Morasch, and W.D. Webber. PNNL-13404, Pacific Northwest National Laboratory, Richland, Washington.

Lindberg, J.W., E.C. Thornton, D.B. Barnett, J.P. McDonald, R.B. Mercer, S.M. Narbutovskih, and M.D. Sweeney. 2002. "Section 2.9. 200 East Area." *Hanford Site Groundwater Monitoring for Fiscal Year 2001*, eds. M.J. Hartman, L.F. Morasch, and W.D. Webber. PNNL-13788, Pacific Northwest National Laboratory, Richland, Washington, pp. 2.219-2.287.

Looney, B.B., and R.W. Falta, eds. 2000. *Vadose Zone: Science and Technology Solutions.* Battelle Press, Columbus, Ohio.

Lundgren, R.E., and A.H. McMakin. 1998. *Risk Communication: A Handbook for Communicating Environmental, Safety, and Health Risks.* Battelle Press, Columbus, Ohio.

MacDonald, J.A. August 1, 1999. "Cleaning Up the Nuclear Weapons." *Environmental Science and Technology* 33(15).

Macilwain, C. 1996. "Science Seeks Weapons Clean-Up Role." *Nature* 385:375-379.

Macilwain, C. 2001. "Out of Sight, Out of Mind?" *Nature* 412(6850):850-852.

Makhijani, A. 1995. "Always the Target?" *The Bulletin of the Atomic Scientists* 51(3)23-27.

Manhattan Project. 1977. "Book IV Pile Project, X-10, Volume 4 Land Acquisition, Hanford Engineers Works." *Manhattan Project: Official History and Documents.* Available: Hanford Technical Library (microfilm), Richland, Washington.

Marceau, T.E., D.W. Harvey, D.C. Stapp, S.D. Cannon, C.A. Conway, D.H. DeFord, B.J. Freer, M.S. Gerger, J.K. Keating, C.F. Noonan, and G. Weisskopf. 2002. *Hanford Site Historic District: History of the Plutonium Production Facilities 1943-1990*. Battelle Press, Columbus, Ohio.

Martin, T. March 3, 2001. "Welcome to Hanford 101: A Course We Must Not Fail." Science and Technology Progress 2001 Edition, *Tri-City Herald*, p. D2, Kennewick, Washington.

Martland, H.S. 1929. "Occupational Poisoning in Manufacture of Luminous Watch Dials, General Review of Hazard Caused by Ingestion of Luminous Paint, With Especial Reference to the New Jersey Cases." *J. Am. Med. Assoc.* 92(466-73):552-559.

Matthias, F. 1942. *Colonel Franklin T. Matthias Diary November 24, 1942-January 29, 1943*. U.S. Department of Energy Public Reading Room, Richland, Washington.

McCullugh, R.W., and J.R. Cartmell. August 30, 1968. *Chronological Record of Significant Events in Separations Operations*. Letter to W. Harty. ARH-780, Atlantic Richfield Hanford Company, Richland, Washington.

McCullough, J., T.C. Hazen, S.M. Benson, F.B. Metting, and A.C. Palmisano. 1999. *Bioremediation of Metals and Radionuclides: What It Is and How It Works*. LBNL-42595, Lawrence Berkeley National Laboratory, Berkeley, California.

McElroy, J.L., W.F. Bonner, H.T. Blair, W.J. Bjorklund, C.C. Chapman, R.D. Dierks, and L.S. Romero. 1976. "Recent Calcination and Vitrification Process Accomplishments." In *Management of Radioactive Wastes from the Nuclear Fuel Cycle*, Vol. 1, pp. 283-302. March 22-26, 1976, Vienna, Austria. International Atomic Energy Agency, Vienna, Austria.

McElroy, J.L., W.J. Bjorklund, and W.F. Bonner. 1982. *Waste Vitrification: A Historical Perspective*. PNL-SA-10261, Pacific Northwest Laboratory, Richland, Washington.

McHenry, J.R. 1954. *Adsorption and Retention of Cesium by Soils of the Hanford Project*. HW-31011, Hanford Atomic Products Operation, Richland, Washington.

McKinley, J.P., C.J. Zeissler, J.M. Zachara, R.J. Serne, R.M. Lindstrom, H.T. Schaef, and R.D. Orr. September 1, 2001. "Distribution and Retention of ^{137}Cs in Sediments at the Hanford Site, Washington." *Environmental Science and Technology* 35(17):3433-3441.

McLean, M. August 23, 1998. "More Than Metal: Pollutants Found in Washington Reach of the Spokane River." *Coeur d' Alene Press*, Coeur d' Alene, Idaho, pp. A1, A3.

McMurray, B.J. 1983. "1976 Hanford Americium Exposure Incident: Accident Description." *Health Physics* 45(4):847-853.

Mercier, P.F., M.D. Wonacott, and C. DeFigh-Price. December 1981. *Survey of the Single-Shell Tank Thermal Histories*. RHO-CD-1172, Rockwell Hanford Company, Richland, Washington.

Michaels, D. June 13, 2000. "DOE Comments on Draft GAO Report 'Low-Level Radiation Standards.'" In *Radiation Standards: Scientific Basis Inconclusive, and EPA and NRC Disagreement Continues.* GAO/RCED-00-152, General Accounting Office, Washington, D.C.

Military Policy Committee. 1943. *Report of August 21, 1943, on Present Status and Future Program on Atomic Fission Bombs.* Records of the Manhattan Engineer District, 1942-1948, Record Group 77, National Archives, Washington, D.C.

Miller, N.R. March 4, 1976. *Directory of ERDA-Owned Nuclear Reactor.* Letter to O.J. Bennett (U.S. Energy Research and Development Administration-Richland Operations Office). United Nuclear Industries, Richland, Washington.

Miller, R.L., and J.M. Steffes. 1987. *Radionuclide Inventory and Source Terms for the Surplus Production Reactors at Hanford.* UNI-3714, Rev. 1, UNC Nuclear Industries, Richland, Washington.

Moeller, D.W. 1990. "Natural Radiation in the Environment." In *Environmental Monitoring, Restoration, and Assessment: What Have We Learned?* Proceedings for the Twenty-Eighth Hanford Symposium on Health and the Environment, ed. R.H. Gray, pp. xix-xxxiii. Pacific Northwest Laboratory, Richland, Washington.

Morgan, R.H. 1959. "Report to the Surgeon General, U.S. Public Health Service on the Control of Radiation Hazards in the United States." Prepared by The National Advisory Committee on Radiation. In U.S. Congress, Joint Committee on Atomic Energy, Special Subcommittee on Radiation. *Hearings.* Vol. 3, pp. 2545-2566. 86th Cong., 1st Session on Fallout from Nuclear Weapons Tests.

Murray, R.L. 1961. *Introduction to Nuclear Engineering.* 2nd ed. Prentice-Hall, Englewood Cliffs, New Jersey.

Murray, R.L. 1989. *Understanding Radioactive Waste.* Battelle Press, Columbus, Ohio.

Napier, B.A. 1992. *Determination of Radionuclides and Pathways Contributing to Cumulative Dose.* BN-SA-3673 HEDR, Pacific Northwest Laboratory, Richland, Washington.

Napier, K. 1998. *Cigarettes: What the Warning Label Doesn't Tell You, The First Comprehensive Guide to Health Consequences of Smoking.* Eds. W.M. London, E.M. Whelan, and A.G. Case, American Council on Science and Health, New York.

Narbutovskih, S.M. 1998. *Results of Phase I Groundwater Quality Assessment for Single-Shell Tank Waste Management Areas B-BX-BY at the Hanford Site.* PNNL-11826, Pacific Northwest National Laboratory, Richland, Washington.

National Academy of Public Administration. 1997. *Deciding for the Future: Balancing Risks, Costs, and Benefits Fairly Across Generations: A Report.* National Academy of Public Administration, Washington, D.C.

National Cancer Institute (NCI). 1997. *Estimated Exposures and Thyroid Doses Received by the American People from Iodine-131 in Fallout Following Nevada Atmospheric Nuclear Bomb Tests.* National Cancer Institute, Bethesda, Maryland.

National Committee on Radiation Protection and Measurements. 1953. *Maximum Permissible Amounts of Radioisotopes in the Human Body and Maximum Permissible Concentrations in Air and Water.* National Bureau of Standards Handbook 52. National Committee on Radiation Protection. U.S. Government Printing Office, Washington, D.C.

National Committee on Radiation Protection and Measurements. 1954. *Permissible Dose From External Sources of Ionizing Radiation: Recommendations from the National Committee on Radiation Protection.* National Bureau of Standards Handbook 59. National Committee on Radiation Protection, U.S. Government Printing Office, Washington, D.C.

National Conference of State Legislatures. 1996. *Environment, Energy and Transportation Program: Assessment of Regulatory and Administrative Streamlining.* Prepared by The Lessons Learned Subcommittee of the Environmental Management Advisory Board and The State and Tribal Government Working Group. Available URL: http://ncsl.org/programs/ESNR/EMREGS.htm.

National Council on Radiation Protection and Measurements (NCRP). 1987a. *Exposure of the Population in the United States and Canada From Natural Background Radiation.* NCRP Report No. 94, Bethesda, Maryland.

National Council on Radiation Protection and Measurements (NCRP). 1987b. *Ionizing Radiation Exposure of the Population of the United States.* NCRP Report No. 93, Bethesda, Maryland.

National Institute for Occupational Safety and Health. 2002. *NIOSH Energy Related Health Research Program: Hanford Environmental Health Foundation Battelle Pacific Northwest Laboratories.* Centers for Disease Control and Prevention, Atlanta, Georgia. Available URL: http://www.cdc.gov/niosh/2001-133h.html.

National Research Council (NRC). 1957. *The Disposal of Radioactive Waste on Land.* National Academy of Sciences, Washington, D.C.

National Research Council (NRC). 1970. *Disposal of Solid Radioactive Waste in Bedded Salt Deposits.* National Academy of Sciences, Washington, D.C.

National Research Council (NRC). 1990a. *Health Effects of Exposure to Low Levels of Ionizing Radiation. Phase 1 Letter Report: BEIR VII.* National Academy Press, Washington, D.C.

National Research Council (NRC). 1990b. *Rethinking High-Level Radioactive Waste Disposal: A Position Statement of the Board on Radioactive Waste Management, Commission on Geosciences, Environment, and Resources.* National Academy Press, Washington, D.C.

National Research Council (NRC). 1994a. *Alternatives for Ground Water Cleanup.* Committee on Ground Water Cleanup Alternatives, National Academy Press, Washington, D.C.

National Research Council (NRC). 1994b. *Building Consensus Through Risk Assessment and Management of the Department of Energy's Environmental Remediation Program.* National Academy Press, Washington, D.C.

National Research Council (NRC). 1995. *Improving the Environment: An Evaluation of DOE's Environmental Management Program.* National Academy Press, Washington, D.C.

National Research Council (NRC). 1996a. *The Hanford Tanks: Environmental Impacts and Policy Choices.* National Academy Press, Washington, D.C.

National Research Council (NRC). 1996b. *Understanding Risk: Informing Decisions in a Democratic Society.* National Academy Press, Washington, D.C.

National Research Council (NRC). 1996c. *Building an Effective Environmental Management Science Program: Final Assessment.* National Academy Press, Washington, D.C.

National Research Council (NRC). 1996d. *Barriers to Science: Technical Management of the Department of Energy Environmental Remediation Program.* National Academy Press, Washington, D.C.

National Research Council (NRC). 1997. *Innovations in Ground Water and Soil Cleanup: From Concept to Commercialization.* National Academy Press, Washington, D.C.

National Research Council (NRC). 1998. *Systems Analysis and Systems Engineering in Environmental Restoration Programs at the Department of Energy Hanford Site.* National Academy Press, Washington, D.C.

National Research Council (NRC). 1998a. *Health Effects of Exposure to Indoor Radon: BEIR VI.* National Academy Press, Washington, D.C.

National Research Council (NRC). 1999a. *An End State Methodology for Identifying Technology Needs for Environmental Management, With an Example From the Hanford Site Tanks.* National Academy Press, Washington, D.C.

National Research Council (NRC). 1999b. *Improving Project Management in the Department of Energy.* National Academy Press, Washington, D.C.

National Research Council (NRC). 1999c. *Alternative High-Level Waste Treatments at the Idaho National Engineering and Environmental Laboratory.* National Academy Press, Washington, D.C.

National Research Council (NRC). 1999d. *Groundwater and Soil Cleanup: Improving Management of Persistent Contaminants.* National Academy Press. Washington, D.C.

National Research Council (NRC). 1999e. *Decision Making in the U.S. Department of Energy's Environmental Management Office of Science and Technology.* National Academy Press, Washington, D.C.

National Research Council (NRC). 1999f. *Exposure of the American People to Iodine-131 from Nevada Nuclear-Bomb Tests: Review of the National Cancer Institute Report and Public Health Implications.* National Academy Press, Washington, D.C.

National Research Council (NRC). 2000a. *Research Needs in Subsurface Science: U.S. Department of Energy's Environmental Management Science Program.* National Academy Press, Washington, D.C.

National Research Council (NRC). 2000b. *Long-Term Institutional Management of U.S. Department of Energy Legacy Waste Sites.* National Academy Press, Washington, D.C.

National Research Council (NRC). 2000c. *Alternatives for High-Level Waste Salt Processing at the Savannah River Site.* National Academy Press, Washington, D.C.

National Research Council (NRC). 2000d. *Review of the Hanford Thyroid Disease Draft Final Report.* National Academy Press, Washington, D.C.

National Research Council (NRC). 2001a. *Disposition of High-Level Waste and Spent Nuclear Fuel: The Continuing Societal and Technical Challenges.* National Academy Press, Washington, D.C.

National Research Council (NRC). 2001b. *Science and Technology for Environmental Cleanup at Hanford.* National Academy Press, Washington, D.C.

National Spent Nuclear Fuel Program. 2002. *Approximate Spent Nuclear Fuel Inventory in Metric Tons of Heavy Metal.* Idaho National Engineering and Environmental Laboratory. Available URL: http://nsnfp.inel.gov/tikal-10-30-00/table32.htm.

Natural Resources Defense Council, Yakama Nation, and Snake River Alliance. February 28, 2002. *Complaint for Declaratory and Injunctive Relief.* Filed in the U.S. District Court for the District of Idaho, Case No. 01-CV-413 (BLW) against Spencer Abraham, Secretary U.S. Department of Energy.

Nature. 1996. "Rethinking on Weapons Clean-Up." *Nature* 383(6599):365.

Nature. 2001. "Decision Time at Yucca Mountain." *Nature* 412(6850):841.

Nelson, I.C., ed. June 1, 1961. *Evaluation of Radiological Conditions in the Vicinity of Hanford for 1960.* HW-68435, Hanford Atomic Products Operation, Richland, Washington.

Neta, R. 2000. "The Promise of Molecular Epidemiology in Defining the Association Between Radiation and Cancer." *Health Physics* 79(1):77-84.

Newcomer, D.R., and M.J. Hartman. 1999. "Hanford Groundwater Monitoring Project." *Hanford Site Environmental Report for Calendar Year 1998.* PNNL-12088, Pacific Northwest National Laboratory, Richland, Washington.

Newcomer, D.R., and M.J. Hartman. 2001. "Hanford Groundwater Monitoring Project." *Hanford Site Environmental Report for Calendar Year 2000.* PNNL-13487, Pacific Northwest National Laboratory, Richland, Washington.

Nies, K.A. 2001. *Marie Curie: The Triumphs & Tragedies of a Scientific Career.* Available URL: http://www.geocities.com/athens/delphi/1836/marie/curie_bio.html.

Norwood, W.D. September 9, 1944. Letter to R.S. Stone. HW-7-623, DuPont de Nemours & Company, Richland, Washington.

Nussbaum, R.H., and C.H. Grossman. March 21, 1999. "Secrets, Lies—A Hanford Health Fantasy Rerun." *Tri-City Herald,* Kennewick, Washington.

O'Farrell, T.P., R.E. Fitzner, and R.O. Gilbert. 1973. *Distribution of Radioactive Jackrabbit Pellets in the Vicinity of the B-C Cribs, 200 East Area, U.S. A.E.C. Hanford Reservation.* BNWL-1794, Pacific Northwest Laboratory, Richland, Washington.

Oakley, D.T. 1972. *Natural Radiation Exposure in the United States.* Report ORP/SID 72-1, U.S. Environmental Protection Agency, Washington, D.C.

Oestreich, D.K., M.A. Casbon, and N.R. Kerr. 1998. *Safety Analysis for the Environmental Restoration Disposal Facility.* BHI-00370, Rev. 3, Bechtel Hanford, Inc., Richland, Washington.

Office of Technology Assessment (OTA). 1989. *Coming Clean: Superfund's Problems Can Be Solved.* OTA-ITE-433, Congress of the United States Office of Technology Assessment, U.S. Government Printing Office, Washington, D.C.

Office of Technology Assessment (OTA). 1991. *Complex Cleanup: The Environmental Legacy of Nuclear Weapon Production.* OTA-O-484, Congress of the United States Office of Technology Assessment, U.S. Government Printing Office, Washington, D.C.

Olson, P.A. November 23, 1948. *Fish and Fish Problems of the Hanford Reservation.* HW-11642, Richland, Washington.

Openness Advisory Panel. November 17, 2000. *Relations Between DOE Facilities and Their Host Communities: A Pilot Review.* Available: U.S. Department of Energy Public Reading Room, Richland, Washington.

Oregon Legislative Assembly. 1999. *House Joint Memorial 11.* HJM 11-A, filed June 17, 1999, with the Oregon Secretary of State.

Oregon Office of Energy. 1999a. *Naval Nuclear Reactor Compartment Shipments on the Columbia River.* Available URL: http://www.energy.state.or.us/nucsafe/Hcleanup.htm.

Oregon Office of Energy. August 1999b. *Hanford Cleanup: The First Ten Years.* Oregon Office of Energy, Salem, Oregon.

Oregon Office of Energy. January 2000a. *Fact Sheet: Cleaning Up Hanford's Nuclear Weapon's Wastes.* Oregon Office of Energy, Salem, Oregon.

Oregon Office of Energy. March 2000b. *Hanford: Issues That Concern Oregon,* Oregon Office of Energy, Salem, Oregon, p. 2.

Oregon Office of Energy. 2000c. *Oregon Hanford Waste Board.* Available URL: http://www.energy.state.or.us/nucsafe/hwboard.htm. Last updated February 28, 2001.

Orlando, T.M., D. Meisel, and D.M. Camaioni. May 1999. "Interfacial Radiolysis Effects in Tank Waste Speciation." *Science to Support DOE Site Cleanup: The Pacific Northwest National Laboratory Environmental Management Science Program Awards,* PNNL-12208, Pacific Northwest National Laboratory, Richland, Washington, pp. 1.1-1.9.

Oughton, D.H. 1996. "Ethical Values in Radiological Protection." *Radiation Protection Dosimetry* 86(3-4):203-208.

Owendoff, J.M. March 12, 1998. Statement of James M. Owendoff, Acting Assistant Secretary for Environmental Management Department of Energy, before the Strategic Forces Subcommittee, Armed Services Committee, United States Senate. Available URL: http://www.fas.org/spp/starwars/congress/1998_h/s980312jo.htm.

Pacific Northwest National Laboratory. 2002. *Tank Site.* Tanks Focus Area. Available URL: http://www.pnl.gov/tfa/sites.stm. Last updated December 14, 2001.

Pajunen, A.L., G.K. Allen, A.B. Carlson, T. Chiao, G.D. Forehand, D.D. Frank, R.W. Harmsen, R.C. Hoyt, G. Jansen, L.J. Johnson, B.J. Knutson, C.N. Krohn, D.C. Lini, J.D. Ludowise, T.R. Lunsford, M.M. McCarthy, L.G. Niccoli, S.E. Seeman, T.L. Waldo, C.E. Worcester, B.D. Zimmerman, R.I. Smith, and R.C. Walling. 1994. *Hanford Strategic Analysis Study.* WHC-EP-0549, 5 vols, Westinghouse Hanford Company, Richland, Washington.

Park, R. 2000. *Voodoo Science.* Oxford University Press, New York.

Parker, G.G. and A.M. Piper. July 1949. *Geologic and Hydrologic Features of the Richland Area, Washington, Relevant to Disposal of Waste at the Hanford Operations of the Atomic Energy Commission*, USGS-W-P-7, U.S. Department of the Interior Geological Survey, Portland, Oregon.

Parker, H.M. August 31, 1944. *Classified Information Required by Operators and Others Exposed to Radiation Hazards.* Letter to D.O. Notman. HW-7-588, Richland, Washington.

Parker, H.M. December 11, 1945a. *Xenon and Iodine Concentrations in the Environs of the T and B Plants.* Letter to S.T. Cantril. HW-7-3005, Richland, Washington.

Parker, H.M. September 11, 1945b. *Status of Problem of Measurement of the Activity of Waste Water Returned to the Columbia River.* Letter to S.T. Cantril. HW-7-2346, Richland, Washington.

Parker, H.M. July 10, 1945c. *Action Taken on Report on Visit to Site W, April 9-April 13, 1945 by G Failla.* Letter to S.T. Cantril. HW-7-1973, Richland, Washington.

Parker, H.M. January 14, 1946. *Tolerable Concentration of Radio-Iodine on Edible Plants.* HW-7-3217, General Electric Company, Hanford Works, Richland, Washington.

Parker, H.M. January 26, 1948a. *Speculations on Long-Range Waste Disposal Hazards.* HW-8674, General Electric Company, Hanford Works, Richland, Washington.

Parker, H.M. July 27, 1948b. *Summary of Hanford Works Radiation Hazards for the Reactor Safeguard Committee.* Declassified letter. HW-10592, Richland, Washington.

Parker, H.M. October 25, 1948c. *Action Taken With Respect to Apparent Enhanced Active Particle Hazard.* HW-11384, Richland, Washington.

Parker, H.M. March 17, 1950. *Columbia River Advisory Group Conference and Cooperation with the U.S. Public Health Service.* Letter to D.G. Sturges. DDTS-GENERATED-557, Richland, Washington.

Parker, H.M. May 4, 1951. *Radioactivity Allowable in A.E.C. Liquid Wastes.* Letter to K.L. Englund of the Atomic Energy Commission. HW-21002. Richland, Washington.

Parker, H.M. May 6, 1952. *Permissible Limits—Release of Reactor Effluent to the Columbia River.* HW-24356, General Electric Company, Hanford Atomic Products Operation, Richland, Washington.

Parker, H.M. June 4, 1954a. *Ground Disposal of Radioactive Wastes at the Hanford Site.* HW-32041, General Electric Company, Hanford Atomic Products Operation, Richland, Washington.

Parker, H.M. September 15, 1954b. *Status of Ground Contamination Problem.* HW-33068, General Electric Company, Hanford Atomic Products Operation, Richland, Washington.

Parker, H.M. 1956. "Radiation Exposure from Environmental Hazards." In *Proceedings of the International Conference on the Peaceful Uses of Atomic Energy*, Vol. 13, pp. 305-310. August 8-20, 1955. United Nations, New York.

Parker, H.M. 1959a. "Statement of Herbert Parker, Manager, Hanford Laboratories." U.S. Congress, Joint Committee on Atomic Energy, Special Subcommittee on Radiation. *Hearings.* Vol. 1, pp. 159-170. 86th Cong., 1st Session on Industrial Radioactive Waste Disposal. U.S. Government Printing Office, Washington, D.C.

Parker, H.M. 1959b. "Radioactive Waste Management Operations at the Hanford Works. Part 1. Hanford Radioactive Waste Management." U.S. Congress, Joint Committee on Atomic Energy, Special Subcommittee on Radiation. *Hearings.* Vol. 1, pp. 202-241. 86th Cong., 1st Session on Industrial Radioactive Waste Disposal. U.S. Government Printing Office, Washington, D.C.

Parker, H.M. 1959c. "Radioactive Waste Management Operations at the Hanford Plant." U.S. Congress, Joint Committee on Atomic Energy, Special Subcommittee on Radiation. *Hearings.* Vol. 1, pp. 171-183. 86th Cong., 1st Session on Industrial Radioactive Waste Disposal. U.S. Government Printing Office, Washington, D.C.

Parker, M.B. 1986. *Tales of Richland, White Bluffs & Hanford 1805-1943.* Ye Galleon Press, Fairfield, Washington.

Patton, G.W. 1999. "Surface Water and Sediment Surveillance." *Hanford Site Environmental Report for Calendar Year 1998.* PNNL-12088, Pacific Northwest National Laboratory, Richland, Washington.

Patton, G.W. 2001. "Appendix B: Additional Monitoring Results for 2000." *Hanford Site Environmental Report for Calendar Year 2000.* PNNL-13487, Pacific Northwest National Laboratory, Richland, Washington.

Patton, G.W., and T.M. Poston. 2000. "Appendix A: Additional Monitoring Results for 1999." *Hanford Site Environmental Report for Calendar Year 1999.* PNNL-13230, Pacific Northwest National Laboratory, Richland, Washington.

Pearce, D.W. 1959. "Radioactive Waste Management Operations at the Hanford Works. Part 5. Disposal of Intermediate and Low-Level Wastes to the Ground." U.S. Congress, Joint Committee on Atomic Energy, Special Subcommittee on Radiation. *Hearings.* Vol. 1, pp. 308-318. 86th Cong., 1st Session on Industrial Radioactive Waste Disposal. U.S. Government Printing Office, Washington, D.C.

Perez, J.M., et al. July 2001. *High-Level Waste Melter Study Report.* PNNL-13582, Pacific Northwest National Laboratory, Richland, Washington.

Petersen, S.W., R.J. Cameron, M.D. Johnson, and M.J. Truex. March 2001. *Technology Alternatives Baseline Report for the 618-10 and 618-11 Burial Grounds, 300-FF-2 Operable Unit.* BHI-01484, Rev. 0, Bechtel Hanford Company, Richland, Washington.

Peterson, R.E., and T.M. Poston. May 2000. *Strontium-90 at the Hanford Site and Its Ecological Implications.* PNNL-13127, Pacific Northwest National Laboratory, Richland, Washington.

Phillips, J.E., and C.E. Easterly. 1980. *Sources of Tritium.* ORNL/TM-6402, Oak Ridge National Laboratory, Oak Ridge, Tennessee.

Pilkey, O.H., A.M. Platt, and C.A. Rohrmann. 1958. "The Storage of High-Level Radioactive Wastes: Design and Operating Experience in the United States." In *Proceedings of the Second United Nations International Conference on the Peaceful Uses of Atomic Energy,* Vol. 18. pp. 7-18. September 1-13, 1958, Geneva, Switzerland. United Nations, Geneva.

Plum, R.L. 1950. *Report on Columbia River Symposium: A Joint Meeting of the Atomic Energy Commission, U.S. Public Health Service, Columbia River Advisory Group, and General Electric Company, June 19-21, 1950.* HO-1, Richland, Washington.

Plum, R.L. June 11, 1968. *Classification of River Temperature Data.* Letter to J.P. Derouin. HAN-100651, Richland, Washington.

Porter, J.C. August 11, 2000. *Potential Waste Inventory Report.* U.S. Department of Energy, Richland Operations Office, Richland, Washington.

Poston, T.M., R.W. Hanf, and R.L. Dirkes, eds. 2000. *Hanford Site Environmental Report for Calendar Year 1999.* PNNL-13230, Pacific Northwest National Laboratory, Richland, Washington.

Poston, T.M., R.W. Hanf, R.L. Dirkes, and L.F. Morasch, eds. 2001. *Hanford Site Environmental Report for Calendar Year 2000.* PNNL-13487, Pacific Northwest National Laboratory, Richland, Washington.

Presidential/Congressional Commission on Risk Assessment and Risk Management. 1997. *Risk Assessment and Risk Management in Regulatory Decision-Making.* The Presidential/ Congressional Commission on Risk Assessment and Risk Management, Final Report, Vol. 2, Superintendent of Document, Pittsburgh, Pennsylvania. Available URL: http:// www.riskworld.com.

Pritikin, T. 1994/1995. "Hanford: Where Traditional Common Law Fails." From Hanford Symposium Part I "Issues Involving the Hanford Nuclear Reservation." *Gonzaga Law Review*, 1994/95, 30(3):523-572, Gonzaga University, Spokane, Washington.

Probst, K.N. 1999. "Institutional Controls: The Next Frontier." *Center for Risk Management Newsletter*, Issue No. 15.

Probst, K.N., and A.I. Lowe. 2000. *Cleaning Up the Nuclear Weapons Complex: Does Anyone Care?* Center for Risk Assessment, Resource for The Future, Washington, D.C.

Probst, K.N., and M.H. McGovern. 1998. *Long-Term Stewardship and the Nuclear Weapons Complex: The Challenge Ahead.* Center for Risk Management, Resource for the Future, Washington, D.C.

Reeves, M. March 4, 2000. "Hanford Advisory Board Urges Timely Waste Cleanup." *Tri-City Herald*, p. D15, Kennewick, Washington.

Reilly, M.A. 1998. *Spent Nuclear Fuel Project Technical Databook.* HNF-SD-SNF-TI-015, Rev. 6, Duke Engineering and Services Hanford, Richland, Washington.

Resnik, D.B. 1997. "Ethical Problems and Dilemmas in the Interaction Between Science and Media." *Proceedings From the Workshop on Ethical Issues in Physics.* M. Thomsen and B. Wylo eds., Eastern Michigan University, Ypsilanti, Michigan.

Resource Conservation and Recovery Act (RCRA) of 1976. 42 USC 6901 et seq., as amended.

Revised Code of Washington (RCW). 1985. "Hazardous Waste Management." RCW 70.105, Olympia, Washington.

RHO. See Rockwell Hanford Operations.

Rhodes, R. 1986. *The Making of the Atomic Bomb.* Simon & Schuster, New York.

Rhodes, R. 1995. *Dark Sun: The Making of the Hydrogen Bomb:* Simon & Schuster, New York.

Richanbach, P.E. D.R. Graham, J.P. Bell, and J.D. Silk. 1997 *The Organization and Management of the Nuclear Weapons Complex.* Institute for Defense Analysis, Alexandria, Virginia.

Rizzo, K. October 11, 2000. "House Approves Compensation for Sick Atomic Workers." Associated Press.

Roberson, J.H. November 19, 2001. *Environmental Management Priorities.* Letter "Memorandum for Director, Office of Management, Budget and Evaluation, Chief Financial Officer." U.S. Department of Energy, Washington, D.C.

Roberson, J. February 4, 2002a. *Memorandum for the Secretary.* Letter to the Secretary of Energy, S. Abraham. U.S. Department of Energy, Washington, D.C.

Roberson, J.H. February 14, 2002b. "Hanford Cleanup Is Capability Under Way." *Tri-City Herald*, p. A-10, Kennewick, Washington.

Roberts, R.E. 1958. *History of Airborne Contamination and Control—200 Areas.* HW-55569, General Electric, Hanford Atomic Products Operation, Richland, Washington.

Rockwell Hanford Operations (RHO). August 1985. *200 Areas Fact Book.* Research and Engineering, Rockwell Hanford Operations, Richland, Washington.

Rodovsky, T.J., and S.L. Bond. 1998. *Final Hazard Classification and Auditable Safety Analysis for the 105-F Building Interim Safe Storage Project.* BHI-01151, Rev. 0, Bechtel Hanford, Inc., Richland, Washington.

Rodovsky, T.J., A.R. Larson, and D. Dexheimer. 1996. *Final Hazard Classification and Auditable Safety Analysis for the 105-C Reactor Interim Safe Storage Project.* BHI-00837, Rev. 0, Bechtel Hanford, Inc., Richland, Washington.

Rohay, V.J. 1993. *Carbon Tetrachloride Evaporative Losses and Residual Inventory Beneath 200 West Area at the Hanford Site.* WHC-SD-EN-TI-101, Westinghouse Hanford Company, Richland, Washington.

Rohay, V.J. 2002. "Section 3.2.3, Carbon Tetrachloride Monitoring and Remediation." *Hanford Site Groundwater Monitoring for Fiscal Year 2001*, eds. M.J. Hartman, L.F. Morasch, and W.D. Webber, PNNL-13788, Pacific Northwest National Laboratory, Richland, Washington, pp. 3.43-3.46.

Rohay, V.J. and L.C. Swanson. 2001. "Section 3.2.4, Carbon Tetrachloride Monitoring and Remediation." *Hanford Site Groundwater Monitoring for Fiscal Year 2000*, eds. M.J. Hartman, L.F. Morasch, and W.D. Webber, PNNL-13404, Pacific Northwest National Laboratory, Richland, Washington.

Ropeik, D. August 6, 2000. "Let's Get Real about Risk." *The Washington Post*, Washington, D.C.

Ross, J.F. 1999. *The Polar Bear Strategy: Reflections on Risk in Modern Life.* Perseus Books, Reading, Massachusetts.

Rothstein, L. 1995. "Nothing Clean About 'Cleanup'." *The Bulletin of the Atomic Scientists* 51(3):34-41.

Rumer, R.R., and J.K. Mitchell, eds. 1996. *Assessment of Barrier Containment Technologies, A Comprehensive Treatment for Environmental Remediation Applications.* Publication #PB96-180583. U.S. Department of Energy, U.S. Environmental Protection Agency, and DuPont Company, National Technical Information Service, Springfield, Virginia.

Russell, M. 1995. *Contamination or Risk: Cost Implications of Alternative Superfund Configurations.* Paper presented at the American Economics Meeting, Washington, D.C., January 7, 1995.

Russell, M., E.W. Colglazier, and M.R. English. 1991. *Hazardous Waste Remediation: The Task Ahead.* R01-2534-19-001-92, University of Tennessee, Waste Management Research and Education Institute, Knoxville, Tennessee.

Russell, D., S. Lewis, and B. Keating. May 1992. *Inconclusive by Design: Waste, Fraud, and Abuse in Federal Environmental Health Research.* Environmental Health Network, Harvey, Louisiana.

Ryker, S.J., and J.L. Jones. 1995. *Nitrate Concentrations in Ground Water of the Central Columbia Plateau.* Open File Report 95-445, U.S. Geological Survey, Reston, Virginia.

Sagan, C. 1996. *The Demon-Haunted World: Science as a Candle in the Dark.* Random House, New York.

Saleska, S., and A. Makhijani. October 1990. "Hanford Cleanup: Explosive Solution." *The Bulletin of the Atomic Scientists* 46(8):14-20.

Sanger, S.L. 1995. *Working on the Bomb: An Oral History of WWII Hanford.* Portland State University, Portland, Oregon.

Sanneh, L. 1989. *Translating the Message: The Missionary Impact on Culture.* Orbis Books, Maryknoll, New York.

Schein, E. 1999. *Corporate Culture Survival Guide: Sense and Nonsense About Culture Change.* Prentice Hall, San Francisco.

Schlemmer, F.C. January 4, 1950. *Columbia River Advisory Group Conference.* Letter to W.J. Williams, DDTS-GENERATED-560, Richland, Washington.

Schwartz, S.I., ed. 1998. *Atomic Audit: The Costs and Consequences of U.S. Nuclear Weapons Since 1940.* Brookings Institution Press, Washington, D.C.

Science. May 1, 1959. U.S. Congress, Joint Committee on Atomic Energy, Special Subcommittee on Radiation. *Hearings.* Vol. 3, pp. 2567. 86th Cong., 1st Session on Fallout from Nuclear Weapons Tests.

Seattle Post-Intelligencer. June 27, 1999. "Hanford Cleanup Needs Technology." Editorial. *Seattle Post-Intelligencer,* Seattle, Washington.

Secretary of Energy Advisory Board. 1993. *Earning Public Trust and Confidence: Requisites for Managing Radioactive Wastes Final Report of the Secretary of Energy Advisory Board Task Force on Radioactive Waste Management.* Task Force on Radioactive Waste Management, Department of Energy, Washington, D.C.

Secretary of Energy Advisory Board. 1995. *Alternative Futures for the Department of Energy National Laboratories.* Task Force on Alternative Futures for the Department of Energy National Laboratories, U.S. Department of Energy, Washington, D.C.

Serber, R. 1992. *The Los Alamos Primer: The First Lectures on How to Build an Atomic Bomb.* R. Rhodes, ed. University of California Press, Berkeley, California.

Serne, R.J., and V.L. LeGore. January 1996. *Strontium-90 Adsorption-Desorption Properties and Sediment Characterization at the 100 N-Area.* PNL-10899, Pacific Northwest National Laboratory, Richland, Washington.

Serne, R.J., H.T. Schaef, B.N. Bjornstad, D.C. Lanigan, G.W. Gee, C.W. Lindenmeier, R.E. Clayton, V.L. LeGore, M.J. O'Hara, C.F. Brown, R.D. Orr, G.V. Last, I.V. Kutnyakov, D.S. Burke, T.C. Wilson, and B.A. Williams. 2002a. *Geologic and Geochemical Data Collected From Vadose Zone Sediments From Borehole 299 W23-19 [SX -115] in the S/ SX Waste Management Area and Preliminary Interpretations.* PNNL-13757-2, Pacific Northwest National Laboratory, Richland, Washington.

Serne, R.J., G.V. Last, G.W. Gee, H.T. Schaef, D.C. Lanigan, C.W. Lindenmeier, R.E. Clayton, V.L. LeGore, R.D. Orr, M.J. O'Hara, C.F. Brown, D.S. Burke, A.T. Owen, I.V. Kutnyakov, and T.C. Wilson. 2002b. *Geologic and Geochemical Data Collected From Vadose Zone Sediments From Borehole SX 41-09-39 in the S/SX Waste Management Area and Preliminary Interpretations.* PNNL-13757-3, Pacific Northwest National Laboratory, Richland, Washington.

Serne, R.J., G.V. Last, H.T. Schaef, D.C. Lanigan, C.W. Lindenmeier, C.C. Ainsworth, R.E. Clayton, V.L. LeGore, M.J. O'Hara, C.F. Brown, R.D. Orr, I.V. Kutnyakov, T.C. Wilson, K.B. Wagnon, B.A. Williams, and D.B. Burke. 2002c. *Geologic and Geochemical Data and Preliminary Interpretations of Vadose Zone Sediment From Slant Borehole SX-108 in the S-SX Waste Management Area.* PNNL-13757-4, Pacific Northwest National Laboratory, Richland, Washington.

Seymour, A.H., and V.A. Nelson. 1973. "Decline of ^{65}Zn in Marine Mussels Following the Shutdown of Hanford Reactors." *Radioactive Contamination of the Marine Environment, Proceedings of a Symposium on the Interaction of Radioactive Contaminants with the Constituents of the Marine Environment,* Seattle, Washington, July 10-14, 1972. International Atomic Energy Agency, Vienna, Austria.

Silverman, M.J. May 2000. *No Immediate Risk: Environmental Safety in Nuclear Weapons Production, 1942-1985.* Ph.D. Dissertation, Carnegie Mellon University. Pittsburgh, Pennsylvania.

Simpson, B.C., R.A. Corbin, and S.F. Agnew. March 2001. *Groundwater/Vadose Zone Integration Project: Hanford Soil Inventory Model.* BHI-01496, Rev. 0, Los Alamos National Laboratory for Bechtel Hanford, Inc., Richland, Washington.

Sivula, C. January 30, 1988a. "Community Rallied to Save N in 1971." *Tri-City Herald,* Kennewick, Washington.

Sivula, C. January 30, 1988b. "Reactor's Future Hinges on Need for Plutonium." *Tri-City Herald,* Kennewick, Washington.

Slovic, P. 1993. "Perceived Risk, Trust, and Democracy." *Risk Analysis* 13(6): 675-682.

Slovic, P. 1996. "Perception of Risk From Radiation." *Radiation Protection Dosimetry* 68(3/4):165-180.

Slovic, P. February 2001. "Risk and Trust." *Risk Excellence Notes* 3(1):4. U.S. Department of Energy, Argonne, Illinois. Available URL: http://riskcenter.doe.gov.

Slovic, P, J.H. Flynn, and M. Layman. December 13, 1991. "Perceived Risk, Trust, and the Politics of Nuclear Waste." *Science* 254:1603-1607.

Smyth, H. August 1945. *A General Account of the Development of Methods of Using Atomic Energy for Military Purposes Under the Auspices of the United States Government, 1940-1945.* War Department. (Reprinted Princeton University Press, New Jersey, September 1945).

Soldat, J.K., and T.H. Essig, eds. September 1966. *Evaluation of Radiological Conditions in the Vicinity of Hanford for 1965.* BNWL-316, Pacific Northwest Laboratory, Richland, Washington.

Soldat, J.K., K.R. Price, and W.D. McCormack. 1986. *Offsite Radiation Doses Summarized From Hanford Environmental Monitoring Reports for the Years 1957-1984.* PNL-5795, Pacific Northwest Laboratory, Richland, Washington.

Stang, J. July 5, 1999a. "Cleanup Funds Expected to Dry Up Nationwide." *Tri-City Herald*, Kennewick, Washington.

Stang, J. June 14, 1999b. "Hanford Faces Rising Tide of Funding Needs." *Tri-City Herald,* Kennewick, Washington.

Stang, J. March 30, 1999c. "Disinterest Spells End of Hanford Watchdog." *Tri-City Herald*, Kennewick, Washington.

Stang, J. December 10, 1999d. "Congress Holds Fate of Glassification Contract." *Tri-City Herald,* Kennewick, Washington.

Stang, J. December 3, 2000a. "Hanford Plumes Bigger Than Solutions." *Tri-City Herald,* Kennewick, Washington.

Stang, J. September 8, 2000b. "Hanford May Need More Waste Tanks, State Says." *Tri-City Herald,* Kennewick, Washington.

Stang, J. March 12, 2000c. "Several Factors Determine Waste Shipments." *Tri-City Herald,* Kennewick, Washington.

Stang, J. May 4, 2000d. "Report Gives N.M. Pause Over Hanford's Transuranic Wastes." *Tri-City Herald,* Kennewick, Washington.

Stang, J. March 9, 2000e. "Worries Over Hanford 'Burping' Tank Appear to Be at End." *Tri-City Herald,* Kennewick, Washington.

Stang, J. April 25, 2000f. "BNFL Submits Project Estimate." *Tri-City Herald,* Kennewick, Washington.

Stang, J. April 12, 2000g. "K Basins Project Could Be Finished a Year Early." *Tri-City Herald,* Kennewick, Washington.

Stang, J. October 11, 2000h. "Officials Debate Hastening DOE's River Shore Cleanup." *Tri-City Herald,* Kennewick, Washington.

Stang, J. October 18, 2000i. "Gregoire Fed Up With DOE." *Tri-City Herald,* Kennewick, Washington.

Stang, J. August 3, 2000j. "4 More Sites Turn Up Small Increases in Plutonium Tied to Fire." *Tri-City Herald,* Kennewick, Washington.

Stang, J. March 24, 2001a. "Gregoire Threatens DOE With Lawsuit." *Tri-City Herald,* Kennewick, Washington.

Stang, J. March 7, 2001b. "Hanford Budget Plans Still Hazy." *Tri-City Herald,* Kennewick, Washington.

Stang, J. December 21, 2001c. "DOE Pushes Ahead With Hanford Plan." *Tri-City Herald,* Kennewick, Washington.

Stang, J. December 21, 2001d. "Hanford Project Looks to Recover." *Tri-City Herald,* Kennewick, Washington.

Stang, J. December 12, 2001e. "Hanford Budget OK for Cleanup." *Tri-City Herald,* Kennewick, Washington.

Stang, J. March 7, 2002a. "Hanford Budget to Get $433 Million Boost for Cleanup." *Tri-City Herald,* Kennewick, Washington.

Stang, J. August 7, 2002b. "DOE Memo Puts Ohio Waste Before Cleanup." *Tri-City Herald,* Kennewick, Washington.

Stang, J. June 7, 2002c. "HAB Mulls Cleanup Proposal." *Tri-City Herald,* Kennewick, Washington.

Stang, J. October 9, 2002d. "Hanford Plan Would Accelerate Tank Work." *Tri-City Herald,* Kennewick, Washington.

Stang, J., and A. Cary. December 7, 2001. "DOE Proposes Vitrification Cuts." *Tri-City Herald,* Kennewick, Washington.

Stannard, J.N. 1988. *Radioactivity and Health: A History.* DOE/RL/01830-T59, Pacific Northwest Laboratory, Richland, Washington.

Stapp, D.C. 2002a. "Chapter 2: Section 3: Reactor Operations." *Hanford Site Historic District: History of the Plutonium Production Facilities 1943-1990*, Battelle Press, Columbus, Ohio.

Stapp, D.C. 2002b. "Chapter 2: Section 12: History of Workers at the Hanford Site." *Hanford Site Historic District: History of the Plutonium Production Facilities 1943-1990*, Battelle Press, Columbus, Ohio.

Stenner, R.D., K.H. Cramer, K.A. Higley, S.J. Jette, D.A. Lamar, T.J. McLaughlin, D.R. Sherwood, and N.C. Van Houten. 1988. *Hazard Ranking System Evaluation of CERCLA Inactive Waste Sites at Hanford.* PNL-6456, 3 vols., Pacific Northwest Laboratory, Richland, Washington.

Stewart, C.W. 2000. "Physical Characteristics of Hanford Tank Waste Relating to Flammable Gas Retention and Release." Proceedings of *Waste Management 2000*, Wednesday Session 54, February 27-March 2, 2000, Tucson, Arizona.

Stewart, C.W., M.E. Brewster, P.A. Gauglitz, L.A. Mahoney, P.A. Meyer, K.P. Recknagle, and H.C. Reid. December 1996. *Gas Retention and Release Behavior in Hanford Single-Shell Waste Tanks.* PNNL-11391, Pacific Northwest National Laboratory, Richland, Washington.

Stewart, D. 1996. *A Radiological Safety Assessment for Disposal of Dredged Material from Lake Wallula.* Master of Science Thesis, Oregon State University.

Stiffler, L. April 19, 2002. "Dark Cloud of Illness Lingers Over Many." *Seattle Post-Intelligencer*, p. A13, Seattle, Washington.

Stock, L.M. September 27, 2000. *The Chemistry of Flammable Gas Generation.* RPP-6664 Rev 0, CH2MHILL Hanford Group, Inc., Richland, Washington.

Straub, C.P., and J.H. Fooks. 1963. "Cooperative Field Studies on Environmental Factors Influencing I^{131} Levels in Milk." *Health Physics* 9:1187-1195.

Strom, D.J., and C.R. Watson. 2002. "On Being Understood: Clarity and Jargon in Radiation Protection." *Health Physics* 82(3):373-386.

Sula, M.J., and P.J. Blumer. April 1981. *Environmental Surveillance at Hanford for CY-1980.* PNL-3728, Pacific Northwest Laboratory, Richland, Washington.

Sutherland, B.M., P.V. Bennett, H. Schenk, O. Sidorkina, J. Laval, J. Trunk, D. Monteleone, and J. Sutherland. 2001. "Clustered DNA Damages Induced by High and Low LET Radiation, Including Heavy Ions." *Physica Medica* 17:202-204, Supplement 1.

Taylor, L.S. 1956a. "The Achievement of Radiation Protection by Legislative and Other Means." In *Proceedings of the International Conference on the Peaceful Uses of Atomic Energy*, Vol. 13, pp. 15-21. August 8-20, 1955. United Nations, New York.

Taylor, L.S. 1956b. "Permissible Exposure to Ionization Radiation." In *Proceedings of the International Conference on the Peaceful Uses of Atomic Energy*, Vol. 13, pp. 196-197. August 8-20, 1955. United Nations, New York.

Terrill, J.G. January 1958. "Some Public Health Aspects of Radioactive Wastes." *Bulletin of Atomic Scientists* 14(1):44-45.

Thayer, H. 1996. *Management of the Hanford Engineer Works in World War II.* American Society of Chemical Engineers (ASCE) Press, New York.

Thompson, M., R. Ellis, and A. Wildavsky. 1990. *Cultural Theory*. Westview Press, Boulder, Colorado.

Thompson, R.C., H.M. Parker, and H.A. Kornberg. 1956. "Validity of Maximum Permissible Standards for Internal Exposure." In *Proceedings of the International Conference on the Peaceful Uses of Atomic Energy*, Vol. 13, pp. 201-204. United Nations, Geneva.

Till, J.E., et al. 2002. *A Risk-based Screening Analysis for Radionuclides Released to the Columbia River From Past Activities at the U.S. Department of Energy Nuclear Weapons Site in Hanford, Washington.* Risk Assessment Corporation, Neeses, South Carolina.

Tiller, B.L., G.E. Dagle, and L.L. Cadwell. 1997. "Testicular Atrophy in a Mule Deer Population." *Journal of Wildlife Diseases* 33(3):420-429.

Tomlinson, R.E. 1959a. "Radioactive Waste Management Operations at the Hanford Works. Part 4. Storage of High-Level Fission Product Wastes." U.S. Congress, Joint Committee on Atomic Energy, Special Subcommittee on Radiation. *Hearings.* Vol. 1, pp. 282-307. 86th Cong., 1st Session on Industrial Radioactive Waste Disposal. U.S. Government Printing Office, Washington, D.C.

Tomlinson, R.E. 1959b. "Radioactive Waste Management Operations at the Hanford Works. Part 3. Release of Gases, Vapors, and Particles to the Atmosphere." U.S. Congress, Joint Committee on Atomic Energy, Special Subcommittee on Radiation. *Hearings.* Vol. 1, pp. 270-281. 86th Cong., 1st Session on Industrial Radioactive Waste Disposal. U.S. Government Printing Office, Washington, D.C.

Treaty. 1855. *Treaty with the Yakama 1855.* 12 Stat. 951.

Treaty. 1855. *Treaty with the Nez Perces.* 12 Stat. 957.

Tri-City Herald. February 14, 2002. "Bush's Hanford Budget Inadequate, Insulting." *Tri-City Herald*, p. A-10, Kennewick, Washington.

Tsivoglou E.C., and M.W. Lammering. March 1961. *A Report by the Columbia River Advisory Group of an Evaluation of the Pollutional Effects of Effluents from Hanford Works.* Available: U.S. Department of Energy Public Reading Room, Richland, Washington.

Tyree, G. November 12, 2001. "Once Infamous Hanford Tank Returned to Service." *Hanford Reach*, p. 9, U.S. Department of Energy, Richland, Washington.

United Nations. 1956. *Proceedings of the International Conference on the Peaceful Uses of Atomic Energy*, Vol. 13. August 8-20, 1955. United Nations, Legal, Administrative, Health and Safety Aspects of Large-Scale Use of Nuclear Energy, New York.

United Nations. 1958. *Proceedings of the Second United Nations International Conference on the Peaceful Uses of Atomic Energy*, Vol. 18. September 1-13, 1958, Geneva, Switzerland. United Nations, Waste Treatment and Environmental Aspects of Atomic Energy, Geneva.

United Nations Scientific Committee on the Effects of Atomic Radiation (UNSCEAR). 1964. "Report of the United Nations Scientific Committee on the Effects of Atomic Radiation." XIX Session, Supplement No.14 (A/5814), United Nations, New York.

United Nations Scientific Committee on the Effects of Atomic Radiation (UNSCEAR). 2000. "Sources and Effects of Ionizing Radiation." 2 vols. No. E.00.IX.3, United Nations, New York.

Upton A.C. February 2001. "The Meaning of Risk." *Risk Excellence Notes* 3(1):1. U.S. Department of Energy, Argonne, Illinois. Available URL: http://riskcenter.doe.gov.

Usdin, S. February 9, 1996. "DOE Lifts the Plutonium Veil." *Nuclear Remediation Week* 3(7).

U.S. Army Corps of Engineers. 2000. *U.S. Army Corps of Engineers Formerly Utilized Sites Remedial Action Program.* Available URL: http://www.hq.usace.army.mil/cecw/fusrap/index.htm. Last updated December 3, 1999.

U.S. Congress. 1959. Joint Committee on Atomic Energy, Special Subcommittee on Radiation. *Hearings.* Vol. 3, pp. 2573-2577. 86th Cong., 1st Session on Fallout from Nuclear Weapons Tests. U.S. Government Printing Office, Washington, D.C.

U.S. Congress. 1959a. Joint Committee on Atomic Energy, Special Subcommittee on Radiation. *Hearings.* Vol. 1 pp. 159-425. 86th Cong., 1st Session on Industrial Radioactive Waste Disposal. U.S. Government Printing Office, Washington, D.C.

U.S. Congressional Budget Office. 1994. *Analyzing the Duration of Cleanup at Sites on Superfund's National Priorities List.* Washington, D.C.

U.S. Department of Commerce. 1953. *Maximum Permissible Amounts of Radioisotopes in the Human Body and Maximum Permissible Concentrations in Air and Water.* National Bureau of Standards Handbook 52, Washington, D.C.

U.S. Department of Defense (DOD). March 12, 1998. *Evaluation of DOD Waste Site Groundwater Pump-and-Treat Operations.* Report Number 98-090, Department of Defense, Office of Inspector General, Office of Audit Services, Arlington, Virginia.

U.S. Department of Energy (DOE). 1987. *Final Environmental Impact Statement: Disposal of Hanford Defense High-Level, Transuranic and Tank Waste, Hanford Site, Richland, Washington.* DOE/EIS-0113, 5 vols., U.S. Department of Energy, Washington, D.C.

U.S. Department of Energy (DOE). 1988a. *Consultation Draft Site Characterization Plan: Reference Repository Location, Hanford Site, Washington.* DOE/RW-0164, 9 vols., U.S. Department of Energy, Office of Civilian Radioactive Waste Management, Washington, D.C.

U.S. Department of Energy (DOE). 1988b. *Defense Waste and Transportation Management: Program Implementation Plan.* DOE/DP-0059, U.S. Department of Energy, Office of Defense Waste and Transportation Management, Washington, D.C.

U.S. Department of Energy (DOE). January 10, 1989. *Investigation of PUREX Limiting Condition for Operation (LOC) Violation Reported on December 7, 1988.* U.S. Department of Energy, Richland Operations Office, Richland, Washington. Available: Department of Energy Public Reading Room, Accession #8124, Richland, Washington.

U.S. Department of Energy (DOE). 1990. *Environmental Restoration and Waste Management: Five-Year Plan Fiscal Years 1992-1996.* DOE/S-0078P, U.S. Department of Energy, Washington, D.C.

U.S. Department of Energy (DOE). 1990a. "Radiation Protection of the Public and the Environment." DOE Order 5400.5, U.S. Department of Energy, Washington, D.C.

U.S. Department of Energy (DOE). 1991a. *Environmental Restoration and Waste Management: Five-Year Plan Fiscal Years 1993-1997.* DOE/S-0090P, U.S. Department of Energy, Washington, D.C.

U.S. Department of Energy (DOE). September 1991b. *Overview of the Hanford Cleanup Five-Year Plan.* Brochure, U.S. Department of Energy, Richland, Washington.

U.S. Department of Energy (DOE). 1992a. *200 East Groundwater Aggregate Area Management Study Report.* DOE/RL-92-19 (draft), U.S. Department of Energy-Richland Operations Office, Richland, Washington.

U.S. Department of Energy (DOE). 1992b. *200 West Groundwater Aggregate Area Management Study Report.* DOE/RL-92-16 (draft), U.S. Department of Energy-Richland Operations Office, Richland, Washington.

U.S. Department of Energy (DOE). 1992c. *Addendum (Final Environmental Impact Statement): Decommissioning of Eight Surplus Production Reactors at the Hanford Site, Richland, Washington.* DOE/EIS-0119, U.S. Department of Energy, Washington, D.C.

U.S. Department of Energy (DOE). 1993. *618-11 Burial Ground Expedited Response Action Proposal.* DOE/RL-93-49 (draft A), U.S. Department of Energy-Richland Operations Office, Richland, Washington.

U.S. Department of Energy (DOE). 1993a. *DOE Management of High-Level Waste at the Hanford Site.* DOE/IG-0325, U.S. Department of Energy, Office of Inspector General, Office of Audit Services, Washington, D.C.

U.S. Department of Energy (DOE). 1994a. *Plutonium Working Group Report on Environmental, Safety and Health Vulnerabilities Associated With the Department's Plutonium Storage.* DOE/EH-0415, U.S. Department of Energy, Office of Environment, Safety, and Health, Washington, D.C.

U.S. Department of Energy (DOE). 1994b. *Strategic Plan for Hanford Site Information Management.* DOE/RL-94-69, U.S. Department of Energy-Richland Operations, Richland, Washington. Available URL: http://www.hanford.gov/irm/irmdoc.htm.

U.S. Department of Energy (DOE). 1995a. *Closing the Circle on the Splitting of the Atom.* DOE/EM-0266, U.S. Department of Energy, Office of Environmental Management, Washington, D.C.

U.S. Department of Energy (DOE). 1995b. *Estimating the Cold War Mortgage: The 1995 Baseline Environmental Management Report.* DOE/EM-0232, U.S. Department of Energy, Office of Environmental Management, Washington, D.C.

U.S. Department of Energy (DOE). 1996. *Hanford Announces Major Progress with Tank Safety Issues.* RL 96-105. Press Release. U.S. Department of Energy, Richland Operations Office, Richland, Washington. Available URL: http://www.hanford.gov/press/1996/96-105.htm.

U.S. Department of Energy (DOE). 1996a. *The 1996 Baseline Environmental Management Report.* DOE/EM-0290, 3 vols., U.S. Department of Energy, Office of Environmental Management, Washington, D.C.

U.S. Department of Energy (DOE). 1996b. *Hanford Strategic Plan.* DOE/RL-96-92, U.S. Department of Energy-Richland Operations Office, Richland, Washington.

U.S. Department of Energy (DOE) 1996c. *Addendum (Final Environmental Impact Statement) Management of Spent Nuclear Fuel from the K Basins at the Hanford Site, Richland, Washington.* DOE/EIS-0245F, U.S. Department of Energy, Washington, D.C.

U.S. Department of Energy (DOE). 1996d. *Plutonium: The First 50 Years, United States Plutonium Production, Acquisition, and Utilization From 1944 Through 1994.* U.S. Department of Energy, Washington, D.C. Available URL: http://apollo.osti.gov/html/osti/opennet/document/pu50yrs/pu50y.html.

U.S. Department of Energy (DOE). 1997. *Environmental Assessment, Management of Hanford Site Non-Defense Production Reactor Spent Nuclear Fuel, Hanford Site, Richland, Washington.* DOE/EA-1185, U.S. Department of Energy-Richland Operations, Richland, Washington.

U.S. Department of Energy (DOE). 1997a. *Linking Legacies: Connecting the Cold War Nuclear Weapons Production Processes to Their Environmental Consequences.* DOE/EM-0319, U.S. Department of Energy, Office of Environmental Management, Washington, D.C.

U.S. Department of Energy (DOE). 1997b. *Waste Site Groupings for 200 Areas Soil Investigations.* DOE/RL-96-81, Rev. 0, U.S. Department of Energy-Richland Operations Office, Richland, Washington.

U.S. Department of Energy (DOE). 1998a. *Screening Assessment and Requirements for a Comprehensive Assessment: Columbia River Comprehensive Impact Assessment.* DOE/RL-96-16, Rev. 1 (final), U.S. Department of Energy, Richland Operations Office, Richland, Washington. Available URL: http://www.hanford.gov/crcia/reports/doe-rl_96-16/crcia03_98.htm.

U.S. Department of Energy (DOE). 1998b. *Accelerating Cleanup: Paths to Closure.* DOE-EM-0362, U.S. Department of Energy, Office of Environmental Management, Washington, D.C.

U.S. Department of Energy (DOE). 1998c. *Accelerating Cleanup: Paths to Closure, Hanford Site.* DOE/RL-97-57, U.S. Department of Energy-Richland Operations Office, Richland, Washington.

U.S. Department of Energy (DOE). 1998d. *Retrieval Performance Evaluation Methodology for the AX Tank Farm.* DOE/RL-98-72 (draft), Jacobs Engineering Group Inc., U.S. Department of Energy-Richland Operations Office, Richland, Washington.

U.S. Department of Energy (DOE). 1998e. *Draft Risk Prospectus: Hanford Site.* U.S. Department of Energy, Chicago Operations Office Center for Risk Excellence, Chicago, Illinois.

U.S. Department of Energy (DOE). 1998f. *100-HR-3 and 100-KR-4 Operable Units Interim Action Performance Evaluation Report.* DOE/RL-97-96, U.S. Department of Energy-Richland Operations Office, Richland, Washington.

U.S. Department of Energy (DOE). 1998g. *Hanford's "C" Reactor Transformed.* Press release, U.S. Department of Energy-Richland Operations Office, Richland, Washington.

U.S. Department of Energy (DOE). June 1998h. *Hanford Site Environmental Management Specifications: Section 4.2.2 Waste Management Project.* DOE/RL-97-55, Rev. 1, U.S. Department of Energy-Richland Operations Office, Richland, Washington.

U.S. Department of Energy (DOE). 1998i. *Richland Environmental Restoration Project: Long Range Plan, FY 1999 Compliance Case.* Rev. 1, U.S. Department of Energy-Richland Operations Office, Richland, Washington.

U.S. Department of Energy (DOE). January 26, 1999. *DOE Resumes Plutonium Stabilization at Hanford Plutonium Finishing Plant.* Press release, U.S. Department of Energy-Richland Operations Office, Richland, Washington.

U.S. Department of Energy (DOE). 1999a. *Hanford: Making Progress…Getting Results.* U.S. Department of Energy-Richland Operations Office, Richland, Washington.

U.S. Department of Energy (DOE). October 1999b. *From Cleanup to Stewardship: A Companion Report to Accelerating Cleanup: Paths to Closure, and Background Information to Support the Scoping Process Required for the 1998 PEIS Settlement Study.* DOE/EM-0466, U.S. Department of Energy, Office of Environmental Management, Washington, D.C.

U.S. Department of Energy (DOE). September 1999c. *Final Hanford Comprehensive Land-Use Plan Environmental Impact Statements.* DOE/EIS-0222-F, U.S. Department of Energy-Richland Operations Office, Richland, Washington.

U.S. Department of Energy (DOE). 1999d. *Groundwater/Vadose Integration Project.* DOE/RL-98-48, 3 vols., U.S. Department of Energy-Richland Operations Office, Richland, Washington.

U.S. Department of Energy (DOE). 1999e. *Hanford Site Cleanup Objectives Inconsistent with Projected Land Uses.* DOE/IG-0446, U.S. Department of Energy, Office of Inspector General, Office of Audit Services, Washington, D.C.

U.S. Department of Energy (DOE). 2000a. *Hanford 2012: Accelerating Cleanup and Shrinking the Site.* DOE/RL-2000-62, Rev. 1, U.S. Department of Energy-Richland Operations Office, Richland, Washington.

U.S. Department of Energy (DOE). February 7, 2000b. "Office of River Protection: Criticality Safety Issue Resolved." *Hanford Progress,* U.S. Department of Energy-Richland Operations, Richland, Washington.

U.S. Department of Energy (DOE). February 7, 2000c. "Cleanup of 'Canyon' Facilities Moved Forward." *Hanford Progress,* U.S. Department of Energy-Richland Operations, Richland, Washington.

U.S. Department of Energy (DOE). February 2, 2000d. *Low-Level Burial Grounds.* Information brochure, REG-0229, U.S. Department of Energy-Richland Operations, Richland, Washington.

U.S. Department of Energy (DOE). 2000e. *The Management of Tank Waste Remediation at the Hanford Site.* DOE/IG-0456, U.S. Department of Energy, Office of Inspector General, Office of Audit Services, Washington, D.C.

U.S. Department of Energy (DOE). June 2000f. *Buried Transuranic-Contaminated Waste Information for U.S. Department of Energy Facilities.* U.S. Department of Energy-Headquarters, Washington, D.C.

U.S. Department of Energy (DOE). March 2000g. *Status of Report on Paths to Closure.* DOE-EM-0526, U.S. Department of Energy-Headquarters, Washington, D.C.

U.S. Department of Energy (DOE). 2000h. "Waste Analysis Plan for PUREX Storage Tunnels." *Dangerous Waste Portion of the Resource Conservation and Recovery Act Permit for the Treatment, Storage, and Disposal of Dangerous Waste, Hanford Facility Dangerous Waste permit Application, PUREX Storage Tunnels.* DOE/RL-90-24, Rev 4, U.S. Department of Energy, Richland, Washington. Available URL: http://www.hanford.gov/docs/hnf-sd-en-wap-007/rl90-24wap.html. Last updated March 28, 2000.

U.S. Department of Energy (DOE). January 2000i. *Fact Sheet: Waste Encapsulation and Storage Facility.* REG-0225, U.S. Department of Energy-Richland Office, Richland, Washington.

U.S. Department of Energy (DOE). February 2000j. *Remedial Design Report and Remedial Action Work Plan for the K Basins Interim Remedial Action.* DOE/RL-99-89, U.S. Department of Energy-Richland Office, Richland, Washington.

U.S. Department of Energy (DOE). 2001. *Hanford Site Cleanup Challenges and Opportunities for Science and Technology: A Strategic Assessment.* DOE/RL-2001-03, Rev. 0, U.S. Department of Energy-Richland Operations Office, Richland, Washington.

U.S. Department of Energy (DOE). October 2001a. *Long-Term Stewardship Study: Volume I-Report.* U.S. Department of Energy, Office of Environmental Management, Washington, D.C.

U.S. Department of Energy (DOE). April 2001b. *Summary Data on the Radioactive Waste, Spent Nuclear Fuel, and Contaminated Media Managed by the U.S. Department of Energy.* U.S. Department of Energy, Office of Environmental Management, Washington, D.C.

U.S. Department of Energy (DOE). February 4, 2002. *A Review of the Environmental Management Program.* Presented to the Assistant Secretary for Environmental Management by the Top-To-Bottom Review Team, U.S. Department of Energy, Washington, D.C. Available online: http://www.em.doe.gov/ttbr.pdf.

U.S. Department of Energy (DOE). 2002a. *Draft Hanford Site Solid (Radioactive and Hazardous) Waste Program Environmental Impact Statement Richland Washington, Vol. 1-2.* DOE/EIS-0286D, U.S. Department of Energy, Richland, Washington.

U.S. Department of Energy (DOE). 2002b. *Human Radiation Studies: Remembering the Early Years: Oral History of Health Physicist Carl C. Gamertsfelder, Ph.D.* DOE/EH-0467, U.S. Department of Energy, Office of Human Radiation Experiments, Washington, D.C. Available URL: http://tis.eh.doe.gov/ohre/roadmap/histories/0467/0467a.html.

U.S. Department of Energy (DOE). 2002c. *Human Radiation Studies: Remembering the Early Years: Oral History of John W. Healy Conducted November 28, 1994.* DOE/EH-0455, U.S. Department of Energy, Office of Human Radiation Experiments, Washington, D.C. Available URL: http://tis.eh.doe.gov/ohre/roadmap/histories/0455/0455toc.html.

U.S. Department of Energy (DOE). 2002d. *Performance Management Plan for the Accelerated Cleanup of the Hanford Site.* DOE-RL-2002-47 Rev. D, U.S. Department of Energy, Richland, Washington.

U.S. Department of Energy (DOE). 2002e. *FY2001 Hanford Technology Deployment Accomplishments*, RL-D02-003, U.S. Department of Energy, Richland, Washington.

U.S. Department of Energy Grand Junction Office. January 1998. *Vadose Zone Characterization Project at the Hanford Tank Farms: TX Tank Farm Report,* GJO-97-30-TAR (GJO-HAN-16), U.S. Department of Energy Grand Junction Office, Grand Junction, Colorado. Available URL: http://www.doegjpo.com/programs/hanf/TXReport/REPORT/content.htm.

U.S. Department of Energy and State of Oregon. August 1997. *Memorandum of Understanding Between the U.S. Department of Energy and State of Oregon.* Signed by J. Wagoner (DOE Hanford Manager) and J. Kitzhaber (Governor of Oregon).

U.S. Department of Energy and Washington State Department of Ecology. 1996. *Tank Waste Remediation System, Hanford Site, Richland, Washington, Final Environmental Impact Statement.* DOE/EIS-0189, U.S. Department of Energy, Washington, D.C.

U.S. Department of Health, Education, and Welfare; Public Health Service; Bureau of State Services, Division of Sanitary Engineering Services; and Robert A. Taft Sanitary Engineering Center. 1954. *Water Quality Studies on the Columbia River.* GEH-21328. Hanford Engineer Works, Richland, Washington.

U.S. Department of Navy. April 1996. *Final Environmental Impact Statement on the Disposal of Decommissioned, Defueled Cruiser, Ohio Class, and Los Angeles Class Naval Reactor Plants.* DOE/EIS-0259, U.S. Department of Navy, Bremerton, Washington.

U.S. District Court for the Eastern District of Tennessee. April 13, 1984. *Legal Environmental Assistance Foundation, Inc., and Natural Resources Defense Council, Inc. Plaintiffs, State of Tennessee Plaintiff-Intervener v. Donald Hodel, Secretary, United States Department of Energy, et al.* No. Civ. 3-83-562, 586 F. Supp.1163.

U.S. District Court for the Eastern District of Washington. June 28, 2000. Ned Charles Lumpkin et al. vs. E.I. DuPont de Nemours and Co., et al. *Class Action Complaint for Injunctive Relief and Damages and Complaint for Individual Claims.* CT-00-5052-EFS, Spokane, Washington.

U.S. Energy Research and Development Administration (ERDA). 1975. *Final Environmental Statement: Waste Management Operations, Hanford Reservation.* ERDA-1538, 2 vols., U.S. Energy Research and Development Administration, Richland, Washington.

U.S. Environmental Protection Agency (EPA). 1985. *High-Level and Transuranic Radioactive Wastes: Background Information Document for Final Rule.* EPA-520/1-85-035, U.S. Environmental Protection Agency, Washington, D.C.

U.S. Environmental Protection Agency (EPA). 1988. *Guidance for Conducting Remedial Investigations and Feasibility Studies Under CERCLA.* OSWER Directive 9355.3-01, EPA/540/G-89/004. U.S. Environmental Protection Agency, Office of Emergency and Remedial Response, Washington, D.C.

U.S. Environmental Protection Agency (EPA). 1991. *Report and Recommendations of the Technology Innovation and Economics Committee: Permitting and Compliance Barriers to U.S. Environmental Technology Innovation.* EPA/101/N-91/001, U.S. Environmental Protection Agency, Office of Cooperative Environmental Management, Washington, D.C.

U.S. Environmental Protection Agency (EPA). 1992. *Framework for Ecological Risk Assessment.* EPA/630/R-92/001, Risk Assessment Forum, Washington, D.C.

U.S. Environmental Protection Agency (EPA). July 1994. *A Review of Ecological Assessment Case Studies From a Risk Assessment Perspective,* Vol. 2. EPA/630/R-94/003, U.S. Environmental Protection Agency, Washington, D.C.

U.S. Environmental Protection Agency (EPA). 1998. *Guidelines for Ecological Risk Assessments.* EPA 630/R-95/002F, U.S. Environmental Protection Agency, Washington, D.C.

U.S. Environmental Protection Agency (EPA). 2000. *Report on the Tank Waste Remediation System (TWRS) Program for the Hanford Federal Facility.* Report No. WAD 99-000421-2000-P-00012, U.S. Environmental Protection Agency, Office of Inspector General for Audits, San Francisco, California.

U.S. Environmental Protection Agency (EPA). April 2001a. *USDOE Hanford Site First Five Year Review Report.* U.S. Environmental Protection Agency, Region 10, Hanford Project Office, Richland, Washington.

U.S. Environmental Protection Agency (EPA). January 4, 2001b. *EPA Response to Consensus #113, Hanford 2012.* Letter from M.F. Gearheard (EPA) to M. Reeves (Hanford Advisory Board). U.S. Environmental Protection Agency, Seattle, Washington.

U.S. Environmental Protection Agency (EPA). 2002. *Columbia River Basin Fish Contaminant Survey 1996-1998.* EPA 910-R-02-006, U.S. Environmental Protection Agency, Region 10, Seattle, Washington.

U.S. House of Representatives. October 2000. *Incinerating Cash: The Department's of Energy's Failure to Develop and Use Innovative Technologies to Clean Up the Nuclear Waste Legacy.* Committee on Commerce, U.S. House of Representatives, Washington, D.C.

U.S. Nuclear Regulatory Commission. 1980. *Three Mile Island, A Report to the Commissioners and to the Public.* Report of Special Inquiry Group, M. Rogovin, director. Vol II, part 2. NUREG/CR-1250, U.S. Nuclear Regulatory Commission, Washington, D.C.

U.S. Public Health Service. September 4, 1964. *Minutes of the Meeting Between Richland Operations Office and United States Public Health Service on Columbia River Contamination at Richland, Washington on July 16, 1964.* HAN-89378, Hanford Operations Office, Richland, Washington.

Van Middlesworth, L. 1954. "Radioactivity in Animal Thyroids From Various Areas." *Nucleonics* 12:56-57.

Van Middlesworth, L. 1990. "Environmental Radioiodine in Thyroids of Grazing Animals." In *Environmental Monitoring, Restoration, and Assessment: What Have We Learned?* Proceedings for the Twenty-Eighth Hanford Symposium on Health and the Environment, ed. R.H. Gray, pp. 15-24. Pacific Northwest Laboratory, Richland, Washington.

Vargo, G.J., ed., V. Poyarkov, V. Bar'yakhtar, V. Kukhar', I. Los', V. Kholosha, and V. Shestopalov. 2000. *Chornobyl Accident: A Comprehensive Risk Assessment.* Battelle Press, Columbus, Ohio.

The Villager. August 22, 1945. "It's Atomic Bombs." *The Villager*, Richland, Washington.

WAC. See Washington Administrative Code.

Waite, J.L. 1991. *Tank Wastes Discharged Directly to the Soil at the Hanford Site.* WHC-MR-0227, Westinghouse Hanford Company, Richland, Washington.

Walker, S. February 2001. "Risk-Based Cleanup Levels Under Superfund." *Risk Excellence Notes* 3(1):8-9. U.S. Department of Energy, Argonne, Illinois Available URL: http://riskcenter.doe.gov.

Ward, J.F. 1994. "The Complexity of DNA Damage: Relevance to Biological Consequences." *International Journal of Radiation Biology* 66:427-432.

Washington Advisory Group. 1999. *Managing Subsurface Contamination: Improving Management of the Department of Energy's Science and Engineering Research on Subsurface Contamination.* The Washington Advisory Group, Washington, D.C.

Washington Administrative Code (WAC). 1982. "Dangerous Waste Regulations." WAC 173-303, Olympia, Washington.

Washington State Department of Ecology. February 1991. *Responsiveness Summary on the Amendments to the Model Toxics Control Act Cleanup Regulation Chapter 173-340 WAC.* Washington State Department of Ecology, Olympia, Washington.

Washington State Department of Ecology. July 20, 1993. *Hanford Waste Management and Environmental Restoration Policy Guidance.*

Washington State Department of Ecology. March 25, 1995. *U.S. Department of Energy Request to Change N-Springs Action Memorandum.* Letter to S.H. Wisness. CCN 012354, U.S. Department of Energy, Richland Operations, Richland, Washington.

Washington State Department of Ecology. 1996. *Focus: Model Toxics Control Act.* Publication F-TC-94-129, Washington State Department of Ecology, Olympia, Washington.

Washington State Department of Ecology. January 5, 2001a. *Response to Hanford Advisory Board Consensus #113, Hanford 2012: Accelerating Cleanup and Shrinking the Site (Hanford 2012).* Letter from T. Fitzimmons (WSDOE) to M. Reeves (HAB). Washington State Department of Ecology, Olympia, Washington.

Washington State Department of Ecology. February 2001b. *Adopted Amendments, The Model Toxics Control Act Cleanup Regulation Chapter 173-340 WAC.* Washington State Department of Ecology, Olympia, Washington.

Washington State Department of Ecology, U.S. Environmental Protection Agency, and U.S. Department of Energy. 1998. *Hanford Federal Facility Agreement and Consent Order.* Document No. 89-10, Rev. 5 (The Tri-Party Agreement), Olympia, Washington.

Washington State Department of Health. 1997. *Hanford Guidance for Radiological Cleanup.* Washington State Department of Health, Olympia, Washington.

Washington State Department of Health. 1998. *1998 Cancer in Washington, Annual Report of the Washington State Cancer Registry.* Available URL: http://www3.doh.wa.gov/WSCR/.

Washington State Department of Health. 2000. *Hanford Site Shoreline Vegetation Study 1999.* WDOH-320-023, Washington State Department of Health, Olympia, Washington. Available URL: http://www.doh.wa.gov/ehp/rp/rp-publ.htm.

Way, K. January 20, 1944. *Time Interval Between Discharge of Metal From the Pile and Processing in Canyon.* Letter to D.F. Babcock, DUH-160, U.S. Department of Energy Public Reading Room, Richland, Washington.

Weber, W.J., L.R. Corrales, et al. 1999. "Radiation Effects in Nuclear Waste Materials." *Science to Support DOE Site Cleanup: The Pacific Northwest National Laboratory Environmental Management Science Program Awards, Section 1.161.* PNNL-12208, Pacific Northwest National Laboratory, Richland, Washington.

Wells, D. 1994. *Special Report: Radioactivity in the Columbia River Sediments and Their Health Effects.* Washington State Department of Health, Olympia, Washington.

Wells, G.W. August 31, 1967. *Chemicals Discharged to the Columbia River From DUN Facilities: Fiscal Year 1967.* DUN-3032, Douglas United Nuclear, Inc., Richland, Washington.

Wenning, R.J. August 2001. "Focus on Sediments: The Different Roles of Ecological Risk Assessment in Regulatory Decision-Making." *Contaminated Soil, Sediment & Water, The Magazine of Environmental Assessment & Remediation* 5:67-70.

Werner, J. 1999. "Long-Term Stewardship." Presented at the Energy Communities Alliance Fall Conference, Richland, Washington.

Western Governors' Association, Southern States Energy Board, and Interstate Technology and Regulatory Cooperation Work Group. November 2000. *Approaches to Improve Innovative Technology Development at the U.S. Department of Energy: Lessons Learned and Recommendations for a Path Forward.* Pacific Rim Enterprise Center, Seattle, Washington.

Westinghouse Hanford Company (WHC). 1995. *B Plant/Waste Encapsulation and Storage Facility: Briefing Booklet.* Westinghouse Hanford Company, Richland, Washington.

Whitman, D. April 17, 2000. "It's a Breath of Fresh Air: Thirty Years After Earth Day, America is Getting Its Environmental Act Together." *U.S. News and World Report* 128(15):16-19.

Wicker, B., and J. Sherwood. November 17, 1999. *Energy Secretary Bill Richardson Announces Action to Help Sick Workers.* Press release, U.S. Department of Energy, Washington, D.C.

Wiggins, W.D., G.P. Ruppert, R.R. Smith, L.L. Reed, L.E. Hubard, and M.L. Courts. 1996. *Water Resources Data, Washington Water Year 1995.* U.S. Geological Survey, Tacoma, Washington.

Wildman, R.D. April 2, 1963. *Statement for Conference on Pollution of the Upper Columbia River.* Letter to N.H. Woodruff. HAN-84700, Richland, Washington.

Wilkinson, G.S., N. Trieff, R. Graham, and R. Priore. 2000. *Final Report: Study of Mortality Among Female Nuclear Weapons Workers.* National Institute for Occupational Safety and Health, Washington, D.C.

Williams, S.J., chairman. April 2, 1948. *Report of the Safety and Industrial Health Advisory Board.* U.S. Atomic Energy Commission, Safety and Industrial Health Advisory Board, Washington, D.C.

Wills, G. 1999. *A Necessary Evil: A History of American Distrust of Government.* Simon & Schuster, New York.

Wilson, M. March 3, 2001. "Ecology Department Serious About Role in Hanford Cleanup." *Tri-City Herald*, p. D2, Kennewick, Washington.

Wilson, R.H., ed. February 24, 1964. *Evaluation of Radiological Conditions in the Vicinity of Hanford for 1963.* HW-80991, Hanford Atomic Products Operation, Richland, Washington.

Wing, S.B. 1998. *Case Control Study of Epidemiology of Multiple Myeloma Among Workers Exposed to Ionizing Radiation and Other Physical and Chemical Agents: Final Technical Report.* University of North Carolina, School of Public Heath for the National Institute for Occupational Safety and Health, Cincinnati, Ohio.

Wodrich, D. 1991. *Historical Perspective of Radioactivity Contaminated Liquid and Solid Wastes Discharged or Buried in the Ground at Hanford (Draft).* TRAC-0151-VA, Westinghouse Hanford Company, Richland, Washington.

Wolman, A. 1959. "The Nature and Extent of Radioactive Waste Disposal Problems." U.S. Congress, Joint Committee on Atomic Energy, Special Subcommittee on Radiation. *Hearings.* Vol. 1, pp. 7-22. 86th Cong., 1st Session on Industrial Radioactive Waste Disposal. U.S. Government Printing Office, Washington, D.C.

Woodcock, G. March 12, 2000. "DOE Health Announcement Needs Context." *Tri-City Herald,* Kennewick, Washington.

Woodruff, R.K., R.W. Hanf, M.G. Hefty, and R.E. Lundgren, eds. December 20, 1991. *Hanford Site Environmental Report for Calendar Year 1990.* PNL-7930, Pacific Northwest Laboratory, Richland, Washington.

Work, J.B. 1951. *Efficiency Evaluation of the Dissolver Cell Silver Reactor and Fiberglass Filter.* HW-19898, General Electric, Hanford Atomic Products Operation, Richland, Washington.

World Health Organization. 1956. "The General Problem of Protection Against Radiations from the Public Health Point of View." In *Proceedings for the International Conference on the Peaceful Uses of Atomic Energy*, Vol. 13, pp. 10-14, United Nations, New York.

World Wide Web. 2002a. *Approximate Spent Nuclear Fuel Inventory in Metric Tons of Heavy Metal.* National Spent Nuclear Fuel Program, Idaho National Engineering and Environmental Laboratory. Available URL: http://nsnfp.inel.gov/tikal-10-30-00/table32.htm. Last updated April 11, 2000.

Wu, R. 1994. *Determination of Effective Doses From Radionuclides in the Columbia River Sediments.* Master of Science Thesis, Oregon State University, Corvallis, Oregon.

Young, A.L. February 2001. "A New Administration, An Opportunity to Strengthen Radiation Policy," *Risk Assessment Notes* 3(1):2-3. U.S. Department of Energy, Argonne, Illinois. Available URL: http://riskcenter.doe.gov.

Zachara, J.M. (Principal Investigator). June 2002. "The Influence of Calcium Carbonate Grain Coatings on Contaminant Reactivity in Vadose Zone Sediments." *Science to Support DOE Site Cleanup: The Pacific Northwest National Laboratory Environmental Management Science Program Awards.* PNNL-13928, Pacific Northwest National Laboratory, Richland, Washington, pp. 3.25-3.29.

Zachara, J.M., S.C. Smith, C. Liu, J.P. McKinley, R.J. Serne, and P.L. Gassman. 2002. "Sorption of Cs+ to Micaceous Subsurface Sediments From the Hanford Site, USA." *Geochim. Cosmochim. Acta* 66(2):193-211.

Zorpette, G. 1996. "Hanford's Nuclear Wasteland." *Scientific American* 274(5):88-97.

Appendix A
About Radionuclides

Atoms are the building blocks of the physical world. They combine in simple to complex patterns to form all solids, liquids, and gases.

Atoms such as calcium, oxygen, silicon, aluminum, carbon, and iron are the building blocks for the chemical elements of the physical world. Matter is made of different types of atoms that contain a central nucleus and negatively charged (-) electrons spinning around the nucleus. The nucleus is composed of positively charged (+) protons and electrically neutral (no charge) neutrons.[1] The number of protons and neutrons in an element's nucleus gives it unique nuclear properties, while the encircling electrons impart unique chemical properties. (The burning of wood or rusting of iron are examples of chemical reactions.)

One hundred and eighteen natural or human-made elements exist. The simplest and lightest element is hydrogen, which has a single proton. The heaviest natural element is uranium. The most abundant type of uranium contains 92 protons and 146 neutrons. This form of uranium is referred to as uranium-238. The total number of protons and neutrons is called an element's atomic weight. Elements can contain different numbers of neutrons creating different forms (isotopes) of that element. For example, there are 22 short- to long-lived isotopes of uranium, 17 isotopes of sodium, and 3 isotopes of hydrogen.

The energies holding nuclei together are millions of times stronger than the energies binding electrons to atoms. When a nucleus splits (fissions), some of this binding energy is released. This is why nuclear reactions, such as in atomic weapons, are much more powerful than conventional chemical explosives or why far greater energy is available from a pound of uranium irradiated inside a reactor than a pound of coal burned in a furnace.

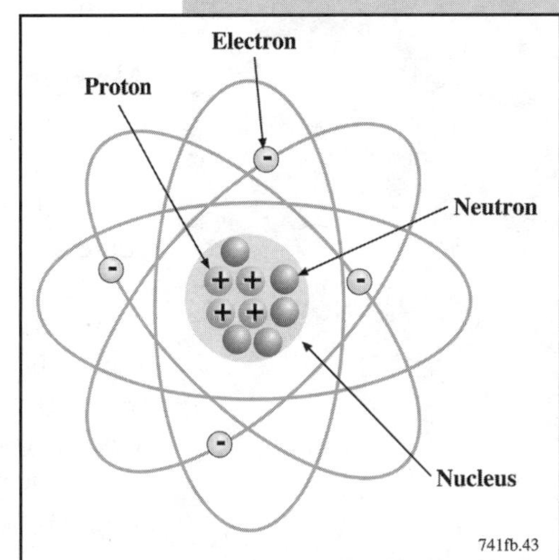

Atoms. Matter is composed of atoms. Atoms are made from combinations of protons, neutrons, and electrons. Some atoms contain a ratio of protons to neutrons that makes the atom stable (nonradioactive). Other atoms contain excess energy and need to release this energy as radiation. They are radioactive.

The nucleus of an atom is held together by a small amount of its mass being converted into binding energy. When all internal forces are well matched, the nucleus is stable (nonradioactive) and remains unchanged over time. If there are too many neutrons or protons, the nucleus has too much energy and wants to release some of it to become stable.

Unstable elements are radioactive. They emit radiation in the form of energy (gamma rays) or particles (alpha, beta, or neutrons) during radioactive decay to become more stable. This transforms the original radioactive element (parent) into new elements (daughter products) that may be stable or radioactive.

Here is one example of radioactive decay. Natural uranium-238 in rocks eventually changes to lead-206. This happens over about 16 decay steps (in a decay series) that begins with naturally radioactive uranium-238 and

(1) In addition to the proton, neutron, and electron, more than 100 other fundamental particles have been discovered or hypothesized (Hammond in Lide 1995).

Half-lives of radionuclides can vary from fractions of a second to billions of years.

ends with nonradioactive lead-206.[2] During each step, one half of the atoms in a radioactive sample changes into another more stable radionuclide or into a nonradioactive element. Each step takes a certain amount of time called a half-life. It takes 4.8 billion years for uranium-238 and the resulting daughter products to change to lead-206. However, nearly 4.5 billion years of that time is spent in the very first step in the decay series—changing uranium-238 to thorium-234. The longest half-life is the decay series of the original parent. After 15 more steps and 325,000 more years, thorium-234 (and its daughter products) will decay to nonradioactive lead-206. Various types of radiation are released during these changes.

About 280 stable and 67 naturally radioactive isotopes exist (Lide 1995). In addition, some heavier isotopes (containing more protons and neutrons than occur naturally) have been produced artificially, at places such as Hanford.

More than 99% of the radioactivity currently at Hanford consist of two radionuclides: cesium-137 and strontium-90. Eventually, they will decay to the nonradioactive elements of barium-137 and zirconium-90.

Different radionuclides have different half-lives. These half-lives can vary from fractions of a second to billions of years. The two longest lasting fission products are exotic radionuclides of neodymium and indium. The half-life of neodymium-144 is 2 million billion years (a 2 followed by 15 zeros). The half-life of indium-115 is 400,000 billion years (a 4 followed by 14 zeros). Examples of the shortest-lived fission products, lasting less than 1 second, are xenon-143 and krypton-94.

More than 99% of the nearly 400 million curies of radioactivity currently at Hanford consist of two radionuclides: cesium-137 and strontium-90. These isotopes have half-lives of 30 and 29 years, respectively. Eventually, they will decay to the nonradioactive elements of barium-137 and zirconium-90. Afterwards, the other longer-lived radionuclides of plutonium, iodine, technetium, and americium will dominate the radionuclide inventory at Hanford.

Natural Radioactivity and Radiation Dose

Naturally occurring radionuclides come from two primary sources: primordial (terrestrial) and extraterrestrial (cosmogenic). Both contribute to external radiation dose. Other sources are inhaled, ingested, or naturally internal to the human body.

Primordial radionuclides are those remaining from materials that formed the earth some 5 billion years ago. These include daughter products of original radionuclides. Most primordial radionuclides are isotopes of heavy elements (those with large atomic weights) from the three radioactive decay series headed by uranium-238, uranium-235, and thorium-232 in addition to potassium-40 and rubidium-87, which decay directly to stable elements.[3]

---

(2) One of these steps produces radon-222. Short-lived daughter products of radon are responsible for most of the natural radiation dose received by people.

(3) Radionuclides are common in granite, coal, phosphate-bearing rocks, shale, and extractable ores.

How Radioactive Decay Works

This table demonstrates how radioactive decay decreases the amount of original material (parent atoms) present while increasing the number of new elements (daughter atoms) present. Let's start with 1000 atoms of a radioactive element. At the beginning (zero half-lives), no daughter atoms exist. After one half-life, 500 parent atoms remain while 500 daughter atoms have accumulated. After two half-lives, 250 parent atoms are left while 750 daughter products exist—and so on. The number of parent atoms is reduced by 50% as each half-life passes. At the same time, the number of daughter atoms increases. By the end of ten half-lives, 1 atom of the original parent remains while 999 daughter atoms exist. In other words, at the end of 10 half-lives, one-tenth of 1% of the original parent remains. Eventually, all parent atoms will disappear because of radioactive decay. The daughter atoms may or may not be radioactive or hazardous.

Half-Lives	Number of Parent Atoms	Number of Daughter Atoms
0	1000	0
1	500	500
2	250	750
3	125	875
4	62	938
5	31	969
6	16	984
7	8	992
8	4	996
9	2	998
10	1	999

Few radionuclides of an isotope remain after 10 half-lives. Radionuclides with half-lives greater than a billion years remain in rocks and soil.

It's natural to wonder, how much natural radioactivity is found in a chunk of earth's surface? Based on David Bodansky's (1996) listing of the average abundance of natural radionuclides in the upper part of the earth plus their radioactivity, one can estimate that 1 curie of radioactivity exists in every cube of the shallow crust measuring about 100 feet on a side. Most of this radioactivity (82%) comes from potassium-40, while the rest is emitted from rubidium-87, thorium-232, and uranium-238.

Just as there are substantial variations in the concentrations of chemical elements from place to place on earth, variations in naturally occurring radionuclides also exist. For example, the abundance of radioactive uranium and potassium is higher in granite than in limestone, marble, or basalt. The average annual radiation dose from these terrestrial radionuclides varies from 16 millirem along the Atlantic Coast to 63 millirem in the Rocky Mountain states (NCRP 1987).

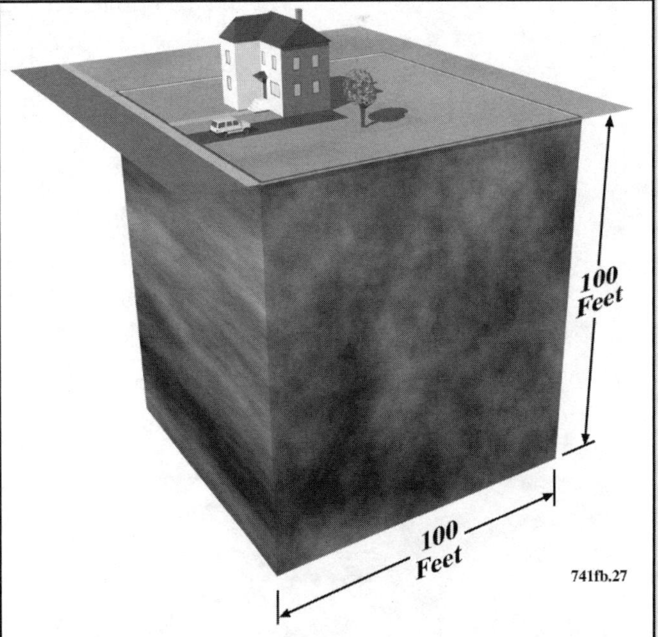

One-Curie Cube of Earth. A cube of earth 100 feet on a side contains about 1 curie of naturally occurring radioactivity, mostly from potassium-40. A typical home smoke detector holds 1 millionth of one curie (1 microcurie) of human-made americium-241.

Because many of the half-lives of cosmic-derived radionuclides vary from less than an hour to a few hundred thousand years, they must be continuously replenished to remain in the environment.

The radiation dose received from background sources is dominated by primordial radionuclides such as potassium-40 or the radioactive series headed by uranium-238 and thorium-232.

Naturally occurring radionuclides with half-lives shorter than perhaps 100 million years are undetectable because they have gone through 50 half-lives since their creation. Few radionuclides of an isotope remain after about 10 half-lives. Radionuclides with half-lives greater than a billion years remain in rocks and soil.

Most radionuclides of cosmic origin are lower in atomic weight and are produced when high-energy cosmic rays, from the sun or outside the solar system (galactic), interact with atoms in our atmosphere. Their intensity depends upon the occurrence of solar flares and adjustments in interplanetary magnetic fields. Because many of the half-lives of cosmic-derived radionuclides vary from less than an hour to a few hundred thousand years, they must be continuously replenished to remain in the environment. Three radioisotopes are primary contributors to background radiation doses to humans: tritium, carbon-14, and sodium-22.

The major contributors to radiation dose from inhaled radionuclides of natural origin are the short-lived decay products of radon-222. These radioisotopes originate from uranium and seep into the basements of homes and buildings as a colorless and odorless gas.

Natural radionuclides found in the human body come from three primary sources: food, air, and water. Once inside, they follow the chemical metabolism of that element. Dose is dominated by primordial radionuclides such as potassium-40 or the radioactive series headed by uranium-238 and thorium-232. The concentration of long-lived radionuclides slowly increases with body age while short-lived radionuclides either disappear or are continually renewed by fresh uptake.

How much natural radioactivity exists in the human body? About 100 billionths of 1 curie, contributed mostly by potassium-40.

According the to the National Council on Radiation Protection and Measurements (NCRP 1987, p. 148), the breakdown of average radiation dose from all natural background sources is as follows:

Origin	Average Radiation Dose (millirem per year)
Cosmic[4]	28
Terrestrial	28
Inhaled	200
Internal to body	40
Total (rounded)	300

(4) This includes radiation dose from cosmic rays and cosmogenic radionuclides such as carbon-14. Radiation dose from cosmic radiation is greater at high elevations than in valleys or at sea level. The average annual cosmic-derived radiation dose increases about 4 millirem for each 1000 feet in elevation. This dose is 26 millirem a year at sea level.

The Half-Lives of Radionuclides

This is a list of the natural and human-made radionuclides used in this book. Their half-lives are also given (Lide 1995). The number listed after each radionuclide is its atomic weight (for example, americium-241 has 241 neutrons plus protons within its nucleus).

Radionuclide	Half-Life	Radionuclide	Half-Life
Americium-241	432 years	Polonium-210	138 days
Antimony-125	2.8 years	Potassium-40	1.26 billion years
Argon-41	1.8 hours	Radium-226	1599 years
Arsenic-76	26 hours	Radon-222	3.8 days
Beryllium-7	53.3 days	Rubidium-87	49 billion years
Carbon-14	5715 years	Ruthenium-103	39 days
Cesium-137	30 years	Ruthenium-106	1.0 year
Cerium-144	285 days	Sodium-22	2.6 year
Cobalt-60	5.3 years	Sodium-24	15 hours
Europium-154	8.6 years	Strontium-90	29 years
Iodine-129	17 million years	Technetium-99	213,000 years
Iodine-131	8.0 days	Thorium-232	14 billion years
Iodine-135	6.6 hours	Thorium-234	24.1 days
Iron-55	2.7 year	Tritium (H-3)	12.3 years
Krypton-85	10.7 years	Uranium-234	245,000 years
Krypton-94	0.2 second	Uranium-235	704 million years
Manganese-56	2.6 hours	Uranium-238	4.46 billion years
Neptunium-239	2.4 days	Xenon-133	5.2 days
Nickel-63	100 years	Xenon-135	9.1 hours
Nickel-65	2.5 hours	Xenon-143	0.3 second
Phosphorus-32	14 days	Yttrium-90	2.7 days
Plutonium-238	87.7 years	Zinc-65	244 days
Plutonium-239	24,000 years	Zinc-95	64 days
Plutonium-240	6537 years		

References

Bodansky D. 1996. *Nuclear Energy: Principles, Practices, and Prospects.* American Institute of Physics Press, Woodbury, New York.

Lide DR, ed. 1995. *CRC Handbook of Chemistry and Physics: A Ready Reference Book of Chemical and Physical Data.* 76th ed. CRC Press, Boca Raton.

National Council on Radiation Protection and Measurements (NCRP). 1987. *Ionizing Radiation Exposure of the Population of the United States.* NCRP Report No. 93, Bethesda, Maryland.

Appendix B
Cultural Legacies
by Stuart Harris
Confederated Tribes of the Umatilla Indian Reservation[1]

Tots Maywe. Good morning.

I am Stuart Harris. I am a staff scientist for the Confederated Tribes of the Umatilla Indian Reservation, or CTUIR. My job is to analyze the risks to our people from pollution impacts. My primary work effort is focused on addressing nuclear pollution and associated cleanup at the Department of Energy's Hanford Site. The CTUIR is a sovereign government that has a legal interest in the natural resources upon which the CTUIR's treaty rights are based. This includes lands of the Hanford Site.

The Umatilla Indian Reservation, located near Pendleton, Oregon, is occupied by descendants of three Columbia Plateau Tribes – the Cayuse, the Walla Walla, and the Umatilla Tribes. Under the Treaty of 1855 [12 Stat. 945], the Tribes ceded lands to the United States yet retained rights to perform many activities on those lands, including but not limited to fishing, hunting, gathering roots and berries, and pasturing livestock. Effective exercise of these treaty rights depends on the health of the natural resources. The CTUIR government does not want the people exercising their treaty rights to be placed at risk.

We, the Tetokin, have been impacted through the encroachment of the Indo-European Americans. Our tribal population has been affected by biological warfare, ecological warfare, economic warfare, and downright attempts at genocide. Yet, we have endured, with our culture intact even through a 600-year holocaust. Our struggle is not over. With each successive generation we are forced to react to numerous environmental, cultural, and health impacts from members from the American society who still perceive us as characters in a Louis L'Amour novel or a Hollywood movie.

My people have to deal with this complex set of problems, complete with numerous entrenched interests such as agribusiness, mining, and government, and competing value sets vying for ever decreasing resources. We use a process that includes education, law, analysis, research, and planning. Each of these processes is filtered and translated through our culture.

The CTUIR culture, which has co-evolved with nature providing thousands of years of ecological education, has provided its people with their unique and valid version of holistic environmental management. Throughout the year, when the Umatilla Indian

(1) This paper was originally given by Stuart Harris at the Plenary Address, Society of Risk Analysis Annual Meeting, December 7, 1998, Phoenix, Arizona. I'm indebted to Stuart and John Burke of the Confederated Tribes of the Umatilla Indian Reservation for their kind permission to use this text. The photograph on page B.7 was added by the book's author.

Reservation traditional American Indian participates in activities such as hunting and gathering for foods, medicines, ceremonies, and subsistence, the associated activities are as important as the end product. In the Judeo-Christian tradition, an analogy would be "kosher" dietary practices.

> All of the foods and implements gathered and manufactured by the traditional American Indian are interconnected in at least one, but more often in many ways. The people of the CTUIR community follow cultural teachings or lessons brought down through history from the elders. Our individual and collective well being is derived, nurtured, and enhanced through membership in a healthy community that has access to ancestral lands and traditional resources. Our health is also enhanced from having the ability to satisfy the personal responsibility to participate in traditional family and community activities and to help maintain the spiritual quality of our resources.

These interconnected behaviorisms are ancient oral traditions within our cultural norms. The material or fabric of this tradition is unique, and is woven into a single tapestry that extends from far in the past to long into the future. In order to encompass the wide range of factors directly to the traditional American Indians of the CTUIR, a risk assessment has to be designed and scaled appropriately.

Only our people, the Tetokin, know what is good for the Tetokin people. We don't give our knowledge away as the cost of buying improved risk assessments that may or may not result in lowered exposures and better resource protection. Within the decision-making context of CERCLA [Comprehensive Environmental Response, Compensation, and Liability Act] and NEPA [National Environmental Policy Act], risk assessment as it now stands is woefully inadequate for addressing Native American concerns. In effect, a re-structuring of the risk assessment process must occur in order to address the overwhelming problems including but not limited to:

- Lack of breadth of coverage,
- Lack of integration and deficiencies related to not addressing the CTUIR traditional American Indian's quality of life,
- The interrelated eco-culture and their unique exposure parameters and pathways.

Unfortunately, the processes, the approach, and even the necessity to account for traditional American Indian lifestyles have gone unnoticed in classical risk assessments that typically focus on suburban lifestyles. A risk from nuclear, hazardous waste, or infrastructure planning that potentially affects one person of the CTUIR community may have lasting impacts throughout all of the community. In other words, a wave of risk and related impacts can ripple outwards, affecting all of the individuals in our culture, just like a wave generated and propagated in a tapestry. If a culture dies, the only remnants are the material artifacts. In the event of the unthinkable happening, a continuously sustainable natural resource-based material culture such as the CTUIR would rapidly disperse into the natural environment leaving no trace of the living CTUIR culture.

The need for understanding the pathways that directly involve the traditional American Indian cannot be overstated. Our ties to the environment are much more complex and intense than is generally understood. Because our tribal culture and religion are essentially synonymous with and inseparable from the land, the quality of the socio-cultural and eco-cultural landscapes is as important as the quality of individual natural resources or ecosystem integrity.

Many of our cultural and religious activities also occur in natural settings, and this increases exposures over suburban factors. The difference in mean exposures between an average suburban resident and an average traditional-subsistence person co-located in a particular contaminated area are due to differences in assumptions such as the percentage of plant material grown locally. The actual percentile will vary with the contaminants that are present in particular media, the pathways that are applicable, and other habitat- and climate-specific factors.

Our average subsistence traditional lifestyle is equivalent to at least a 90th percentile of the average suburban exposure. Initial sensitivity analyses show that the difference between means of the two types of lifestyles ranges from twofold to a hundredfold. The magnitude of the difference is due to the fact that the traditional of life as it is currently practiced is more than just a suburban lifestyle with extra fish consumption.

There are certain exposures that are potentially underestimated for a broad cross section of tribal members. For example, animal parts have many non-food uses that could contribute to personal exposure: teeth and bones are used for decoration and whistles, skin is made into clothing, fish belly fat is rendered and used as a base for body paint, and so on.

As with game, plants are used for more than just nutrition. Daily cleaning, preparation and ingestion of stored plants, and crafting of plant materials into household goods occur throughout the year. The cattail provides an example: in the spring the shoots are eaten, the roots are consumed, and the fibrous stalks are split, woven, or twisted into baskets, mats, or cook-hole layers. Later in the year the pollen is used for breads. Each of these activities involves the selection and gathering the plants from marshy areas, sorting, cleaning, stripping, peeling, splitting, chewing, and using various parts of the plant. Our basket weavers typically hold plant materials in their mouths during separation of the inner and outer bark. In addition to the plant itself, they contact sediment and water, and generally there will be cuts on the hands from the sharp edges that could facilitate dermal absorption during gathering, preparation, and weaving.

Likewise, the scenarios for children and other segments of our populations with greater exposures or greater sensitivities are not explicitly called out. These factors are part of both the uncertainty analysis and the risk characterization.

There are additional co-risk factors that could modify both exposure and sensitivity, such as:

- Individual cancer and non-cancer risk using the subsistence exposure scenario,
- Exposure to future populations,
- Community-level exposure burdens,

- Additional background exposures such as evidenced by fish advisories,
- Underlying health problems using health statistics where available,
- Influence of underlying nutritional status, dietary quality, and the physiological effects of substituting a modern diet if the traditional diet is not available,
- Socioeconomic status and access to health care,
- Potential differences in biochemical genetics and ethnopharmacology.

It is clear that evaluating impacts to a traditional way of life would include environmental quality and community quality of life in addition to personal exposures to contaminants.

Given this complex set of factors, along with the sovereign standing that our government has, I have found it useful to express my work within general guiding principles. There are some overriding principles, such as stewardship, trusteeship, and treaty compliance that apply to every aspect of my job. When we combine those principles with sustainable environmental management and a naturally integrated and holistic perspective, this leads us to some core performance measures that are reflected in my tribal mission statement. The things that are important to my tribal policy makers are:

1. Continuity and well-being of our people and our land,
2. Treaty rights, sovereignty, and the ability of my fellow tribal members to safely exercise their treaty rights,
3. Restoration of environmental conditions for cultural wellness and subsistence rights based on traditional environmental knowledge,
4. Individual and community health over time,
5. Equity within this generation and between generations,
6. Trusteeship of cultural and natural resources and landscapes,
7. Sustainability of cultures within ecosystems, and
8. Protection of the ethno-ecosystem or ecocultural health.

There are no assessments to which those principles do not apply. The following is a short list of projects that I am working on to fix or modify to meet these information needs.

First of all, I have developed a Native American subsistence scenario, which is being used in the CERCLA process and other assessments. But ultimately CERCLA assessments need to be based on more than just human exposure. They must include cultural risk as well. There are two NEPA EISs [environmental impact statements] at Hanford that have some unique features such as recognizing treaty rights and using the Native American subsistence exposure scenario (NASS).

Second, one of the EISs is also using the ethnohabitat concept and is substantially improving many of the components within the environmental justice section.

Third, the Natural Resource Damage Assessment process at Hanford is now including the cultural use of natural resources with new metrics for evaluating cultural use.

Fourth, I have borrowed terminology from EPA's [U.S. Environmental Protection Agency's] comparative risk methodology, which has three components – human health, ecological health, and quality of life. The quality of life component was modified to reflect cultural well-being specifically for the tribes.

Fifth, I have recommended certain modifications to the Hazard Ranking System so that it would be suitable for use as a tribal hazard ranking system, including incorporation of a tribal cultural perspective.

The point of these examples is that if you are well-grounded in the values and perspectives of the people you are trying to protect, then you should be able to find a way to apply those principles to any situation or tool. I believe that there is no risk assessment tool that cannot be made to follow those principles.

Now I am going to shift topics and discuss how risk managers can use cultural risk information. As an example, in any particular cleanup situation, there could be three bases for selecting a remedy:

1. If only human health is evaluated, and if the only cleanup goal is to reduce human risk, then institutional controls might appear to be the most cost-effective remedy. The most permanent remedy might be complete excavation regardless of the environmental damage this causes.
2. If the cleanup goal is to protect both human health and the environment, then the remedy might consist of limited excavation, a cap, a fence, and natural attenuation. This remedy ignores the cultural effects of lost access and use.
3. If, however, the cleanup goal is to protect human health, the environment, *and cultural use*, then the most cost-effective remedy might be a more expensive but less intrusive remedial technology combined with ALARA (a process to reduce concentrations to **as** **l**ow **as** **r**easonably **a**chievable) plus mitigation for impaired cultural use.

While protecting cultural resources and cultural health might seem to be an obstacle to achieving site closure because it might prevent excavation in culturally sensitive areas, I think that this is a great opportunity for innovative and credible and acceptable negotiated closures. More thoughtful scholarship needs to be applied to this area. More scholarship is also needed in the neglected discipline of risk ethics.

Please note that the current EPA guidance for environmental justice fails to capture tribal concerns and does not deal fairly with the science of traditional environmental management. It does not adequately describe how to evaluate the distribution of risk between population groups such as tribes compared to suburbia. It completely omits evaluation of differences in impacts between cultures and the resources on which those cultures depend. For example, we know that traditional members with subsistence lifestyles receive 2 to 100 times more exposure than a suburban resident might receive at identical environmental concentrations. We also know that tribal members typically have a larger burden of co-risk factors such as poor nutritional status, loss of natural diet, poorer access to health care, differences in metabolism, and so on. This means that tribal members might hypothetically not only receive more exposure but might also be more sensitive. Therefore, the cumulative impacts could be greatly magnified for tribal populations versus suburban populations. We as members of the risk community need to consider the concept that risk to people and their culture is composed of both exposure and sensitivity.

I have talked about cultural risk as one of the three types of risk, and have described why it should be evaluated just as rigorously and systematically as human and ecological risk. Risk characterization needs to include those three types of risk integrated into a more holistic summary that tells a more useful story about all the impacts to my culture that contamination causes. The current approach to risk characterization is to determine a probability of developing adverse human health effects, and sometimes to describe some ecological effects completely separately from human health effects. From the perspective of an exposed community, what is needed is a more complete story that describes everything that is at risk from the particular contamination incident, including a cultural way of life. Risk characterization is another neglected part of risk assessment.

Ultimately, the complete story about the long-term impacts of pollution on our culture needs to be incorporated within the oral histories because of the long-lived and/or persistent nature of some of the contaminants. This relates to many stewardship issues that are gaining attention. As the original managers of sustainable environmental systems, I believe tribal scientists can contribute a great deal to stewardship programs.

In closing, I want to review the conventional scientific method because our tribal religion is based on an observational and applied science that has proved its worth over thousands of years through survival of the Tetokin (people). I want to briefly review the process for moving from observation, to hypothesis, to theory, to law. Tribal science has followed this path also.

Science is the observation, identification, description, experimental investigation, and theoretical explanation of phenomena. The scientific method is a general term for the lines of reasoning that scientists follow in attempting to explain natural phenomena. It typically includes observation, analysis, synthesis, classification, and inductive inference, in order to arrive at a hypothesis that seems to explain the phenomenon or solve the problem.

Remember that a hypothesis becomes theory if it withstands repeated testing and application. A hypothesis is a conception of proposition that is tentatively assumed, and then tested for validity by comparison with observed facts and by experimentation. A theory is a hypothesis that is supported to some extent by experimentation or factual evidence but that has not been so conclusively proven as to be generally accepted as law. Deductive use of the theory may then explain additional problems.

Science is a product of the society that develops it, and it is formed to serve the needs of that society. American Indians have been observing natural phenomena, describing them experimentally investigating them, and explaining natural phenomena and natural resources for thousands of years. This tribal environmental knowledge forms the basis of traditional environmental management.

The reasoning that led to the determination of how to behave in the environment, based on what the environment consists of, is transferred to members of the tribe. Therefore, when a tribal member is gathering cultural materials, whether it is food or something else, he or she does it in a manner that reflects the principles of the science of traditional environmental management. This is the application of science, traditional tribal science, distilled into daily practice for the survival of a people.

The principles of traditional environmental management have been codified into law. There are some things you can do out in the environment and other things that you cannot do. The results of an action affect many things. As we know from the first law of thermodynamics, energy is conserved. Yet the entropy of reactions, especially in complex ecosystems, is difficult to determine, but has been observed by our elders through the noting of occurrences of the most probable reactions. For example, the type, quality, quantity, and occurrence of food or other natural resources has been noted and is related to young people through oral histories. Attention to the knowledge passed down means immediate survival and continuation of our people. Disregarding the knowledge can result in eating a poison, starvation, or poor health.

For countless generations our elders have told us about environmental conditions, and that our behavior is a product of rigorous and proven methodology that has guaranteed our survival through all types of natural cycles. Our lifestyle is resilient and has persisted through floods, droughts, cataclysms, upheavals, and warfare. We carry the unique and individual genes specifically adapted to and modified by our homelands.

Therefore, when I am asked, "What is cultural risk?" my answer is:

> "Because our people, the Tetokin, have been genetically modified by the ecology for thousands upon thousands of years, and have had their behavior modified as a result of responding to the flux of the ecology of our land for thousands upon thousands of years, and have produced a viable holistic environmental management system designed for continuously sustainable enhancement of our culture, and because the fabric of our very existence, including our sounds, medicine, science, art, music, and lifestyle is a reflection of thousands upon thousands of years of site-specific environmental shaping, any impact to those resources of which we are an inseparable part, is a risk to our culture."

I was asked by an educated man once, "How can a culture be irradiated?" He thought that only tangible things can be irradiated and therefore only tangible things can be at risk. My answer was, and still is: "If our Tribal people are kept from a sacred site because that piece of mother Earth has been contaminated, then I cannot transmit traditional teaching to future generations about the life significance of that site and therefore a significant part of our culture will be irreversibly altered."

How can you put a price on a sacred song that is derived from a landscape feature and is significant to the survival to our people and therefore our unique gene pool?

U.S. Department of Energy 741fb.106

Tribal Nations. This is a 1928 picture of a Native American family living at the Horn Rapids fishing camp now on the southern edge of Hanford. Members of several Native American Tribes have lived in the Hanford area for thousands of years. The Yakama, Umatilla, and Nez Perce nations retain rights to hunt, fish, and gather natural foods and medicines on lands ceded to the U.S. government in the Treaties of 1855.

Impacts to the ecology directly impact the health of our people and put our culture at risk. Through time, our sovereign tribal genetic characteristics may be adversely affected, thus destroying a multi-thousand year-long fabric of blood. When an organism interacts and specializes within a finite set of environmental factors for thousands and thousands of years, that organism becomes the ecology. Within an ecological system all parts are important and all parts interact. Eventually the parts become mutually dependent, and neither part can be removed without harming or killing the whole.

The Tetokin have developed the science of holistic environmental management and have evolved with the ecology in our homelands, driving it towards a sustainable, aesthetically nurturing environment that mutually enhances our culture through time. This is why I come to you today, to bring forth the concept of the reality of cultural risk and how I have, through my work, been able to develop it in terms that are easily understood by risk managers and as a process that can be used by risk assessors so that we, as risk professionals, will have the tools to provide more complete and satisfactory answers, and make better environmental decisions.

Thank you.

Glossary

actinide—an element having an atomic number (number of protons in nucleus) between 89 and 103. Examples include uranium, neptunium, plutonium, and americium.

activation—the process of causing radioactivity in a nonradioactive element by bombarding it with neutrons or other types of radiation.

alpha particle—a positively charged particle given off during radioactive decay. It consists of two protons (positive charge) and two neutrons (no charge) emitted from the nucleus of some long-lived radioactive elements such as uranium-238, plutonium-239, and radium-226. It is identical to the nucleus of a helium atom.

Applicable or Relevant and Appropriate Requirements—the Comprehensive Environmental Response, Compensation, and Liability Act requires organizations to select cleanup standards that comply with other laws and regulations. Such standards are called applicable or relevant and appropriate requirements. Applicable requirements are standards that apply to a specific substance, remedial action, location, or other circumstance found at a site. Relevant and appropriate requirements are standards not legally applicable, but they do address similar problems or situations.

aquifer—a permeable underground layer of sediments or fractured rock that can hold and transmit large quantities of groundwater.

ARAR—see Applicable or Relevant and Appropriate Requirements.

atom—the basic component of all matter. Most atoms consist of a nucleus of neutrons and protons surrounded by a cloud of electrons.

atomic number—the number of protons in the nucleus of an atom. For example, an atom of iodine contains 53 protons.

atomic weight—the number of protons and neutrons within the nucleus of an atom. For example, an atom of iodine-129 contains 129 protons and neutrons.

beta particle—an electron (negative charge) or positron (positively charged electron) emitted from the nucleus of an unstable (radioactive) atom. Cesium-137 and strontium-90 are two radionuclides that emit beta particles.

breeder reactor—a nuclear reactor that can produce both power and nuclear fuel in the form of plutonium-239. A breeder reactor produces more fuel than it consumes.

calcine—a general term for the granular, dehydrated ceramic powder created when high-level radioactive waste and certain chemical additives are heated at a high temperature.

canning—the placing of a protective metal covering around a cylindrical slug of uranium metal. This coating (cladding) prevents uranium from corroding as well as releasing radionuclides produced during nuclear fission.

cascade—the process of overflowing liquid waste from one underground tank to another. A series of two to four tanks were used in each tank farm overflow system. Solid particles settled in the first tank while the liquids (tank supernatant) overflowed or were pumped into succeeding tanks. Liquids were sometimes discharged to the soil after the last tank was filled.

CERCLA—see Comprehensive Environmental Response, Compensation, and Liability Act.

chain reaction—a rapidly repeating series of nuclear fission events caused by atoms absorbing neutrons, splitting into fragments, and releasing more neutrons that cause additional atoms to split.

cladding—the outer protective metal coating covering and bonded to a uranium fuel slug.

Clean Air Act—this federal law regulates air emissions addressing such national problems as smog, acid rain, and ozone depletion. The act gives the U.S. Environmental Protection Agency authority to establish National Ambient Air Quality Standards. The original act was passed in 1963 (then called the Air Quality Act). Today's air pollution program is based on the 1970 version of the law as amended in 1977 and 1990.

Clean Water Act—a 1977 amendment to the Federal Water Pollution Control Act of 1972 that sets the basic structure for regulating discharges of pollutants to surface waters. The law gave the U.S. Environmental Protection Agency the authority to set effluent standards. In its 1987 reauthorization, the Clean Water Act included citizen suit provisions and funded the construction of sewage treatment plants.

cleanup—all actions taken to contain, treat, store, or dispose of unwanted radioactive and chemical materials, whether in buildings, tanks, on or below the ground, or in surface waters.

collective effective dose equivalent—the total of the effective dose equivalents for all individuals living in a select population or area (such as within 50 miles of a nuclear power plant or a waste site).

colloids—collections of very small particles (one millionth to one billionth of a meter in size) that might assist in moving otherwise immobile contaminants.

committed dose equivalent—radiation dose (in rem) to a specific organ or tissue received from the intake of radioactive material by an individual during the 50-year period following intake. Commonly, this value is attributed to the single year intake was received.

Comprehensive Environmental Response, Compensation, and Liability Act—this act, enacted by Congress in 1980 and amended in 1986, governs the cleanup of hazardous, toxic, and radioactive substances at abandoned or uncontrolled waste sites. A trust fund, known as the Superfund, was created to finance the investigation and cleanup of some waste sites.

contaminant—a unwanted chemical, physical, biological, or radioactive substance that degrades the natural or desired quality of the environment.

cosmic rays—high energy particles (electrons and nuclei of atoms, mostly hydrogen) that are found in space and filter through as well as interact with the earth's atmosphere. They come from the sun and outside of the solar system. Most cosmic rays originate from solar flares.

crib—an underground box-like structure with an open bottom or a gravel-filled tile field designed to receive intermediate-level radioactive liquids. Cribs normally consisted of a series of buried perforated pipes or connected tiles. Cribs use the filtration and chemical exchange capability of the soil to retard the movement of contaminants.

criticality—the condition in which a nuclear chain reaction, involving neutrons, can support itself. It normally takes place slowly inside a nuclear reactor or rapidly when a nuclear weapon detonates.

curie—a unit describing the rate of decay of a radioactive material. One curie equals 37 billion particle emissions (disintegrations) a second. The greater the number of curies, the more energy released. Radiation hazards depend upon many variables including the type of radiation emitted, energy of radiation received, tissue exposed, intensity of exposure, and age of person.

daughter product—the product of radioactive decay of an element. Also known as a decay product. A radionuclide may have one or more daughter products before it transforms into a stable (nonradioactive) atom.

decay chain—the process that unstable (radioactive) elements pass through to become stable (nonradioactive).

decay product—see daughter product.

DOE—see U.S. Department of Energy.

dose—general term used to describe the amount of radiation or chemicals absorbed.

dose equivalent—a value (in rem) for comparing potential biological hazards that takes into account the different types of radiation, their intensity, and the organs/tissues exposed. The dose equivalent is found by multiplying the energy received (absorbed dose expressed in rad) by a radiation-weighting factor (quality factor). This value is commonly used to assess radiation risk over a limited region of the body.

EDTA—see ethylenediaminetetraacetic acid.

effective dose equivalent—the weighted sum of all committed dose equivalents for every organ and tissue in the body. This gives an "effective" whole-body radiation dose. Weighting factors are used to place doses to different organs and tissues on an equal footing so that potential whole-body health risks are compared. Commonly used when comparing risks between individuals.

end point—the products or environmental quality expected when a cleanup project ends.

end state—the final or long-term condition of a site (for example, an area of land, a body of water) following cleanup.

engineered barrier—a human-made barrier built above or below ground to stabilize a waste site and control contaminant release as well as entry by moisture, plants, and animals. Surface barriers are built using multiple layers of sand, gravel, clay, cement, asphalt, or plastic.

environmental impact statement—a document written to describe the effects of proposed activities on the land, water, air, and living organisms. In addition, it addresses environmental values and the social, cultural, and economic impacts of those activities.

ethylenediaminctctraacetic acid (EDTA)—this organic compound is capable of reacting with less mobile metals; for example, plutonium or cobalt, to form more soluble and environmentally mobile metals.

Federal Facilities Compliance Act—this act amends the Resource Conservation and Recovery Act by requiring all federal agencies to meet the legal requirements of federal, state, and local laws governing waste management and cleanup in the same manner as private industry.

fertile—material that becomes fissile upon absorbing a neutron.

FFCA—see Federal Facilities Compliance Act.

fissile—material that will undergo fission, that is, split into fragments upon absorbing a neutron.

fission—the process of an atom splitting into two or more lighter atoms. This is accompanied by the release of neutrons, other particles (called fission products), and high-energy gamma rays. The energy available for fission comes from the combined effects of (1) the repulsive force between protons and (2) the attractive nuclear force between neutrons or protons inside the nucleus.

fission products—the nuclei pieces produced during fission of a more massive nucleus.

Formerly Utilized Sites Remedial Action Program—Congress established this program in 1974 to identify and clean up contaminated sites and buildings previously used for nuclear weapons research and production. Many of these sites operated between the 1940s and the 1960s. Contaminants are primarily low levels of uranium, thorium, and radium, plus their decay products. Mixed waste (radionuclides and chemicals) is sometimes present. Management of this program was transferred from the U.S. Department of Energy to the U.S. Army Corps of Engineers in 1997.

French drains—vertical concrete pipes, buried about 10 feet deep and filled with coarse gravel, used to dispose of low volumes of cooling water and steam condensate from the reprocessing plants.

fuel—short, cylinder-shaped pieces of uranium manufactured to exact specifications, enclosed inside a metal coating (cladding), and inserted into a nuclear reactor. Inside the reactor, a portion of the fuel is converted into energy, plutonium, and other radionuclides.

fuel-grade plutonium—plutonium produced inside a reactor is divided into grades distinguished by the amount of plutonium-239. Fuel-grade plutonium contains 94% to 82% by mass plutonium-239. The rest is made of other plutonium isotopes, especially plutonium-240. See also weapons-grade plutonium.

FUSRAP—see Formerly Utilized Sites Remedial Action Program.

gamma ray—a high-energy, deeply penetrating form of radiation, originating from the nucleus of an unstable (radioactive) atom. Gamma rays are similar to x-rays but have more energy.

General Accounting Office—the investigative arm of Congress. The office examines the use of public funds, evaluates federal activities, and provides analyses, recommendations, and other assistance to help Congress make more informed decisions. For information on this office and documents published, see http://www.gao.gov/.

groundwater—water existing beneath the land surface. Most groundwater originates from past precipitation entering the ground or from surface water bodies such as lakes and rivers. It naturally contains dissolved elements. Human-made contamination, such as chemicals, can also be found in groundwater.

half-life—the average time for a radioactive sample to lose half of its activity because of radioactive decay. At the end of one half-life, 50% of the original radioactive material has decayed away forming a stable or unstable daughter product. Half-lives range from less than a second to billions of years.

Hanford Federal Facility Agreement and Consent Order—the legally enforceable agreement signed by the U.S. Department of Energy, U.S. Environmental Protection Agency, and the Washington State Department of Ecology in 1989 to clean up the Hanford Site. To learn more about the agreement, see http://www.hanford.gov/tpa/tpahome.htm.

hazardous waste—nonradioactive waste such as metals and chemical compounds that pose a risk to the environment and human health.

high-level waste—chemical solutions created during the initial reprocessing of used reactor fuel; this waste contains most of the unwanted and concentrated amounts of radionuclides. High-level waste can be liquids or solids created from those liquids.

injection wells—narrow boreholes that contaminated liquids were pumped or poured into as a disposal method (also known as reverse wells).

ion—an atom or group of atoms that carries a positive or negative electrical charge as a result of gaining or losing one or more electrons.

ionizing radiation—radiation that has enough energy to remove an electron from an atom, leaving a positively charged particle (positive ion) behind. High doses of ionizing radiation can break chemical bonds, causing damage to living cells. The splitting or decay of radioactive atoms emits ionizing radiation. There are four basic types of ionizing radiation: alpha particles, beta particles, neutrons, and gamma rays.

irradiated—to expose something to radiation such as when uranium is exposed to neutrons inside a nuclear reactor.

isotopes—different forms of the same element distinguished by different numbers of neutrons in the nucleus. A single element may have many isotopes; some isotopes are radioactive, others are not. For example, 17 isotopes of argon exist. Many are radioactive such as argon-41. It was released from Hanford reactors after naturally occurring argon-40 captured an extra neutron.

low-level waste—a general term describing any radioactive waste containing low amounts of radioactivity such that minimal or no protective containment is needed. Low-level waste is not irradiated fuel or high-level waste. Further, it does not contain high concentrations of transuranic elements. It can include liquids or contaminated solids such as clothing, tools, and equipment. The U.S. Nuclear Regulatory Commission has divided low-level radioactive waste generated by commercial power reactors and other sources into the following categories:

- **Class A**: waste having the lowest concentrations of radionuclides that can be disposed with the least stringent requirements on waste forms and disposal packaging. Class A waste is intended to be safe after 100 years.
- **Class B**: waste containing higher concentrations of shorter-lived radionuclides. This waste must be packaged more stringently than Class A materials. Intended to be safe for 300 years.
- **Class C**: waste that must meet the form and stability requirements applicable to Class B waste as well as actions taken at the disposal site to protect against inadvertent human intrusion for 500 years.
- **Greater-Than-Class-C**: waste is not typically disposed of in shallow land burial sites. May require disposal in a geologic repository.

Manhattan Project—the U.S. government project that produced the first nuclear weapons during World War II. This project began in 1942 and ended in 1946.

maximally exposed individual—a hypothetical member of the public who could receive the highest possible radiation dose from radionuclide releases.

millirem—a unit of radiation equal to one-thousandth of a rem. Abbreviated mrem.

mixed waste—low-level radioactive or transuranic contaminated material combined with hazardous chemicals or metals.

Model Toxics Control Act—a human health and environmental protection law passed in 1994 in the state of Washington establishing the standards to identify, investigate, and clean up facilities where hazardous substances are found. Its provisions may also be applied to potential and ongoing releases of hazardous substances.

moderator—a substance, such as graphite, that slows down neutrons so that they are more likely captured by atoms of uranium to cause nuclear fission.

mrem—see millirem.

National Environmental Policy Act—this federal law, passed in 1970, requires the U.S. government to consider the potential environmental impacts from major actions when it makes decisions.

National Priorities List—a list of waste sites in the United States that are potentially the most hazardous and warrant further investigations including possible cleanup. This list is maintained by the U.S. Environmental Protection Agency using a screening analysis model called the Hazard Ranking System.

National Research Council—the principal operating agency for both the National Academy of Sciences and the National Academy of Engineering. It was created in 1916 to provide high-quality scientific and technological responses to questions raised by government agencies, as requested.

natural attenuation—degradation of contaminants in the environment using naturally occurring physical, chemical, or biological processes.

Natural Resource Damage Assessment—a provision under the Comprehensive Environmental Response, Compensation, and Liability Act providing a process for collecting, compiling, and analyzing information to calculate the cost of repairing the natural environment from injuries or compensating the public for such injuries.

NEPA—see National Environmental Policy Act.

neutron—one of the basic particles of all atoms except elemental hydrogen. Neutrons are located in an atom's nucleus. They are electrically neutral and have a mass slightly heavier than a proton. The fission process emits neutrons.

Nuclear Waste Policy Act—originally passed in 1982 and amended in 1987, this act established the Office of Civilian Radioactive Waste Management within the U.S. Department of Energy to develop, construct, and operate a system for spent nuclear fuel and high-level radioactive waste disposal including a permanent geologic repository, interim storage, and supporting transportation network.

Office of the Inspector General—this office within the various federal agencies conducts, supervises, monitors, and initiates investigations relating to the programs of that agency.

organic compound or chemical—a substance containing mostly carbon, hydrogen, and oxygen.

outrage—people's feelings and beliefs causing fear, anger, or frustration about something they consider valuable. Outrage is the silent partner to risk.

permeability—a measure of the ease with which liquids move through sediment or fractured rocks. Liquid flows rapidly through material of high permeability.

pitchblende—a dark-colored mineral rich in uranium as well as a source of radioactive radium and polonium.

proton—one of the basic particles of an atom. Protons are located in an atom's nucleus, have a positive electrical charge, and have a mass slightly lighter than a neutron.

rad—see radiation absorbed dose.

radiation—particles (for example, alpha, beta, or neutrons) and energy waves (for example, gamma rays) released from a radioactive element. Often used as the shorthand form for ionizing radiation.

radiation absorbed dose—a unit that measures the amount of ionizing radiation (energy) absorbed by material, such as human tissue. One rad equals an energy absorption of 100 ergs per gram of material. (An erg is a metric unit of energy.)

radioactive—giving off energy in the form of particles or rays during radioactive decay. Often used to modify a word or phrase, such as radioactive material.

radioactive decay—the spontaneous change in the nucleus of an atom by the release of one or more particles or other forms of energy. Radioactive decay occurs as atoms change from an unstable state to a more stable state.

radioactivity—property possessed by some atoms of emitting radiation spontaneously from their nucleus during radioactive decay.

radionuclide—any type of an atom that is radioactive; a radioactive form (isotope) of an element. Radioactive elements emit particles or energy to become more stable.

RCRA—see Resource Conservation and Recovery Act.

recharge—the process of replenishing groundwater lying below the land's surface.

Record of Decision—an official document that states the decision made and describes the environmental factors considered, the preferred cleanup approach, and the alternatives considered in an environmental impact statement or remedial action.

rem—see roentgen equivalent man.

reprocessing—the chemical and mechanical treatment of spent nuclear fuel to separate usable products, such as plutonium and uranium, from chemical solutions.

Resource Conservation and Recovery Act (RCRA)—a federal law enacted in 1976 to address the cradle-to-grave treatment, storage, and disposal of hazardous waste.

reverse wells—see injection wells.

risk—the chance of injury, harm, or loss. Estimation of risk is based on the likelihood and consequence of an event happening.

risk assessment—an organized, scientific process used to describe and estimate the likelihood of adverse health or environmental impacts from exposure to contamination or performing an activity. Risk assessments take into consideration the types of hazards, extent of exposure to hazards, relationship between exposure and receptor response, and characterization of risk.

risk management—deciding what to do when an outcome is uncertain. Risk management is the process of using risk assessment results and other information to make more informed decisions for reducing or controlling risk.

Rivers and Harbors Act—the first federal legislation, passed in 1899, protecting the nation's navigable waters to safeguard commerce. The act requires permits for the construction of any structure in or over any federally listed water way, the excavation from or disposal of materials in those waters, or the accomplishment of other work affecting the course, location, condition, or capacity of such waters.

roentgen equivalent man— a unit of ionizing radiation dose that indicates the potential for biological damage. It takes into account the amount of radiation received plus other factors such as the type of radiation, how radiation was delivered, the radionuclide in question, and the tissue exposed.

Safe Drinking Water Act—a federal law established by Congress in 1974 to protect human health from contaminants in drinking water and to prevent contamination of existing water supplies. Contaminants are defined as any potentially harmful levels of physical, chemical, biological, or radiological substances. Provisions in this act also protect groundwater aquifers from contaminants injected into the ground.

saltcake—a moist material (sometimes like wet beach sand) created from the crystallization of chemicals after tank waste liquid was evaporated. Saltcake is usually made of water-soluble chemicals.

science—knowledge about the physical, chemical, or biological properties and processes taking place in the natural world or engineered systems.

sludge—a thick layer of water-insoluble chemicals precipitated or settled to the bottom of a tank.

slug—a short cylindrical rod of uranium metal, coated with aluminum or zirconium, inserted inside the core of a nuclear reactor. After enough uranium is inserted, a chain reaction begins. Synonymous with fuel rod or fuel element.

sluicing—the use of high-velocity water and chemical sprays to break up chunks of waste inside a tank so it can be removed.

slurry—a mixture of small solid particles suspended in a liquid.

solvent—a liquid in which solids will dissolve. In nature, water is the most common solvent. In the reprocessing of spent nuclear fuel, organic solvents, such as hexone or tributyl phosphate, are used to separate select radionuclides from unwanted chemicals. This is possible because some radionuclides and other metals are more soluble in one solvent than in another; thus, selective extraction is possible.

sorption—a general term used to describe the retention of chemicals, metals, or radionuclides into or on the surface of a solid or liquid.

specific retention trench—see trench.

spent fuel—uranium fuel that was irradiated inside a nuclear reactor and has been removed or is ready for removal from the reactor.

stewardship—all activities required to maintain an acceptable level of protection to human health and the environment posed by contaminants or end products remaining after active cleanup is completed. It includes physical controls, institutional constraints, land-use restraints, information management, and monitoring.

supernatant liquid—a liquid easily pumped from a waste tank; generally floats above a layer of settled solids. Also known as supernate.

technology—a capability that solves a problem. This capability is based upon knowledge (science).

Toxic Substances and Control Act—this federally managed law was enacted in 1976 and gives the U.S. Environmental Protection Agency broad authority to protect human health through regulating the manufacture, use, distribution in commerce, and disposal of substances containing toxic chemicals.

transuranic element—an element with atomic number greater than 92. All transuranic elements are radioactive. Examples include plutonium, neptunium, and americium.

transuranic waste—radioactive waste containing alpha-emitting transuranic elements, such as plutonium or neptunium, with half-lives greater than 20 years and present in concentrations of more than 100 billionths of a curie (100 nanocuries) per gram of waste. (Twenty-eight grams makes 1 ounce.)

trench—a long open ditch dug in the ground for the disposal of select batches of low-level to intermediate-level radioactively contaminated liquids that would otherwise interfere with the chemical absorption of contaminants in the soil if repeatedly discharged to a crib. A trench is covered with soil after use. Also known as a specific retention trench.

Tri-Party Agreement—the legally enforceable agreement (formally known as the Hanford Federal Facility Agreement and Consent Order) signed by the U.S. Department of Energy, U.S. Environmental Protection Agency, and the Washington State Department of Ecology in 1989 to clean up the Hanford Site.

UMTRA—see Uranium Mill Tailings Remedial Action Project.

uranium—the heaviest element normally found in nature. Uranium is a hard, silvery, metallic radioactive element.

Uranium Mill Tailings Remedial Action Project—in 1978, Congress passed this act directing the U.S. Department of Energy to stabilize, dispose, and control uranium mill tailings in a safe and environmentally acceptable manner. Tailings remain from the mining of uranium used in manufacturing nuclear fuel.

U.S. Department of Energy—this cabinet-level department, created in 1977, focuses on national security, energy, environment, and science and technology. The department is responsible for cleaning up the nuclear weapons complex. For more information about the Energy Department, see http://www.energy.gov/.

vadose zone—the layer of sediments lying between ground level and the water table. The open pore spaces between sediment grains are filled with a mixture of water and air. Sometimes referred to as the "partially saturated" or "unsaturated" zone.

valence—the capacity of an atom, expressed in the number of electrons it gives up or accepts, to combine with other atoms. An atom's valence state is the number of electrons it can donate (positive valence state) or accept (negative valence state) to achieve a chemical bond.

vitrification—a high-temperature glass melting process that mixes glass-forming materials with radioactive waste. This mixture hardens into solid amorphous glass, immobilizing the waste and making it more suitable for long-term disposal.

Watch List—Public Law 101-510 (Section 3137), passed in 1990, required the U.S. Department of Energy to treat underground radioactive waste storage tanks at Hanford in such a way as to avoid any potential releases of unwanted materials to the environment.

water table—the upper boundary of an aquifer where the pore spaces between soil grains or rock fractures are filled with water.

weapons-grade plutonium—plutonium containing 94% or more by mass of plutonium-239. The remaining is mostly the isotope plutonium-240. Weapons-grade plutonium is used in nuclear weapons. See also fuel-grade plutonium.

zircaloy—a type of metal cladding for reactor fuel. It is made of zirconium combined with tin, and small amounts of iron, chromium, and nickel.

Index

The following page number notations are used in this index: t after the page number means the information is in the table; f, figure; q, chapter quote; b, side discussion boxes.

J

K

L

M

N

S

T

U

V

W